Physical Metallurgy of Metals and Alloys

Physical Metallurgy of Metals and Alloys

Editors

Pan Gong
Maojun Li
Guangchao Han
Xin Wang

MDPI • Basel • Beijing • Wuhan • Barcelona • Belgrade • Manchester • Tokyo • Cluj • Tianjin

Editors
Pan Gong
Huazhong University of Science and Technology
China

Maojun Li
Hunan University
China

Guangchao Han
China University of Geosciences
China

Xin Wang
Hebei University of Technology
China

Editorial Office
MDPI
St. Alban-Anlage 66
4052 Basel, Switzerland

This is a reprint of articles from the Special Issue published online in the open access journal *Materials* (ISSN 1996-1944) (available at: https://www.mdpi.com/journal/materials/special_issues/metallurgy_metal_alloy).

For citation purposes, cite each article independently as indicated on the article page online and as indicated below:

LastName, A.A.; LastName, B.B.; LastName, C.C. Article Title. *Journal Name* **Year**, *Volume Number*, Page Range.

ISBN 978-3-0365-7804-0 (Hbk)
ISBN 978-3-0365-7805-7 (PDF)

Contents

About the Editors

Pan Gong

Pan Gong, Ph.D., is an Associate Professor of Materials Processing Engineering at Huazhong University of Science and Technology in Wuhan, China. His current research interests broadly centre on bulk metallic glasses, high-entropy alloys, titanium alloys, metallic composites, precision metal plastic forming, powder metallurgy, and incremental sheet forming. He teaches professional courses such as Materials Forming Equipment and Automation, Fundamentals of Precision Plastic Forming, and Stamping Technology and Mold Design at the School of Materials Science and Engineering. He obtained a PhD in Mechanical Engineering from Tsinghua University, and both BSc and MSc degrees from Huazhong University of Science and Technology.

Maojun Li

Maojun Li, Ph.D., is an Associate Professor of Precision Manufacturing at Hunan University in Changsha, China. His current research interests broadly centre on the machinability of advanced aerospace alloys and composite materials, microscale machining, the development of novel/innovative processes (e.g., hybrid machining, high-power fibre laser cutting, ultrasonic-assisted machining, etc.), CFRP/Al joints, and computer numerically controlled (CNC) operation of machine tools including five-axis machining and industrial robots. His work has a strong materials element, particularly in relation to studying the effect of cutting variables on workpiece surface integrity and their subsequent influence on the in-service performance of parts. Many of his research activities involve strong industrial collaboration and relevance. He obtained a PhD in Mechanical Engineering from University of Birmingham, UK, and both BSc and MSc degrees from Huazhong University of Science and Technology.

Guangchao Han

Guangchao Han, Ph.D., is a Professor of Advanced Manufacturing at China University of Geosciences in Wuhan, China. His research focusses on studying ultrasonic-assisted micro-forming processing, multi-energy field composite processing technology and multi-material 3D printing technology. He has developed a multi-mode ultrasonic vibration platform and a multi-mode ultrasonic composite forming process. He teaches "Interchangeability and Measuring Technology" at the department of Mechanical Engineering at China University of Geosciences. He obtained a PhD in Materials Processing Engineering from Huazhong University of Science and Technology, and both BSc and MSc degrees in Materials Processing Engineering from Wuhan University of Science and Technology.

Xin Wang

Xin Wang, Ph.D, is an Associate Professor of Metal Matrix Composites at Hebei University of Technology, in Tianjin, China. His research interests focus on the pressure infiltration molding of amorphous composite materials, functional amorphous alloys for wastewater treatment, and inoculation treatment techniques for aluminium, magnesium, and steel materials. He is the Deputy Director of Hebei Key Laboratory of New Functional Materials and the Deputy Director of the Department of Metal Materials Engineering. He teaches professional courses such as Fundamentals of Materials Science, Mechanical Engineering Materials, and Material Surface and Interface at the School of Materials Science and Engineering. He obtained a PhD in Materials Science and Engineering (2013) from Tsinghua University, and a bachelor's degree in Materials Science and Engineering (2001) from Hebei University of Technology.

materials

MDPI

Article

Effect of Cooling Rate on the Microstructure Evolution and Mechanical Properties of Iron-Rich Al–Si Alloy

Xiao Shen [1], Shuiqing Liu [1,2,*], Xin Wang [3,*], Chunxiang Cui [3], Pan Gong [4], Lichen Zhao [3], Xu Han [1,2] and Zirui Li [1]

1 School of Mechanical Engineering, Hebei University of Technology, Tianjin 300401, China; sx961216@163.com (X.S.); xhan@hebut.edu.cn (X.H.); lizirui@gmail.com (Z.L.)
2 State Key Laboratory for Reliability and Intelligence of Electrical Equipment, School of Materials Science and Engineering, Hebei University of Technology, Tianjin 300401, China
3 Key Laboratory for New Type of Functional Materials in Hebei Province, School of Materials Science and Engineering, Hebei University of Technology, Tianjin 300401, China; hutcui@hebut.edu.cn (C.C.); zhlch@hebut.edu.cn (L.Z.)
4 State Key Laboratory of Materials Processing and Die & Mould Technology, School of Materials Science and Engineering, Huazhong University of Science and Technology, Wuhan 430074, China; pangong@hust.edu.cn
* Correspondence: sqliu@hebut.edu.cn (S.L.); ahaxin@hebut.edu.cn (X.W.)

Abstract: The mechanical properties of iron-rich Al–Si alloy is limited by the existence of plenty of the iron-rich phase (β-Al_5FeSi), whose unfavorable morphology not only splits the matrix but also causes both stress concentration and interface mismatch with the Al matrix. The effect of the cooling rate on the tensile properties of Fe-rich Al–Si alloy was studied by the melt spinning method at different rotating speeds. At the traditional casting cooling rate of ~10 K/s, the size of the needle-like β-Al_5FeSi phase is about 80 μm. In contrast, the size of the β-Al_5FeSi phase is reduced to 500 nm and the morphology changes to a granular morphology with the high cooling rate of ~10^4 K/s. With the increase of the cooling rate, the morphology of the β-Al_5FeSi phase is optimized, meanwhile the tensile properties of Fe-rich Al–Si alloy are greatly improved. The improved tensile properties of the Fe-rich Al-Si alloy is attributed to the combination of Fe-rich reinforced particles and the granular silicon phase provided by the high cooling rate of the melt spinning method.

Keywords: Fe-rich Al–Si alloy; β-Al_5FeSi phase; cooling rate; melt spinning; microstructure; strengthening mechanisms

Citation: Shen, X.; Liu, S.; Wang, X.; Cui, C.; Gong, P.; Zhao, L.; Han, X.; Li, Z. Effect of Cooling Rate on the Microstructure Evolution and Mechanical Properties of Iron-Rich Al–Si Alloy. *Materials* **2022**, *15*, 411. https://doi.org/10.3390/ma15020411

Academic Editor: Amir Mostafaei

Received: 29 November 2021
Accepted: 4 January 2022
Published: 6 January 2022

1. Introduction

Recycled aluminum is obtained through the recycling of waste aluminum products. The production energy consumption is only 3% to 5% of the energy consumption of primary aluminum [1]. It has obvious advantages in energy saving and emission reduction, thus it has become one of the research hotspots in the field of lightweight structural materials. Iron is considered to be the most harmful impurity element in recycled aluminum. The main reason is that the solid solubility of iron in cast aluminum–silicon alloy is low (only 0.05 wt.%) and a variety of iron-rich intermetallics can be formed during solidification, such as α-AlFeSi, β-Al_5FeSi, π-$Al_8Mg_3FeSi_6$, and δ-Al_4FeSi_2. Among them, β-Al_5FeSi is long and needle-like on the two-dimensional optical micrographs and the three-dimensional space is a lath structure [2–4].

The size and morphology of the secondary phase are the key factors affecting the properties of the alloy. The needle-like β-Al_5FeSi phase is easy to cause stress concentration and split the matrix, which is unfavorable to the mechanical properties of recycled aluminum [5,6]. In addition, excessive iron content will lead to poor melt fluidity and reduce casting quality [7]. Therefore, many researchers are working to modify the morphology of β-Al_5FeSi phases, such as by adding the alloying elements Mn, Nb, Co, Sc, Er, etc. [8–12]. In addition, increasing the melt superheat and heat treatment are also effective ways to improve the morphology of the iron-rich phase. In our previous work, it was found that the combination addition of Mg and La elements can promote the transition from β-Al_5FeSi to the script-like π-$Al_8Mg_3FeSi_6$ phase, resulting in a 65% increase in the yield strength of Al–Si–Mg–Fe alloy [13]. However, the addition of the above elements only changes the morphology of the iron-rich phase, brings some technical problems, and further increases the difficulty of the secondary resource recovery of Al–Si alloys. Therefore, it is a key issue to develop high performance Al–Si alloys with a high Fe content.

Increasing the cooling rate is an effective way to change the morphology of an intermetallic. Melt spinning is a typical rapid solidification method which is used in the preparation of inoculants due to its chemical uniformity and large solid solubility [14–16]. Additionally, the microstructure evolution has been reported to be dependent on the cooling rate and Fe content in the Al–Si alloy [17,18]. The cooling rates range from 1 to 10 K/s in traditional mold casting to 10 K/s–100 K/s in high-pressure die casting. It can be seen that the morphology control of the cooling rate on the β-Al_5FeSi phase is generally limited to the sub-rapid solidification (cooling rates < 10^3 K/s) condition and there is a need to explore the microstructure under the rapid solidification condition with cooling rates higher than 10^3 K/s. Moreover, although some studies have analyzed the growth behavior and morphology of the needle-like β-Al_5FeSi phases in specific recycled Al–Si alloy, a systematic way to understand the formation or suppression of the needle-like phase is not available.

In this work, we achieve different cooling rates by adjusting the speed of the copper roller during melt spinning. Through in-depth analysis of the phase microstructure morphology of β-Al_5FeSi under different cooling rates, the solidification path and formation conditions are discussed. Meanwhile, the effect of the cooling rates on the microstructure and tensile properties using the rapid solidification technology is further investigated.

2. Materials and Methods

The Al-10Si-1.5Fe (all compositions cited in this work are in wt.%) alloy investigated in this work was prepared from the as-cast ingots of commercial Al-12Si and Al-20Fe alloys, and its chemical composition is shown in Table 1. The raw and processed materials were melted in a vacuum arc furnace under an argon atmosphere. For a low cooling rate sample, the ingots were remelted by the induction smelting method, during which it was poured into the preheated 200 °C steel mold with a cavity diameter of 20 mm and a height of 120 mm, which was denoted as the C1 alloy. For high cooling rate samples, secondary melting of intermediate alloy was carried out in a high-frequency induction smelting furnace and blown on a copper roller with a diameter of 22 cm and rotating speed of both 2000 rpm and 4000 rpm to produce Al-10Si-1.5Fe ribbons (abbreviated as C2 and C3 alloy, respectively). According to the results of the previous studies [19–23], the cooling rates were calculated as 6.7×10^2 K/s, 2.4×10^4 K/s, respectively.

Table 1. Composition of the experimental alloy.

Element	Al	Si	Fe	Mg	Cu	Zn
Content	Balance	9.96	1.52	0.33	0.09	0.07

The specimens were cut from the middle of the casting rod and the spun ribbon for metallographic examination, and then were carefully ground by sandpaper in ethanol as well as dried with air. After mechanical polishing, the specimens were corroded by Poulton's reagent and observed by optical microscope. The microstructure of the samples was etched with 0.5% vol. HF solution before scanning electron microscope (SEM, Hitachi, Tokyo, Japan) observation. The chemical composition of the phases in the prepared alloys were examined by energy dispersive spectroscopy (EDS, Hitachi, Tokyo, Japan), which was performed on at least 3 parts of the phases. The tensile tests for the as-cast rods were carried out at a constant strain rate of 5×10^{-4} s^{-1} and loaded parallel to the axis of the specimens following standard ASTM B209. Based on the same experimental scheme, the C2 and C3 ribbon samples were tensile-tested with a width of 2 mm and length of 120 mm. Three tests were conducted for each sample condition. The samples were characterized by Bruker D8 Discover X-ray diffraction (XRD, Bruker AXS, Karlsruhe, Germany), Cu Ka radiation, the S4800 scanning electron microscope (SEM, TESCAN, Czech Republic), Nova Nano SEM450 (SEM, FEI Company, Portland, America) and the optical microscope (OM, Olympus, Jiang dong ou yi testing instrument, Ningbo, China).

3. Results and Discussion

Figure 1 shows the typical XRD spectra of Al-10Si-1.5Fe alloys with different cooling rates. It can be seen that at the cooling rate of ~30 K/s, the C1 alloy had three groups of sharp diffraction peaks corresponding to Al, Si, and β-Al5FeSi phases, respectively. It is clear that the δ-Al_3FeSi_2 phase appeared in the C2 alloy with the cooling rate increase to 6.7×10^2 K/s. It is noteworthy that the iron-rich phase disappeared in the matrix as the cooling rate continued to increase to 2.4×10^4 K/s.

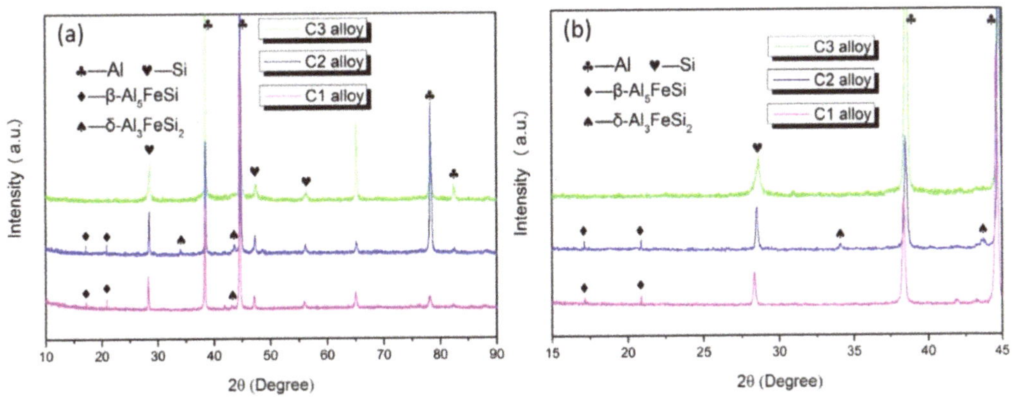

Figure 1. (**a**) XRD spectra of the C1, C2, and C3 alloys. (**b**) Shows the partially amplified spectra of (**a**).

Figure 2 shows the metallography of Al–Si–Fe alloy with different cooling rates. A large number of needle-like phases about 80 μm in length and the acicular-like phase existed in the aluminum matrix. Take the XRD spectra into consideration; they are β-Al_5FeSi phase and eutectic Si phase, respectively. It can be seen that the β-Al5FeSi phase was randomly distributed among the dendrites, as shown in Figure 2a, and its unfavorable morphology is particularly unfavorable to the strength and ductility of the matrix. It is noteworthy that the morphology of the Fe-rich phase in C2 changed significantly, that is, from needle-like to short rod-like with a length of 10 μm with the increase of the cooling rate to ~10^2 K/s. In addition, increasing the cooling rate not only changes the morphology of the Fe-rich phase but also the eutectic Si phases transformed from coarse acicular-like to fine particle-like in the C2 alloy. As the cooling rate increased to ~10^4 K/s, the needle-like or fibrous phases were no longer visible in the field of view and a large number of dispersed particles were replaced, as shown in Figure 2e,f. Compared with the C1 and C2 alloy, the phases distributed in the C3 alloy matrix were more uniform and dispersed.

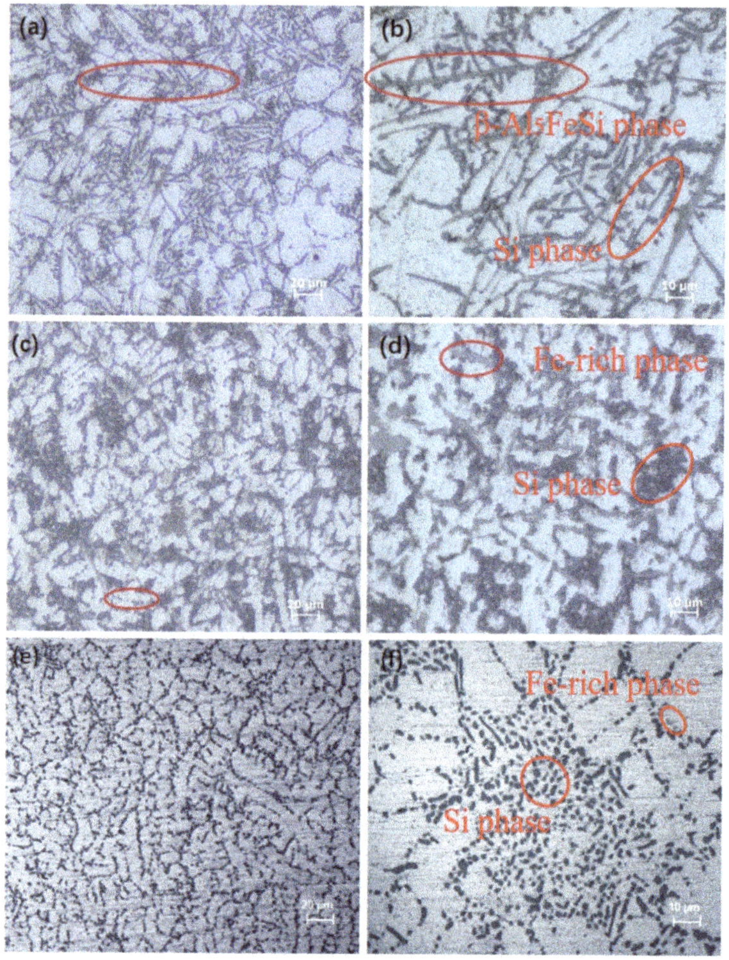

Figure 2. (**a**,**c**,**e**) are optical micrographs of C1, C2, and C3 alloy. (**b**,**d**,**f**) are optical micrographs showing locally amplified spectra of (**a**,**c**,**e**), respectively.

Figure 3 shows the SEM images of Al-10Si-1.5Fe alloys at different cooling rates. Figure 3a,c are the SEM images of the C1 and C2 samples, among which Figure 3b,d are the local enlarged views of Figure 3a,c. In Figure 3a,b, the presence of long needle-like phases and sharp morphology in the die casting sample is more clearly observed than in Figure 3c–f. Combined with the EDS analysis in Figure 3e,f, the phase can be judged as β-Al_5FeSi. The comparison of Figure 3a,c show that the Fe-rich phase transformed from a sharp long needle-like structure to a short rod-like morphology, wherein the distribution state was more diffuse and there was almost no obvious interweaving connection. Figure 3d clearly shows that the short rod-shaped phase in the C2 alloy with the cooling rate of 6.7×10^2 K/s was a fishbone-like shape, which can be determined as the δ-Al_3FeSi_2 phase. The fishbone-like morphology phase had little effect on the fracture of the Al matrix, thus the mechanical properties of the matrix will be improved accordingly. Based on the above analysis, the increase in the cooling rate does have a regulatory effect on the morphology of the Fe-rich phase in the aluminum alloy.

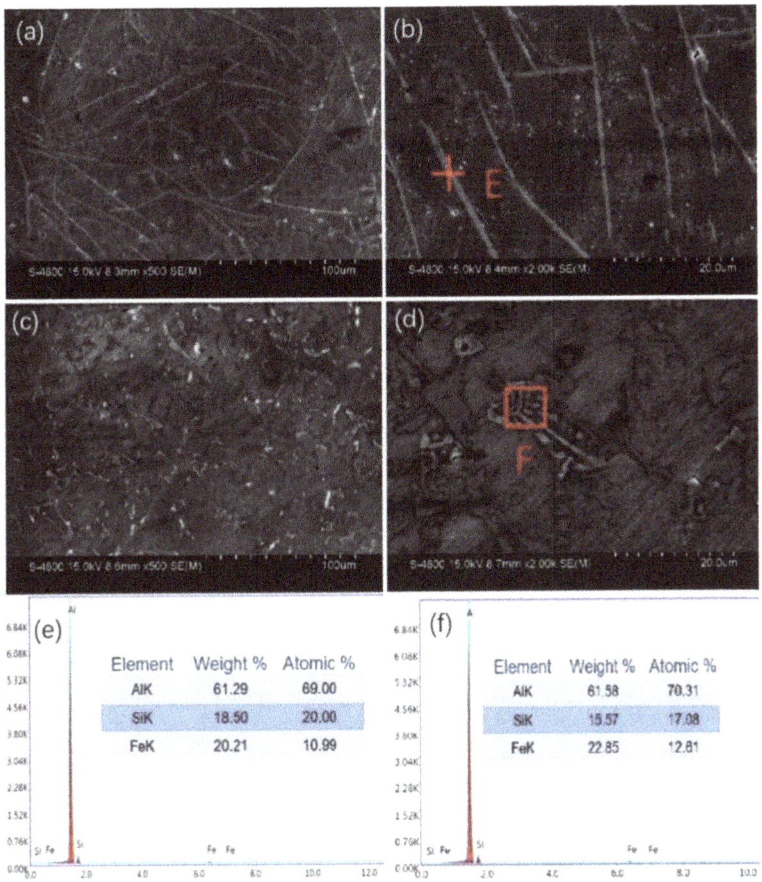

Figure 3. SEM images of the (**a**,**b**) C1 alloy and (**c**,**d**) C2 alloy. (**e**,**f**) are the EDS results of the (**b**,**d**) corresponding areas.

Figure 4 reveals the SEM images of the C3 alloy with the cooling rate of 2.4×10^4 K/s. It can be seen that two kinds of particles with different brightness were distributed in the matrix and the particle size ranged from 200 nm to 500 nm. From the EDS inset in Figure 4b, it can be seen that the content of iron in this region was about 1 wt.%, which is lower than the nominal composition of the Al–Si–Fe alloy. The cooling rate can obviously change the morphology of iron-rich phase because at a high cooling rate, Fe does not have enough time to precipitate in the form of the intermetallic phase. Instead, Fe is trapped in the Al matrix to form a supersaturated solid solution. In order to clearly describe the change of the Fe-rich phase size with the cooling rate, the particle length and width statistical distribution of the three alloys is shown in Figure 5. It can be seen that with the increase of the cooling rate, the length of the Fe-rich phase decreased sharply and the length and width tended to be consistent, which tends to form spherical particles. The change is beneficial to the improvement of the tensile properties of the Fe-rich Al alloy.

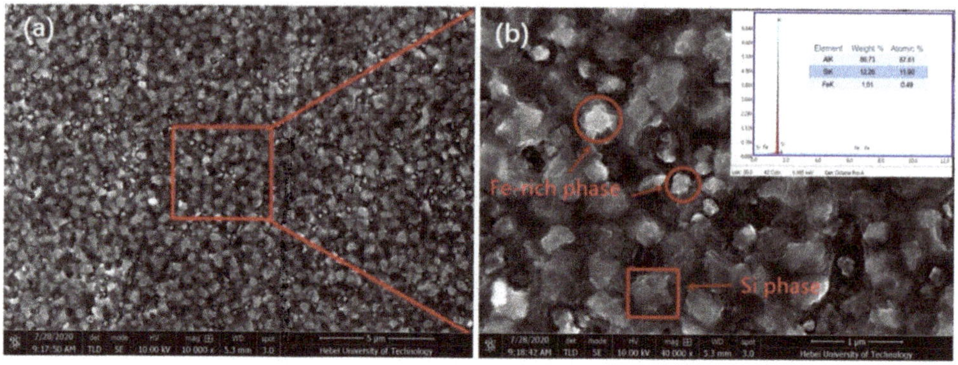

Figure 4. (**a**) SEM images of the C3 alloy. (**b**) is the magnification of the region in (**a**), while the inset is the EDS results of the region in (**a**).

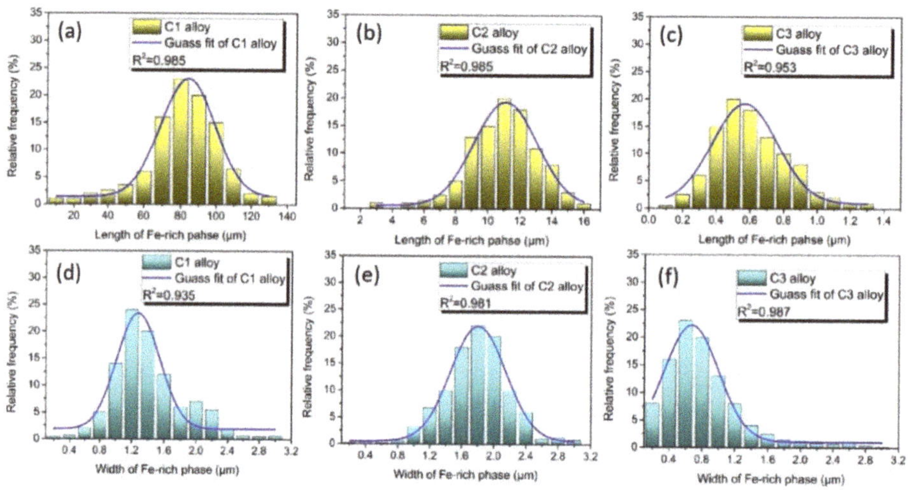

Figure 5. Statistical distribution of the length of Fe-rich phase of the (**a**) C1, (**b**) C2, and (**c**) C3 alloys, (**d**,**e**,**f**) are the statistical distribution of the width of Fe-rich phase of the C1, C2 and C3 alloys, respectively.

Figure 6 reveals the Differential Thermal Analysis (DTA)curves of the alloy at different cooling rates, where the peak of the DTA curve represents the eutectic temperature of the alloy. As the cooling rate increased, the eutectic temperature gradually decreased from 577.2 °C to 571.7 °C. According to the binary phase diagram of Al–Si alloy, the eutectic zone moves towards the non-metallic phase under the non-equilibrium solidification condition. The growth morphology of the silicon phase is not only restricted by the concentration distribution of the silicon atom at the growth interface but also is affected by the growth rate of α-Al. When the cooling rate is high, the α-Al phase is the first to nucleate and grow during solidification, and silicon atoms are disposed to the front of the solid–liquid interface, which increases the concentration of the silicon element in liquid phase. At this time, the silicon phase changes its growth direction in staggered or twin mode. It can be seen that from the C2 and C3 alloys, they had a subtle derivative curve peak in the eutectic region corresponding to the formation of the δ-Al_3FeSi_2 phase. That is, an increase in the cooling rate can promote the precipitate of the fishbone-like δ-Al_3FeSi_2 phase to replace the harmful β-Al_5FeSi phase.

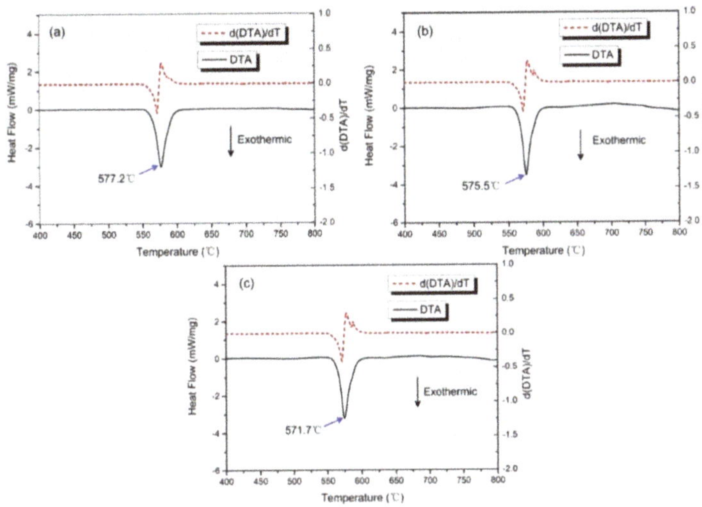

Figure 6. (**a**–**c**) DTA thermograph of C1 alloy, C2 alloy, and C3 alloy, respectively.

Figure 7 shows the tensile properties of the alloys with different cooling rates. As the cooling rates increased, the tensile strength and elongation of the alloys were significantly improved, as shown in Figure 7a. The quality index Q can be expressed as follows [24]:

$$Q = \text{UTS(MPa)} + 150 \times \log(\%\text{EI}) \tag{1}$$

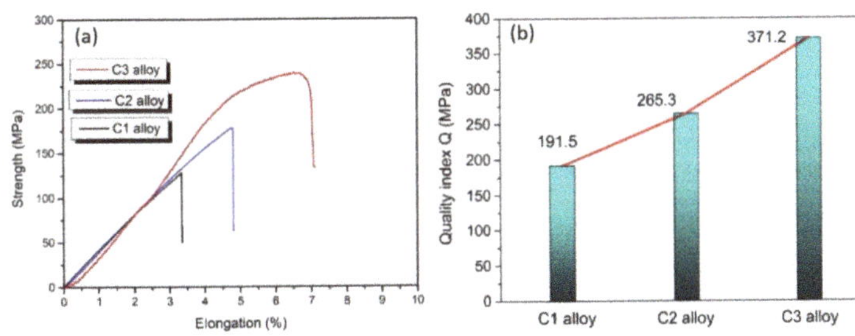

Figure 7. (**a**) Tensile properties of the alloys with different cooling rates. (**b**) The quality index Q of the three alloys.

Mass coefficient Q includes both tensile strength (UTS) and elongation after fracture (EI), which is a more widely used and reliable parameter in engineering applications. As can be seen from Figure 7, there was no significant difference between the tensile properties of the C2 alloy and the as-cast C1 alloy, which is consistent with the results of Figures 2 and 3. The morphology of the needle-like β-phase did not change completely at the moderate cooling rate but the aspect ratio decreased. However, the tensile properties of the C3 alloy with the high cooling rate are obviously improved compared with those of the C1 alloy and C2 alloy. It is mainly attributed to the transformation of the coarse needle-like β-phase into the δ-Al_3FeSi_2 granular phase. In addition, the quality index of the C3 alloy is obviously higher than those of the other two alloys. The other main reason is that the acicular eutectic silicon phase transforms into fine fibers, which significantly reduces the stress concentration around the acicular silicon phase during the tensile process. As the cooling rates increased, this remarkably reduced the aspect ratio of the silicon phase. Overall, the strengthening mechanism is considered to be associated with the morphology change of the needle-like β-phase and with the decrease in aspect ratio of the eutectic silicon phase. The optimization of the phase morphology can significantly reduce the stress concentration and improve the load transfer capacity so as to inhibit the initiation and propagation of cracks during the tensile process [25–27].

In order to better understand the failure mechanism of the alloy at different cooling rates, fracture studies were carried out on each tensile fracture sample. Figure 8a,c show the cleavage mode of the fracture with evidence of many brittle cleavage planes separated by tearing ridges. It can be observed in the high-magnification SEM fractography of Figure 8a,c that coarse secondary phases caused the initiation and fast propagation of cracks. This can be responsible for the low elongation value of the alloys. The plastic deformation of α-Al grains are inhibited because the dislocation cannot pass through the β phase by the shear or bypass mechanism [28,29]. As a result, dislocations continue to accumulate at the interface between the β-phase and aluminum matrix and result in severe stress concentrations, which lead to a higher tendency of brittle fractures. The existence of these cleavage planes indicates that these alloys exhibit brittle fracture and need sufficient modification for practical applications. It is worth noting that the fracture mode of the C3 alloy was different from that of the C1 and C2 alloys, and a large number of dimples appeared on the fracture surface, as shown in Figure 8e,f. The ductile fracture was determined by the size of the dimples, meaning that more homogenous and deeper dimples produce a higher ductility of the alloys. The existence of deep and conical dimples in the C3 alloy shows the higher ductility, which is consistent with the results shown in Figure 7. The realization of better mechanical properties is attributed to the combined effect of the harmful morphology of the β phase and to the spheroidization of the silicon phase.

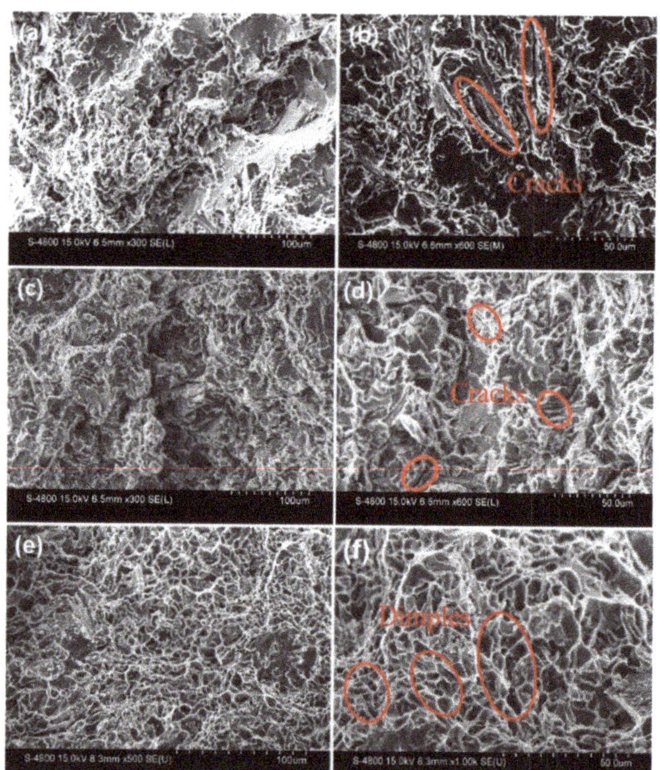

Figure 8. Fracture surface of the alloys with different cooling rates: (**a**,**c**,**e**) represent C1, C2, and C3 alloy, respectively. (**b**,**d**,**f**) are the enlarged images of the corresponding areas.

4. Conclusions

(1) The morphology of the Fe-rich phase in Al–Si alloy is greatly affected by the cooling rate. The needle-like β-Al_5FeSi phase had the size of ~80 μm at the conventional casting cooling rate. As the cooling rate increases to ~10^2 K/s, the needle-like β-phase changes into the fishbone-like δ-Al_3FeSi_2 phase. At a high cooling rate of ~10^4 K/s, the size of the β-Al_5FeSi phase decreases to ~500 nm and the morphology changes to a granular form.

(2) The change of cooling rate reduces the eutectic temperature of Fe-rich Al–Si alloy and promotes the phase transformation of eutectic silicon. The DTA analysis results showed that the increase in the cooling rate contributes to forming the fishbone-like δ-Al_3FeSi_2 phase and plays an important role in modifying the structure of the eutectic Si.

(3) The platelet-like β phase is indeed the initiation point of the crack, which can be confirmed by the fracture morphology analysis. In addition, the platelet-like β-phase accelerates crack propagation. On the contrary, the fine Fe-rich particles are relatively helpful to improve the tensile strength and elongation.

(4) The melt spinning method transforms the brittle failure of Fe-rich Al–Si alloy into ductile failure, which is of great significance to the recovery and reuse of Fe-rich Al–Si alloy and has broad industrial application prospects.

Author Contributions: Conceptualization, S.L.; methodology, S.L.; validation, S.L. and X.W.; writing—original draft preparation, X.S.; writing—review and editing, X.H. and Z.L.; visualization, P.G. and L.Z.; software, C.C.; supervision, S.L. and X.W. All authors have read and agreed to the published version of the manuscript.

Funding: This research study was funded the Natural Science Foundation of Hebei Province (no. E2021202140), Hebei Provincial Department of Human Resources and Social Security of China (grant no. B2020005006), and Postdoctoral Innovation Fund of State Key Laboratory for Reliability and Intelligence of Electrical Equipment (EERIPD 2018003).

Institutional Review Board Statement: Not applicable.

Informed Consent Statement: Not applicable.

Data Availability Statement: The data used to support the findings of this study are available from the corresponding author upon request.

Conflicts of Interest: The authors declare no conflict of interest. We declare that we have no financial and personal relationships with other people or organizations that can inappropriately influence our work.

References

1. Zhu, X.; Blake, P.; Dou, K.; Ji, S. Strengthening die-cast Al-Mg and Al-Mg-Mn alloys with Fe as a beneficial element. *Mater. Sci. Eng. A* **2018**, *732*, 240–250. [CrossRef]
2. Puncreobutr, C.; Lee, P.D.; Karech, K.M.; Connolley, T.; Fife, J.L.; Phillion, A.B. Influence of Fe-rich intermetallics on solidification defects in Al-Si-Cu alloys. *Acta Mater.* **2014**, *68*, 42–51. [CrossRef]
3. Cameron, M.D.; John, A.T.; Arne, K.D. As-cast morphology of iron-intermetallics in Al-Si foundry alloys. *Scripta Mater.* **2005**, *53*, 955–958.
4. Terzi, S.; Taylor, J.A.; Cho, Y.H.; Salvo, L.; Suery, M.; Boller, E.; Dahle, A.K. In situ study of nucleation and growth of the irregular α-Al/β-AlFeSi eutectic by 3D synchrotron X-ray microtomography. *Acta Mater.* **2010**, *58*, 5370–5380. [CrossRef]
5. Feng, S.K.; Liotti, E.; Lui, A.; Matthew, D.; Connolley, T.; Mathiesen, R.; Patrick, S. In-situ X-ray radiography of primary Fe-rich intermetallic compound formation. *Acta Mater.* **2020**, *196*, 759–769. [CrossRef]
6. Que, Z.P.; Wang, Y.; Zhang, F. Formation of the Fe-containing intermetallic compounds during solidification of Al-5Mg-2Si-0.7Mn-1.1Fe alloy. *Metall. Trans. A.* **2018**, *49*, 2173–2181. [CrossRef]
7. Gao, T.; Li, Z.Q.; Zhang, Y.X.; Qin, J.Y.; Liu, X.F. Evolution of Fe-rich phases in Mg melt and a novel method for separating Al and Fe from Al-Si-Fe alloys. *Mater. Des.* **2017**, *134*, 71–80. [CrossRef]
8. Li, X.F.; Xia, C.J.; Wu, Y.; Chen, D.; Wang, M.L.; Ma, N.H.; Wang, H.W. Effect of Er addition on the high temperature strength of Al-Si-Cu-Ni-Mg-Fe piston alloys by T5 and T6 heat treatment. *Mater. Sci.* **2019**, *25*, 376–382.
9. Wu, X.Y.; Zhang, H.R.; Zhang, F.X.; Ma, Z.; Yang, B.; Tao, T.X.; Zhang, H. Effect of cooling rate and Co content on the formation of Fe-rich intermetallics in hypoeutectic Al7Si0.3Mg alloy with 0.5%Fe. *Mater. Charact.* **2018**, *139*, 116–124. [CrossRef]
10. Tang, Q.; Zhao, J.H.; Chen, J.; He, K. The effects of neodymium addition on the intermetallic microstructure and mechanical properties of Al-7Si-0.3Mg-0.3Fe alloys. *J. Alloys Compd.* **2018**, *741*, 161–173. [CrossRef]

11. Zhang, J.Y.; Feng, J.; Zuo, L.J.; Ye, B.; Kong, X.Y.; Jiang, H.Y.; Ding, W.J. Effect of Sc microalloying addition on microstructure and mechanical properties of as-cast Al-12Si alloy. *Mater. Sci. Eng. A* **2019**, *24*, 138343. [CrossRef]
12. Khan, M.H.; Das, A.; Li, Z.; Kotadia, H.R. Effects of Fe, Mn, chemical grain refinement and cooling rate on the evolution of Fe intermetallics in a model 6082 Al-alloy. *Intermetallics* **2021**, *132*, 107132. [CrossRef]
13. Li, D.F.; Cui, C.X.; Wang, X.; Wang, Q.Z.; Chen, C.; Liu, S.Q. Microstructure evolution and enhanced mechanical properties of eutectic Al-Si die cast alloy by combined alloying Mg and La. *Mater. Des.* **2016**, *90*, 820–828. [CrossRef]
14. Easton, M.A.; StJohn, D.H. Improved prediction of the grain size of aluminum alloys that includes the effect of cooling rate. *Mater. Sci. Eng. A* **2008**, *486*, 8–13. [CrossRef]
15. Verma, A.; Kumar, S.; Grant, P.S.; O'Reilly, K.A.Q. Influence of cooling rate on the Fe intermetallic formation in an AA6063 Al alloy. *J. Alloys Compd.* **2013**, *555*, 274–282. [CrossRef]
16. Kilicaslan, M.F.; Altaib, S.S.; Vurdu, C.D. Effect of Ni Addition on the Morphology and Microstructure of Both Conventional Cast and Melt-Spun of Al-Si-Fe-Nb (at wt%) Alloy. *Met. Mater. Int.* **2019**, *25*, 1457–1466. [CrossRef]
17. Rajabi, M.; Simchia, A.; Davamia, P. Microstructure and mechanical properties of Al-20Si-5Fe-2X (X=Cu, Ni, Cr) alloys produced by melt-spinning. *Mater. Sci. Eng. A* **2008**, *492*, 443–449. [CrossRef]
18. Liu, Z.T.; Wang, B.Y.; Wang, C.; Zha, M.; Liu, G.J.; Yang, Z.Z.; Wang, J.G.; Li, J.H.; Wang, H.Y. Microstructure and mechanical properties of Al-Mg-Si alloy fabricated by a short process based on sub-rapid solidification. *J. Mater. Sci. Technol.* **2020**, *41*, 178–186. [CrossRef]
19. Liu, S.Q.; Wang, X.; Cui, C.X.; Zhao, L.C.; Liu, S.J.; Chen, C. Fabrication, microstructure and refining mechanism of in situ CeB6/Al inoculant in aluminum. *Mater. Des.* **2015**, *65*, 432–437. [CrossRef]
20. Liu, S.Q.; Wang, X.; Cui, C.X.; Zhao, L.C. Enhanced grain refinement of in situ CeB_6/Al composite inoculant on pure aluminum by microstructure control. *J. Alloys Compd.* **2017**, *701*, 926–934. [CrossRef]
21. Liu, S.Q.; Wang, X.; Cui, C.X.; Han, X.; Cui, C. Significantly improved particle strengthening of Al-Sc alloy by high Sc composition design and rapid solidification. *Mater. Sci. Eng. A* **2021**, *800*, 140304. [CrossRef]
22. Liu, S.Q.; Cui, C.X.; Wang, X.; Li, N.; Cui, S. Effect of cooling rate on microstructure and grain refining behavior of in situ CeB_6/Al composite inoculant in aluminum. *Metals* **2017**, *7*, 204. [CrossRef]
23. Han, B.H.; Liu, S.Q.; Wang, X.; Cui, C.X. Simultaneously improving strength and ductility of hybrid Al-Si matrix composite with polyphasic and multi-scale ceramic particles reinforced. *Mater. Sci. Eng. A* **2021**, *804*, 140517. [CrossRef]
24. Wu, X.F.; Wang, Z.C.; Wang, K.Y.; Zhao, R.D.; Wu, F.F. Microstructural refinement and tensile properties enhancement of Al-10Mg2Si cast alloys by copper addition. *J. Alloys Compd.* **2022**, *896*, 163058. [CrossRef]
25. Azimi, H.; Nourouzi, S.; Jamatti, R. Effects of Ti particles and T6 heat treatment on the microstructure and mechanical properties of A356 alloy fabricated by compocasting. *Mater. Sci. Eng. A* **2021**, *818*, 141443. [CrossRef]
26. Liu, T.; Pei, Z.R.; Barton, D.; Brewer, L.N. Characterization of nanostructures in a high pressure die cast Al-Si-Cu alloy. *Acta Mater.* **2022**, *224*, 117500. [CrossRef]
27. Bogdanoff, T.; Lattanzi, L.; Merlin, M.; Ghassemali, E.; Jarfors, A.E.W.; Senfeddine, S. The complex interaction between microstructural features and crack evolution during cyclic testing in heat-treated Al–Si–Mg–Cu cast alloys. *Mater. Sci. Eng. A* **2021**, *825*, 141930. [CrossRef]
28. Kim, D.; Kim, J.H.; Jeon, J.Y.; Kim, Y.W.; Kobayashi, E. Local elongation of a high Fe-containing Al-Si-Cu-Mg alloy by a deformation-semisolid extrusion process. *Mater. Lett.* **2021**, *309*, 131337. [CrossRef]
29. Que, Z.P.; Mendis, C.L. Heterogeneous nucleation and phase transformation of Fe-rich intermetallic compounds in Al–Mg–Si alloys. *J. Alloys Compd.* **2020**, *836*, 155515. [CrossRef]

materials MDPI

Article

Fabrication and Properties of Zn-3Mg-1Ti Alloy as a Potential Biodegradable Implant Material

Shuo Zhang, Pengkai Yuan, Xin Wang, Tiebao Wang, Lichen Zhao * and Chunxiang Cui *

Key Laboratory for New Type of Functional Materials of Hebei Province, School of Materials Science and Engineering, Hebei University of Technology, Tianjin 300400, China; zskln1996@163.com (S.Z.); ypk1079420647@163.com (P.Y.); ahaxin@hebut.edu.cn (X.W.); wtb@hebut.edu.cn (T.W.)
* Correspondence: zhlch@hebut.edu.cn (L.Z.); hutcui@hebut.edu.cn (C.C.)

Abstract: A Zn-3Mg-1Ti alloy was fabricated by ultrasonic treatment of Zn-Mg alloy melt using a Ti ultrasonic radiation rod. The microstructure, phase structure, mechanical properties, degradation property, and in vitro cytotoxicity were investigated systematically. The obtained Zn-3Mg-1Ti alloy is composed of the Zn, Mg_2Zn_{11}, and $TiZn_{16}$. Owing to the grain refinement and second phase reinforcement, the mechanical properties of Zn-3Mg-1Ti alloy is improved. In addition, the Zn-3Mg-1Ti alloy exhibits minimal cytotoxicity compared to pure Zn and Zn-1Ti alloy. Electrochemical tests show that the Zn-3Mg-1Ti alloy has an appropriate degradation rate in Hank's solution.

Keywords: Zn-Mg-Ti alloy; biodegradable metals; mechanical properties; cytotoxicity

Citation: Zhang, S.; Yuan, P.; Wang, X.; Wang, T.; Zhao, L.; Cui, C. Fabrication and Properties of Zn-3Mg-1Ti Alloy as a Potential Biodegradable Implant Material. *Materials* **2022**, *15*, 940. https://10.3390/ma15030940

Academic Editor: Halina Krawiec

Received: 23 December 2021
Accepted: 18 January 2022
Published: 26 January 2022

1. Introduction

Biodegradable metal implants have gained much attention owing to their good mechanical properties, excellent biocompatibility, and degradability [1]. After completing the purpose of repair and treatment, it can be completely dissolved in the human body, avoiding injury to the patient from a second operation [2,3]. Recently, biodegradable metals such as Mg, Fe, and Zn have been increasingly accepted as implant materials [4–11]. However, Fe-based alloys and Mg-based alloys have not been able to solve the problem of too slow and too rapid degradation. Among the three candidate metals, Zn and Zn-based alloys have drawn increasing concern due to their moderate degradation rate [12–14]. Zinc is an essential nutrient element that participates in the metabolic processes of the human body and is also a constituent element of a variety of enzymes [13–15]. Zn affects the regulation of taste, vision immune function, sexual function, and plays a crucial role in protein synthesis, deoxyribonucleic acid synthesis. In addition, Zn can also induce Ca-P deposition, stimulate bone formation, growth, and mineralization. [16–20]. As a biodegradable material with good biocompatibility and an acceptable degradation rate, pure zinc has been intensively investigated by researchers. However, its poor mechanical properties severely limit its application [14,15]. As we all know, alloying can usually improve mechanical properties. Up to now, more than ten kinds of Zn alloy systems have been developed for biomedical applications. Considering biocompatibility, most biomedical zinc alloys contain nutrient elements [21]. For example, Mg [22], Sr [22], Ca [22], Cu [23], Mn [24], Ag [25], Fe [26], and Li [27]. Considering the biocompatibility and mechanical properties, Mg is the preferred alloying element. Mg is an indispensable trace element in the normal life activities and metabolic processes of the human body, and participates in various metabolic activities of the human body, including the formation of bone cells, activation of various enzymes, and protein synthesis [28,29]. So far, many Zn-Mg alloys have been reported, such as Zn-0.8Mg [30], Zn-1Mg [22], Zn-1.2Mg [31], Zn-1.6Mg [32], Zn-3Mg [33], Zn-0.8Li-0.4Mg [34], Zn-1Mg-0.5Ca [35], Zn-3Cu-1Mg [36] alloys. The results show that the Zn-Mg alloy can improve mechanical properties and biocompatibility, but has little effect on the corrosion properties. In particular, the Zn-3Mg alloy has excellent biocompatibility [33]. In addition,

Ti and some of its alloys are widely used as implant materials in vivo due to their excellent biocompatibility and mechanical properties [37]. Zn-Ti alloy can improve the mechanical properties and biocompatibility, and promote the degradation of the alloy [38,39]. To date, no researchers have systematically investigated Zn-Mg-Ti alloys. In addition, the ultrasonic treatment process can effectively refine the grains and improve the mechanical properties of the alloy [40].

This paper aims to prepare a Zn-Mg-Ti alloy by ultrasonic treatment with a Ti ultrasonic radiation rod, and the microstructure, phase structure, mechanical properties, degradation property, and cytotoxicity were systematically investigated.

2. Materials and Methods

2.1. Alloy Preparation

Commercially pure zinc (purity, 99.995%) and pure magnesium (purity, 99.995%) ingots were used as raw materials. A pure titanium rod with a diameter of 30 mm and a length of 130 mm was used as the ultrasonic radiation rod. The ultrasonic radiation rod was first preheated in pure zinc melt (660 °C) for 60 s. At the same time, a certain amount of Mg was dissolved in another zinc melt (600 °C) and 99.999% argon was used as a protective gas to prevent oxidation. The Zn-Mg melt was next ultrasonically treated for 60 s at a frequency of about 19,000 Hz. After ultrasonic processing, the melt was poured into a steel mold (φ20 mm) to obtain a Zn-Mg-Ti alloy. For comparison, pure zinc melt (600 °C) was also ultrasonically treated to obtain a Zn-Ti alloy. The actual chemical compositions of the zinc alloys determined by X-ray fluorescence spectrometer (XRF, ZSX Primus 2) are given in Table 1. Hereafter, the obtained Zn-Mg-Ti alloy is designated as Zn-3Mg-1Ti, the obtained Zn-Ti alloy is designated as Zn-1Ti.

Table 1. Chemical compositions of the specimens.

Specimens	Mg (wt.%)	Ti (wt.%)	Zn (wt.%)
Zn-1Ti	0	1.01	Bal.
Zn-3Mg-1Ti	3.03	0.94	Bal.

2.2. Microstructure and Phase Characterization

The microstructure and morphology of the specimens were characterized by using an optical microscope (Zeiss AxioCam ICc5, Jena, Germany) and a scanning electron microscope (SEM, S-4800, Hitachi Ltd., Tokyo, Japan) with an EDAX energy dispersive spectrometer (EDS). Phase structures of alloys were determined by X-ray diffraction (XRD, Bruker D8 Discover, Rheinstetten, Germany, 6°/min) using Cu K_α radiation. The high-resolution transmission electron microscopy (HRTEM) images of the specimens were obtained by transmission electron microscopy (TEM, Titan Themis G2, The Thermo Fisher Scientific, Waltham, MA, USA). The high-angle annular dark-field scanning transmission electron microscopy (HAADF-STEM) images of the specimens were obtained by spherical aberration correction scanning transmission electron microscope (ac-STEM, FEI-Tecnil-Talos-F200, The Thermo Fisher Scientific). Overall TEM foil specimens were treated by a FIB-SEM crossbeam workstation (Helios Nanolab 600i, The Thermo Fisher Scientific).

2.3. Mechanical Test

The mechanical property tests of the specimens include hardness test and compressive test. The HMV-2T microhardness tester was used to test the Vickers hardness of the specimens. The load applied for the test was 100 g, and the holding time was 15 s. Under the same condition, each specimen was tested at least five times, and the average value was taken as the microhardness (HV) of the specimen. The compression test was carried out by using a WDW-300 universal tester with a crosshead speed of 1 mm/min at room temperature. Cylindrical specimens 4 mm in diameter and 8 mm long prepared by wire electrical discharge machining were used for compression tests. Since the compliance of the testing machine significantly affects the accuracy of the elastic modulus, the elastic moduli of the specimens were not obtained by the stress-strain curves.

2.4. Electrochemical Test

The electrochemical test was carried out in Hank's solution at 37 °C with an electrochemical workstation (CHI660E, Corrtest Instruments, Wuhan, China). A three-electrode cell method was used for electrochemical measurements. The specimen with 1.0 cm^2 exposed area was used as working electrode, the reference electrode was a saturated calomel electrode (SCE) and the counter electrode was a graphite rod. The pH value of Hank's solution [15] was adjusted to 7.4 with HCl solution and $NaHCO_3$ solution. The open-circuit potential was first measured in Hank's solution for 2400 s to reach stability. Then, the potentiodynamic polarization curve was determined at a scanning rate of 1.0 mV/s. The corrosion current density (I_{corr}) was estimated by linear fit and Tafel extrapolation using cathodic and anodic branches of the polarization curve. The corrosion rate of the specimen was calculated as follows [23,24,41]:

$$CR = K\frac{I_{corr}M}{n\rho} \quad (1)$$

where *CR* is corrosion rate, *K* is 3.27×10^{-3} mm g μA^{-1} cm^{-1} yr^{-1}, I_{corr} is the corrosion current density in $\mu A/cm^2$, *M* is the atomic mass fraction, *n* is the number of transfer electron, ρ is the density of specimens.

2.5. Cytotoxicity Test

Mouse fibroblasts (L-929) were used to evaluate the cytotoxicity of the specimens. The cytotoxicity test was conducted by an indirect contact method. First, the specimen was immersed in a sterile Petri dish containing 75% absolute ethyl alcohol for 1 h with ultraviolet lamp irradiation. After that, the specimen was air-dried under ultraviolet irradiation, and leached in Roswell Park Memorial Institute (RPMI) 1640 culture solution at 37 °C for 24 h, with the leaching ratio of 0.2 g/mL. Then, L929 cells in the logarithmic phase were digested with 0.25% trypsin and blown into a single cell suspension. The cell concentration was adjusted to approximately 2×10^4 mL^{-1} with RPMI 1640 medium containing 10% calf serum, and 100 µL per well was inoculated into a 96-well plate. The plate was incubated at 37 °C for 24 h in a 5% CO_2 incubator. After the incubation, the supernatant in each well was cleaned up, and 100 µL of extracts with concentrations of 100%, 50%, 25%, 12.5%, and 6.25% were added to the specimen group, 5% or 10% dimethyl sulfoxide (DMSO) solution was added to the positive control group, and fresh culture solution was added to the cell control group. In, addition, the culture plates were cultured in 37 °C and 5% CO_2 incubator for 1, 3, and 5 days, respectively. After the culture, 10 µL thiazole blue (MTT) solution of 5 mg mL^{-1} was added to each well, and continued to incubate at 37 °C for 4 h. Then, 180 µL DMSO was added to each well after aspirating the supernatant. Finally, the absorbance (OD) was measured at 570 nm with a microplate reader, the formula of cell proliferation rate (RGR) is as follows [42,43]:

$$RGR = \frac{OD_{specimen}}{OD_{control}} \times 100\% \quad (2)$$

3. Results

3.1. Microstructures and Phase Structures of the Alloys

The metallographic microstructure of the Zn-3Mg-1Ti alloy is shown in Figure 1a. The optical micrograph exhibits that the alloy is mainly composed of laminar and rod-like eutectic mixtures. In addition, some bulks are distributed in the eutectic structures. The SEM images of the Zn-3Mg-1Ti alloy are shown in Figure 1b,c. It can be clearly seen that the morphology of the alloy is basically the same as the result of the optical micrograph. In addition, the eutectic structure has not only a layered structure, but also a rod-like structure.

For comparison, the SEM image of the Zn-1Ti alloy is shown in Figure 1d. The grains of α-Zn are separated by white discontinuous network precipitates. Compared with the as-cast pure zinc, the alloy crystal grains are refined [24]. Similar to Zn-3Mg-1Ti alloy, some blocks with regular shapes are embedded in the alloy. Combined with the EDS results (Figure 1e,f), the eutectic structure is rich in Mg and the block-like phase is a Ti-rich phase. Besides, the atomic ratio of Zn and Ti is approximately 16:1. According to the Zn-Mg phase diagram [44], the Zn-Ti phase diagram [45], and the XRD pattern (Figure 2), it can be concluded that the eutectic mixtures are composed of α-Zn and Mg_2Zn_{11}, the Ti-rich phase may be $TiZn_{16}$.

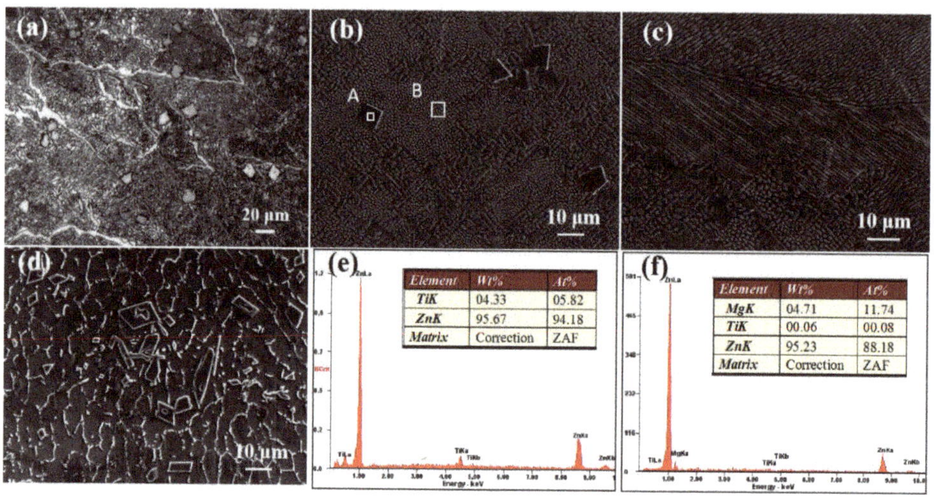

Figure 1. (**a**) Optical micrograph of the Zn-3Mg-1Ti alloy; (**b**,**c**) SEM images of Zn-3Mg-1Ti alloy; (**d**) SEM image of Zn-1Ti alloy; (**e**,**f**) the EDS results of the area A and B in (**b**), respectively.

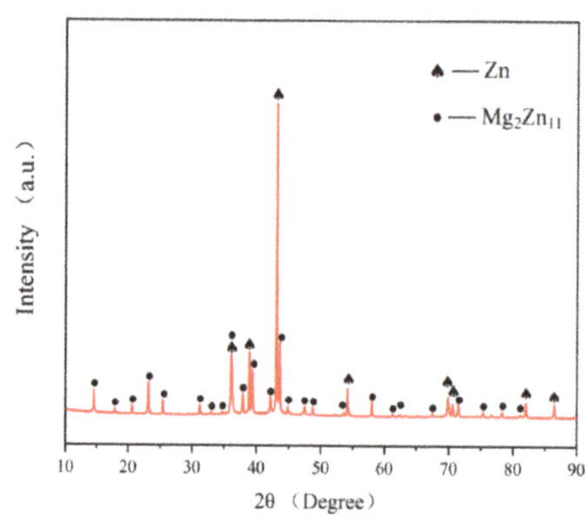

Figure 2. XRD pattern of Zn-3Mg-1Ti alloy.

To further investigate and qualitative analysis of the Ti-rich phase, the Zn-3Mg-1Ti alloy was observed by TEM. The TEM image of the alloy is shown in Figure 3a. It can be found that the image includes dark areas and bright areas. Figure 3b is the HRTEM image of the specimen, and the image was further filtered by fast Fourier transform (FFT). The FFT electron diffraction patterns of zone C and zone B as shown in Figure 3c,d, respectively. The characterization results show that the HRTEM image contains Mg_2Zn_{11} phase (zone A), $TiZn_{16}$ phase (zone C), and the $TiZn_{16}$/ Mg_2Zn_{11} interface (zone B). In addition, the $TiZn_{16}$ phase has an ideal interface matching with the Mg_2Zn_{11} phase, which is (330)$TiZn_{16}$//(223)Mg_2Zn_{11}. The orientation relationship is a typical coherent interface. It should be noted that some dislocations can also be observed in the interface (Figure 3e).

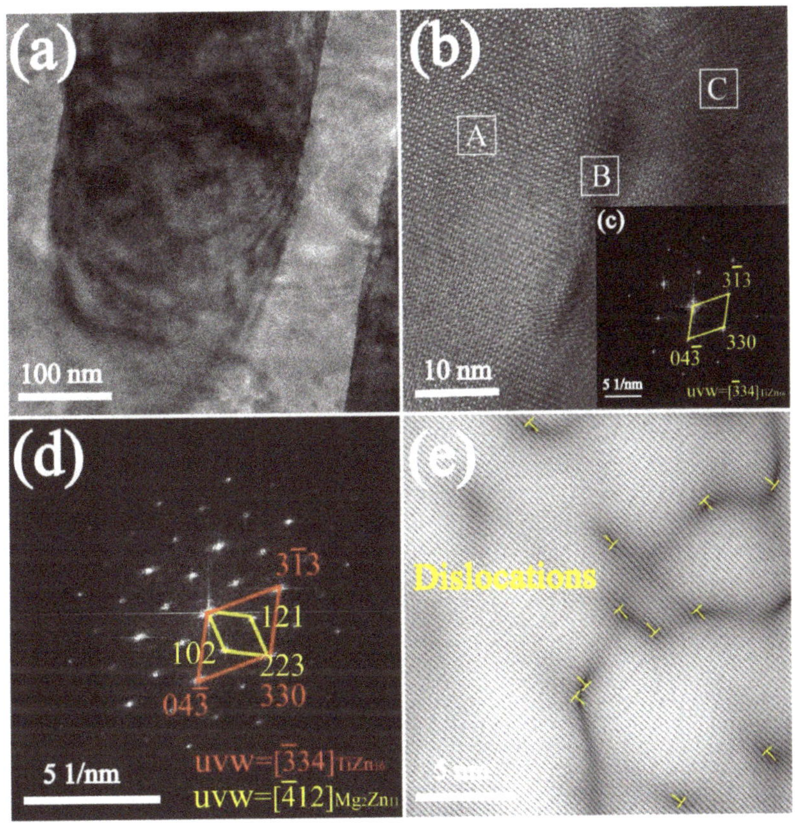

Figure 3. (**a**) TEM image of Zn-3Mg-1Ti alloy; (**b**) HRTEM image of the alloy; (**c**,**d**) FFT electron diffraction patterns of zone C and zone B in (**b**), respectively; (**e**) IFFT image of zone B in (**b**).

Figure 4a,b are the HAADF-STEM images of Zn-3Mg-1Ti alloy. It can be clearly seen that Figure 4a has no lattice distortion. According to the measured results, the interplanar spacing of 2.11 Å and 2.31 Å corresponds to $(3\bar{1}3)$ and $(04\bar{3})$ planes of $TiZn_{16}$ (65-2261), respectively. The measured angle between $(3\bar{1}3)$ and $(04\bar{3})$ planes is approximately 68.6 °, which is also consistent with the theoretical value. Figure 4b exhibits that the part lattices of $TiZn_{16}$ phase are strongly distorted.

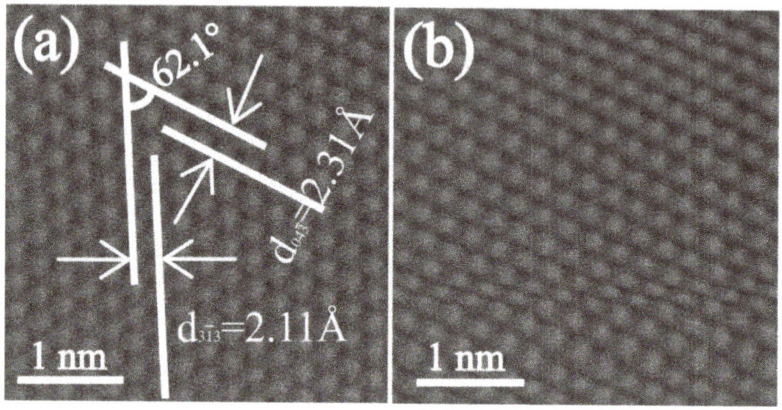

Figure 4. HAADF-STEM images of Zn-3Mg-1Ti alloy: (**a**) no lattice distortion; (**b**) lattice distortion.

3.2. Mechanical Properties of the Alloys

As we all know, the compressive strength (<22 MPa) and microhardness (41 HV) of pure zinc are very low, which cannot meet the requirements of biodegradable implants [15,23]. Figure 5a,b show the compression curves and microhardness histogram of the specimens. It can be clearly seen that after ultrasonic treatment and alloying with Ti, the compressive strength and microhardness of the Zn-1Ti alloy reach 265.1 MPa and 71.3 HV. The Zn-3Mg-1Ti alloy reaches 625.1 MPa and 226.4 HV. The results show that Zn-3Mg-1Ti alloy has more excellent compressive strength and microhardness. However, the plasticity of the Zn-3Mg-1Ti alloy is poor. This may be related to the formation of the intermetallic compound (Mg_2Zn_{11}) in the alloy.

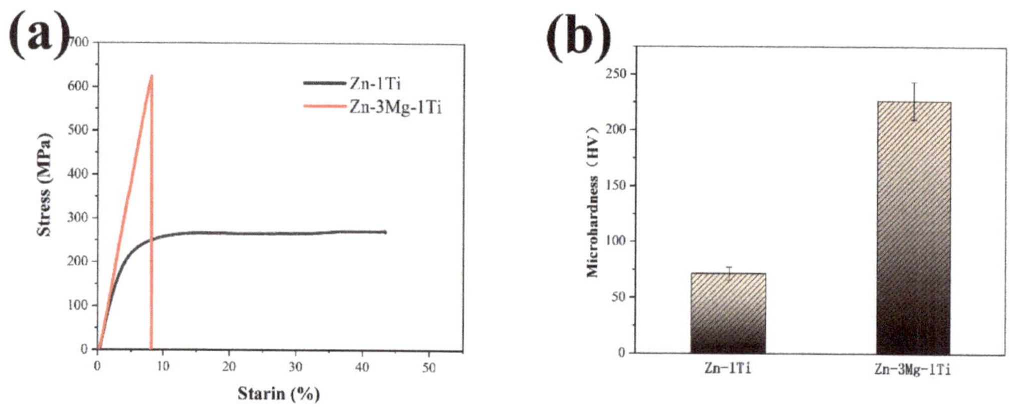

Figure 5. (**a**) Compressive stress–strain curves of specimens; (**b**) microhardness of specimens.

3.3. Electrochemical Characterization of the Alloys

Figure 6 shows the potentiodynamic polarization curves of the specimens. The corrosion potentials (E_{corr}), corrosion current densities (I_{corr}), and Tafel slopes (b_a and b_c) directly derived from the polarization curves are listed in Table 2. The calculated polarization resistance (R_p) and corrosion rate are also presented in Table 2. It can be seen that the corrosion potential of Zn-3Mg-1Ti alloy is close to that of Zn-1Ti alloy, but lower than that of pure Zn. However, the corrosion current density of the Zn-3Mg-1Ti alloy is lower than the Zn-1Ti alloy and higher than the pure Zn. After calculation, the corrosion rate of the Zn-3Mg-1Ti alloy (78.5 μm/y) is higher than pure Zn (35.1 μm/y) and lower than the Zn-1Ti alloy (145.9 μm/y).

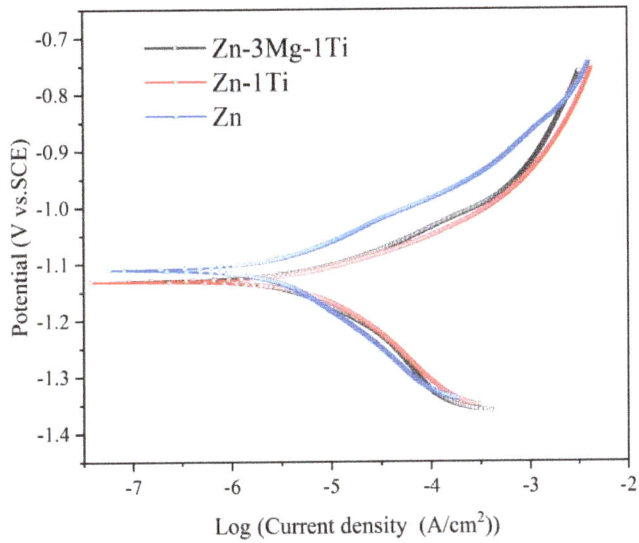

Figure 6. Potentiodynamic polarization curves of the specimens.

Table 2. Electrochemical parameters of Zn-3Mg-1Ti, Zn-Ti alloy and pure Zn.

Specimens	E_{corr}/V_{SCE}	I_{corr} ($\mu A/cm^2$)	b_a (mV/dec)	b_c (mV/dec)	R_p ($k\Omega/cm^2$)	Corrosion Rate(μm/y)
Zn	−1.110	2.35	70.0	130.9	8.44	35.1
Zn-1Ti	−1.131	9.77	79.4	178.4	2.44	145.9
Zn-3Mg-1Ti	−1.130	5.26	80.7	125.9	4.06	78.5

3.4. Cytotoxicity of the Alloys

Figure 7a–c shows the RGR of L-929 cells cultured in 6.25%, 12.5%, 25%, 50%, and 100% extracts of pure Zn, Zn-1Ti alloy and Zn-3Mg-1Ti alloy, respectively. The results show that Zn-3Mg-1Ti alloy, Zn-1Ti alloy, and pure Zn cultured in 6.25%, 12.5%, 25% extracts have good cell viability, and the cytotoxicity is basically non-toxic (cytotoxicity grade 0). However, with the increase of cell extract culture concentration and culture time, the RGR in pure Zn decreased dramatically and the cytotoxicity increased to severe toxicity (cytotoxicity grade 3–4). After being cultured in 50% extracts for 5 days, the Zn-1Ti alloy showed a slight decrease in cell activity (cytotoxicity grade 1–2), but when cultured in 100% extracts, the cytotoxicity increased significantly (cytotoxicity grade 4). Nevertheless, after being cultured in 100% extract for 5 days, the cytotoxicity of the Zn-3Mg-1Ti alloy is still slightly toxic (cytotoxicity grade 2). The results suggest that the biocompatibility of Zn-3Mg-1Ti alloy is better than that of Zn-1Ti alloy and pure Zn. In addition, the measured Zn^{2+} ion concentration in the extract of the Zn-3Mg-1Ti alloy is (3.1792 μg/mL), which is the lowest among these specimens (Figure 7d).

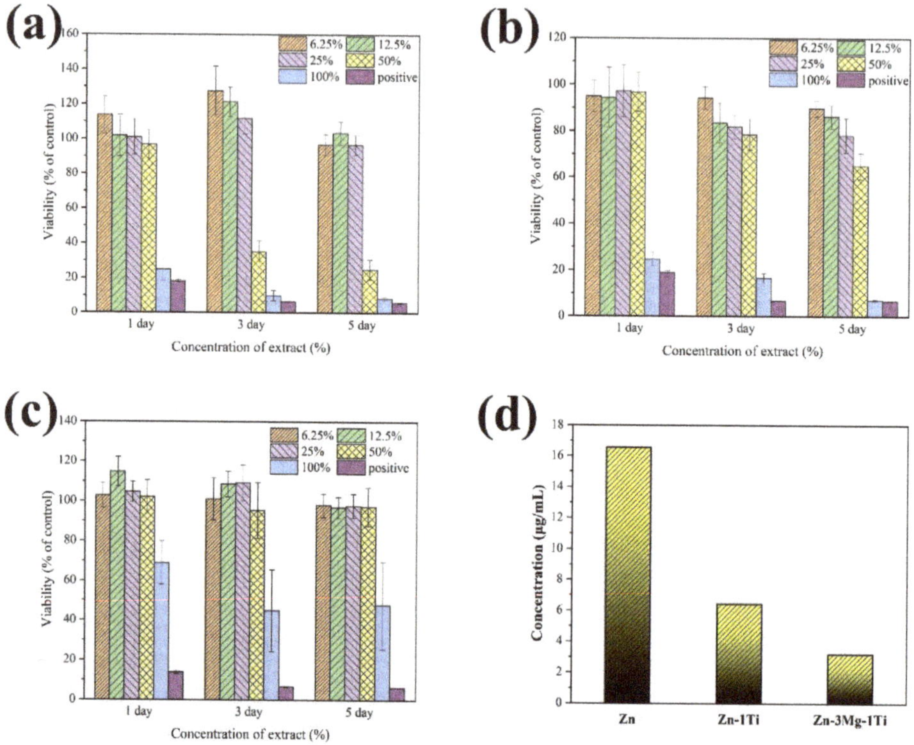

Figure 7. Cell viability after culturing with (**a**) pure Zn, (**b**) Zn-1Ti, (**c**) Zn-3Mg-1Ti extracts for 1, 3, and 5 d; (**d**) Zn^{2+} ion concentration in the extract.

4. Discussion

4.1. Influence of Ultrasonic Treatment

After ultrasonic treatment, Ti element was introduced into the Zn and Zn-Mg melts by diffusion. This can solve the problem that Zn alloy containing Ti is not easy to be prepared because the melting point of Zn (419.53 °C) is too different from that of Ti (1668 °C). Besides, ultrasonic treatment can not only introduce Ti element, but also reduce the casting defects and specific gravity segregation [40,44]. Not only that, as shown in Figure 8, when the zinc melt is ultrasonically processed, the sound waves propagating into the melt medium will generate alternating high-pressure (compression) and low-pressure (rarefaction) cycles, and the cavitation bubbles will undergo a series of dynamic processes of nucleation, growth, and collapse. The tremendous energy and high-pressure shock wave generated by this cavitation effect will break the primary crystals in the zinc alloy melt and increase the number of heterogeneous crystal nuclei. These nucleation particles spread to all areas of the melt driven by the sound flow, which greatly increases the probability of equiaxial crystal nuclei in the melt [45–47].

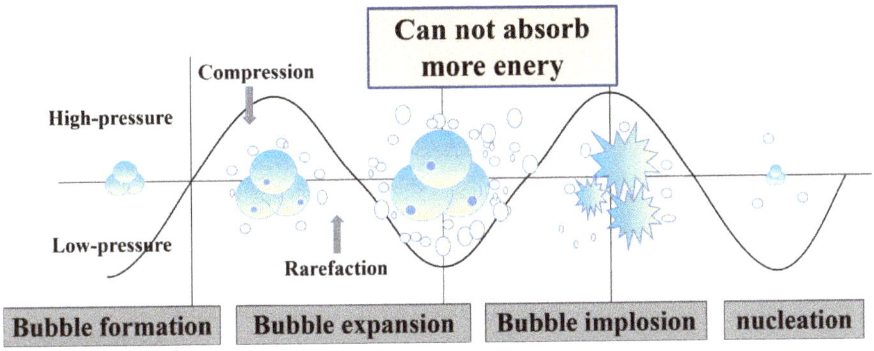

Figure 8. Schematic diagram of ultrasonic cavitation principle.

4.2. Microstructures and Mechanical Properties of the Alloys

According to the Zn-Ti phase diagram [45], when the melt temperature reaches 418.6 °C, the eutectic reaction of $L \rightarrow \alpha - Zn + TiZn_{16}$ will occur in the melt. Zn and Ti will form the $TiZn_{16}$ phase, and some eutectic structures are distributed on the grain boundaries, which significantly refine the grains [38,48]. Compared with the Zn-1Ti alloy, the morphology of Zn-3Mg-1Ti alloy changed correspondingly with the addition of Mg. Zn-3Mg-1Ti alloy is mainly composed of bulky $TiZn_{16}$ phase and eutectic mixture of α-Zn and Mg_2Zn_{11}. The phase diagram of Zn-Mg shows that eutectic reaction will occur when the temperature is 364 °C and the mass fraction of Mg is 3 wt.% ($L \rightarrow \alpha - Zn + Mg_2Zn_{11}$). Therefore, the diffraction peaks of Zn and Mg_2Zn_{11} can be observed in the XRD pattern of the Zn-3Mg-1Ti alloy (Figure 2). However, due to the low content of the Ti element and fast scanning speed, $TiZn_{16}$ was not shown in the XRD pattern. In addition, the TEM test confirmed the existence of $TiZn_{16}$ phase and found that Mg_2Zn_{11} and $TiZn_{16}$ had an excellent orientation relationship.

The compressive strength and microhardness of the Zn-1Ti alloy (265.1 MPa, 71.3 HV) far exceed pure Zn. This is because the addition of Ti refines the grains of the alloy and produces the second phase of $TiZn_{16}$. According to the Hall-Petch relationship [49], the smaller the grains of the alloy, the higher its strength. In addition, the second phase strengthening also plays a positive role. When the Mg element is introduced to form a ternary alloy, large amounts of eutectic structures are produced. The compression strength and microhardness of Zn-3Mg-1Ti alloy (625.1 MPa, 226.4 HV) are higher than that of Zn-1Ti alloy, which might be attributed to the increase of the eutectic mixtures. Mg_2Zn_{11} is a hard and brittle phase, which will significantly improve the hardness and strength of the alloy, but inevitably sacrifice a part of plasticity. Furthermore, based on the edge-by-edge matching model [50,51], the mismatch between Mg_2Zn_{11} phase and $TiZn_{16}$ phase is calculated to be 3.7%, which is less than 6%. In addition, the Mg_2Zn_{11} phase and $TiZn_{16}$ phase have a coherent orientation relationship, resulting in a strong combination between Mg_2Zn_{11} and $TiZn_{16}$. The load can be efficiently transferred from one phase to another reinforcing phase so that the hard phase with higher strength can share the load and improve the strength and hardness of the alloy [52]. The lattice distortion is also one of the factors for the improvement of mechanical properties. Therefore, the synergistic effects of grain refinement strengthening, second phase strengthening, lattice distortion and load transfer jointly improve the mechanical properties of the alloy.

4.3. Corrosion Behavior of the Alloys

After adding Ti, the corrosion rate of the alloy is significantly increased. This is due to the potential difference between $TiZn_{16}$ phase and α-Zn in the Zn-1Ti alloy, and galvanic corrosion occurred in Hank's solution, thereby accelerating the corrosion process of the Zn alloy. The potential of the $TiZn_{16}$ phase is lower than that of α-Zn, and it is usually regarded as the anode of micro-battery to be preferentially corroded. Both the Ti-rich phase and the Mg-rich phase will undergo galvanic corrosion with the zinc matrix and be preferentially corroded, thereby accelerating the degradation rate of the alloy. However, Yao et al. [53] and Yang et al. [54] reported that the Mg_2Zn_{11} phase will reduce the electrode potential difference and inhibit the progress of corrosion. Therefore, the degradation rate of Zn-3Mg-1Ti alloy with Mg_2Zn_{11} phase is lower than that of Zn-Ti.

4.4. Cytotoxicity Assessment of the Alloys

Cytotoxicity evaluation showed that the cell activity of the Zn-3Mg-1Ti alloy was better than that of the Zn-1Ti alloy, while pure zinc has the worst cell activity. The results can be attributable to the Zn^{2+} concentration of the extract. It is well-known that excessive local Zn^{2+} will promote cell death [55]. Our experimental results show that the order of the Zn^{2+} concentration in the leaching solution is as follows: Zn-3Mg-1Ti alloy < Zn-1Ti alloy < pure Zn. Although the degradation rate of Zn-3Mg-1Ti alloy is higher than that of pure zinc and lower than that of Zn-1Ti, it releases the least Zn^{2+} ions. This is due to the $TiZn_{16}$ phase acting as a sacrificial anode to protect the cathode, which reduces the release rate of Zn^{2+} ion, and a large amount of Mg_2Zn_{11} precipitates in the alloy which will reduce the degradation rate of the alloy [53]. Not only that, both Ti and Mg elements are beneficial to the human body and have important physiological functions in the human body. Combining the above reasons, Zn-3Mg-1Ti alloy has very low cytotoxicity and excellent biocompatibility.

5. Conclusions

In this study, the Zn-3Mg-1Ti alloy was fabricated by ultrasonic treatment with a Ti ultrasonic radiation rod. The microstructure, phase structure, compressive strength, microhardness, degradation rate, and cytotoxicity of the alloy were studied. The main conclusions are as follows:

1. Ultrasonic treatment with Ti rod can introduce Ti element, refine the grains of the alloy, and promote the number of heterogeneous nucleation cores;
2. Zn-3Mg-1Ti alloy is mainly composed of Mg_2Zn_{11}, $TiZn_{16}$, and α-Zn. The compressive strength and microhardness of the alloy are excellent. But the alloy is relatively brittle and has poor plasticity;
3. The electrochemical test shows that the Zn-3Mg-1Ti alloy has a suitable degradation rate and is a very promising biodegradable implant material;
4. The Zn-3Mg-1Ti alloy exhibits minimal cytotoxicity and excellent biocompatibility.

In a word, the Zn-3Mg-1Ti alloy has high application potential as a biodegradable implant. We will further improve the plasticity of the alloy and optimize its comprehensive properties in future research.

Author Contributions: Conceptualization, S.Z. and L.Z.; methodology, S.Z. and L.Z.; formal analysis, S.Z.; investigation, X.W. and P.Y.; data curation, X.W.; supervision, T.W.; writing—original draft preparation, S.Z. and L.Z.; writing—review and editing, S.Z. and L.Z. and C.C.; Funding acquisition: L.Z. and C.C; All authors have read and agreed to the published version of the manuscript.

Funding: This work was supported by Science and Technology Research Project of Hebei Province Colleges and Universities with No. ZD2021034.

Institutional Review Board Statement: Not applicable.

Informed Consent Statement: Not applicable.

Data Availability Statement: The data presented in this study are available on request from the corresponding author.

Conflicts of Interest: The authors declare no conflict of interest.

References

1. Hermawan, H.; Mantovani, D. Degradable metallic biomaterials: The concept, current developments and future directions. *Minerva Biotecnol.* **2009**, *21*, 207–216.
2. Hernández-Escobar, D.; Champagne, S.; Yilmazer, H.; Yilmazer, H.; Dikici, B.; Boehlert, C.J.; Hermawan, H. Current status and perspectives of zinc-based absorbable alloys for biomedical applications. *Acta Biomater.* **2019**, *97*, 1–22. [CrossRef] [PubMed]
3. Yuan, G.; Zhang, J.; Ding, W. Research progress of Mg-based alloys as degradable biomedical materials. *Mater. China* **2011**, *30*, 44–50.
4. Witte, F. The history of biodegradable magnesium implants: A review. *Acta Biomater.* **2010**, *6*, 1680–1692. [CrossRef] [PubMed]
5. Zheng, Y.F.; Gu, X.N.; Witte, F. Biodegradable metals. *Mater. Sci. Eng. R.* **2014**, *77*, 1–34. [CrossRef]
6. Hermawan, H. *Biodegradable Metals from Concept to Applications*; Springer: Berlin/Heidelberg, Germany, 2012.
7. Seitz, M.; Durisin, M.; Goldman, J.; Drelich, J.W. Recent advances in biodegradable metals for medical sutures: A critical review. *Adv. Healthc. Mater.* **2015**, *4*, 1915–1936. [CrossRef]
8. Heiden, M.; Walker, E.; Stanciu, L. Magnesium, iron and zinc alloys, the trifecta of bioresorbable orthopaedic and vascular implantation—A review. *J. Biotechnol. Biomater.* **2015**, *5*, 178.
9. Li, H.; Zheng, Y.; Qin, L. Progress of biodegradable metals. *Proc. Nat. Sci.-Mater. Int.* **2014**, *24*, 414–422. [CrossRef]
10. Im, S.H.; Jung, Y.; Kim, S.H. Current status and future direction of biodegradable metallic and polymeric vascular scaffolds for next-generation stents. *Acta Biomater.* **2017**, *60*, 3–22. [CrossRef] [PubMed]
11. Hermawan, H. Updates on the research and development of absorbable metals for biomedical applications. *Prog. Biomater.* **2018**, *7*, 93–110. [CrossRef]
12. Zhao, L.C.; Zhang, Z.; Song, Y.T.; Liu, S.J.; Wang, X.; Cui, C.X. Mechanical properties and in vitro biodegradation of newly developed porous Zn scaffolds for biomedical applications. *Mater. Des.* **2016**, *108*, 136–144. [CrossRef]
13. Uddin, M.S.; Hall, C.; Murphy, P. Surface treatments for controlling corrosion rate of biodegradable Mg and Mg-based alloy implants. *Sci. Technol. Adv. Mat.* **2015**, *16*, 053501. [CrossRef] [PubMed]
14. Bowen, P.K.; Drelich, J.; Goldman, J. Zinc exhibits ideal physiological corrosion behavior for bioabsorbable stents. *Adv. Mater.* **2013**, *25*, 77–82. [CrossRef] [PubMed]
15. Vojtěch, D.; Kubásek, J.; Šerák, J. Mechanical and corrosion properties of newly developed biodegradable Zn-based alloys for bone fixation. *Acta Biomater.* **2011**, *7*, 3515–3522. [CrossRef] [PubMed]
16. Fraga, C.G. Relevance, essentiality and toxicity of trace elements in human health. *Mol. Aspects Med.* **2005**, *26*, 235–244. [CrossRef]
17. Tubek, S. Selected zinc metabolism parameters in women with arterial hypotension. *Biol. Trace Elem. Res.* **2007**, *116*, 73–79. [CrossRef]
18. Lin, W.; Qin, L.; Qi, H.; Zhang, D.; Zhang, G.; Gao, R.; Qiu, H.; Xia, Y.; Cao, P.; Wang, X. Long-term in vivo corrosion behavior, biocompatibility and bioresorption mechanism of a bioresorbable nitrided iron scaffold. *Acta Biomater.* **2017**, *54*, 454–468. [CrossRef]
19. Huan, Z.G.; Zhou, J.; Leeflang, M.A.; Fratila, L.E.; Duszczyk, J. In vitro degradation behavior and cytocompatibility of Mg–Zn–Zr alloys. *J. Mater. Sci. Mater. Med.* **2010**, *21*, 2623–2635. [CrossRef]
20. Tapiero, H.; Townsend, D.M.; Tew, K.D. Trace elements in human physiology and pathology: Zinc and metallothioneins. *Biomed. Pharmacother.* **2003**, *57*, 386–398. [CrossRef]
21. Venezuela, J.; Dargusch, M.S. The influence of alloying and fabrication techniques on the mechanical properties, biodegradability and biocompatibility of zinc: A comprehensive review. *Acta Biomater.* **2019**, *87*, 1–40. [CrossRef]
22. Li, H.F.; Xie, X.H.; Zheng, Y.F.; Cong, Y.; Zhou, F.Y.; Qiu, K.J.; Wang, X.; Chen, S.H.; Huang, L.; Tian, L. Development of biodegradable Zn-1X binary alloys with nutrient alloying elements Mg, Ca and Sr. *Sci. Rep.* **2015**, *5*, 10719. [CrossRef]
23. Claudia, G.M.; Laura, C.C.; Judit, B.P.; Andrea, M.; Emilio, J.P.; Ginebra, M.P.; José, L.C.; Marta, P. Zn-Mg and Zn-Cu alloys for stenting applications: From nanoscale mechanical characterization to In Vitro degradation and biocompatibility. *Bioact. Mater.* **2021**, *6*, 4430–4446.
24. Liu, X.; Sun, J.; Zhou, F.; Yang, Y.; Chang, R.; Qiu, K.; Pu, Z.; Li, L.; Zheng, Y. Micro-alloying with Mn in Zn-Mg alloy for future biodegradable metals application. *Mater. Des.* **2016**, *94*, 95–104. [CrossRef]
25. Li, P.; Christine, S.; Ernst, S.; Frank, R.; Alexander, H.; Claudia, L.; Klotz, U.E.; Jürgen, G.G.; Lutzet, S. Mechanical characteristics, in vitro degradation, cytotoxicity, and antibacterial evaluation of Zn-1.0Ag alloy as a biodegradable material. *Int. J. Mol. Sci.* **2018**, *19*, 755. [CrossRef]
26. Kafri, A.; Ovadia, S.; Yosafovich-Doitch, G.; Aghion, E. In vivo performances of pure Zn and Zn–Fe alloy as biodegradable implants. *J. Mater. Sci. Mater. Med.* **2018**, *29*, 94. [CrossRef]
27. Shan, Z.; Seitz, J.M.; Eifler, R.; Maier, H.J.; Roger, J.; Earley, E.J.; Drelich, A.; Goldman, J.; Drelich, J.W. Zn-Li alloy after extrusion and drawing: Structural, mechanical characterization, and biodegradation in abdominal aorta of rat. *Mater. Sci. Eng. C* **2017**, *76*, 301–312.
28. Vormann, J. Magnesium: Nutrition and metabolism. *Mol. Asp. Med.* **2003**, *24*, 27–37. [CrossRef]
29. Touyz, R.M. Magnesium in clinical medicine. *Front Biosci.* **2004**, *9*, 1278–1293. [CrossRef] [PubMed]
30. Liu, S.Y.; Kent, D.; Doan, N.; Dargusch, M.; Wang, G. Effects of deformation twinning on the mechanical properties of biodegradable Zn-Mg alloys. *Bioact. Mater.* **2019**, *4*, 8–16. [CrossRef]

31. Shen, C.; Liu, X.; Fan, B.; Lan, P.; Zhou, F.Y.; Li, X.; Guo, Z.; Pu, Z.; Zheng, Y.F. Mechanical properties, in vitro degradation behavior, hemocompatibility and cytotoxicity evaluation of Zn-1.2Mg alloy for biodegradable implants. *RSC Adv.* **2016**, *6*, 86410–86419. [CrossRef]
32. Kubásek, J.; Vojtěch, D.; Jablonská, E.; Lipov, J.; Ruml, T. Structure, mechanical characteristics and in vitro degradation, cytotoxicity, genotoxicity and mutagenicity of novel biodegradable Zn-Mg alloys. *Mater. Sci. Eng. C* **2016**, *58*, 24–35. [CrossRef] [PubMed]
33. Murni, N.S.; Dambatta, M.S.; Yeap, S.K.; Froemming, G.R.A.; Hermawan, H. Cytotoxicity evaluation of biodegradable Zn–3Mg alloy toward normal human osteoblast cells. *Mater. Sci. Eng. C* **2015**, *49*, 560–566. [CrossRef] [PubMed]
34. Yang, H.T.; Jia, B.; Zhang, Z.C.; Qu, X.; Zheng, Y.F. Alloying design of biodegradable zinc as promising bone implants for load-bearing applications. *Nat. Commun.* **2020**, *11*, 401. [CrossRef] [PubMed]
35. Tang, Z.; Huang, H.; Niu, J.; Zhang, L.; Zhang, H. Design and characterizations of novel biodegradable Zn-Cu-Mg alloys for potential biodegradable implants. *Mater. Des.* **2017**, *117*, 84–94. [CrossRef]
36. Katarivas Levy, G.; Leon, A.; Kafri, A.; Ventura, Y.; Drelich, J.W.; Goldman, J.; Vago, R. Evaluation of biodegradable Zn-1%Mg and Zn-1%Mg-0.5%Ca alloys for biomedical applications. *J. Mater. Sci. Mater. Med.* **2017**, *28*, 174. [CrossRef]
37. Geetha, M.; Singh, A.K.; Asokamani, R.; Gogia, A.K. Ti based biomaterials, the ultimate choice for orthopaedic implants—A review. *Mater. Sci.* **2009**, *54*, 397–425. [CrossRef]
38. Wang, K.; Tong, X.; Lin, J.X.; Wang, K.; Tong, X.; Lin, J.X.; Wei, A.; Li, Y.; Dargusch, M.; Wen, C. Binary Zn–Ti alloys for orthopedic applications: Corrosion and degradation behaviors, friction and wear performance, and cytotoxicity. *J. Mater. Sci. Technol.* **2021**, *74*, 216–229. [CrossRef]
39. Zhang, L.; Liu, X.Y.; Huang, H.; Zhan, W. Effects of Ti on microstructure, mechanical properties and biodegradation behavior of Zn-Cu alloy. *Mater Lett.* **2019**, *244*, 119–122. [CrossRef]
40. Liao, L.; Yao, J.; Lei, Z.; Su, S. Effects of ultrasonic treatment on solidification structure and mechanical properties of AS31 magne-sium alloy. *Spec. Cast. Nonferr. Alloy.* **2017**, *21*, 1241–1246.
41. *ASTM G102-89*; Standard Practice for Calculation of Corrosion Rates and Related Information from Electrochemical Measurements. ASTM Standards: West Conshohocken, PA, USA, 2015; 1.
42. Zhao, L.C.; Xie, Y.; Cui, C.X.; Zhang, Z.; Wang, X. Fabrication and properties of biodegradable ZnO nano-rods/porous Zn scaffolds. *Mater. Charact.* **2018**, *144*, 227–238. [CrossRef]
43. Gu, X.N.; Zhou, W.R.; Zheng, Y.F.; Liu, Y.; Li, Y.X. Degradation and cytotoxicity of lotus-type porous pure magnesium as potentialtissue engineering scaffold material. *Mater Lett.* **2010**, *64*, 1871–1874. [CrossRef]
44. Eskin, G.I. Cavitation mechanism of ultrasonic melt degassing. *Ultrason. Sonochem.* **1995**, *95*, S137–S141. [CrossRef]
45. Suslick, K.S. Applications of ultrasound to materials chemistry. *Annu. Rev. Mater. Sci.* **1999**, *29*, 295–326. [CrossRef]
46. Jian, X.; Xu, H.; Meek, T.T. Effect of power ultrasound on solidification of aluminum A356 alloy. *Mater. Lett.* **2005**, *59*, 190–193. [CrossRef]
47. Han, Y.; Li, K.; Wang, J.; Da, S.; Sun, B. Influence of high-intensity ultrasound on grain refining performance of Al-5Ti-1B master alloy on aluminium. *Mater. Sci. Eng. A* **2005**, *405*, 306–312. [CrossRef]
48. Yin, Z.Y. Microstructural Evolution and Mechanical Properties of Zn–Ti Alloys for Biodegradable Stent Applications. Master's Thesis, Michigan Technological University, Houghton, MI, USA, 2017.
49. Hall, E.O. The deformation and ageing of mild steel: III discussion of results. *Proc. Phys. Soc. Lond. B* **1951**, *64*, 747–752. [CrossRef]
50. Zhang, M.X.; Kelly, P.M.; Easton, M.A.; Taylor, J.A. Crystallographic study of grain refinement in aluminum alloys using the edge-to-edge matching model. *Acta Mater.* **2005**, *53*, 1427–1438. [CrossRef]
51. Celotto, S. TEM study of continuous precipitation in Mg-9 wt%Al-1 wt% Zn alloy. *Acta Mater.* **2000**, *48*, 1775–1787. [CrossRef]
52. Khodabakhshi, F.; Arab, S.M.; Svec, P.; Gerlichd, A.P. Fabrication of a new Al-Mg/graphene nanocomposite by multi-pass friction-stir processing: Dispersion, microstructure, stability, and strengthening. *Mater. Charact.* **2017**, *132*, 92–107. [CrossRef]
53. Yao, C.Z.; Wang, Z.C.; Gao, W.; Chen, D.; Zhu, D.; Dai, K. Effects of Mg on microstructure and corrosion properties of Zn-Mg alloy. *J. Alloys Compd.* **2014**, *602*, 101–107. [CrossRef]
54. Yang, H.; Qu, X.; Lin, W.; Chen, D.; Zhu, D.; Dai, K. Enhanced osseointegration of Zn-Mg composites by tuning the release of Zn ions with sacrificial Mg-rich anode design. *ACS Biomater. Sci. Eng.* **2019**, *5*, 453–467. [CrossRef] [PubMed]
55. Shearier, E.R.; Bowen, P.K.; He, W.; Drelich, A.J.; Zhao, F. In vitro cytotoxicity, adhesion, and proliferation of human vascular cells exposed to zinc. *ACS Biomater. Sci. Eng.* **2016**, *2*, 634–642. [CrossRef] [PubMed]

materials

MDPI

Article

The Microstructure and Corrosion Resistance of Fe-B-W-Mn-Al Alloy in Liquid Zinc

Zixiang Luo [1,2], Ke Liu [3], Zizhen Cui [1,2], Xuemei Ouyang [1,2,*], Chen Zhang [3] and Fucheng Yin [1,2]

[1] School of Material Science and Engineering, Xiangtan University, Xiangtan 411105, China; 201921001459@smail.xtu.edu.cn (Z.L.); czz2012nino@sina.com (Z.C.); fuchengyin@xtu.edu.cn (F.Y.)
[2] Key Laboratory of Materials Design and Preparation Technology of Hunan Province, School of Material Science and Engineering, Xiangtan University, Xiangtan 411105, China
[3] Jianglu Machinery and Electronics Group Co., Ltd., Xiangtan 411100, China; liuke820x@126.com (K.L.); ZC719500@163.com (C.Z.)
* Correspondence: ouyangxuemei@xtu.edu.cn; Tel.: +86-731-5829-8428

Abstract: The microstructure, interfacial characteristics, and corrosion resistance of Fe-W-Mn-Al-B alloys in molten zinc at 520 °C have been investigated using scanning electron microscopy (SEM), X-ray diffractometry (XRD), and electron probe micro-analysis (EPMA). The experimental results indicate that the Fe-B alloy with 11 wt.% W, 7 wt.% Mn, and 4 wt.% Al addition displays a lamellar eutectic microstructure and excellent corrosion resistance to molten zinc. The toughness of M_2B-type borides in the hyper-eutectic Fe-4.2B-11W-7Mn-4Al alloy can be more than doubled, reaching 10.5 $MPa{\cdot}m^{1/2}$, by adding Mn and Al. The corrosion layer of the Fe-3.5B-11W-7Mn-4Al alloy immersed in molten zinc at 520 °C comprises Fe_3AlZn_x, δ-$FeZn_{10}$, ζ-$FeZn_{13}$, and η-Zn. The lamellar borides provide the mechanical protection for α-(Fe, Mn, Al), and the thermal stability of borides improves as the fracture toughness of the borides increases, which jointly contribute to the improvement of the corrosion resistance to the molten zinc.

Keywords: Fe-B-W alloy; manganese; aluminum; molten zinc; microanalysis; corrosion

Citation: Luo, Z.; Liu, K.; Cui, Z.; Ouyang, X.; Zhang, C.; Yin, F. The Microstructure and Corrosion Resistance of Fe-B-W-Mn-Al Alloy in Liquid Zinc. *Materials* **2022**, *15*, 1092. https://doi.org/10.3390/ma15031092

Academic Editor: Pavel Novák

Received: 19 November 2021
Accepted: 27 January 2022
Published: 30 January 2022

1. Introduction

Hot-dip galvanizing is one of the more effective and economical methods to treat steel against atmospheric corrosion, and it has been widely used [1–4]. Nevertheless, the severe corrosion in continuous galvanizing line (CGL) processing is a challenge for the long-term submersed hardware [5–8]. The corrosion seriously decreases the service life of the equipment immersed in the molten zinc bath, which further affects the production efficiency and product quality. Hence, it is necessary to search for excellent corrosion-resistant materials under elevated temperatures from 450 to 600 °C in molten zinc.

In recent years, some studies on corrosion-resistant materials in CGL have been conducted, and many kinds of materials have been used in the immersed hardware, including ceramics, Stellite alloys, intermetallics, coatings, and other composite materials [9,10]. Among these materials, because of the unique non-wetting characteristic of the Fe_2B phase in molten zinc and their reticular structure, the Fe-B alloys display remarkable anti-corrosion properties and have drawn much research interest [11–15]. Ma et al. [11] investigated the effect of B on the corrosion resistance of the Fe-B alloy, finding that the Fe-B alloy containing 3.5 wt.% B showed the optimal corrosion resistance. They also found that the thermal stability of the Fe_2B phase decreased significantly once the temperature of molten zinc reached or went above 520 °C, promoting the spalling of the borides and corrosion failure [12]. Alloying methods are widely used to improve the toughness of borides and the corrosion resistance of alloys. The effects of Cr, Mn, W, Mo, Ti, rare earth, and other elements on the fracture toughness and hardness of Fe_2B in Fe-B-C cast alloy have been investigated [16,17], while the effect of Cr, Ni, W, Mo, and other elements on the corrosion resistance of Fe-B alloys in molten zinc baths was reported upon [18–21]. Recently, Jian et al. [16] investigated the effect of Mn addition on the microstructure, mechanical properties, and wear behavior of an Fe-3.0B alloy, finding that proper Mn addition could improve the toughness of Fe_2B. Liu et al. [20] reported that adding 11 wt.% W to the Fe-3.5B alloy could significantly increase its corrosion resistance to molten zinc by the improvement of boride stability and through the grain refining of eutectic borides. However, the brittle characteristics of the boride phase and the preferential corrosion of the α-Fe phase remain dominant reasons for corrosion failure, which cause difficulties in meeting (CGL) industry requirements. In addition, based on the previous studies [22,23], the Fe-Al intermetallic layers could obstruct the diffusion of Zn atoms, and thereby delay the reaction between the Fe substrate and molten zinc. Al can improve the hardness and red hardness of steel, which is beneficial for improving the wear resistance of steel at elevated temperatures. Al can form a highly stable and protective oxide scale, which can improve the corrosion resistance of steel at elevated temperatures [24]. Therefore, Mn and Al were chosen as the alloying elements in the present work. In this study, we investigate the further improvement of the corrosion resistance of an Fe-3.5B-11W alloy in molten zinc by adding proper Al and Mn. The microstructural evolution of the Fe-B alloys with proper Mn and Al addition was characterized. Special attention was paid to the synergistic effects of Mn and Al on the corrosion behaviors of the Fe-3.5B-11W alloy in the molten zinc.

2. Experimental Procedures

2.1. Specimens Preparation

The raw materials are 99.9 wt.% iron ingot, 99.9 wt.% pure W, 99.9 wt.% pure Mn, 99.9 wt.% pure Al, and an FeB alloy which contains 19.0–21.0 wt.% B, less than 0.15 wt.% C, and less than 1.0 wt.% Si according to the GB/T 5682-2015 Ferro Boron. Firstly, the Al-Mn master alloy is prepared by vacuum melting to reduce the evaporation of Al. As known by the Al-Mn phase diagram [25], a large amount of Al can be dissolved in βMn. The melting point of the Mn-Al alloys increases with the addition of Al compared with that of pure Al. Then, pure Fe, W, FeB alloy, and Al-Mn master alloy were melted together by vacuum melting. Four cast Fe-B alloys containing about 3.5 wt.% B, 11 wt.% W, 7 wt.% Mn, and 0–6 wt.% Al were prepared through orthogonal tests. The compositions of these four Fe-B alloys are given in Table 1.

Table 1. The nominal chemical composition of Fe-3.5B alloys (wt.%).

Specimens	W	Mn	B	Al	Balance
Fe-3.5B-11W	11	0	3.5	0	Fe
Fe-3.5B-11W-7Mn-1Al	11	7	3.5	1	Fe
Fe-3.5B-11W-7Mn-4Al	11	7	3.5	4	Fe
Fe-3.5B-11W-7Mn-6Al	11	7	3.5	6	Fe
Fe-4.2B-11W-7Mn-4Al	11	7	4.2	4	Fe

2.2. Morphology and Phase Characterization of the Specimen

All specimens embedded in resin were polished to a mirror finish using Al_2O_3 and etched with 4 vol.% nital solution. The surface and cross-sections of the as-cast and corroded alloys were observed using an optical microscope and JSM-6360LV scanning electron microscope (JEOL, Tokyo, Japan) equipped with an energy dispersive spectrometer. An JXA-8230 electron probe microanalyzer (JEOL, Tokyo, Japan) and a wavelength-dispersive X-ray spectrometer (WDX) were employed to determine the phase chemical composition. The constituent phases were further identified with an X-ray diffractometer (XRD, Rigaku Ultima IV, Tokyo, Japan) with Cu Kα radiation as the X-ray source, and performed in continuous scanning mode at 40 kV and 40 mA. The corroded samples were ground carefully to remove the pure zinc layer.

2.3. Corrosion Test in Molten Zinc

The corrosion depth profiles were used to characterize the average corrosion rate. The original thickness of each specimen was measured with a microscale micrometer 10 times at multiple locations across the cross-section of the specimen to obtain average values. The corrosion tests were executed in a graphite crucible placed in a vertical resistance furnace with the temperature maintained at 520 ± 3 °C for 24, 48, 72, 96, and 120 h. Then, these specimens were removed from the molten zinc and cooled quickly by water quenching to preserve the microstructure at that temperature. The average corrosion depth and corrosion rate of the tested specimen were calculated with the following Equations (1) and (2) [18–21]:

$$H = (a - b)/2 \quad (1)$$

$$R = h/t \quad (2)$$

where h is the average corrosion depth (μm), a is the original thickness (μm), b is the final thickness (μm), R is the corrosion rate ($\mu m \cdot h^{-1}$), and t is the corrosion time (h).

2.4. Vickers Micro-Indentation Fracture Toughness Test

The micro-hardness of the boride phase in the Fe-B alloy was measured using an MH-5L Vickers hardness tester with a load of 200 g, according to the ASTM E384 standard [26]. Each value was obtained from the average of five measured values. The fracture toughness can be calculated with Equation (3) [14–17]:

$$K_c = XP/c^{3/2} \quad (3)$$

$$X = 0.064(E/H)^{1/2} \quad (4)$$

where X is the residual indentation coefficient, which depends on the hardness-to-modulus ratio (E/H) of the boride. E and H are the Young's modulus and micro-hardness, respectively. Additionally, the value of E is approximately 336 GPa [27] for the fracture toughness calculation. P is the applied load and c is the indentation half crack length.

3. Results

3.1. Microstructural Characteristics

Four as-cast Fe-B alloys containing approximately 3.5 wt.% of B, 11 wt.% of W, 0–7 wt.% of Mn, and 0–6 wt.% of Al, were prepared through orthogonal testing. The microstructure and corrosion resistance of the four alloys are described and discussed below.

The microstructure of the four as-cast alloys and the constituent phases based on the WDX and XRD results is shown in Figure 1, and the XRD patterns are shown in Figure 2b. The microstructure of alloy Fe-3.5B-11W is a typical hypoeutectic morphology consisting of the primary solid solution, a gray, net-like M_3B and white FeWB, as shown in Figure 1a. Compared with alloy Fe-3.5B-11W, the gray, net-like M_3B-type boride changed to a net-like M_2B-type boride, and the amount of FeWB was reduced. As the addition of Al further increases, the microstructure of the alloy Fe-3.5B-11W-7Mn-4Al changes to the almost pure finer eutectic structure without the primary solid solution. Alloy Fe-3.5B-11W-7Mn-6Al containing 6 wt.% Al has a typical hypoeutectic morphology consisting of the primary solid solution, a gray, net-like M_2B. Firstly, the white, binary eutectic FeWB disappears (L→FeWB+α), which can eliminate the effect of destroying the integrity of the reticular borides. The size of the primary solid solution becomes small, and the primary solid solution disappears in alloy Fe-3.5B-11W-7Mn-4Al. Compared with other alloys, the eutectic borides in alloy Fe-3.5B-11W-7Mn-4Al exhibit a lamellar structure, and the microstructure is significantly finer. The software Image-Pro Plus was used to calculate the phase fraction, and the results are shown in Table 2. The microstructure treated by three different colors is shown in Figure 1e–h, in which the red phase is the α solid solution, the green one is M_3B and M_2B, and the blue one is FeWB.

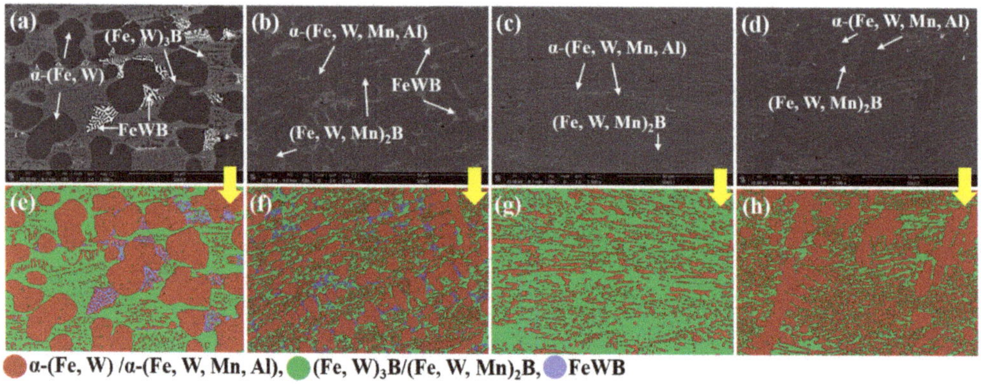

Figure 1. The as-cast microstructures and phase distribution of alloys: (**a**,**e**) alloy Fe-3.5B-11W; (**b**,**f**) alloy Fe-3.5B-11W-7Mn-1Al; (**c**,**g**) alloy Fe-3.5B-11W-7Mn-4Al; (**d**,**h**) alloy Fe-3.5B-11W-7Mn-6Al.

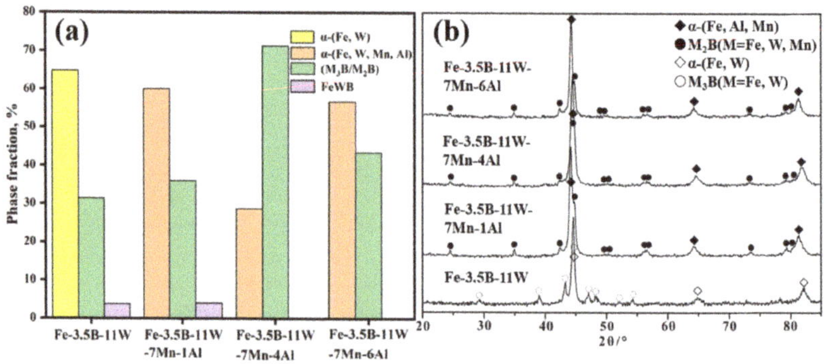

Figure 2. (**a**) The calculated phase fractions of the alloys; (**b**) the XRD patterns of alloys.

Table 2. Average composition of the phases in alloys by WDX (at.%) and calculated phase fraction.

Alloy	Phase	Fe	W	B	Mn	Al	Calculated Phase Fraction
Fe-3.5B-11W	α-(Fe, W)	98.7	1.3	-	-	-	64.8%
	$(Fe, W)_3B$	75.3	3.0	21.7	-	-	31.4%
	FeWB	34.1	32.3	33.6	-	-	3.8%
Fe-3.5B-11W-7Mn-1Al	α-(Fe, W, Mn, Al)	87.8	0.8	-	5.2	6.2	60.0%
	$(Fe, W, Mn)_2B$	52.4	2.6	32.4	12.2	0.4	36.0%
	FeWB	32.2	35.2	32.6	-	-	4.0%
Fe-3.5B-11W-7Mn-4Al	α-(Fe, W, Mn, Al)	82.0	0.7	-	5.5	11.8	28.7%
	$(Fe, W, Mn)_2B$	50.7	3.8	32.4	11.6	1.5	71.3%
Fe-3.5B-11W-7Mn-6Al	α-(Fe, W, Mn, Al)	80.2	0.6	-	5.3	13.9	56.6%
	$(Fe, W, Mn)_2B$	56.3	3.5	25.7	11.8	2.7	43.4%

The calculated phase content in the alloys is shown in Figure 2a. The results indicate that the content of M_2B increases first and then decreases, and that of α decreases first and then increases with the addition of Fe-3.5B-11W. Alloy Fe-3.5B-11W-7Mn-4Al has most M_2B borides and less of the primary α-(Fe, Mn, Al) phase. According to the composition results of the WDX analysis (see in Table 2), the Fe-3.5B-11W atoms are mainly distributed in the α solid solution, while most Mn and W atoms are present in the borides. Compared with alloy A1, the addition of Fe-3.5B-11W and Mn can promote the formation of M_2B-type borides and restrain the formation of M_3B-type borides and FeWB. As a consequence, the addition of Fe-3.5B-11W and Mn can affect the solidified microstructure and promote the formation of dense borides, which benefits the improvement of corrosion resistance to the molten zinc.

3.2. Micro-Hardness Testing

The toughness is a key factor in the stability of borides. We investigated the effects of Mn and Al addition on the fracture toughness and hardness of Fe_2B; however, the borides in alloy Fe-3.5B-11W-7Mn-4Al were too small for fracture toughness testing via Vickers micro-indentation. Therefore, alloy Fe-4.2B-11W-7Mn-4Al, with a hypereutectic composition, was designed to obtain the block Fe_2B phase. A schematic of the indentation mark and crack length, as well as an optical micrograph of the micro-cracks in the M_2B, are shown in Figure 3. The average micro-hardness of Fe_2B in alloy Fe-4.2B-11W-7Mn-4Al is 1565 $HV_{0.1}$, somewhat higher than that of Fe_2B (~1552 $HV_{0.1}$) and (Fe, Mn)$_2$B (1560 $HV_{0.1}$), and lower than that of (Fe, W)$_2$B (>1570 $HV_{0.1}$). The average value of the fracture toughness is approximately 10.5 MPa·m$^{1/2}$, compared to the fracture toughness values of Fe_2B at 3.8 MPa·m$^{1/2}$, (Fe, W)$_2$B at 6.9 MPa·m$^{1/2}$, and (Fe, Mn)$_2$B at 4.7 MPa·m$^{1/2}$ [15,16]. It can be concluded that the addition of Mn and W to Fe_2B benefits the improvement of the fracture toughness. The radius of the Mn atom (1.32 Å) and W atom (1.41 Å) is larger than the Fe atom (1.27 Å), which may introduce more impact to the surrounding B atoms and bond energy in the [002] B-B. The primary weak B-B bond is expected to be strengthened naturally by the addition of the Mn and W elements, which is consistent with the results calculated by Zhou et al. [28].

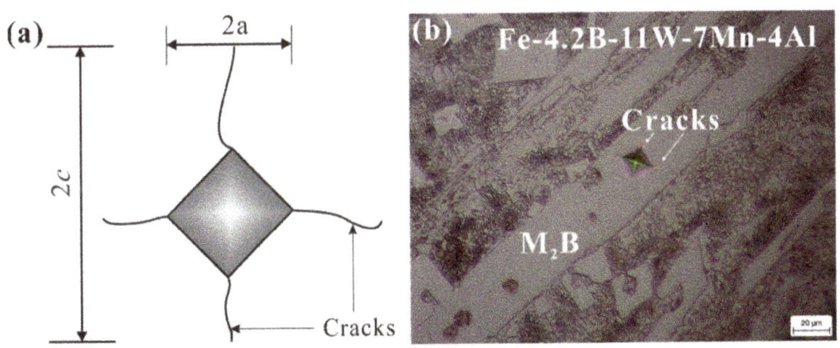

Figure 3. (**a**) Schematic of indentation mark and crack length; (**b**) micrographs of indentation marks and cracks on M_2B in Fe-4.2B-11W-7Mn-4Al.

3.3. Corrosion Kinetics

The corrosion behaviors of these four as-cast alloys in the molten zinc were investigated with a corrosion test. Alloy Fe-3.5B-11W was selected as the standard against which to compare corrosion resistance. Figure 4 shows the corrosion depths and corrosion rates as a function of corrosion time based on Equations (1) and (2). From Figure 4a, it can be seen that the corrosion depth of all the Fe-B alloys increases with increasing immersion time. The Fe-3.5B-11W-7Mn-4Al alloy shows the best corrosion resistance among the various alloys, including alloys Fe-3.5B-11W, Fe-3.5B-11W-7Mn-1Al, Fe-3.5B-11W-7Mn-4Al, Fe-3.5B-11W-7Mn-6Al, Fe-3.5B, and Fe-3.5B-5Cr. From Figure 4b, it can be seen that all investigated samples show declining corrosion rates with increased immersion time. Meanwhile, alloy Fe-3.5B-11W-7Mn-4Al shows a comparatively low corrosion rate (<2 μm·h^{-1}) among these four kinds of Fe-B alloys.

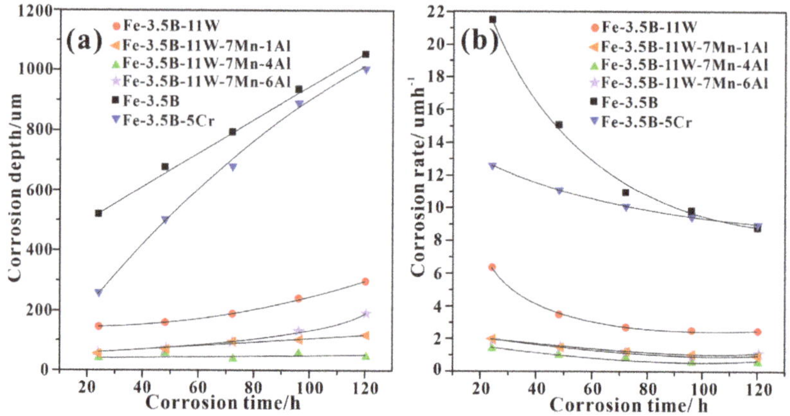

Figure 4. Corrosion depth and corrosion rate as a function of corrosion time at 520 °C: (**a**) corrosion depth vs. corrosion time; (**b**) corrosion rate—corrosion time.

3.4. Corrosion Layer Characterization

The cross-sectional morphologies of the corrosion interface of the samples corroded by the molten zinc at 520 °C for 48 h are illustrated in Figure 5. The cross-sections of these corroded samples can be divided into three different regions: The un-corroded matrix, the corrosion layer, and the Fe-Zn compound layer. Generally, the corrosion interfaces are not uniform, and the reticular structure is destroyed by molten zinc, except in alloy Fe-3.5B-11W-7Mn-4Al. The corrosion layer mainly comprises the borides and Fe-Zn compounds. The WDX analysis results of the Fe-Zn and Fe-Al-Zn intermetallic com pounds, measured at each point 10 times, are given in Table 3.

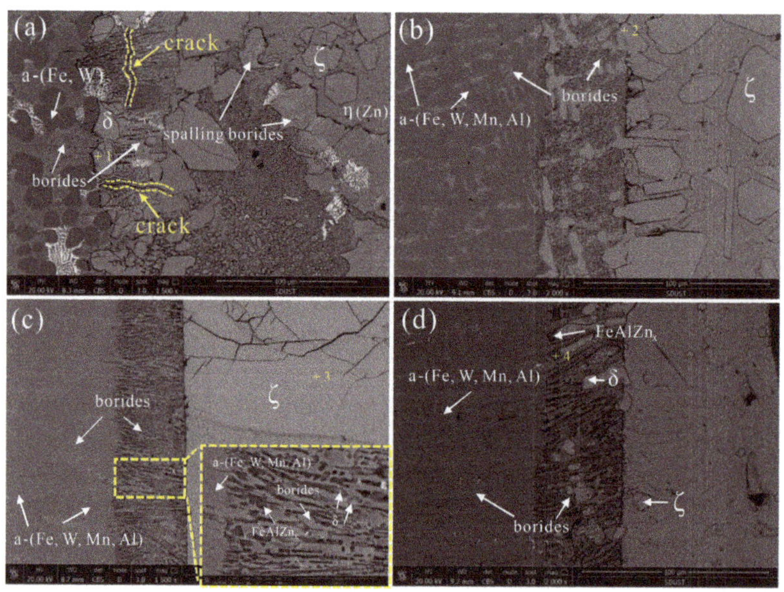

Figure 5. The morphology of the corrosion cross-section after corrosion testing at 520 °C for 48 h: (**a**) alloy Fe-3.5B-11W; (**b**) alloy Fe-3.5B-11W-7Mn-1Al; (**c**) alloy Fe-3.5B-11W-7Mn-4Al; (**d**) alloy Fe-3.5B-11W-7Mn-6Al.

Table 3. The chemical composition of the phases in the corroded layer of the alloys of different compositions after 48 h of corrosion under the zinc solution at 520 °C by EPMA.

Position	Fe		Al		Zn		W		Mn		Phase
	wt.%	at.%	wt.%	at.%	wt.%	at.%	wt.%	at.%	wt.%	at.%	
+1 in Fe-3.5B-11W	10.9	13.1	-	-	81.4	84.0	7.7	2.5	-	-	δ-$FeZn_{10}$
+2 in Fe-3.5B-11W-7Mn-1Al	11.7	13.7	0.4	1.0	84.1	83.7	3.3	1.2	0.5	0.4	δ-$FeZn_{10}$
+3 in Fe-3.5B-11W-7Mn-4Al	7.1	8.2	-	-	92.9	91.8	-	-	-	-	ζ-$FeZn_{13}$
+4 in Fe-3.5B-11W-7Mn-6Al	31.6	24.1	38.7	61.2	17.7	11.6	11.6	2.7	0.4	0.4	Fe_3AlZn_x

The Fe_3AlZn_x alloy was found to be close to the matrix, as shown in Figure 5d, which was not confirmed by the XRD because the content is too small. The Fe-Al-Zn intermetallic compounds were observed at the corrosion interface of the Fe-B alloy to be immersed in liquid 0.25 wt.% Al-Zn [12]. The main phases in the corrosion layer are identified to be δ-$FeZn_{10}$, ζ-$FeZn_{13}$, and η-(Zn), which are identified by the XRD shown in Figure 6. The δ-$FeZn_{10}$ phase is distributed on the right side of the corroded layer, near to the Fe-Zn compound layer, while the ζ-$FeZn_{13}$ phase presents dispersed blocks in the molten zinc. There are many cracks in the δ-$FeZn_{10}$ phase, which provide the channel for the diffusion of zinc atoms. Moreover, little spalling of the borides is found in the three alloys, which demonstrates that both alloys have good corrosion resistance to the molten zinc, but alloy Fe-3.5B-11W-7Mn-4Al is superior to alloy Fe-3.5B-11W.

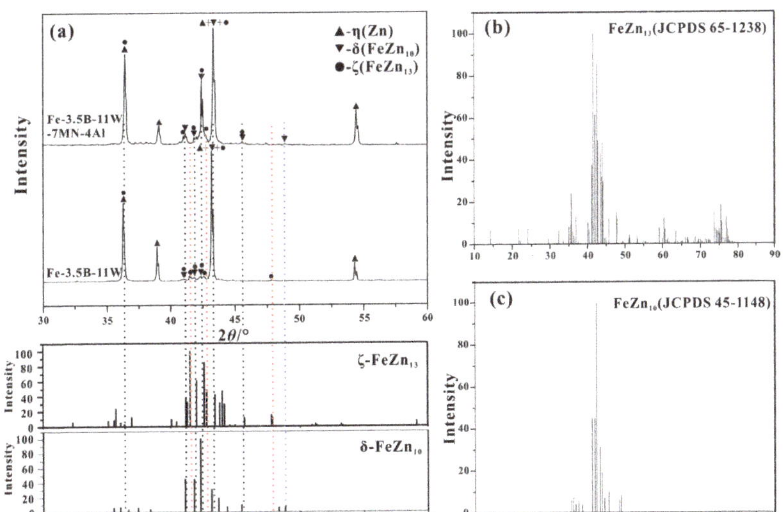

Figure 6. The XRD patterns of the corrosion products in alloys Fe-3.5B-11W and Fe-3.5B-11W-7Mn-4Al.

The element distribution of the corrosion layer near the matrix in alloy Fe-3.5B-11W-7Mn-1Al is shown in Figure 7. It can be seen that the Fe-3.5B-11W element is indeed enriched at the interface between the matrix and the corrosion diffusion layer. Mn atoms mainly enter into the borides because of the similar atomic radius (R_{Mn} = 0.132 nm, R_{Fe} = 0.127 nm) and electro-negativities ($X_P^{Mn} = 1.55$, $X_P^{Fe} = 1.83$) of Mn as a replacement for Fe in borides. It is related to the smaller W diffusion coefficient and smaller solubility in the molten zinc pool. It can be seen from the distribution of Zn in the boride phase that the solubility of Zn is very low, which is the reason why the boride resists molten zinc corrosion. Due to the greater diffusion coefficient of zinc ($D_{Zn} > D_{Fe}$) in the zinc bath, zinc atoms easily diffuse to the corrosion interface along the grain boundaries of Fe-Zn compounds.

Subsequently, a few zinc atoms enter the (Fe, W, Mn)$_2$B lattice. The fracture toughness of (Fe, W, Mn)$_2$B is higher than (Fe, W)$_2$B, which delays the formation of the crack. Moreover, Al diffuses in preference to Fe atoms, and dissolves at the interface between α-Fe and the borides, leading to the formation of Fe_3AlZn_x.

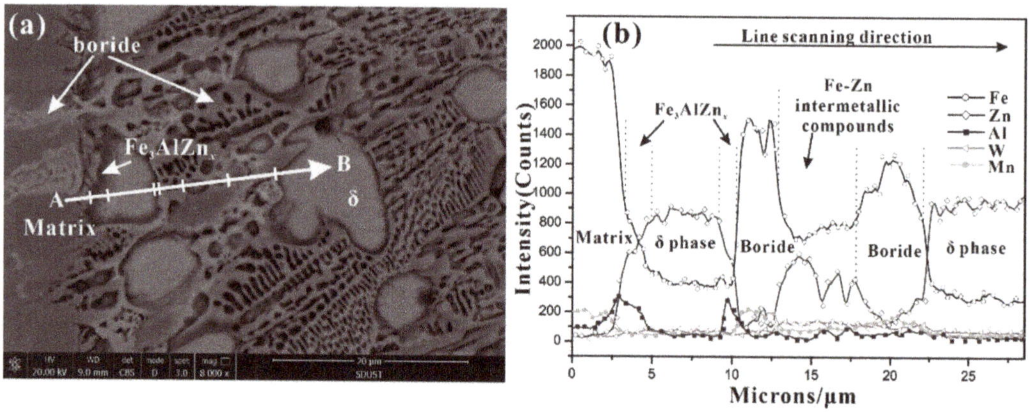

Figure 7. Line scanning results of element distribution near the matric and corrosion layer in alloy Fe-3.5B-11W-7Mn-1Al after corrosion testing at 520 °C for 48 h: (**a**) line scan SEM image; (**b**) line scan element distribution results.

4. Corrosion Mechanism

The microstructure of the corrosion layer of the Fe-3.5B-11W-7Mn-4Al alloy immersed in the zinc for 48 h, 72 h, and 120 h is shown in Figure 8. The alloy Fe-3.5B-11W is enriched at the interface, forming Fe_3AlZn_x, and promoting the formation of the bulky ζ-$FeZn_{13}$, as shown in Figure 8a. No microcracks were found in the alloy corroded for 48 h, and the chunky ζ-$FeZn_{13}$ with a lot of cracks forms at the front end of the interface, and the interface close to the η-Zn is uniform, as shown in Figure 8b. With the increase in corrosion time, the size of the ζ-$FeZn_{13}$ become small and the number of the cracks at the corrosion layer increases, as shown in Figure 8c,d. Meanwhile, most cracks grow perpendicular to the corrosion interface, which can delay the fracture of boride. In addition, few cracks were found on the δ-$FeZn_{10}$ alloy following the addition of the Al element compared with that of Fe-B alloys reported [11,12]. With the increase in the number of cracks on the boride, the alloy eventually fails with the continuous spalling of the boride. Since the spalling boride moves with the flowing zinc, it is difficult to observe the spalling boride at the interface of the samples. In order to visually describe the entire corrosion process, the whole corrosion behavior of the Fe-W-Mn-Al-B as-cast alloy in liquid zinc can be represented schematically in Figure 9.

Based on the observations above, the main reasons for this good corrosion resistance should be the lamellar eutectic structure, the increase in fracture toughness and volume of the borides, and the preferred growth direction of the Fe_2B perpendicular to the corrosion interface. Alloy Fe-3.5B-11W-7Mn-4Al comprises the eutectic structure with no primary solid solution phase, which corrodes first and transforms to the brittle δ-$FeZn_{10}$. The lamellar structure of the borides provides the mechanical protection for α-(Fe, W, Mn, Al). The alloying elements dissolve into molten zinc to promote the growth of the bulky ζ-$FeZn_{13}$ and delay the diffusion of Zn atoms to the matrix [27]. With the increase in time, the size of ζ-$FeZn_{13}$ becomes small and the effect slows down. The high Al dissolution in solid solution can reduce the activity of Zn and then retard the reaction of Fe and Zn [12,29]. Because the fracture toughness of borides is increased with the Mn and Al addition, the thermal stability of the borides improves, the cracks in the borides decrease, and boride spalling is caused by the transformation stress of Fe-Zn intermetallics and the attack from molten zinc is efficiently reduced [30–32]. In addition, the orientation of M_2B has a great effect on the corrosion behavior of Fe-B alloys in molten zinc. It was found that the Fe-B alloy with Fe_2B oriented vertically to the corrosion interface possessed the best corrosion resistance to the molten zinc [33,34]. It can be seen that the biggest angle between the interface and the orientation of the borides in the corrosion layer is approximately 90°. The result is consistent with previous studies indicating that the sample with the preferred Fe_2B growth direction oriented perpendicular to the corrosion interface has the best corrosion resistance.

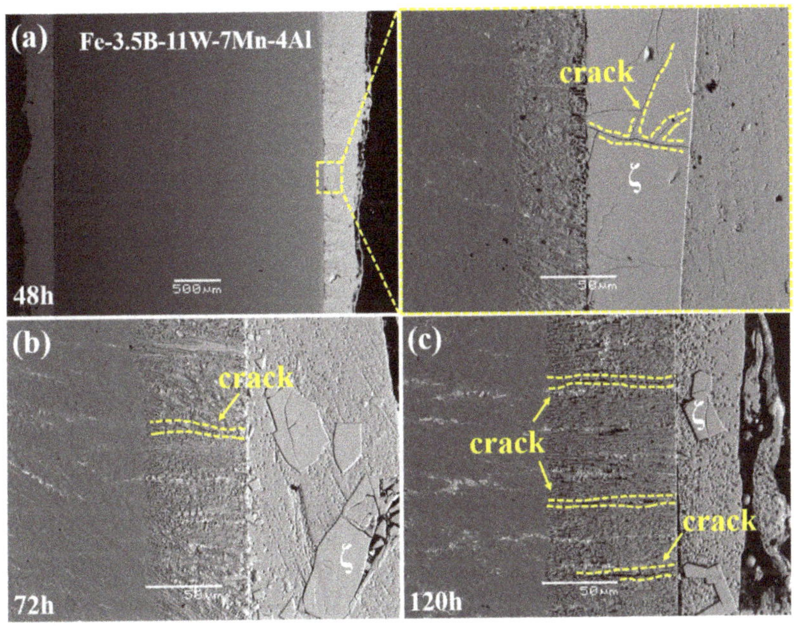

Figure 8. The microstructure of the corrosion layer of the alloy Fe-3.5B-11W-7Mn-4Al immersed in the zinc for different times: (**a**) 48 h; (**b**) 72 h; (**c**) 120 h.

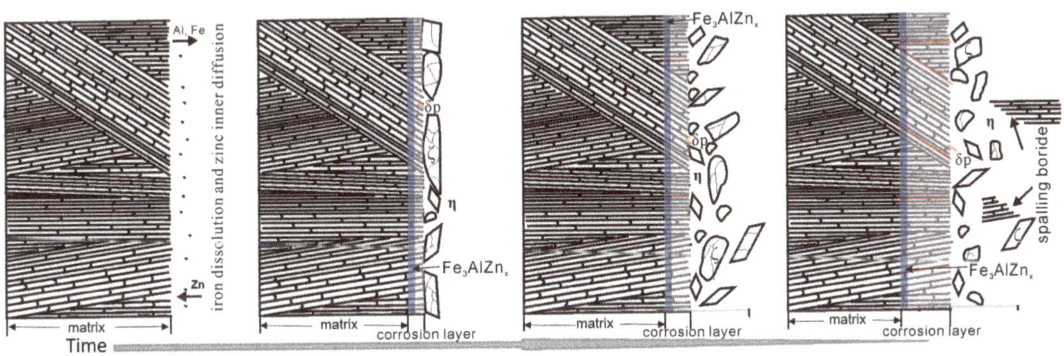

Figure 9. The schematic diagram of the whole corrosion behavior of alloy Fe-3.5B-11W-7Mn-4Al in liquid zinc.

5. Conclusions

The effects of Mn and Al addition on the microstructure and corrosion behaviors of the Fe-3.5B-11W alloys in molten zinc have been investigated in the present work. The results obtained are summarized below:

(1) With the addition of Mn to Fe-3.5B-11W, the microstructures of the Fe-3.5B-11W alloys change from hypoeutectic to eutectic, facilitating the formation of M_2B-type boride and restraining the formation of M_3B-type borides. In the meantime, the fracture toughness obviously increases.

(2) Proper Mn and Fe-3.5B-11W contents can stabilize the Fe_2B phase and improve the corrosion resistance in a zinc bath. The Fe-3.5 wt.% B alloy containing 11 wt.% W, 7wt. % Mn, and 4 wt.% Fe-3.5B-11W, with few microcracks, has the best corrosion resistance.

(3) The lamellar borides provide the mechanical protection for α-(Fe, W, Mn, Al). With the fracture toughness of the borides increasing, the thermal stability of the borides improves, thereby suppressing boride spalling and corrosion failure.

Author Contributions: Conceptualization, Z.L.; Data curation, Z.C.; Funding acquisition, K.L. and C.Z.; Investigation, Z.L. and Z.C.; Methodology, K.L. and X.O.; Project administration, K.L. and C.Z.; Validation, X.O.; Writing—original draft, Z.L.; Writing—review and editing, X.O. and F.Y. All authors have read and agreed to the published version of the manuscript.

Funding: This work was supported by the National Science Foundation of China (No. 51471141, 2015), the Key Project of the Education Department of Hunan Province (No. 15A179), and the Scientific Research Fund of Hunan Provincial Educational Department (No.2016JC2005, 2016), and Scientific Research Foundation of Hunan Provincial Education Department (19B550).

Institutional Review Board Statement: Not applicable.

Informed Consent Statement: Not applicable.

Data Availability Statement: The raw/processed data required to reproduce these findings cannot be shared at this time as the data also forms part of an ongoing study.

Acknowledgments: The authors gratefully acknowledge the support provided by Fucheng Yin and his team members for X-ray diffraction, metallography, and electronic microscopy.

Conflicts of Interest: The authors declare no conflict of interest.

References

1. Peng, B.; Wang, J.; Su, X.; Li, Z.; Yin, F. Effects of zinc bath temperature on the coatings of hot-dip galvanizing. *Surf. Coat. Technol.* **2008**, *202*, 1785–1788.
2. Verma, A.R.B.; van Ooij, W.J. High-temperature batch hot-dip galvanizing. Part 1. General description of coatings formed at 560 °C. *Surf. Coat. Technol.* **1997**, *89*, 132–142. [CrossRef]
3. Rudnik, E. Hydrometallurgical recovery of zinc from industrial hot dipping top ash. *Trans. Nonferrous Met. Soc. China* **2020**, *30*, 2239–2255. [CrossRef]
4. Lu, J.-t.; Wang, X.-h.; Che, C.-s.; Kong, G.; Chen, J.-h.; Xu, Q.-y. Crystallographic research of spangle on hot dip galvanized steel sheets. *Trans. Nonferrous Met. Soc. China* **2007**, *17*, 351–356. [CrossRef]
5. Townsend, H.E.; Steinbicker, R.N.; Yau, Y.H. Corrosion of Stainless Steel Conductor Rolls in a Continuous Sheet Electrogalvanizing Line. *Corrosion* **1990**, *46*, 418–423. [CrossRef]
6. Wang, W.J.; Lin, J.P.; Wang, Y.L.; Chen, G.L. The corrosion of intermetallic alloys in liquid zinc. *J. Alloys Compd.* **2007**, *428*, 237–243. [CrossRef]
7. Wang, W.J.; Lin, J.P.; Wang, Y.L.; Chen, G.L. The corrosion of Fe_3Al alloy in liquid zinc. *Corros. Sci.* **2007**, *49*, 1340–1349. [CrossRef]
8. Peng, Z.Y.; Wang, X.M.; Yin, F.C.; Ouyang, X.M.; Jing-Xian, H.U. Phase equilibria of CoMoZn ternary system. *Trans. Nonferrous Met. Soc. China* **2020**, *30*, 417–427. [CrossRef]
9. Yuan-Tao, H.U.; Zheng, L.; Yan, H.J.; Lian-Kui, W.U.; Jiang, M.Y. Improving hot corrosion resistance of aluminized TiAl alloy by anodization and pre-oxidation. *Trans. Nonferrous Met. Soc. China* **2021**, *31*, 193–206.
10. Liu, X.; Barbero, E.; Xu, J.; Burris, M.; Chang, K.-M.; Sikka, V. Liquid metal corrosion of 316L, Fe_3Al, and FeCrSi in molten Zn-Al baths. *Metall. Mater. Trans. A* **2005**, *36*, 2049–2058. [CrossRef]

11. Ma, S.; Xing, J.; Fu, H.; Yi, D.; Zhi, X.; Li, Y. Effects of boron concentration on the corrosion resistance of Fe–B alloys immersed in 460 °C molten zinc bath. *Surf. Coat. Technol.* **2010**, *204*, 2208–2214. [CrossRef]
12. Ma, S.; Xing, J.; Fu, H.; Yi, D.; Li, Y.; Zhang, J.; Zhu, B.; Gao, Y. Microstructure and interface characteristics of Fe–B alloy in liquid 0.25 wt.% Al–Zn at various bath temperatures. *Mater. Chem. Phys.* **2012**, *132*, 977–986. [CrossRef]
13. Makuch, N. Nanomechanical properties and fracture toughness of hard ceramic layer produced by gas boriding of Inconel 600 alloy. *Trans. Nonferrous Met. Soc. China* **2020**, *30*, 428–448. [CrossRef]
14. Huang, Z.; Xing, J.; Guo, C. Improving fracture toughness and hardness of Fe_2B in high boron white cast iron by chromium addition. *Mater. Des.* **2010**, *31*, 3084–3089. [CrossRef]
15. Huang, Z.; Xing, J.; Lv, L. Effect of tungsten addition on the toughness and hardness of Fe_2B in wear-resistant Fe–B–C cast alloy. *Mater. Charact.* **2013**, *75*, 63–68. [CrossRef]
16. Jian, Y.; Huang, Z.; Xing, J.; Guo, X.; Wang, Y.; Lv, Z. Effects of Mn addition on the two-body abrasive wear behavior of Fe-3.0 wt% B alloy. *Tribol. Int.* **2016**, *103*, 243–251. [CrossRef]
17. Fu, H.; Xiao, Q.; Kuang, J.; Jiang, Z.; Xing, J.D. Effect of rare earth and titanium additions on the microstructures and properties of low carbon Fe-B cast steel. *Mater. Sci. Eng. A* **2007**, *466*, 160–165. [CrossRef]
18. Ma, S.; Xing, J.; Fu, H.; Yi, D.; Zhang, J.; Li, Y.; Zhang, Z.; Zhu, B.; Ma, S. Interfacial morphology and corrosion resistance of Fe–B cast steel containing chromium and nickel in liquid zinc. *Corros. Sci.* **2011**, *53*, 2826–2834. [CrossRef]
19. Ma, S.; Xing, J.; Yi, D.; Fu, H.; Zhang, J.; Li, Y.; Zhang, Z.; Liu, G.; Zhu, B. Effects of chromium addition on corrosion resistance of Fe-3.5B alloy in liquid zinc. *Surf. Coat. Technol.* **2011**, *205*, 4902–4909. [CrossRef]
20. Liu, X.; Wang, M.; Yin, F.; Ouyang, X.; Li, Z. Effects of Tungsten Addition on the Microstructure and Corrosion Resistance of Fe-3.5B Alloy in Liquid Zinc. *Materials* **2017**, *10*, 399. [CrossRef]
21. Yin, F.; Ouyang, X.; Chen, G.; Zhao, M.; Liu, Y. Effect of Molybdenum on the Microstructures of As-Cast Fe-B Alloys and Their Corrosion Resistance in Molten Zinc. *Corrosion* **2017**, *73*, 942–952.
22. Kato, T.; Nunome, K.; Kaneko, K.; Saka, H. Formation of the ζ phase at an interface between an Fe substrate and a molten 0.2 mass% Al–Zn during galvannealing. *Acta Mater.* **2000**, *48*, 2257–2262. [CrossRef]
23. Akdeniz, M.V.; Mekhrabov, A.O. The effect of substitutional impurities on the evolution of Fe-Al diffusion layer. *Acta Mater.* **1998**, *46*, 1185–1192. [CrossRef]
24. Liu, J.; Chen, W. Microstructure, mechanical properties and corrosion behavior of an Fe-10Cr-2.7B-5.5Al-13Mn alloy prepared by spark plasma sintering. *J. Alloys Compd.* **2018**, *741*, 348–359. [CrossRef]
25. Liu, X.J.; Ohnuma, I.; Kainuma, R.; Ishida, K. Thermodynamic assessment of the Aluminum-Manganese (Al-Mn) binary phase diagram. *J. Phase Equilibria* **1999**, *20*, 45. [CrossRef]
26. Hetzner, D. Microindentation Hardness Testing of Materials Using ASTM E384. *Microsc. Microanal.* **2003**, *9*, 708–709. [CrossRef]
27. Zhang, K.; Tang, N.-Y.; Goodwin, F.E.; Sexton, S. Reaction of 316L stainless steel with a galvanizing bath. *J. Mater. Sci.* **2007**, *42*, 9736–9745. [CrossRef]
28. Li, J.; Li, J.; Li, C.; Liu, Y. Reactive synthesis of FeWB powders and preparation of bulk materials. *Int. J. Refract. Met. Hard Mater.* **2014**, *46*, 80–83. [CrossRef]
29. Qiu, K.; Wang, R.-c.; Peng, C.-q.; Wang, N.-g.; Cai, Z.-y.; Zhang, C. Effects of Mn and Sn on microstructure of Al–7Si–Mg alloy modified by Sr and Al–5Ti–B. *Trans. Nonferrous Met. Soc. China* **2015**, *25*, 3546–3552. [CrossRef]
30. Ma, S.; Xing, J.; Fu, H.; He, Y.; Bai, Y.; Li, Y.; Bai, Y. Interface characteristics and corrosion behaviour of oriented bulk Fe_2B alloy in liquid zinc. *Corros. Sci.* **2014**, *78*, 71–80. [CrossRef]
31. Kong, G.; Lu, J.-t.; Xu, Q.-y. Dissolution mechanism of solid nickel in liquid zinc saturated with Fe. *Trans. Nonferrous Met. Soc. China* **2007**, *17*, 564–569. [CrossRef]
32. Esmaily, H.; Habibolahzadeh, A.; Tajally, M. Improving pulsed laser weldability of duplex stainless steel to 5456 aluminum alloy via friction stir process reinforcing of aluminum by BNi-2 brazing alloy. *Trans. Nonferrous Met. Soc. China* **2019**, *29*, 1401–1412. [CrossRef]
33. Wang, Y.; Xing, J.; Ma, S.; Liu, G.; He, Y.; Yang, D.; Bai, Y. Effect of Fe_2B orientation on erosion–corrosion behavior of Fe–3.5 wt.% B steel in flowing zinc. *Corros. Sci.* **2015**, *98*, 240–248. [CrossRef]
34. Ma, S.; Xing, J.; He, Y.; Fu, H.; Li, Y.; Liu, G. Effect of orientation and lamellar spacing of Fe_2B on interfaces and corrosion behavior of Fe-B alloy in hot-dip galvanization. *Acta Mater.* **2016**, *115*, 392–402. [CrossRef]

materials

MDPI

Article

Effect of Radiofrequency Plasma Spheroidization Treatment on the Laser Directed Energy Deposited Properties of Low-Cost Hydrogenated-Dehydrogenated Titanium Powder

Min Liu [1,2,3,†], Liufei Huang [3,4,†], Congcong Ren [3], Dou Wang [3], Qiang Li [1,2,*] and Jinfeng Li [3,*]

1 Xinjiang Key Laboratory of Solid State Physics and Devices, Xinjiang University, Urumqi 830046, China; liuminchn@foxmail.com
2 School of Physics and Technology, Xinjiang University, Urumqi 830046, China
3 Institute of Materials, China Academy of Engineering Physics, Mianyang 621908, China; lyyhuangliufei@126.com (L.H.); rc13572802927@163.com (C.R.); wangd0005@126.com (D.W.)
4 School of Mechanical Engineering, Xinjiang University, Urumqi 830047, China
* Correspondence: qli@xju.edu.cn (Q.L.); lijinfeng305@126.com (J.L.)
† These authors contributed equally to this work.

Abstract: Titanium for additive manufacturing presents a challenge in the control of costs in the fabrication of products with expanding applications compared with cast titanium. In this study, hydrogenated–dehydrogenated (HDH) titanium powder with a low cost was employed to produce spherical Ti powder using the radiofrequency plasma (RF) technique. The spherical Ti powder was used as the raw material for laser directed energy deposition (LDED) to produce commercially pure titanium (CP-Ti). Microstructural analyses of the powder revealed that RF treatment, not only optimized the shape of the titanium powder, but also benefited in the removal of the residual hydride phase of the powder. Furthermore, the LDED-HDH-RF-produced samples showed an excellent combination of tensile strength and tensile ductility compared to the cast and the LDED-HDH-produced samples. Such an enhancement in the mechanical properties was attributed to the refinement of the α grain size and the dense microstructure. The present work provides an approach for LDED-produced CP-Ti to address the economic and mechanical properties of the materials, while also providing insights into the expanding application of HDH titanium powder.

Keywords: hydrogenated–dehydrogenated titanium powder; laser directed energy deposition; microstructure; mechanical property; radio frequency plasma technique

Citation: Liu, M.; Huang, L.; Ren, C.; Wang, D.; Li, Q.; Li, J. Effect of Radiofrequency Plasma Spheroidization Treatment on the Laser Directed Energy Deposited Properties of Low-Cost Hydrogenated-Dehydrogenated Titanium Powder. *Materials* **2022**, *15*, 1548. https://doi.org/10.3390/ma15041548

Academic Editor: Jan Haubrich

Received: 22 January 2022
Accepted: 15 February 2022
Published: 18 February 2022

1. Introduction

Titanium and its alloys, as engineering materials, are employed in the aerospace, military, and medical fields due to their high specific strength, high specific modulus, and corrosion resistance [1]. Commercially pure titanium (CP-Ti) has better biocompatibility, weldability, and osseointegration than other titanium alloys, which makes it extremely useful [2]. However, it is difficult to melt and process owing to CP-Ti's notoriously active chemical activity, which limits its large-scale applicability [3].

Components with complex internal structures can be directly produced using additive manufacturing (AM) technology, which provides a powerful tool in the application of titanium and its alloys. Recently, additive manufacturing (AM) fabrication of CP-Ti has received considerable research attention [4]. Many researchers have suggested that, among the large number of metallic materials, CP-Ti is well suited to AM due to AM-produced titanium materials usually presenting attractive mechanical properties, which enables the precision forming of large-sized complex components designed for specific uses [5]. Santos et al. found that adjusting the power of a laser beam and the laser scanning hatch spacing in the AM process helps to improve the density of CP-Ti samples, from ~92% to ~98%, which is beneficial to improving the torsional fatigue strength [6]. Hasib et al. found that that AM-produced CP-Ti samples generally yielded a higher tensile strength and fatigue crack growth resistance compared to wrought material [7,8]. AM technology provides a simple tool to obtain titanium components that are expected to have broad application in medical or chemical-plant industries [9,10].

Previous studies on CP-Ti using AM have focused on parameter optimization to achieve high-density parts and to evaluate their mechanical properties [11]. However, there are limitations to producing corresponding AM-produced titanium components of economical utility for practical applications as the current preparation of high-quality spherical metal powers relies mainly on the atomization and rotating electrode methods, which need high-purity raw materials [12]. The development of the radio frequency plasma (RF) technique provides a powerful tool to reduce the costs of spherical metal powders [13]. This approach has been intensively researched for the manufacturing of cost-effective titanium spherical powders using low-cost hydrogenated–dehydrogenated (HDH) titanium powder [14]. Combining the advantages of the manufacturing capacity of AM technology and spherical HDH titanium powder with RF treatment, in this study, AM-HDH-RF-produced Ti components achieved superior mechanical properties and were prepared at a low-cost compared to those prepared using traditional processing techniques [15]. The HDH titanium powder used in this work was made of sponge titanium, machining residues, etc., as raw materials, and could be used for laser deposition after the radio frequency plasma (RF) process. Comparing with AM using titanium wire, this method omits the process of suspension melting, or arc melting, and the process of preparing wire, which has great advantages in terms of improvements to energy conservation and efficiency [16–18].

2. Experimental Section

The experimental apparatus for the HDH titanium powder treatment consisted of a radio frequency (RF), an inductively coupled plasma torch, a water-cooled steel chamber, a powder feeder, and a vacuum system. The plasma torch had a two-part structure, a water-cooled three-layer quartz confinement tube and a four-turn copper coil with water cooling. The plasma operated at an oscillating frequency of 3.5 MHz and used a 40 kW plasma plate. The central plasma and sheath gas used high-purity argon at flow rates of 20~30 L·min^{-1} and 80~90 L·min^{-1}, respectively. The HDH-Ti powder was supplied by Jiangxi Weila Metal Materials Co., Ltd., Ganzhou, Jiangxi, China and had an oxygen content of 723.87 ppm and a nitrogen content of 9.38 ppm. The purity of the HDH-Ti powder was over 99.9 wt%. The average size of the HDH powder was about 135 μm, and was measured using a particle size analyzer (model Microtrac-S3500). The HDH-Ti powder entered the plasma arc via the carrier gas (argon). The flow rates of the carrier gas were 30 g/min, 45 g/min, and 60 g/min. Powder particles rapidly absorbed heat and spheroidized in the plasma arc, and finally entered the cooling chamber to rapidly condense and form spherical powder.

Laser directed energy deposition (LDED) has the characteristics of net-forming and rapid-forming, and is one of the most promising AM methods in the preparation of complex metal components. In this work, the AM process was performed using LDED in an LDM6050 (YuChen Tech. Ltd., Nanjing, Jiangsu, China). The HDH titanium powder with RF treatment was processed in a drying oven at 393 k for 4 h under vacuum conditions. Specimens were produced on a pure Ti substrate under an argon atmosphere with an oxygen content less than 10 ppm. The parameters of the optimized process were as follows: a laser beam wavelength of 1070 nm, laser beam spot size of ~2 mm, laser power of 800 W, laser scanning speed of 600 mm/min, powder feeding rate of 9~14 g/min, coaxial argon gas flow of 12~20 L/min, and a sheath gas flow rate of 6~10 L/min. Figure 1 shows a schematic of the RF technique and the LDED for fabricating the CP-Ti sample. The cast Ti samples were fabricated via arc melting in a copper mold under a high-purity argon atmosphere. The raw material was at least 99.9 wt%.

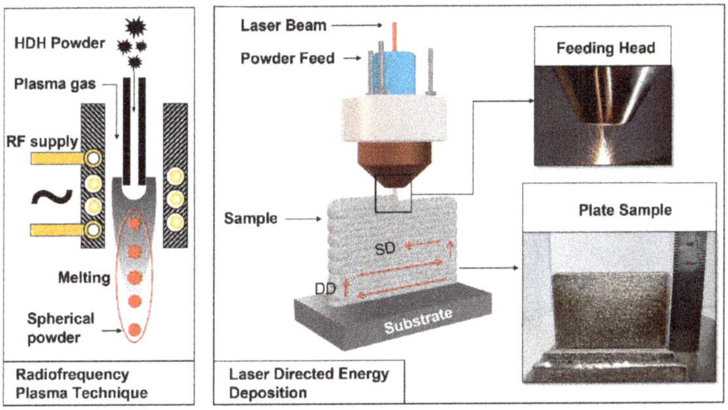

Figure 1. Schematic of the RF technique and the LDED fabrication process for Ti samples.

Metallographic characterizations were performed using an ECLIPSE LV 150N optical microscope (OM, NIKON CORPORATION, Shinogawa Intercity Tower C, Konan, Minato-ku, Tokyo, Japan). The metallographic samples were mechanically polished and then etched in a solution of 10% HF, 5% HNO_3, and 85% deionized water. The microstructures of the metallographic samples and the powder samples were characterized using scanning electron microscopy (SEM, LEO1530, LEO Electron Microscopy Ltd., Krefeld, Germany). A Microtrac-S3500 laser particle size analyzer was used to measure the powder size. Phase types were investigated using a Rigaku X-ray diffractometer(XRD, D/max-RB, Rigaku, Japan) using Cu-Kα radiation. The tensile properties of samples were examined at an ambient temperature using an Instron 5982 static testing machine with a strain rate of 10^{-3} s^{-1}. To confirm that the results were reliable, three tensile samples from each group were tested.

3. Results and Discussion

The morphology of the HDH titanium powders, before and after induced RF treatment, were studied using SEM and the granularity of the powders were measured with a laser granularity analyzer. Figure 2 shows the microstructures of HDH titanium powders, before and after induced RF treatment. It can be seen in Figure 2a that the HDH titanium powder precursor has a rough surface in terms of morphology. There are some scraps and cracks located on the surface of the powders. For the HDH titanium powders with an RF treatment with a powder feeding rate of 30 g/min (Figure 2b), most of the irregular powders were changed into the spherical powders. However, with an increase of powder feeding rate from 30 g/min to 60 g/min, the proportion of non-spherical powders increased, as shown in Figure 2c,d. Based on the above results, it was revealed that RF treatment can markedly improve the morphology of HDH titanium powders. When powders are injected into a high-temperature plasma region, they rapidly absorb energy, resulting in the surfaces of the powders being heated to melt and are then subsequently quenched into a spherical shape. In the process of RF treatment, a lower powder feeding rate benefits the contact between powders and plasma. Therefore, the production efficiency of spherical titanium powders will be reduced with an increase in the powder feeding rate.

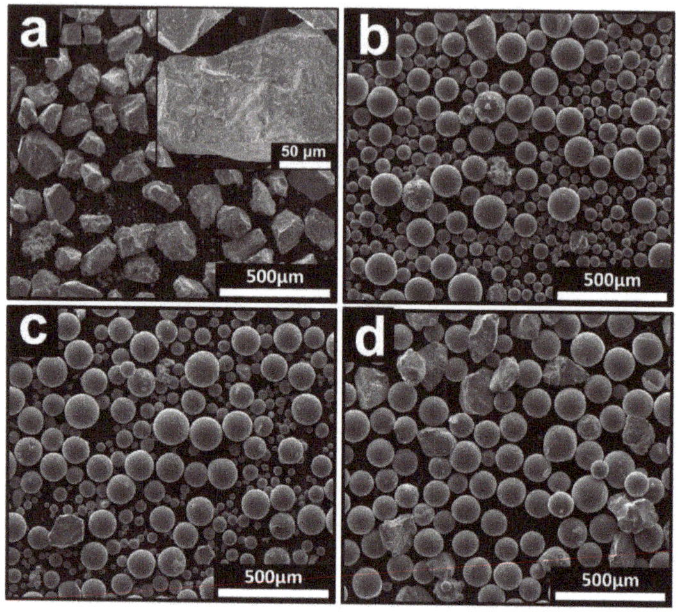

Figure 2. The micrograph show HDH titanium powder under different parameters: (**a**) without RF, (**b**) RF with a feeding rate of 30 g/min, (**c**) RF with a feeding rate of 45 g/min and (**d**) RF with a feeding rate of 60 g/min.

The particle size distributions of the HDH titanium powders, before and after RF treatment, are shown in Figure 3. Compared to powders without RF treatment, the peaks of the particle size distributions after RF treatment showed a lower position, indicating that the RF treatment has the ability to refine the HDH titanium powders. Furthermore, the peaks of the particle size distributions moved to a lower position with a reduction in the powder feeding rate. Some powders with a loose structure that are fragile cannot be subjected to the pressure caused by the thermal expansion and the residual titanium hydride dehydrogenation reaction, leading to the powders being broken up into a finer powder [19]. Therefore, there was more fine powder with the lower powder feeding rate.

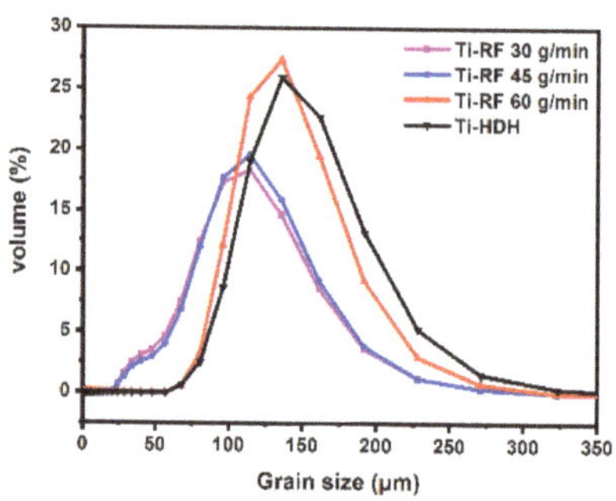

Figure 3. The particle size distributions of HDH titanium powders, before and after radio-frequency plasma treatment.

Figure 4 shows the XRD spectrum of HDH titanium powders, before and after RF treatment. Peaks that represent the close-packed hexagonal (HCP) Ti crystal structure were observed in the XRD spectra of the original powder without RF treatment. At the same time, weak peaks belonging to the Ti_2H phase were also observed in the enlargement of the XRD spectra of the original powder, as shown in Figure 4b, revealing that the process of dehydrogenated of the HDH Ti powder was incomplete. For the HDH titanium powders with RF treatment, the peaks showed that the Ti_2H phases faded away in the XRD spectrum along with the reduction in the powder feeding rate. The above results indicated that the radio frequency plasma treatment, not only optimized the shape of the titanium powder, but also benefitted the removal of the residual Ti_2H phase in the powder.

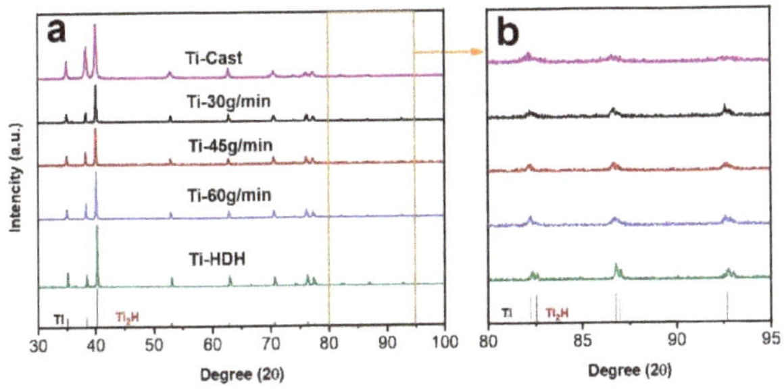

Figure 4. (**a**) XRD pattern of HDH titanium powders before and after radio frequency plasma treatment; (**b**) zoom-in image of the XRD patterns with 2θ ranging from 80° to 95°.

In order to analyze the effect of HDH powder and HDH-RF powder on the LDED-produced CP-Ti wall, the microstructures of the samples were observed with OM. The OM images in Figure 5a show that the morphology of the cast sample exhibited a specific dendritic grain microstructure. In Figure 5b, many blowhole defects with diameters of ~300 μm were left on the LDED samples prepared using HDH powder. Figure 5c shows an OM image of the LMD samples prepared using HDH-RF powder with a feeding rate of 30 g/min, which presents an almost-dense microstructure without porosity, revealing that HDH-RF powder is more suitable for the LDED technique than HDH powder. Furthermore, Figure 5d shows an enlarge image of the square in the Figure 5c, presenting that the microstructures of the LDED samples prepared using HDH-RF powder were composed of a mix of coarse and lath-shaped grains. Diverse microstructural features between the cast and LDED samples could be explained by the heating history, where significantly higher cooling rates were used in the process of LDED, leading to martensitic β transformation in CP-Ti solidification to an α phase with lath-shaped grains [20].

The tensile stress–strain curves of the titanium samples are shown in Figure 6. Compared with the LDED-HDH-produced samples and the cast samples, the LDED-HDH-RF-produced samples possessed an excellent combination of tensile strength and tensile ductility. It is well-known that defects in the samples cause performance degradation. The blowhole defects in the LDED-HDH-produced samples resulted in lesser mechanical properties compared to those of the LDED-HDH-RF-produced samples with an almost-dense microstructure. A fine microstructure results in a high yield strength, which is determined according to the Hall–Petch relation law, providing an increase in the yield stress of a polycrystalline material as its interplanar spacing decreases [21]. For CP-Ti, the yield strength is inversely proportional to the α grain size. The fine α lath shape obtained in LDED-HDH-RF-produced samples may be related to the high cooling rate during the LDED process, which promoted the β phase to α phase, with a fine net microstructure transformation [15]. Therefore, LDED-HDH-RF-produced samples with a lath shape showed a higher yield strength than the cast samples with equiaxed grains. The improved yield strength by employing the LDED-HDH-RF technique can enhance an implant CP-Ti material's resistance to permanent shape change, which may increase its effectiveness for implant applications.

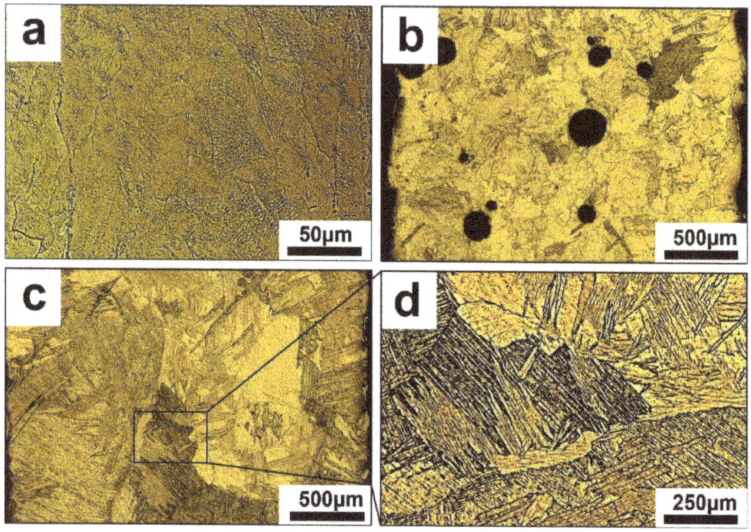

Figure 5. The optical microscope of CP-Ti using different prepared methods: (**a**) cast, (**b**) LDED-HDH titanium, (**c**) LDED-HDH-RF titanium, (**d**) zoom-in image of the square in LDED-HDH-RF titanium.

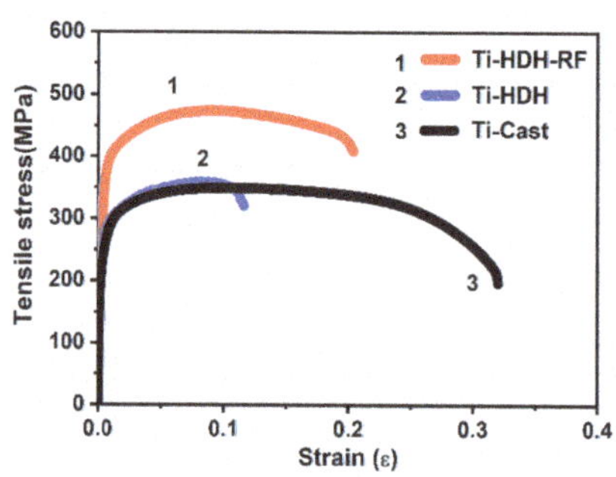

Figure 6. The tensile stress–strain curves of CP-Ti produced using different prepared methods.

4. Conclusions

Hydrogenated–dehydrogenated (HDH) titanium powder has been employed to produce a spherical Ti powder using the RF technique. The spherical Ti powder was also used as the raw materials for LDED to create the CP-Ti. Spherical Ti and CP-Ti samples were further analyzed by means of XRD and SEM. The main conclusions were described as follows:

(1) Microstructural analyses of the powder revealed that the radio frequency plasma treatment, not only optimizes the shape of titanium powder, but also benefits the removal of the residual Ti_2H phase in the powder.

(2) Microstructural analyses of the CP-Ti samples present that the LDED-HDH-RF-produced samples have a denser microstructure than the LDED-HDH-RF-produced samples.

(3) Tensile testing indicated that LDED-HDH-RF-produced samples possess an excellent combination of tensile strength and tensile ductility. Compared to the cast samples, the higher yield strength of the LDED-HDRHT-produced samples should be attribute to the fine α lath shape.

(4) The present work provides an approach for LDED-fabricated CP-Ti to accommodate the economic and mechanical properties of CP-Ti, while providing insights into the expanding application of HDH titanium powder.

Author Contributions: Conceptualization, J.L. and Q.L.; methodology, M.L.; software, L.H.; validation, M.L., L.H. and C.R.; formal analysis, M.L.; investigation, L.H. and D.W.; resources, D.W.; data curation, M.L. and D.W.; writing—original draft preparation, M.L.; writing—review and editing, J.L.; visualization, M.L.; supervision, Q.L.; project administration, J.L.; funding acquisition, Q.L. All authors have read and agreed to the published version of the manuscript.

Funding: This research was funded by the Leading Talents of Tianshan Cedar Program of Xinjiang Uygur Autonomous Region (No. 2019XS02), the Tianshan Innovation Team Program of Xinjiang Uygur Autonomous Region (No. 2020D14038) and China Academy of Engineering Physics Foundation (Grant No. CX2019020 and Grant No. YZJJLX2019010), and National Natural Science Foundation of China (Grant No: 52001288).

Institutional Review Board Statement: Not applicable.

Informed Consent Statement: Not applicable.

Data Availability Statement: The data used to support the findings of this study are available from the corresponding author upon request.

Conflicts of Interest: The authors declare no conflict of interest. We declare that we have no financial and personal relationships with other people or organizations that can inappropriately influence our work.

References

1. Na, T.-W.; Kim, W.R.; Yang, S.-M.; Kwon, O.; Park, J.M.; Kim, G.-H.; Jung, K.-H.; Lee, C.-W.; Park, H.-K.; Kim, H.G. Effect of laser power on oxygen and nitrogen concentration of commercially pure titanium manufactured by selective laser melting. *Mater. Charact.* **2018**, *143*, 110–117. [CrossRef]
2. Zhao, X.; Niinomi, M.; Nakai, M.; Hieda, J.; Ishimoto, T.; Nakano, T. Optimization of Cr content of metastable β-type Ti–Cr alloys with changeable Young's modulus for spinal fixation applications. *Acta Biomater.* **2012**, *8*, 2392–2400. [CrossRef] [PubMed]
3. Froes, F.; Mashl, S.; Hebeisen, J.; Moxson, V.; Duz, V. The technologies of titanium powder metallurgy. *JOM* **2004**, *56*, 46–48. [CrossRef]
4. Uhlmann, E.; Kersting, R.; Klein, T.B.; Cruz, M.F.; Borille, A.V. Additive manufacturing of titanium alloy for aircraft components. *Procedia CIRP* **2015**, *35*, 55–60. [CrossRef]
5. Kovacı, H. Comparison of the microstructural, mechanical and wear properties of plasma oxidized Cp-Ti prepared by laser powder bed fusion additive manufacturing and forging processes. *Surf. Coat. Technol.* **2019**, *374*, 987–996. [CrossRef]
6. Kang, N.; Yuan, H.; Coddet, P.; Ren, Z.; Bernage, C.; Liao, H.; Coddet, C. On the texture, phase and tensile properties of commercially pure Ti produced via selective laser melting assisted by static magnetic field. *Mater. Sci. Eng. C* **2017**, *70*, 405–407. [CrossRef] [PubMed]
7. Hasib, M.T.; Ostergaard, H.E.; Liu, Q.; Li, X.; Kruzic, J.J. Tensile and fatigue crack growth behavior of commercially pure titanium produced by laser powder bed fusion additive manufacturing. *Addit. Manuf.* **2021**, *45*, 102027. [CrossRef]
8. Barro, Ó.; Arias-González, F.; Lusquiños, F.; Comesaña, R.; del Val, J.; Riveiro, A.; Badaoui, A.; Gómez-Baño, F.; Pou, J. Improved commercially pure titanium obtained by laser directed energy deposition for dental prosthetic applications. *Metals* **2021**, *11*, 70. [CrossRef]
9. Sercombe, T.; Zhang, L.-C.; Li, S.; Hao, Y. Additive manufacturing of cp-Ti, Ti-6Al-4V and Ti2448. In *Titanium in Medical and Dental Applications*; Woodhead Publishing: Sawston, UK, 2018; pp. 303–324.
10. Amado, J.; Rodríguez, A., Montero, J.; Tobar, M.; Yáñez, A. A comparison of laser deposition of commercially pure titanium using gas atomized or Ti sponge powders. *Surf. Coat. Technol.* **2019**, *374*, 253–263. [CrossRef]
11. Kusuma, C.; Ahmed, S.H.; Mian, A.; Srinivasan, R. Effect of laser power and scan speed on melt pool characteristics of commercially pure titanium (CP-Ti). *J. Mater. Eng. Perform.* **2017**, *26*, 3560–3568. [CrossRef]
12. Attar, H.; Bermingham, M.; Ehtemam-Haghighi, S.; Dehghan-Manshadi, A.; Kent, D.; Dargusch, M. Evaluation of the mechanical and wear properties of titanium produced by three different additive manufacturing methods for biomedical application. *Mater. Sci. Eng. A* **2019**, *760*, 339–345. [CrossRef]
13. Bolzoni, L.; Ruiz-Navas, E.M.; Gordo, E. Powder metallurgy CP-Ti performances: Hydride-dehydride vs. sponge. *Mater. Des.* **2014**, *60*, 226–232. [CrossRef]
14. Balog, M.; Viskic, J.; Krizik, P.; Schauperl, Z.; Snajdar, M.; Stanec, Z.; Catic, A. CP TI fabricated by low temperature extrusion of HDH powder: Application in dentistry. *Key Eng. Mater.* **2016**, *704*, 351–359. [CrossRef]

15. Tao, Q.; Wang, Z.; Chen, G.; Cai, W.; Cao, P.; Zhang, C.; Ding, W.; Lu, X.; Luo, T.; Qu, X. Selective laser melting of CP-Ti to overcome the low cost and high performance trade-off. *Addit. Manuf.* **2020**, *34*, 101198. [CrossRef]
16. Fang, X.; Ren, C.; Zhang, L.; Wang, C.; Huang, K.; Lu, B. A model of bead size based on the dynamic response of CMT-based wire and arc additive manufacturing process parameters. *Rapid Prototyp. J.* **2021**, *27*, 741–753. [CrossRef]
17. Kulkarni, J.D.; Goka, S.B.; Parchuri, P.K.; Yamamoto, H.; Ito, K.; Simhambhatla, S. Microstructure evolution along build direction for thin-wall components fabricated with wire-direct energy deposition. *Rapid Prototyp. J.* **2021**, *27*, 1289–1301. [CrossRef]
18. Tang, S.; Wang, G.; Song, H.; Li, R.; Zhang, H. A novel method of bead modeling and control for wire and arc additive manufacturing. *Rapid Prototyp. J.* **2021**. ahead-of-print. [CrossRef]
19. Wang, J.-J.; Hao, J.-J.; Guo, Z.-M.; Wang, Y.-M. Preparation of spherical tungsten and titanium powders by RF induction plasma processing. *Rare Met.* **2015**, *34*, 431–435. [CrossRef]
20. Lonardelli, I.; Gey, N.; Wenk, H.-R.; Humbert, M.; Vogel, S.; Lutterotti, L. In situ observation of texture evolution during $\alpha\rightarrow\beta$ and $\beta\rightarrow\alpha$ phase transformations in titanium alloys investigated by neutron diffraction. *Acta Mater.* **2007**, *55*, 5718–5727. [CrossRef]
21. Stanford, N.; Carlson, U.; Barnett, M. Deformation twinning and the Hall-Petch relation in commercial purity Ti. *Metall. Mater. Trans. A* **2008**, *39*, 934–944. [CrossRef]

materials

MDPI

Article

Designing High Entropy Bulk Metallic Glass (HE-BMG) by Similar Element Substitution/Addition

Hongyu Ding [1,2,*], Hengwei Luan [2], Hengtong Bu [2], Hongjie Xu [2] and Kefu Yao [2,*]

1 Marine Equipment and Technology Institute, Jiangsu University of Science and Technology, Zhenjiang 212100, China
2 School of Materials Science and Engineering, Tsinghua University, Beijing 100084, China; luanhengwei6770@163.com (H.L.); bht20@mails.tsinghua.edu.cn (H.B.); xuhongjie@mail.tsinghua.edu.cn (H.X.)
* Correspondence: dinghongyu2018@just.edu.cn (H.D.); kfyao@tsinghua.edu.cn (K.Y.)

Abstract: In this paper, we report that two newly designed high entropy bulk metallic glasses (HE-BMGs), $Ti_{20}Hf_{20}Cu_{20}Ni_{20}Be_{20}$ with a critical diameter of 2 mm, and $Ti_{16.7}Zr_{16.7}Nb_{16.7}Cu_{16.7}Ni_{16.7}Be_{16.7}$ with a critical diameter of 1.5 mm, can be fabricated by copper mold casting method. These newly developed HE-BMGs exhibited a high fracture strength over 2300 MPa. The glass forming ability and atomic size distribution characteristics of the HE-BMGs are discussed in detail. Moreover, a parameter δ' was proposed to evaluate the atomic size distribution characteristics in different HEAs. It showed that this new parameter is closely related to the degree of lattice distortion and phase selection of high-entropy alloys. Adjusting the value of δ' parameter by similar element substitution/addition would be beneficial for designing high entropy bulk metallic glasses.

Keywords: high entropy alloy; bulk metallic glass; similar element substitution/addition; glass forming ability; lattice distortion

Citation: Ding, H.; Luan, H.; Bu, H.; Xu, H.; Yao, K. Designing High Entropy Bulk Metallic Glass (HE-BMG) by Similar Element Substitution/Addition. *Materials* **2022**, *15*, 1669. https://doi.org/10.3390/ma15051669

Academic Editor: Filippo Berto

Received: 31 December 2021
Accepted: 15 February 2022
Published: 23 February 2022

1. Introduction

In the past few decades, bulk metallic glasses (BMGs) [1–9] and high entropy alloys (HEAs) [10–18] have attracted much attention, owing to their unique structure and properties, such as high strength/hardness, good corrosion/wear resistance, etc. Previously, BMGs and HEAs were developed separately in most cases, following different composition design and fabrication routes. While recent studies show that intersections exist between these two domains, namely some HEAs with meticulously designed composition could be made into BMGs, and hence the high entropy bulk metallic glasses (HE-BMGs) were developed [19–41]. An investigation into HE-BMGs is beneficial for understanding the phase formation rules of HEAs and fundamental issues of BMGs, so it is very important to develop more HE-BMGs.

In our previous work, a $Ti_{20}Zr_{20}Cu_{20}Ni_{20}Be_{20}$ HE-BMG with a critical diameter of 3 mm was successfully obtained by copper mold casting method [24]. By introducing Hf as the sixth constituent element, $Ti_{16.7}Zr_{16.7}Hf_{16.7}Cu_{16.7}Ni_{16.7}Be_{16.7}$ with a critical diameter of 15 mm [25] and a series of $Ti_{20}Zr_{20}Hf_{20}(Cu_{20-x}Ni_x)Be_{20}$ HE BMGs with critical diameters of larger than 12 mm was developed [26,27]. These results indicate that similar element substitution/addition is an effective way for developing new HE-BMGs, just the same as traditional BMGs. Since Hf is an element chemically similar to Zr while Nb and Zr are also very close in the periodic table of elements, it is reasonable to suppose that by substituting Zr with Hf, or by adding Nb in the $Ti_{20}Zr_{20}Cu_{20}Ni_{20}Be_{20}$ quinary HEA system, new HE-BMG with good properties can be obtained. Accordingly, two new HEAs, namely $Ti_{20}Hf_{20}Cu_{20}Ni_{20}Be_{20}$ and $Ti_{16.7}Zr_{16.7}Nb_{16.7}Cu_{16.7}Ni_{16.7}Be_{16.7}$, were designed to verify this assumption, and their glass-forming ability, atomic size distribution characteristics, lattice distortion, and phase selection rules of HEAs are discussed in detail.

2. Experimental

The master alloy ingots with nominal compositions of $Ti_{20}Hf_{20}Cu_{20}Ni_{20}Be_{20}$ and $Ti_{16.7}Zr_{16.7}Nb_{16.7}Cu_{16.7}Ni_{16.7}Be_{16.7}$ in equal atomic ratio were prepared by arc melting the mixtures of high purity Ti, Hf, Cu, Ni, Zr, Nb plates, and Be granules (purity higher than 99.99 wt.%) within a pure argon gas environment. Cylindrical rods with different diameters were prepared by copper mold injection or suction casting method. Arc melting and casting was conducted on multi-functional high vacuum arc-melting and melt-spinning system, which was produced by SKY Technology Development Corporation, Shenyang, China. The glassy nature of these as-prepared samples was examined by X-ray diffraction (XRD) technique using a Rigaku D/max-RB XRD spectrometry (Rigaku Corporation, Tokyo, Japan) with Cu Kα radiation (λ = 0.15406 nm). Thermal properties of the glassy alloys were examined by a Shimadzu DSC-60 differential scanning calorimeter (Shimadzu Corporation, Kyoto, Japan) instrument under the protection of N_2 gas (flow rate: 50 mL/min). The applied heating rate was set as 20 K/min. The DSC instrument was calibrated with In and Zn standard specimens. The errors are within $\pm$1 K. Compression tests with specimens of Ø2 × 4 mm and Ø1.5 × 3 mm in size were carried out on WDW-100 testing machine (Shanghai Precision Instrument Co., Ltd, Shanghai, China) under a stain rate of 4×10^{-4} s^{-1}.

3. Results

Figure 1 shows the XRD spectra of the as-cast $Ti_{20}Hf_{20}Cu_{20}Ni_{20}Be_{20}$ and $Ti_{16.7}Zr_{16.7}Nb_{16.7}Cu_{16.7}Ni_{16.7}Be_{16.7}$ rods with different diameters. No sharp diffraction peak corresponding to the crystalline phase was observed in the Ø2 mm $Ti_{20}Hf_{20}Cu_{20}Ni_{20}Be_{20}$ and Ø1.5 mm $Ti_{16.7}Zr_{16.7}Nb_{16.7}Cu_{16.7}Ni_{16.7}Be_{16.7}$ samples, indicating that they both possess a fully amorphous structure.

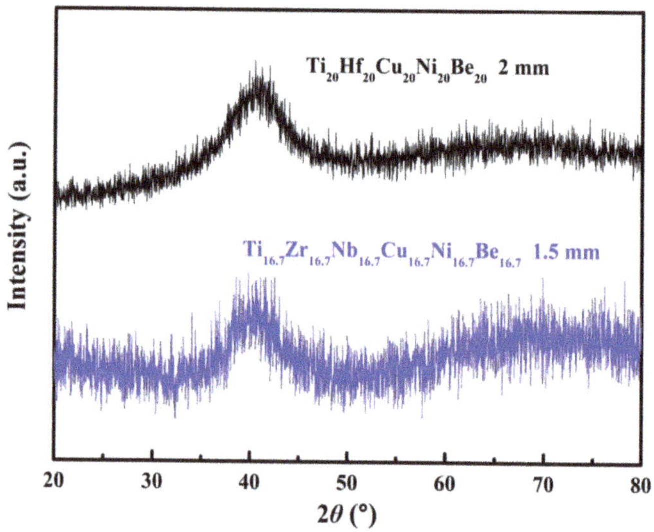

Figure 1. XRD spectra of the Ø2 mm $Ti_{20}Hf_{20}Cu_{20}Ni_{20}Be_{20}$ rod sample and Ø1.5 mm $Ti_{16.7}Zr_{16.7}Nb_{16.7}Cu_{16.7}Ni_{16.7}Be_{16.7}$ rod sample.

The DSC curves of the $Ti_{20}Hf_{20}Cu_{20}Ni_{20}Be_{20}$ and $Ti_{16.7}Zr_{16.7}Nb_{16.7}Cu_{16.7}Ni_{16.7}Be_{16.7}$ samples are shown in Figure 2. The highest test temperature reached 1273 K (1000 °C). However, since the endothermic peak is very high in the high temperature part, glass transition would be very ambiguous in the curve. In order to demonstrate the glass transition phenomenon (which is very important for glasses) clearly, we just cut out temperature less than 1000 K in Figure 2. The glass transition temperature T_g and initial crystallization temperature T_x were marked with arrows. T_g, T_x, T_m (melting temperature) and T_l (liquidus temperature) were measured as 717 K, 760 K, 1095 K, and 1220 K for the $Ti_{20}Hf_{20}Cu_{20}Ni_{20}Be_{20}$ HE-BMG, and 684 K, 739 K, 1066 K, and 1218 K for the $Ti_{16.7}Zr_{16.7}Nb_{16.7}Cu_{16.7}Ni_{16.7}Be_{16.7}$ HE-BMG, respectively. These data were listed in Table 1.

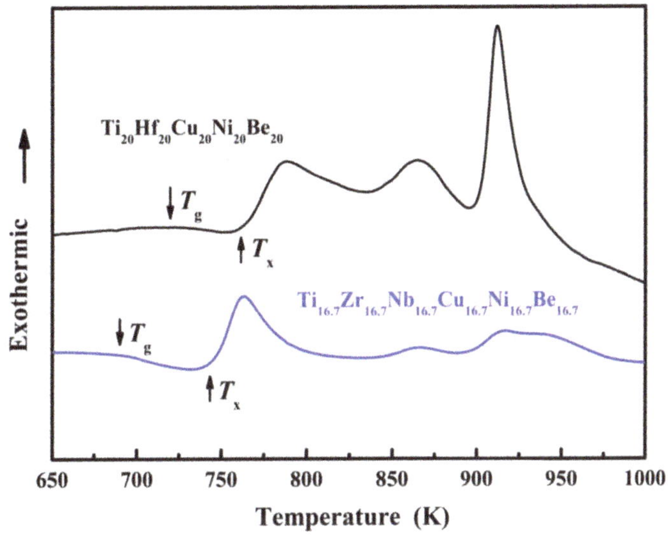

Figure 2. DSC curves of the $Ti_{20}Hf_{20}Cu_{20}Ni_{20}Be_{20}$ and $Ti_{16.7}Zr_{16.7}Nb_{16.7}Cu_{16.7}Ni_{16.7}Be_{16.7}$ HE-BMGs.

Table 1. Thermal and mechanical properties of some HE-BMGs.

Composition	T_g (K)	T_x (K)	T_m (K)	T_l (K)	$\sigma_{0.2}$ (MPa)	σ_b (MPa)	ε_p (%)	Year
$Ti_{20}Zr_{20}Cu_{20}Ni_{20}Be_{20}$	683	729	1076	1161	-	2315	0	2013 [24]
$Ti_{16.7}Zr_{16.7}Hf_{16.7}Cu_{16.7}Ni_{16.7}Be_{16.7}$	681	751	1019	1100	1943	2064	0.6	2014 [25]
$Ti_{20}Hf_{20}Cu_{20}Ni_{20}Be_{20}$	717	760	1095	1220	-	2425	0	This work
$Ti_{16.7}Zr_{16.7}Nb_{16.7}Cu_{16.7}Ni_{16.7}Be_{16.7}$	684	739	1066	1218	2330	2450	0.5	This work

The stress strain curves of Ø2 × 4 mm $Ti_{20}Hf_{20}Cu_{20}Ni_{20}Be_{20}$ and Ø1.5 × 3 mm $Ti_{16.7}Zr_{16.7}Nb_{16.7}Cu_{16.7}Ni_{16.7}Be_{16.7}$ HE-BMG samples in uniaxial compression test were shown in Figure 3. The fracture strength σ_b was 2425 MPa for $Ti_{20}Hf_{20}Cu_{20}Ni_{20}Be_{20}$ HE-BMG, the yield strength $\sigma_{0.2}$, fracture strength σ_b and plasticity ε_p were 2330 MPa, 2450 MPa and 0.5% for $Ti_{16.7}Zr_{16.7}Nb_{16.7}Cu_{16.7}Ni_{16.7}Be_{16.7}$ HE-BMG, respectively, which were also listed in Table 1. The specimens fractured in a shear mode. It is interesting to note that both $Ti_{20}Hf_{20}Cu_{20}Ni_{20}Be_{20}$ and $Ti_{20}Zr_{20}Cu_{20}Ni_{20}Be_{20}$ quinary HE-BMGs fractured without any plasticity [24], while $Ti_{16.7}Zr_{16.7}Nb_{16.7}Cu_{16.7}Ni_{16.7}Be_{16.7}$ and $Ti_{16.7}Zr_{16.7}Hf_{16.7}Cu_{16.7}Ni_{16.7}Be_{16.7}$ senary HE-BMGs exhibited a compressive plasticity of about 0.5%, as well as serration behavior [25]. The reason of this difference remains unclear.

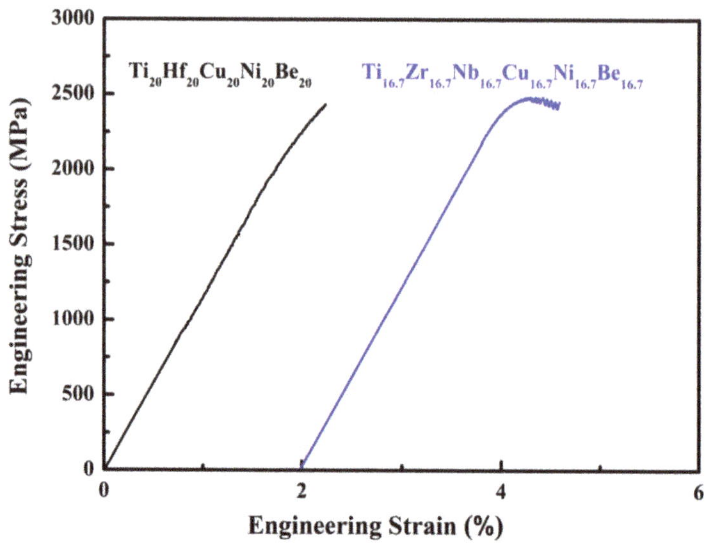

Figure 3. Stress strain curves of the $Ti_{20}Hf_{20}Cu_{20}Ni_{20}Be_{20}$ and $Ti_{16.7}Zr_{16.7}Nb_{16.7}Cu_{16.7}Ni_{16.7}Be_{16.7}$ HE-BMGs.

4. Discussion

4.1. Glass Forming Ability (GFA) of High Entropy Alloys by Element Addition/Substitution

The parameters of supercooled liquid region ΔT (= $T_x - T_g$), reduced glass transition temperature T_{rg} (= T_g/T_l), and γ parameter (= $T_x/(T_g + T_l)$) are calculated as 43 K, 0.588, and 0.392 for $Ti_{20}Hf_{20}Cu_{20}Ni_{20}Be_{20}$, while 55 K, 0.562, and 0.388 for $Ti_{16.7}Zr_{16.7}Nb_{16.7}Cu_{16.7}Ni_{16.7}Be_{16.7}$, respectively. Compared with $Ti_{20}Zr_{20}Cu_{20}Ni_{20}Be_{20}$ alloy (3 mm), the critical diameter of $Ti_{20}Hf_{20}Cu_{20}Ni_{20}Be_{20}$, alloy (2 mm) and $Ti_{16.7}Zr_{16.7}Nb_{16.7}Cu_{16.7}Ni_{16.7}Be_{16.7}$ (1.5 mm) both decreased. It is noticed that by substitution Zr with Hf, although T_{rg} remains the same, ΔT and γ decreased; by addition of Nb as the sixth element, both T_{rg} and γ decreased, although ΔT increased [24]. It implies that the parameter γ is better than T_{rg} and ΔT in judging the GFA in these high-entropy glassy alloys; meanwhile high entropy is not always beneficial to the GFA of the HEAs. The substitution of element Hf and the addition of Nb brings the liquidus temperature T_l higher than that of $Ti_{20}Zr_{20}Cu_{20}Ni_{20}Be_{20}$ alloy [24]. As a result, the GFA of the HEA was slightly deteriorated. On the other hand, by the addition of Hf as the sixth element, liquidus temperature T_l was lowered down. Therefore, the GFA of the $Ti_{16.7}Zr_{16.7}Hf_{16.7}Cu_{16.7}Ni_{16.7}Be_{16.7}$ was greatly improved as compared with $Ti_{20}Zr_{20}Cu_{20}Ni_{20}Be_{20}$ alloy [25]. These results indicate that lowering down liquidus temperature would be helpful for enhancing the GFA.

4.2. Atomic Radius Characteristics of HE-BMG

The atomic size distribution characteristics of existing HE-BMGs were shown in Table 2. Based on the atomic radius of constituent elements, they were divided into five categories, namely super large atom ($r > 0.165$ nm), large atom ($r \approx 0.16$ nm), medium atom ($r \approx 0.14$ nm), small atom ($r \approx 0.12$ nm), and ultra-small atom ($r < 0.12$ nm). It is noticed that most HE-BMGs were comprised of 3 to 4 categories, except for those containing nonmetal element such as Si, P, B, C, etc [29,33,36]. In high entropy alloys, larger atomic radius difference leads to larger lattice distortion. In case that lattice distortion exceeds some degree, the lattice collapse and amorphous structure formed accordingly. This is in agreement with Zhang's work [13].

Table 2. Atomic size distribution characteristics of existing HE-BMGs.

Composition	Super Large Atom $r > 0.165$ nm	Large Atom $r \approx 0.16$ nm	Medium Atom $r \approx 0.14$ nm	Small Atom $r \approx 0.12$ nm	Ultra Small Atom $r < 0.12$ nm	Year
$Ti_{20}Zr_{20}Hf_{20}Cu_{20}Ni_{20}$		Zr, Hf	Ti	Cu, Ni		2002 [19]
$Sr_{20}Ca_{20}Yb_{20}Mg_{20}Zn_{20}$	Sr, Ca, Yb	Mg	Zn			2011 [20,23]
$Er_{20}Tb_{20}Dy_{20}Ni_{20}Al_{20}$	Tb, Dy, Er		Al	Ni		2011 [21]
$Pd_{20}Pt_{20}Cu_{20}Ni_{20}P_{20}$			Pt, Pd	Cu, Ni	P	2011 [22]
$Ti_{20}Zr_{20}Cu_{20}Ni_{20}Be_{20}$		Zr	Ti	Cu, Ni	Be	2013 [24]
$Ti_{16.7}Zr_{16.7}Hf_{16.7}Cu_{16.7}Ni_{16.7}Be_{16.7}$		Zr, Hf	Ti	Cu, Ni	Be	2014 [25]
$Ti_{20}Zr_{20}Hf_{20}(Cu_{20-x}Ni_x)Be_{20}$		Zr, Hf	Ti	Cu, Ni	Be	2015 [26,27]
$Ho_{20}Er_{20}Co_{20}Al_{20}Dy_{20}$	Dy, Ho, Er		Al	Co		2015 [28]
$Fe_{25}Co_{25}Ni_{25}(B, Si)_{25}$				Co, Ni, Fe	Si, B	2015 [29]
$Zr_{40}Hf_{10}Ti_4Y_1Al_{10}Cu_{25}Ni_7Co_2Fe_1$	Y	Zr, Hf	Ti, Al	Cu, Co, Ni, Fe		2015 [30]
$Er_{18}Gd_{18}Y_{20}Al_{24}Co_{20}$	Y, Gd, Er		Al	Co		2018 [31]
$Er_{20}Dy_{20}Co_{20}Al_{20}RE_{20}$ (RE = Gd, Tb, Tm)	Gd/Tb, Dy, Er	Tm	Al	Co		2018 [32]
$Fe_{25}Co_{25}Ni_{25}(P_{0.4}C_{0.2}B_{0.2}Si_{0.2})_{25}$				Co, Ni, Fe	Si, P, B, C	2018 [33]
$La_{25-35}Ce_{25-35}Ni_{5-15}Cu_{5-15}Al_{20}$	La, Ce		Al	Cu, Ni		2018 [34]
$Fe_{25}Co_{25}Ni_{25}Mo_5P_{10}B_{10}$			Mo	Co, Ni, Fe	P, B	2019 [35]
$(Fe_{1/3}Co_{1/3}Ni_{1/3})_{80}(P_{1/2}B_{1/2})_{20}$				Co, Ni, Fe	P, B	2019 [36]
$Zr_{35}Hf_{17.5}Ti_{5.5}Al_{12.5}Co_{7.5}Ni_{12}Cu_{10}$		Zr, Hf	Ti, Al	Cu, Co, Ni		2019 [37]
$Gd_{25}Co_{25}Al_{25}Y_{15}RE_{10}$ (RE = Dy, Ho, Er)	Y, Gd, (Dy, Ho, Er)		Al	Co		2020 [38]
$Fe_{20-35}Ni_{20}Cr_{20-30}Mo_{5-15}(P_{0.6}C_{0.2}B_{0.2})_{20}$			Mo	Cr, Ni, Fe	P, B, C	2020 [39]
$(Gd_{0.2}Dy_{0.2}Er_{0.2}Co_{0.2}Al_{0.2})_{99.5}Si_{0.5}$	Gd, Dy, Er		Al	Co	Si	2021 [40]
$Zr_{33}Hf_8Ti_6Cu_{32}Ni_{10}Co_5Al_6$		Zr, Hf	Ti, Al	Cu, Co, Ni		2021 [41]
$Ti_{20}Hf_{20}Cu_{20}Ni_{20}Be_{20}$		Hf	Ti	Cu, Ni	Be	This work
$Ti_{16.7}Zr_{16.7}Nb_{16.7}Cu_{16.7}Ni_{16.7}Be_{16.7}$		Zr	Ti, Nb	Cu, Ni	Be	This work

4.3. Assessing Degree of Lattice Distortion in High Entropy Alloys by Parameter δ'

The phase formation rule in HEA is of great importance both scientifically and technologically. The formed phase(s) in HEAs (solid solution, intermetallics and amorphous phase) at certain conditions (alloy composition, preparation method, service environment, etc.) remain unknown for most HEAs [10,13,15,18,42]. Many researchers proposed various criteria to solve this problem, such as the δ-ΔH_{mix} diagram proposed by Zhang et al. [13], VEC criteria proposed by Guo et al. [15], electronegativity mismatch D_c proposed by Toda-Caraballo et al. [43], etc. Lattice distortion is a crucial factor in HEAs, and it is also very important in determining phase formation. However, the relationship between lattice distortion and phase formation is still not clear. Much research has been devoted to characterizing the degree of lattice distortion, and to further illustrate its correlation with phase formation, such as the γ parameter proposed by Wang et al. [44], the α_2 parameter proposed by Wang et al. [45], etc. However, it is far from clearly understanding. Further investigation is still required.

It is noticed from Table 2 that in most HE-BMGs, the atomic radius of the constituent elements atomic sizes distribute in a wide range; while for many solid solution forming HEAs, atomic sizes are more concentrated (especially for CuCoCrNiFe [10] and Cantor alloy [11], they both possess FCC structure, meanwhile atomic size difference of the constituent elements are very small). However, this is a qualitative description, and it is somehow ambiguous. As a result, a quantitative exemplification is needed.

Based on Table 2, here we propose a new parameter δ' to assess the degree of lattice distortion in HEAs. Supposing that a HEA contains N elements, the atomic fractions are c_1, $c_2 \ldots \ldots c_N$, respectively, and the atomic radii are $r_1, r_2 \ldots \ldots r_N$ ($r_1 < r_2 < \ldots \ldots < r_N$), respectively (data from ref. [46]). Then, the average atomic size is defined as $\bar{r}$:

$$\bar{r} = \sum_{1}^{N} c_i r_i \quad (1)$$

The lattice distortion parameter δ' is defined as

$$\delta' = 100 \sum_{1}^{N-1} \frac{c_{i+1} + c_i}{2} \frac{r_{i+1} - r_i}{\bar{r}} \tag{2}$$

In particular, for equal atomic alloy, δ' is given as

$$\delta' = \frac{100}{N} \frac{r_N - r_1}{\bar{r}} \tag{3}$$

According to Formula (2), lattice distortion parameter δ' for some typical HEAs were calculated and listed in Table 3. For clarity, the relationship between atomic size distribution, lattice distortion parameter δ', and phase selection is demonstrated in Figure 4. It is noticed that δ' is closely related to phase selection in HEAs: when atomic size difference is relatively small, δ' is also small ($\delta' < 2.2$), FCC solid solution would be formed; when the atomic size difference became larger, δ' increased, FCC + BCC solid solution would tend to form as $2.2 < \delta' < 2.9$; with even larger δ' ($2.9 < \delta' < 4.9$), BCC solid solution would be formed; amorphous phase would be formed as δ' exceeds 4.9.

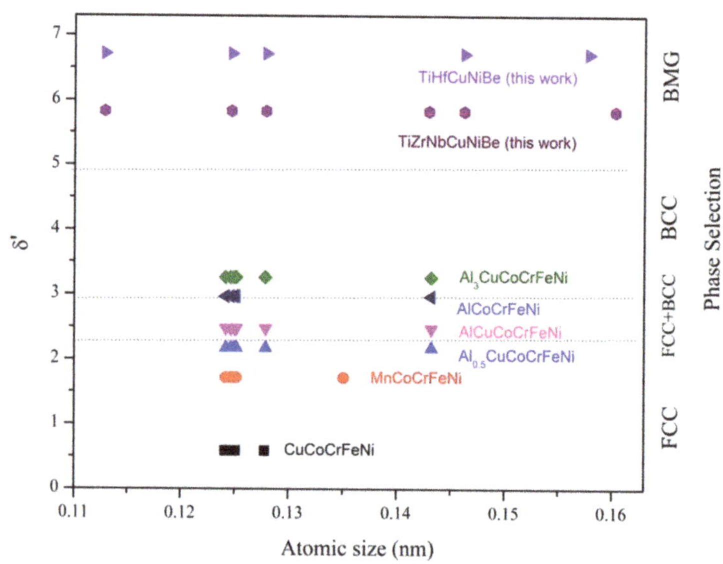

Figure 4. Correlation between atomic size distribution, lattice distortion degree parameter δ' and phase selection in some typical HEAs [10–12].

The parameter δ' can be understood from the point of view of dense atomic packing. Since it is correlated with the degree of lattice distortion, when δ' was small, dense random packing FCC phase formed (its density is about 74% for monolic element); with the increase of lattice distortion, δ' became larger, looser BCC phase (density of about 68% for monolic element) appeared, and its concentration increased accordingly; when lattice distortion became even serious, lattice collapse and amorphous phase would form eventually. Then, adjusting the lattice distortion or the parameter δ' of HEAs, such as by similar element substitution or addition, could be helpful in designing high-entropy metallic glasses.

The new parameter δ' we proposed here is somewhat similar with the δ parameter proposed by Zhang et al. [13]; it is also affected by the number, type, and concentration of the elements, while its value is smaller than δ, as is demonstrated in Table 3. As compared with δ, δ' is more sensitive to addition/substitution of an ultra large/small atom. Taking alloy 11 and 12 in Table 3 for example, it can be seen that by substituting Hf element with much smaller Be element in the Ti-Zr-Hf-Cu-Ni HEA, δ increased from 10.324 to 12.514, the growth rate is 21%; while δ' increased from 4.977 to 7.065, the growth rate is 42%, much larger than that of δ. It indicates that the new parameter δ' is more sensitive than δ in certain circumstance.

Additionally, it is noticed from Formula (3), for equiatomic high entropy alloys, as the number of elements N increased, δ' decreased and lattice distortion is mitigated accordingly. As a result, it is not beneficial for amorphous phase formation, especially for N > 10. This is in consistent with Cantor's result that an alloy with 16 to 20 elements in equiatomic concentration does not form amorphous phase [11].

Table 3. Correlation between atomic size distribution, lattice distortion and phase selection in some typical HEAs.

No.	Composition	$r >$ 0.165 nm	$r \approx 0.16$ nm	$r \approx 0.14$ nm	$r \approx 0.12$ nm	$r < 0.12$ nm	$\bar{r}$	δ [13]	δ'	VEC [15]	Phase
1	CrMnFeCoNi			Mn	Co, Cr, Ni, Fe		1.26744	3.267	1.717	8	FCC [11]
2	CuCoCrNiFe				Cu, Co, Cr, Ni, Fe		1.25304	1.031	0.587	8.8	FCC [10]
3	$Al_{0.3}CuCoCrFeNi$			Al	Cu, Co, Cr, Ni, Fe		1.26315	3.416	2.042	8.472	FCC [10]
4	$A_{0.5}lCuCoCrFeNi$			Al	Cu, Co, Cr, Ni, Fe		1.26928	4.161	2.178	8.273	FCC [10]
5	$Al_{0.8}CuCoCrFeNi$			Al	Cu, Co, Cr, Ni, Fe		1.27768	4.912	2.363	8	FCC + BCC [10]
6	AlCuCoCrFeNi			Al	Cu, Co, Cr, Ni, Fe		1.28281	5.271	2.475	7.833	FCC + BCC [10]
7	$Al_{2.5}CuCoCrFeNi$			Al	Cu, Co, Cr, Ni, Fe		1.31259	6.466	3.106	6.867	FCC + BCC [10]
8	$Al_{2.8}CuCoCrFeNi$			Al	Cu, Co, Cr, Ni, Fe		1.31717	6.554	3.201	6.718	BCC [10]
9	$Al_{3.0}CuCoCrFeNi$			Al	Cu, Co, Cr, Ni, Fe		1.32004	6.598	3.259	6.625	BCC [10]
10	AlCoCrFeNi			Al	Co, Cr, Ni, Fe		1.28378	5.767	2.968	7.2	BCC [12]
11	$Ti_{20}Zr_{20}Hf_{20}Cu_{20}Ni_{20}$		Zr, Hf	Ti	Cu, Ni		1.43308	10.324	4.977	-	BMG [19]
12	$Ti_{20}Zr_{20}Cu_{20}Ni_{20}Be_{20}$		Zr	Ti	Cu, Ni	Be	1.34318	12.514	7.065	-	BMG [24]
13	$Ti_{16.7}Zr_{16.7}Hf_{16.7}Cu_{16.7}Ni_{16.7}Be_{16.7}$		Zr, Hf	Ti	Cu, Ni	Be	1.38223	12.773	5.721	-	BMG [25]
14	$Ti_{20}Hf_{20}Cu_{20}Ni_{20}Be_{20}$		Hf	Ti	Cu, Ni	Be	1.33818	11.993	6.718	-	BMG (this work)
15	$Ti_{16.7}Zr_{16.7}Nb_{16.7}Cu_{16.7}Ni_{16.7}Be_{16.7}$		Zr	Ti, Nb	Cu, Ni	Be	1.35495	11.546	5.826	-	BMG (this work)

5. Conclusions

In this paper, two new high entropy bulk metallic glasses (HE-BMGs) have been successfully fabricated using copper mold casting method, namely $Ti_{20}Hf_{20}Cu_{20}Ni_{20}Be_{20}$ with a critical diameter of 2 mm and $Ti_{16.7}Zr_{16.7}Nb_{16.7}Cu_{16.7}Ni_{16.7}Be_{16.7}$ with a critical diameter of 1.5 mm. These two HE-BMGs exhibit high fracture strength over 2300 MPa. The glass forming ability and atomic size distribution characteristics of the HE-BMGs are discussed, and it is found that atomic radius spans over a wide range in HE-BMGs. Moreover, we propose a new parameter δ' to assess the degree of lattice distortion in high entropy alloys (HEAs). It emphasizes the difference between atoms with adjacent atomic size, and it is closely related to phase selection in HEAs. When δ' is relatively small (δ' < 2.2), FCC solid solution formed; when 2.2 < δ' < 2.9, FCC + BCC phases formed; when 2.9 < δ' < 4.9, BCC phase formed; while δ' > 4.9, amorphous phase would be formed. This new parameter δ' is beneficial for understanding lattice distortion and phase selection in HEAs. The present work suggests that through adjusting the parameter δ' by similar element substitution/addition, that is, adjusting the lattice distortion, is an effective way for designing high entropy bulk glassy alloy.

Author Contributions: Conceptualization, H.D.; Data curation, H.X.; Formal analysis, H.L.; Funding acquisition, H.D. and K.Y.; Investigation, H.B.; Methodology, H.B.; Project administration, K.Y.; Software, H.X.; Validation, H.X.; Visualization, H.L.; Writing–Original draft, H.D.; Writing–review & editing, H.L., H.B. and K.Y. All authors have read and agreed to the published version of the manuscript.

Funding: This research was funded by the National Key Basic Research and Development Program (Grant No. 2016YFB00300500), National Natural Science Foundation of China (Grant Nos. 51571127 and 51871129). and Youth Fund of Jiangsu Natural Science Foundation (Grant No. BK20190979).

Institutional Review Board Statement: Not applicable.

Informed Consent Statement: Not applicable.

Data Availability Statement: The data that support the plots within this paper and other findings of this study are available from the corresponding authors upon reasonable request.

Conflicts of Interest: The authors declare no competing interests.We declare that we have no financial and personal relationships with other people or organizations that can inappropriately influence our work.

References

1. Greer, A.L. Metallic glasses. *Science* **1995**, *267*, 1947–1953. [CrossRef] [PubMed]
2. Johnson, W.L. Bulk glass-forming metallic alloys: Science and technology. *MRS Bull.* **1999**, *24*, 42–56. [CrossRef]
3. Inoue, A. Stabilization of metallic supercooled liquid and bulk amorphous alloys. *Acta Mater.* **2000**, *48*, 279–306. [CrossRef]
4. Wang, W.H.; Dong, C.; Shek, C.H. Bulk metallic glasses. *Mater. Sci. Eng. R* **2004**, *44*, 45–89. [CrossRef]
5. Inoue, A.; Takeuchi, A. Recent development and applications of bulk glassy alloys. *Int. J. Appl. Glass Sci.* **2010**, *1*, 273–295. [CrossRef]
6. Wang, W.H. The elastic properties, elastic models and elastic perspectives of metallic glasses. *Prog. Mater. Sci.* **2012**, *57*, 487–656. [CrossRef]
7. Greer, A.L. Metallic glasses. *Phys. Metall.* **2014**, *35*, 305–385. [CrossRef]
8. Li, H.X.; Lu, Z.C.; Wang, S.L.; Wu, Y.; Lu, Z.P. Fe-based bulk metallic glasses: Glass formation, fabrication, properties and applications. *Prog. Mater. Sci.* **2019**, *103*, 235–318. [CrossRef]
9. Yang, C.; Zhang, C.; Chen, Z.J.; Li, Y.; Yan, W.Y.; Yu, H.B.; Liu, L. Three-dimensional hierarchical porous structures of metallic glass/copper composite catalysts by 3D printing for efficient wastewater treatments. *ACS Appl. Mater. Inter.* **2021**, *13*, 7227–7237. [CrossRef]
10. Yeh, J.W.; Chen, S.K.; Lin, S.J.; Gan, J.Y.; Chin, T.S.; Shun, T.T.; Tsau, C.H.; Chang, S.Y. Nanostructured high-entropy alloys with multiple principle elements: Novel alloy design concepts and outcomes. *Adv. Eng. Mater.* **2004**, *6*, 299–303. [CrossRef]
11. Cantor, B.; Chang, I.; Knight, P.; Vincent, A. Microstructural development in equiatomic multicomponent alloys. *Mater. Sci. Eng. A* **2004**, *375–377*, 213–218. [CrossRef]
12. Zhou, Y.J.; Zhang, Y.; Wang, Y.L.; Chen, G.L. Solid solution alloys of $AlCoCrFeNiTi_x$ with excellent room-temperature mechanical properties. *Appl. Phys. Lett.* **2007**, *90*, 181904. [CrossRef]
13. Zhang, Y.; Zhou, Y.J.; Lin, J.P.; Chen, G.L.; Liaw, P.K. Solid-solution phase formation rules for multi-component alloys. *Adv. Eng. Mater.* **2008**, *10*, 534–538. [CrossRef]
14. Senkov, O.N.; Wilks, G.B.; Miracle, D.B.; Chuang, C.P.; Liaw, P.K. Refractory high-entropy alloys. *Intermetallics* **2010**, *18*, 1758–1765. [CrossRef]
15. Guo, S.; Ng, C.; Lu, J.; Liu, C.T. Effect of valence electron concentration on stability of fcc or bcc phase in high entropy alloys. *J. Appl. Phys.* **2011**, *109*, 103505. [CrossRef]
16. Gludovatz, B.; Hohenwarter, A.; Catoor, D.; Chang, E.H.; George, E.P.; Ritchie, R.O. A fracture-resistant high-entropy alloy for cryogenic applications. *Science* **2014**, *345*, 1153–1158. [CrossRef]

17. Lu, Y.P.; Dong, Y.; Guo, S.; Jiang, L.; Kang, H.J.; Wang, T.M.; Wen, B.; Wang, Z.J.; Jie, J.C.; Cao, Z.Q.; et al. A promising new class of high-temperature alloys: Eutectic high-entropy alloys. *Sci. Rep.* **2014**, *4*, 6200. [CrossRef]
18. Ye, Y.F.; Wang, Q.; Lu, J.; Liu, C.T.; Yang, Y. Design of high entropy alloys: A single-parameter thermodynamic rule. *Scr. Mater.* **2015**, *104*, 53–55. [CrossRef]
19. Ma, L.Q.; Wang, L.M.; Zhang, T.; Inoue, A. Bulk Glass Formation of Ti–Zr–Hf–Cu–M (M = Fe,Co,Ni) Alloys. *Mater. Trans.* **2002**, *43*, 277–280. [CrossRef]
20. Zhao, K.; Xia, X.X.; Bai, H.Y.; Zhao, D.Q.; Wang, W.H. Room temperature homogeneous flow in a bulk metallic glass with low glass transition temperature. *Appl. Phys. Lett.* **2011**, *98*, 141913. [CrossRef]
21. Gao, X.Q.; Zhao, K.; Ke, H.B.; Ding, D.W.; Wang, W.H.; Bai, H.Y. High mixing entropy bulk metallic glasses. *J. Non-Cryst. Solids* **2011**, *357*, 3557–3560. [CrossRef]
22. Takeuchi, A.; Chen, N.; Wada, T.; Yokoyama, Y.; Kato, H.; Inoue, A.; Yeh, J.W. $Pd_{20}Pt_{20}Cu_{20}Ni_{20}P_{20}$ high-entropy alloy as a bulk metallic glass in the centimeter. *Intermetallics* **2011**, *19*, 1546–1554. [CrossRef]
23. Li, H.F.; Xie, X.H.; Zhao, K.; Wang, Y.B.; Zheng, Y.F.; Wang, W.H.; Qin, L. In vitro and in vivo studies on biodegradable CaMgZnSrYb high-entropy bulk metallic glass. *Acta Biomater.* **2013**, *9*, 8561–8573. [CrossRef] [PubMed]
24. Ding, H.Y.; Yao, K.F. High entropy $Ti_{20}Zr_{20}Cu_{20}Ni_{20}Be_{20}$ bulk metallic glass. *J. Non-Cryst. Solids* **2013**, *364*, 9–12. [CrossRef]
25. Ding, H.Y.; Shao, Y.; Gong, P.; Li, J.F.; Yao, K.F. A senary TiZrHfCuNiBe high entropy bulk metallic glass with large glass-forming ability. *Mater. Lett.* **2014**, *125*, 151–153. [CrossRef]
26. Zhao, S.F.; Yang, G.N.; Ding, H.Y.; Yao, K.F. A quinary Ti-Zr-Hf-Be-Cu high entropy bulk metallic glass with a critical size of 12 mm. *Intermetallics* **2015**, *61*, 47–50. [CrossRef]
27. Zhao, S.F.; Shao, Y.; Liu, X.; Chen, N.; Ding, H.Y.; Yao, K.F. Pseudo-quinary $Ti_{20}Zr_{20}Hf_{20}Be_{20}(Cu_{20-x}Ni_x)$ high entropy bulk metallic glasses with large glass forming ability. *Mater. Des.* **2015**, *87*, 625–631. [CrossRef]
28. Huo, J.T.; Huo, L.S.; Men, H.; Wang, X.M.; Inoue, A.; Wang, J.Q.; Chang, C.T.; Li, R.W. The magnetocaloric effect of Gd-Tb-Dy-Al-M (M= Fe, Co and Ni) high-entropy bulk metallic glasses. *Intermetallics* **2015**, *58*, 31–35. [CrossRef]
29. Qi, T.L.; Li, Y.H.; Takeuchi, A.; Xie, G.Q.; Miao, H.T.; Zhang, W. Soft magnetic $Fe_{25}Co_{25}Ni_{25}(B, Si)_{25}$ high entropy bulk metallic glasses. *Intermetallics* **2015**, *66*, 8–12. [CrossRef]
30. Chen, C.; Pang, S.J.; Cheng, Y.Y.; Zhang, T. A centimeter-size $Zr_{40}Hf_{10}Ti_4Y_1Al_{10}Cu_{25}Ni_7Co_2Fe_1$ bulk metallic glass with high mixing entropy designed by multi-substitution. *J. Non-Cryst. Solids* **2015**, *410*, 39–42. [CrossRef]
31. Kim, J.; Oh, H.S.; Kim, J.; Ryu, C.W.; Lee, G.W.; Chang, H.J.; Park, E.S. Utilization of high entropy alloy characteristics in Er-Gd-Y-Al-Co high entropy bulk metallic glass. *Acta Mater.* **2018**, *155*, 350–361. [CrossRef]
32. Li, J.; Xue, L.; Yang, W.M.; Yuan, C.C.; Huo, J.T.; Shen, B.L. Distinct spin glass behavior and excellent magnetocaloric effect in $Er_{20}Dy_{20}Co_{20}Al_{20}RE_{20}$ (RE = Gd, Tb and Tm) high-entropy bulk metallic glasses. *Intermetallics* **2018**, *96*, 90–93. [CrossRef]
33. Xu, Y.Q.; Li, Y.H.; Zhu, Z.W.; Zhang, W. Formation and properties of $Fe_{25}Co_{25}Ni_{25}(P, C, B, Si)_{25}$ high-entropy bulk metallic glasses. *J. Non-Cryst. Solids* **2018**, *487*, 60–64. [CrossRef]
34. Wu, L.; Zhao, Y.; Li, J.J.; Wu, J.L.; Zhang, B. Correlation between mechanical and thermodynamic properties for La–Ce–Ni–Cu–Al high-entropy metallic glasses. *J. Iron Steel Res. Int.* **2018**, *25*, 658–665. [CrossRef]
35. Wu, K.N.; Liu, C.; Li, Q.; Huo, J.T.; Li, M.C.; Chang, C.T.; Sun, Y.F. Magnetocaloric effect of $Fe_{25}Co_{25}Ni_{25}Mo_5P_{10}B_{10}$ high-entropy bulk metallic glass. *J. Magn. Magn. Mater.* **2019**, *489*, 165404. [CrossRef]
36. Li, C.Z.; Li, Q.; Li, M.C.; Chang, C.T.; Li, H.X.; Dong, Y.Q.; Sun, Y.F. New ferromagnetic $(Fe_{1/3}Co_{1/3}Ni_{1/3})_{80}(P_{1/2}B_{1/2})_{20}$ high entropy bulk metallic glass with superior magnetic and mechanical properties. *J. Alloys Compd.* **2019**, *791*, 947–951. [CrossRef]
37. Wada, T.; Jiang, J.; Yubuta, K.; Kato, H.; Takeuchi, A. Septenary Zr–Hf–Ti–Al–Co–Ni–Cu high-entropy bulk metallic glasses with centimeter-scale glass-forming ability. *Materialia* **2019**, *7*, 100372. [CrossRef]
38. Pang, C.M.; Chen, L.; Xu, H.; Guo, W.; Lv, Z.W.; Huo, J.T.; Cai, M.J.; Shen, B.L.; Wang, X.L.; Yuan, C.C. Effect of Dy, Ho, and Er substitution on the magnetocaloric properties of Gd-Co-Al-Y high entropy bulk metallic glasses. *J. Alloys Compd.* **2020**, *827*, 154101. [CrossRef]
39. Li, Y.H.; Wang, S.W.; Wang, X.W.; Yin, M.L.; Zhang, W. New FeNiCrMo(P, C, B) high-entropy bulk metallic glasses with unusual thermal stability and corrosion resistance. *J. Mater. Sci. Technol.* **2020**, *43*, 32–39. [CrossRef]
40. Shao, L.L.; Wang, Q.Q.; Xue, L.; Zhu, M.Y.; Wang, A.D.; Luan, J.H.; Yin, K.B.; Luo, Q.; Zeng, Q.S.; Sun, L.T.; et al. Effects of minor Si addition on structural heterogeneity and glass formation of GdDyErCoAl high-entropy bulk metallic glass. *J. Mater. Res. Technol.* **2021**, *11*, 378–391. [CrossRef]
41. Jalali, A.; Malekan, M.; Park, E.S.; Rashidi, R.; Bahmani, A.; Yoo, G.H. Thermal behavior of newly developed $Zr_{33}Hf_8Ti_6Cu_{32}Ni_{10}Co_5Al_6$ high-entropy bulk metallic glass. *J. Alloys Compd.* **2021**, *892*, 162220. [CrossRef]
42. Luan, H.W.; Shao, Y.; Li, J.F.; Mao, W.L.; Han, Z.D.; Shao, C.L.; Yao, K.F. Phase stabilities of high entropy alloys. *Scr. Mater.* **2020**, *179*, 40–44. [CrossRef]
43. Toda-Caraballo, I.; Rivera-Diaz-del-Castillo, P.E.J. A criterion for the formation of high entropy alloys based on lattice distortion. *Intermetallics* **2016**, *71*, 76–87. [CrossRef]
44. Wang, Z.J.; Huang, Y.H.; Yang, Y.; Wang, J.C.; Liu, C.T. Atomic-size effect and solid solubility of multicomponent alloys. *Scr. Mater.* **2015**, *94*, 28–31. [CrossRef]

45. Wang, Z.J.; Qiu, W.F.; Yang, Y.; Liu, C.T. Atomic-size and lattice-distortion effects in newly developed high-entropy alloys with multiple principal elements. *Intermetallics* **2015**, *64*, 63–69. [CrossRef]
46. Senkov, O.N.; Miracle, D.B. Effect of the atomic size distribution on glass forming ability of amorphous metallic alloys. *Mater. Res. Bull.* **2001**, *36*, 2183–2198. [CrossRef]

materials

MDPI

Article

Metallization on Sapphire and Low-Temperature Joining with Metal Substrates

Jiajun Fang [1], Qiaoxin Zhang [1], Zhou Luo [1], Wei Huang [1], Zhenyu Liu [1], Zhiwen Chen [2], Xueqiang Cao [1] and Li Liu [1,*]

1 School of Materials Science and Engineering, Wuhan University of Technology, Wuhan 430070, China; jjf@whut.edu.cn (J.F.); zhangqx@whut.edu.cn (Q.Z.); lz15571339343@whut.edu.cn (Z.L.); whuthuangwei@whut.edu.cn (W.H.); liu.zy@whut.edu.cn (Z.L.); xcao@whut.edu.cn (X.C.)

2 The Institute of Technological Science, Wuhan University, Wuhan 430072, China; zwchen_lu@163.com

* Correspondence: l.liu@whut.edu.cn

Abstract: To meet the packaging requirements of sapphire in special electronic components, there is an urgent need for a joining process that can realize a good connection between sapphire and dissimilar metals at a low temperature. In this work, the surface of a sapphire substrate was successfully catalytically activated and metallized by an electroless nickel plating process. Moreover, the solderability and interconnection of metallized sapphire with Sn-based solders were evaluated and investigated at 250 °C, and the wetting angle of the Sn-based solders on sapphire on sapphire without and with metallization was 125° and 51°, respectively. The interfacial microscopic morphology and element distribution in the Cu/Sn-Ag solder/sapphire solder joints were analyzed. It was found that the middle solder layer has diffused during the reflow process, inferring good adhesion between sapphire and Cu substrate with the aid of the Ni-P deposition. Thus, a sapphire welding method with a simple process suitable for practical applications is demonstrated.

Keywords: sapphire; surface metallization; deposition mechanism; wettability; low-temperature joining

Citation: Fang, J.; Zhang, Q.; Luo, Z.; Huang, W.; Liu, Z.; Chen, Z.; Cao, X.; Liu, L. Metallization on Sapphire and Low-Temperature Joining with Metal Substrates. *Materials* **2022**, *15*, 1783. https://doi.org/10.3390/ma15051783

Academic Editor: Israel Felner

Received: 31 December 2021
Accepted: 25 January 2022
Published: 26 February 2022

1. Introduction

Sapphire has superior optical and mechanical properties, including high melting point, high hardness, strong corrosion resistance and good thermal conductivity, which makes sapphire highly demanded in electronics and scientific instruments [1,2]. For electronics, sapphire is required to achieve joining with common metals by a simple reflow process. Unfortunately, sapphire is hard to achieve reliable solder joints during the reflow process as it is an Al_2O_3 single crystal. Currently, enormous works have been conducted on sapphire welding by various processes. Metallization on the sapphire surface is an important method. Related research on metallization methods of sapphire is listed in Table 1. Another method is using solders with active additions of Ti, Al and Mg. For instance, Ti addition in solders can significantly promote the solder wettability on sapphire. Mu et al. systematically studied the wetting behaviors of a Sn-Ti alloy on sapphire by the solid drop method [3]. However, Ti-based solders were required to be brazed in a high-temperature vacuum environment, which has been reported by Ning et al. [4]. With the assistance of ultrasonic, the brazing temperature of sapphire solder joints can be greatly reduced. Cui et al. used an Al-4.5Cu-1.5Mg alloy as a filling metal to connect sapphire through ultrasonic-assisted hot dipping [5]. A dense transient layer formed at the metal–sapphire interface to achieve solid connections between sapphire and the aluminum alloy. At present, various related researches on sapphire welding have been conducted in the above aspects. Unfortunately, they all have shortcomings for electronics applications, including high joining temperature, specific active solder, auxiliary ultrasonic brazing and so on.

Table 1. Studies on metallization methods on sapphire.

Methods	Temperature	Conditions	Disadvantage	Reference
Mo-Mn sintering	950~1000 °C	Reducing atmosphere	Pore and microcrack defects	[6]
Magnetron sputtering	<50 °C	Vacuum condition	High cost, low material utilization	[7]
Ultrasonic assisted hot dipping	210~230 °C	Active solders (Ti, Al, Mg)	Ultrasonic damage	[8,9]

Electroless nickel plating is a conventional surface treatment technology for wide applications prospects [10–14], which has the advantages of uniform deposition, dense plating, good conductivity and good solderability [15,16]. Electroless nickel plating on porous alumina ceramics was achieved by Zhang et al. [17]. Dehchar et al. successfully deposited copper films on non-conductive epoxy glass substrates by electroless plating as well [18]. Currently, many works related to electroless nickel plating have been carried out on different ceramic substrates. However, the related research on electroless nickel plating on sapphire substrates has not been reported yet.

In this work, we realized the metallization on sapphire by electroless nickel plating and elaborated its deposition mechanism. Therefore, the Cu/solder/sapphire sandwich solder joints can be fabricated, and then the interfacial morphology of these solder joints can be observed. Moreover, the fracture mechanism of solder joints was also studied. Finally, sapphire substrates can be joined by a simple reflow process at a low temperature and low pressure.

2. Experimental Works

In this work, sapphire substrates (10 mm × 10 mm × 1 mm, α-Al_2O_3 single crystal, purity >99.9%) were used for electroless plating. To achieve electroless nickel plating on the sapphire surface, the influence of substrate pretreatment on sapphire was studied to successfully deposit electroless nickel plating. The surface roughness of the sapphire substrates was measured by an atomic force microscope (AFM, Nanoscope IV, Veeco, Suzhou, China). Different types of chemical etching solutions were chosen to prepare the smooth deposition of the coatings. After grinding, etching, sensitizing, activating and initial plating (25 s) of the sapphire substrates, the surface morphology of sapphire after each process was observed through a scanning electron microscope (SEM, MIRA 3, TESCAN, Shanghai, China). Moreover, the changes in the surface morphology of the substrates were compared and analyzed.

The electroless plating solution is an acidic solution composed of deionized water, nickel sulfate, sodium hypophosphite, sodium citrate and ammonium sulfate. Table 2 lists the chemical reagents and the corresponding electroless plating conditions. The crystal structure of Ni-P coating was also analyzed by X-ray diffraction (XRD, Smart SE, Rigaku, Tokyo, Japan). Afterwards, the surface morphology and chemical composition of the electroless deposition by SEM and EDS (MIRA 3, TESCAN, Shanghai, China), respectively.

Table 2. Chemical plating bath composition and conditions.

Chemical Reagents	Composition (g/L)
$NiSO_4{\cdot}6H_20$	20–35
$NaH_2PO{\cdot}2H_2O$	25–35
$Na_3C_6H_5O_7{\cdot}2H_2O$	60–70
$((NH_4)_2SO_4$	60–80
Plating condition	Temp: 65–75 °C, pH: 5–6, Time: 15–30 min

A contact angle measuring instrument (JC2000D2, Shanghai zhongchen digital technic apparatus Co. Ltd, Shanghai, China) was used to test the wetting angle of the Sn-3Ag solder balls (diameter 760 µm, 1.6 mg weight/ball, Langfang Bangzhuang Electronic Materials Co. Ltd, Langfang, China) on the sapphire with and without metallization in an atmospheric environment at 240 °C. Afterwards, sandwich-structured Cu/solder/sapphire solder joints were prepared by the reflow process. Pure copper substrates (purity 99%,

5 mm × 5 mm × 1 mm) and nickel-plated sapphire substrates were used as the upper and lower substrates, while solder pastes were evenly plated on the sapphire surface by a 100 μm thick stainless mask to connect the substrates. Then, the assembled Cu/solder/sapphire sandwich structure was prepared in a crucible furnace (vacuum degree <10^{-3} Pa) at a peak reflow temperature of 250 °C for 60 min under a pressure of 0.2 Mpa. The macroscopic appearance of sapphire solder joints is shown in Figure 1. The interfacial microstructure, morphology and element distribution of the sapphire solder joints and shear fracture surface were observed by SEM and EDS.

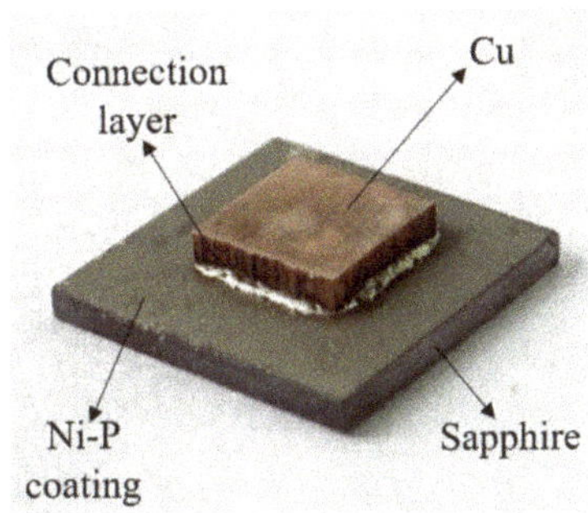

Figure 1. Macroscopic appearance of samples.

3. Results and Discussion

3.1. Metallization Mechanism on Sapphire

The surface activity of sapphire is rather poor and cannot auto catalyze, so the pretreatment is very important for deposition on the sapphire surface. The pretreatment process mainly includes surface degreasing, polishing, chemical etching, chemical sensitization and activation. The roughness of the polished and smooth surface measured by atomic force microscope (AFM) was 84.0 nm and 1.79 nm, respectively. Some works show that substrates with reasonable surface roughness can provide good anchorage for Ni-P films, which can significantly improve deposition adhesion [19,20]. After grinding, the deposition area between the coating and sapphire is greatly increased, thereby improving the bonding force of the deposit. Chemical treatment is also carried out to further improve the bonding force. Since the substrate is α-Al_2O_3, which is the most stable phase in the majority of alumina and is generally insoluble in acids and alkalis. In the subsequent electroless plating process, a mixed solution of HF and HCl was used as the coarsening solution. Therefore, Ni-P coating can be deposited on the sapphire, indicating that the coarsening solution can react with α-Al_2O_3 and form enormous micro-holes on the substrates. Thus, the substrate can be activated for the following electroless plating. During the plating process, a complete Ni-P coating can be directly deposited on the sapphire surface at once. The results show that sapphire can be metallized by a simple electroless plating process with proper pretreatments.

Figure 2 shows the surface morphology of sapphire substrates in each pretreatment process. Figure 2a–d is the substrate surface after grinding, etching, activating and initial plating, respectively. Figure 2a,b has a similar microstructure, while the surface morphology in Figure 2b is smoother. This is contributed to the etching solution that can react with the sapphire substrates. Figure 2c shows the morphology of sapphire substrates after activation by $SnCl_2$/HCl and $PdCl_2$/HCl solution. When the specimens absorbed with Sn^{+2} ions were immersed in the dilute $PdCl_2$ solution, the Pd^{+2} ion was reduced to Pd metal and acted as a catalyst to initiate the deposition reaction according to Equation (1). As a result, some individual Pd patches can be formed on the sapphire surface, as shown in Figure 3.

$$Sn^{+2} + Pd^{+2} \rightarrow Sn^{+4} + Pd \quad (1)$$

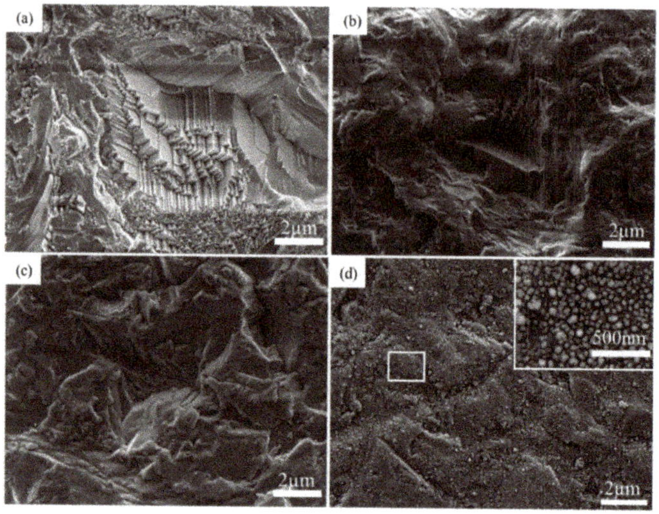

Figure 2. Topography of the substrate in electroless plating pretreatment after (**a**) grinding, (**b**) etching, (**c**) activation and (**d**) initial plating.

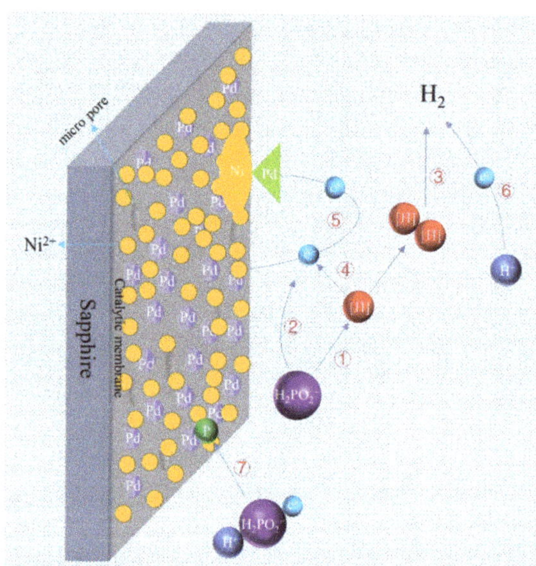

Figure 3. Deposition process and mechanism in the electroless plating solution.

Figure 2d exhibits the surface morphology of the sapphire after pre-plating for 25 s. As can be seen from the high-magnification figure in Figure 2d, enormous small particles of the Ni-P alloy are deposited on the sapphire surface in uniform distribution. Generally, the size range of Ni-P particles is between 50 nm to 100 nm. No area that is not covered with Ni-P coatings can be found even after electroless plating for only 25 s, indicating this work has a relatively fast deposition rate. The results show that by changing the pretreatment, the coating can be deposited on the sapphire substrate, and the metallization of sapphire is successfully realized.

To further explore the deposition mechanism of the sapphire metallized layer, Figure 3 is the reaction diagram simulating the Ni-P depositions on sapphire in the electroless plating solution. Under acidic conditions, the electroless nickel plating process of the phosphite ester system is basically carried out according to the following equation [21,22]. As a reducing agent, sodium hypophosphite first undergoes a dehydrogenation reaction, and the phosphorus–hydrogen bond is split to produce HPO_2^- ions and reducing hydrogen atoms, as shown in Equation (2). However, HPO_2^- will be oxidized to HPO_3^- due to instability, and electrons will be released at the same time according to Equation (3). The reducing hydrogen atoms are oxidized in the acid solution to generate electrons (Equation (4)). Since there are individual catalytic Pd patches on the sapphire surface, Ni^{2+} will accumulate on the sapphire substrates in a large amount and be reduced to metal after obtaining nickel electrons according to Equation (5). At the same time, the hypophosphite also obtains electrons and is reduced to elemental phosphorus (Equation (6)). Since metallic nickel and elemental phosphorus are produced by the reaction at the same time and then deposited on the sapphire substrates. Thus, the deposition of the Ni-P coating can be realized. Meanwhile, many bubbles were generated during the electroless plating process. These bubbles are reduced to H_2 by the combination of two reducing hydrogen atoms or electrons obtained from hydrogen ions in Equations (7) and (8).

$$H_2PO_2^- \rightarrow HPO_2^- + [H] \tag{2}$$

$$HPO_2^- + H_2O \rightarrow H_2PO_3^- + H^+ + e^- \tag{3}$$

$$[H] \rightarrow H^+ + e^- \tag{4}$$

$$Ni^{2+} + 2e^- \rightarrow Ni \tag{5}$$

$$H_2PO_2^- + 2H^+ + e^- \rightarrow 2H_2O + P \tag{6}$$

$$2H^+ + 2e^- \rightarrow H_2 \tag{7}$$

$$[H] + [H] \rightarrow H_2 \tag{8}$$

3.2. Microstructure and Wettability of Ni-P Coatings

The surface metallization on sapphire by electroless plating was successfully achieved. Figure 4a shows the surface morphology of the Ni-P coating by scanning electron microscope (SEM). No defects, including micropores, cracks and undeposit areas, can be found in the coating, which shows good quality. This coating exhibits a nodulus structure with a particle range of 0.45–0.96 μm. Figure 4b shows the cross-section topography of the Ni-P coating. The middle bright layer is the Ni-P coating with a relatively uniform thickness of 6.3 μm. Generally, this metallized layer is evenly distributed on sapphire and shows good quality.

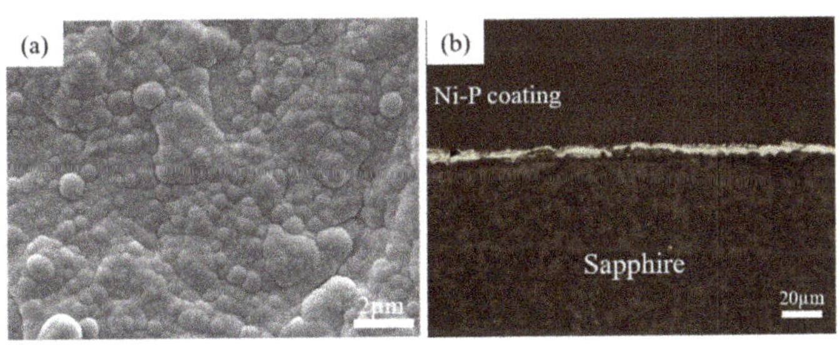

Figure 4. (**a**) Surface morphology, (**b**) section morphology of Ni-P coating.

Figure 5 shows the XRD pattern of Ni-P coating, and there is a wide peak of Ni phase at 46°. In addition, the crystallinity of this coating is 0% calculated by JADE software, indicating that this coating has an amorphous structure. Moreover, the P content of this

Ni-P coating is around 13–14 wt.%. As previously reported, the lattice strain of Ni (111) grains increases when the P content in Ni-P alloy exceeds 8 wt.%, which leads to a Ni-P coating with an amorphous structure [23].

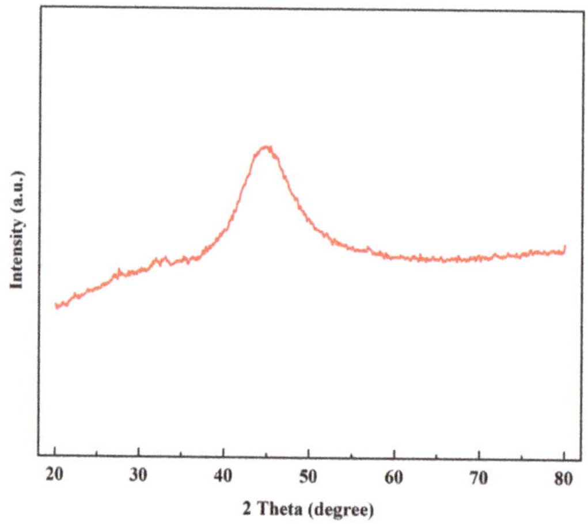

Figure 5. XRD pattern of Ni-P coating.

Moreover, the vital effect of depositions on sapphire is to improve its solderability. Figure 6a,b are the cross-sections of the wetting behaviors of the Sn-3 Ag solder balls on sapphire without and with metallization, respectively. According to the Young's equation [24], the solid–liquid interaction occurring when a droplet contacts a surface is expressed by the following equation:

$$\cos\theta = \frac{\gamma_{sg} - \gamma_{sl}}{\gamma_{gl}} \tag{9}$$

where θ is the contact angle of a smooth surface; γ_{sg}, γ_{sl} and γ_{gl} represent the interfacial tension at solid-gas, solid–liquid and gas–liquid interfaces, respectively.

Figure 6. Wetting angles of Sn-Ag solder balls on (**a**) bare sapphire, (**b**) Ni-P plated sapphire and (**c**) bare Cu substrate.

Wettability is the ability of liquid solders that spread on the surface to be soldered. Many wettability studies have shown that the values of the wetting angle can indicate the degree of wettability, as shown in Table 3 [25]. The contact angle of Sn-Ag solder on a bare copper substrate is 34°, as shown in Figure 6c, which shows good wetting. The contact angle of the sapphire sample without the Ni-P coating is 125°, which exceeds 90° and exhibits unacceptable wettability for Sn-based solder balls. On the contrary, with the aid of Ni-P coatings on sapphire, the contact angle can significantly decrease to 51°, enhancing wettability.

Table 3. The correlation of wetting angle range with wettability.

Wetting Angle Range	Wettability
$0° \leq \theta < 30°$	Excellent wetting
$30° \leq \theta < 40°$	Good wetting
$40° \leq \theta < 55°$	Acceptable wetting
$55° < \theta < 70°$	poor wetting
$70° \leq \theta$	Unacceptable wetting

3.3. Joining and Fracture Mechanism of Sapphire Solder Joints

Afterwards, sapphire substrates plated with Ni-P coatings were joined with bare copper sheets by a simple reflow process. Figure 7 shows the surface topography of the sapphire solder joints and the elemental distribution. The interface microstructure at the Cu/sapphire solder interfaces shows a dense structure without obvious holes and cracks. From the EDS mapping results, the bright gray stripe in the middle is the Sn-based solder, while the upper and lower substrates are copper and sapphire, respectively. Combined with the overall morphology and Ni distribution, it is found that there is an intermittent Ni-P coating between the middle layer and the sapphire, and the coating is deposited on the sapphire with good adhesion. It shows that the coating thickness decreases gradually during the reflow process, which is caused by the reaction between the Ni-P coating and the solder.

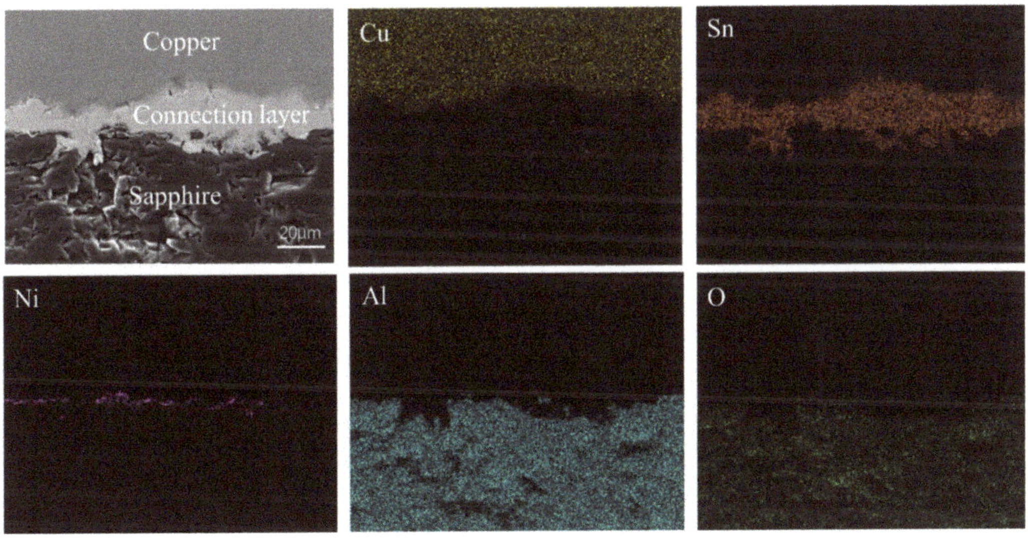

Figure 7. Morphology and EDS elemental distribution at Cu/sapphire solder interfaces.

The fracture surface topography and EDS elemental distribution are shown in Figure 8. It can be seen that there is an obvious bright white curve in the middle, and shear fracture mainly occurs at the interface between the coating and the substrate. The distribution of Sn and Ni indicates that the electroless nickel plating layer is reacted by metallurgy during the joining process. It is speculated that Sn-Ni intermetallic compounds are formed, which is consistent with studies on the interfacial reaction between Ni-P coatings and Sn-based solders [26,27].

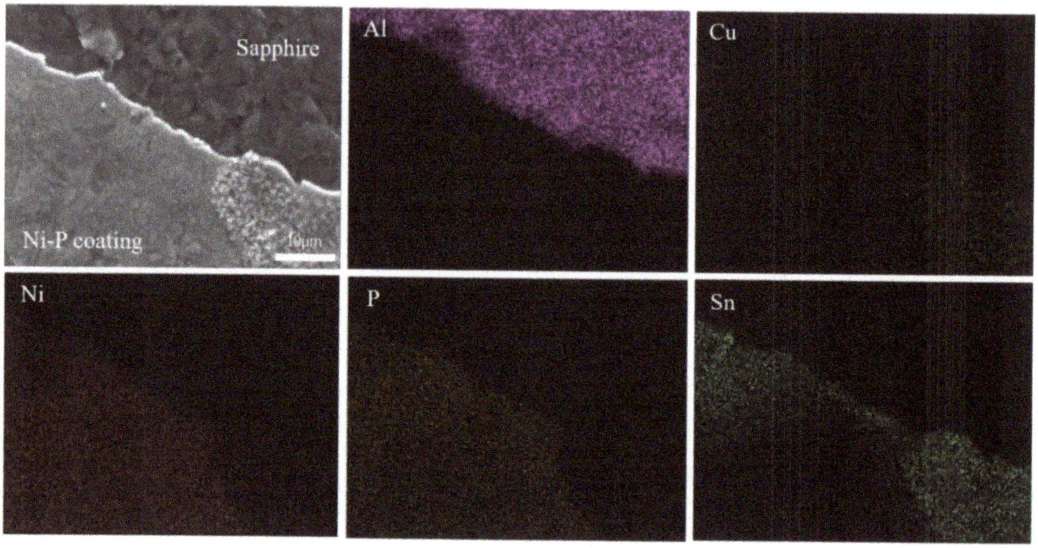

Figure 8. Element analysis diagram of fracture surface morphology.

The fracture diagram of these sapphire solder joints is illustrated in Figure 9. During the connection process, the Sn-based solder reacts with the Ni-P coating to form Ni-Sn IMC layers, leading to a continuous decrease in the thickness of the Ni-P coating. When the thickness of Ni-P coating decreases to a certain value, shear fracture occurs at the coating–sapphire interface.

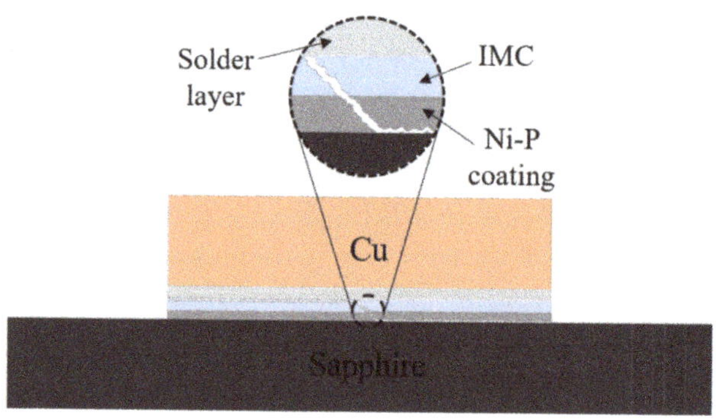

Figure 9. Cu/solder/sapphire fracture schematic diagram.

4. Conclusions

In this work, the sapphire substrates were successfully catalytic activated and metallized by electroless nickel plating to achieve direct joining with copper substrates via the conventional reflow process. The conclusions can be drawn as follows:

1. A proper pretreatment for sapphire substrates was proposed to successfully deposit metallization on sapphire by a simple electroless plating process. Moreover, the deposition mechanism of the metallization on sapphire was also elaborated.
2. The metallization significantly improved the wettability of the substrates as the wetting angle of the Sn-based solders on sapphire with the metallization was reduced from 125° (without metallization) to 51°.

3. The direct joining between sapphire and copper was achieved by the conventional reflow process at 250 °C and reduced pressure. The nickel atoms from the Ni-P metallization interacted with Sn atoms in the solder, resulting in the formation of Ni-Sn IMCs to achieve metallurgical bonding at the joint interfaces.

Author Contributions: Conceptualization, J.F. and L.L.; methodology, Z.L. (Zhou Luo); software, W.H.; validation, J.F., L.L. and Q.Z.; formal analysis, L.L.; investigation, Q.Z.; resources, Z.C.; data curation, J.F.; writing—original draft preparation, Z.L. (Zhenyu Liu); writing—review and editing, L.L.; visualization, Q.Z.; supervision, L.L.; project administration, X.C.; funding acquisition, L.L. All authors have read and agreed to the published version of the manuscript.

Funding: This research was funded by Wuhan University of Technology, grant number 430070. The APC was funded by Natural Science Foundation of China (62004144, 61904127). Additional thanks to the Fundamental Research Funds for the Central Universities (Grant No. 213201006, 213134005, 2042019kf0013).

Data Availability Statement: Data presented in this article are available at request from the corresponding author.

Conflicts of Interest: The authors declare no conflict of interest.

References

1. Dobrovinskaia, E.R.; Litvinov, L.A.; Pishchik, V.V. *Sapphire: Materials, Manufacturing, Applications*; Springer: New York, NY, USA, 2009.
2. Cheng, J.; Wu, J. Experimental investigation of fracture behaviors and subsurface cracks in micro-slot-grinding of monocrystalline sapphire. *J. Mater. Processing Technol.* **2016**, *242*, 160–181. [CrossRef]
3. Mu, D.; Feng, K.; Lin, Q.; Huang, H. Low-temperature wetting of sapphire using Sn–Ti active solder alloys. *Ceram. Int.* **2019**, *45*, 22175–22182. [CrossRef]
4. Ning, H.; Geng, Z.; Ma, J.; Huang, F.; Qian, Z.; Han, Z. Joining of sapphire and hot pressed Al_2O_3 using Ag70.5Cu27.5Ti2 brazing filler metal. *Ceram. Int.* **2003**, *29*, 689–694. [CrossRef]
5. Cui, W.; Li, S.; Yan, J.; He, J.; Liu, Y. Ultrasonic-assisted brazing of sapphire with high strength Al–4.5Cu–1.5Mg alloy. *Ceram. Int.* **2015**, *41*, 8014–8022. [CrossRef]
6. Chen, X.; Yang, J.; Huang, G.; Chen, D.; Xiang, J.; Tiantao, L.; Liu, E. Study on the metallization mechanism of sapphire by active Molybdenum-Manganese method technology. *Vac. Electron.* **2018**, *3*, 53–57.
7. Xin, C.; Li, N.; Yan, J. Microstructural evolution in the braze joint of sapphire to Kovar alloy by Ti-Cu metallization layer. *J. Mater. Processing Technol.* **2017**, *248*, 115–122. [CrossRef]
8. Xu, Y.; Ma, X.; Tang, H.; Yan, J. Mechanism of the interfacial reaction between sapphire and Sn-3.5Ag-4Ti solder at a low temperature in air by ultrasound. *Ceram. Int.* **2020**, *46*, 4435–4443. [CrossRef]
9. Xu, Y.; Ma, X.; Xiu, Z.; Yan, J. Bonding and strengthening mechanism on ultrasonic-assisted soldering of sapphire using Sn-3.5Ag-4Al solder. *J. Mater. Processing Technol.* **2021**, *288*, 116893. [CrossRef]
10. Loto, C.A. Electroless Nickel Plating—A Review. *Silicon* **2016**, *8*, 177–186. [CrossRef]
11. Qin, T.; Ma, L.; Yao, Y.; Ni, C.; Ding, Y. An in situ measure method to study deposition mechanism of electroless Ni-P plating on AZ31 magnesium alloy. *Trans. Nonferrous Met. Soc. China* **2011**, *21*, 2790–2797. [CrossRef]
12. Wang, M.; Yang, Z.; Zhang, C.; Liu, D. Growing process and reaction mechanism of electroless Ni–Mo–P film on SiO_2 substrate. *Trans. Nonferrous Met. Soc. China* **2013**, *23*, 3629–3633. [CrossRef]
13. Georgiza, E.; Novakovic, J.; Vassiliou, P. Characterization and corrosion resistance of duplex electroless Ni-P composite coatings on magnesium alloy. *Surf. Coat. Technol.* **2013**, *232*, 432–439. [CrossRef]
14. Wei, Y.; Li, Y.; Zhang, Y.; Luo, X.; Li, C. Corrosion resistant nickel coating with strong adhesion on AZ31B magnesium alloy prepared by an in-situ shot-peening-assisted cold spray. *Corros. Sci.* **2018**, *138*, 105–115. [CrossRef]
15. Guo, X.; Du, K.; Guo, Q.; Wang, Y.; Wang, F. Experimental study of corrosion protection of a three-layer film on AZ31B Mg alloy *Corros. Sci.* **2012**, *65*, 367–375. [CrossRef]
16. Shang, W.; Zhan, X.; Wen, Y.; Li, Y.; Wang, C. Deposition mechanism of electroless nickel plating of composite coatings on magnesium alloy. *Chem. Eng. Sci.* **2019**, *207*, 1299–1308. [CrossRef]
17. Zhang, P.; Lv, Z.; Liu, X.; Xie, G.; Zhang, B. Electroless nickel plating on alumina ceramic activated by metallic nickel as electrocatalyst for oxygen evolution reaction. *Catal. Commun.* **2021**, *149*, 106238. [CrossRef]
18. Dehchar, C.; Chikouche, I.; Kherrat, R.; Sahari, A.; Zouaoui, A.; Merati, A. Electroless copper deposition on epoxy glass substrate for electrocatalysis of formaldehyde. *Mater. Lett.* **2018**, *228*, 439–442. [CrossRef]
19. Hui, Z.; Huang, Z.; Cui, J. A new method for electroless Ni–P plating on AZ31 magnesium alloy. *Surf. Coat. Technol.* **2007**, *202*, 133–139.

20. Ranjbar, M.; Garavand, N.; Mahdavi, S.; Zad, A. Electroless plating of palladium on WO3 films for gasochromic applications. *Sol. Energy Mater. Sol. Cells* **2010**, *94*, 201–206. [CrossRef]
21. Zhang, B. Mechanism of Electroless Plating. *Amorph. Nano Alloy. Electroless Depos.* **2016**, *4*, 583–627.
22. Lin, J.; Chou, C. The influence of phosphorus content on the microstructure and specific capacitance of etched electroless Ni-P coatings. *Surf. Coat. Technol.* **2019**, *368*, 126–137. [CrossRef]
23. Lewis, D.B.; Marshall, G.W. Investigation into the structure of electrodeposited nickel-phosphorus alloy deposits. *Surf. Coat. Technol.* **1996**, *78*, 150–156. [CrossRef]
24. Liu, Y.; Shi, J.; Lu, Q.; Guo, Y.; Chen, R.; Yin, D. Survey on Calculation of Solid Surface Tension Based on Young's Equation. *Mater. Rev.* **2013**, *27*, 123–129.
25. Moser, Z.; Gąsior, W.; Pstruś, J.; Dębski, A. Wettability Studies of Pb-Free Soldering Materials. *Int. J. Thermophys.* **2008**, *29*, 1974–1986. [CrossRef]
26. Kumar, A.; Min, H.; Zhong, C. Barrier properties of thin Au/Ni–P under bump metallization for Sn–3.5Ag solder. *Surf. Coat. Technol.* **2005**, *198*, 283–286. [CrossRef]
27. Yang, Y.; Yang, C.; Ouyang, F. Interfacial reaction of Ni_3Sn_4 intermetallic compound in Ni/SnAg solder/Ni system under thermomigration. *J. Alloy. Compd.* **2016**, *674*, 331–340. [CrossRef]

materials

MDPI

Article

Improved Anti-Vulcanization and Bonding Performance of a Silver Alloy Bonding Wire by a Cathodic Passivation Treatment with Palladium

Guannan Yang [1,2], Zhiqiang Zhou [1], Haide Zhang [1], Yu Zhang [1,2,*], Zhen Peng [3,*], Pan Gong [4], Xin Wang [5] and Chengqiang Cui [1,2,*]

1 State Key Laboratory of Precision Electronic Manufacturing Technology and Equipment, Guangdong University of Technology, Guangzhou 510006, China; ygn@gdut.edu.cn (G.Y.); 2111901233@mail2.gdut.edu.cn (Z.Z.); 3119000357@mail2.gdut.edu.cn (H.Z.)
2 Jihua Laboratory, Foshan 528225, China
3 School of Materials Science and Engineering, Jiangsu University, Zhenjiang 212013, China
4 State Key Laboratory of Materials Processing and Die & Mould Technology, School of Materials Science and Engineering, Huazhong University of Science and Technology, Wuhan 430074, China; pangong@hust.edu.cn
5 Key Laboratory for New Type of Functional Materials in Hebei Province, School of Materials Science, Hebei University of Technology, Tianjin 300401, China; ahaxin@hebut.edu.cn
* Correspondence: zhangyu@gdut.edu.cn (Y.Z.); peng@ujs.edu.cn (Z.P.); cqcui@gdut.edu.cn (C.C.)

Abstract: As a traditional interconnect material, silver alloy bonding wires are widely used in electronic packaging, but their propensity to vulcanize quickly has not been sufficiently addressed. The current surface anti-oxidation and anti-sulfidation treatments are often accompanied by a decline in bonding performance, which hinders the use of silver alloy bonding wires in new applications. In the present paper, we develop a new cathodic passivation treatment in a Pd-containing solution for silver bonding wires, which not only significantly improves their vulcanization resistance, but also maintains their bonding performance. The surface of the treated wires remains unaffected after vulcanization in 0.3 $\mu g/m^3$ of ammonium sulfide for 60 min. Compared to a Pd-free passivation treatment, the bonding strength of the wire passivated with the Pd-containing solution improves from 0.20 to 0.27 N. XPS analysis confirms the existence of Pd on the surface of the wire. The solder ball formed an obtuse angle instead of a sharp angle on the pad, which is beneficial for bonding strength.

Keywords: silver alloy bonding wire; bonding strength; vulcanization resistance; cathodic passivation

Citation: Yang, G.; Zhou, Z.; Zhang, H.; Zhang, Y.; Peng, Z.; Gong, P.; Wang, X.; Cui, C. Improved Anti-Vulcanization and Bonding Performance of a Silver Alloy Bonding Wire by a Cathodic Passivation Treatment with Palladium. *Materials* **2022**, *15*, 2355. https://doi.org/10.3390/ma15072355

Academic Editor: Chih Chen

Received: 3 March 2022
Accepted: 20 March 2022
Published: 22 March 2022

1. Introduction

As a traditional packaging interconnect technology, wire bonding still occupies an important place in electronic packaging, due to its simple process and low cost [1]. The materials for bonding wires have developed from pure metals to a variety of alloys based on, for example, gold, silver, copper, and aluminum [2–5]. Among these metals, silver has the best electrical and thermal conductivity, and its price is relatively moderate. Therefore, silver bonding wires have been considered as a potential alternative to gold bonding wires. However, silver alloy bonding wires can be easily contaminated when used in an environment without a protective gas [6,7], which becomes an obstacle for extending their use to new applications.

Bonding reliability is another important property of bonding wires. The device will fail, as long as one of the bonding points is damaged. Generally, the failure modes of the solder ball of wire bonding can be divided into intra-ball failure and interfacial failure, which usually start from the intermetallic compounds, interfacial voids and interfacial corrosion points [8,9]. Zhang et al. [10] studied the failure of LED packages and pointed out that the stress concentration generated in the thermal shock test can lead to the neck fracture of the solder balls. On the other hand, interfacial failures are also widely observed. Wu et al. [11]

conducted power cycle life tests on different types of insulated gate bipolar translator (IGBT) modules, and found that the recrystallization and electrothermal migration could lead to the fracture of the bonding interface. Zheng et al. [12] found that the separation of the bonding wire and the pad is the main failure mode of IGBT modules. The mismatch of thermal expansion coefficients between the bonding wire and the chip causes stress concentration and leads to cracks at the bonding interface to form an open circuit.

To address these problems, various technologies, such as alloying and surface treatments, were developed. For instance, Cheng and Hsiao [13] tried to improve the oxidation and interfacial corrosion resistance of Ag wires through the addition of Au and Pd. Guo and Jong-Soo [14] revealed that doping with Pd can improve the bonding force of the wire by increasing the interconnect reliability between the solder ball and the pad interface. Tseng et al. [15] plated silver wires with gold to improve its oxidation resistance and mechanical properties. Fei-Yi et al. [16] galvanized aluminum bonding wires to control the shape of the solder ball. However, new problems appeared with these treatments. After the bonding wire was alloyed, its electrical and thermal conductivity decreased, which would lead to an increase in manufacturing costs [17]. The thickness of the noble metal element electroplated on the bonding wires was relatively large (~100 nm) and also expensive [18]. The application of a surface treatment affected the mechanical performance of the bonding wires, thereby reducing the life of the package structure [16,19]. As a result, finding a new anti-vulcanization processing method that combines low cost, facile operation, and minimal influence on the bonding performance remains a challenge for the use of silver bonding wires.

In this study, we develop a new cathodic passivation solution for the anti-vulcanization treatment of silver bonding wires. Through the addition of palladium nitrate in solution, the vulcanization resistance of the passivated wires is significantly improved. On the other hand, the treatment also improved the surface wettability of the bonding wires on Au pads. Therefore, the solder ball tends to form an obtuse angle instead of a sharp angle on the pad, which is beneficial to reducing the interface stress concentration and improving the bonding strength. This study could provide a way to improve the sulfidation resistance of bonding wires with less impact on the bonding performance, and shows that wettability modulation may be an effective way to improve the reliability of bonding interfaces.

2. Experiments

2.1. Preparation of the Materials

Silver alloy bonding wires with a diameter of 20 μm were purchased from Xinqipai Electronic Technology Co., Ltd., Chongqing, China. The compositions of the silver alloy bonding wires are shown in Table 1. A hydrotropic solution with 0.010 mol/L of chromium acetate, 0.008 mol/L of trisodium citrate, 0.010 mol/L of sodium tartrate, 0.015 mol/L of crystalline sodium acetate, and 0.140 mol/L of sodium hydroxide was used as the Pd-free passivation solution. Another Pd-containing passivation solution was prepared by adding 0.001 mol/L of palladium nitrate to the Pd-free passivation solution.

Table 1. Compositions of the silver alloy bonding wires.

Element	Ag (%)	Fe (ppm)	Pd (ppm)	Mg (ppm)	Si (ppm)
Content	≥99	≤5	≤1	≤2	≤1

2.2. Experimental Methods

Figure 1a provides an illustration of the cathodic passivation device. The silver alloy bonding wire was sequentially immersed into the passivation solution and deionized water. A current of 500 A/m^2 was applied to the wires and the passivation solution. The wire is the cathode. The wire was processed in the passivation solution for 5 s at 25 °C.

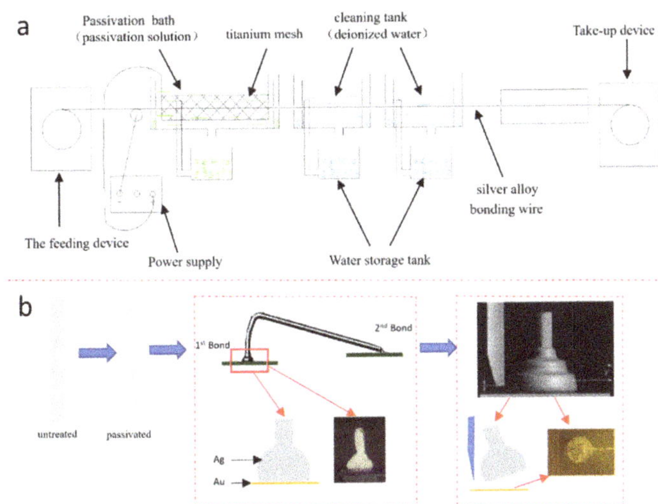

Figure 1. (**a**) Illustration of the cathodic passivation device. (**b**) Illustrations of the wire bond and the shear force test.

Figure 1b shows a flow chart of the experimental procedures after the passivation treatment. The original silver wires and the passivated wires in the Pd-free and Pd-containing solutions were bonded onto Au pads (diameter = 50 μm) using an automatic bonding tool (AW386, Ada Intelligent Equipment Co., Ltd., Foshan, China) for the ball shear force test. The wires formed two bond points, the first being the ball bond and the second being the wedge bond. The detailed bonding parameters are shown in Table 2. After the bonding wires were attached on the Au pads, the cross-sections of the ball bonds on the Au pads were observed by a scanning electron microscope (SEM, HITACHI SU8220, Tokyo, Japan). Using a thrust tester (Condor Sigma, XYZTEC), the ball bonds were pushed away from the Au pads, and the bonding strength was recorded. The fracture morphologies of the Au pads after the shear tests were observed by the SEM.

Table 2. Bonding parameters for the silver bonding wires.

Parameters	1st Bond	2nd Bond
Bonding time (ms)	14	14
Power (mW)	55	80
Pressure (gf)	24	40
Bonding temperature (°C)	150	150
Electronic flame-off current (mA)	28	0
Electronic flame-off time (μs)	530	0

The original silver wires and the passivated wires in the Pd-free and Pd containing solutions were placed in a sealed 3 × 3 × 3 cm^3 box filled with ammonium sulfide gas with a concentration of 0.3 $\mu g/m^3$. The surface condition of the wires was recorded by an optical microscope. Due to the small size of the silver alloy bonding wires, silver plates with a thickness of 10 μm were passivated in the solutions using the same parameters. The surfaces of the passivated silver plates were analyzed by X-ray photoelectron spectroscopy (XPS, Escalab 250Xi).

3. Results and Discussions

3.1. Vulcanization Resistance of the Passivated Silver Alloy Bonding Wires in Different Solutions

Figure 2 shows the surface of the passivated silver alloy bonding wires in different solutions, before and after vulcanization in ammonium sulfide for 20 and 60 min. The surface of the untreated silver alloy bonding wire turned yellow after 20 min of vulcanization, and the surface was further vulcanized to dark red after 60 min. The surface of the silver alloy bonding wire treated with the standard, Pd-free passivation solution turned pale yellow after vulcanization for 20 min. After 60 min of vulcanization, the surface of the bonding wire darkened further. The surface of the silver alloy bonding wire treated with the new Pd-containing passivation solution after vulcanization for 60 min had the same appearance and silver-white color as the surface of the wire before vulcanization. This result indicate that the Pd-containing passivation solution provides better vulcanization resistance than the Pd-free passivation solution.

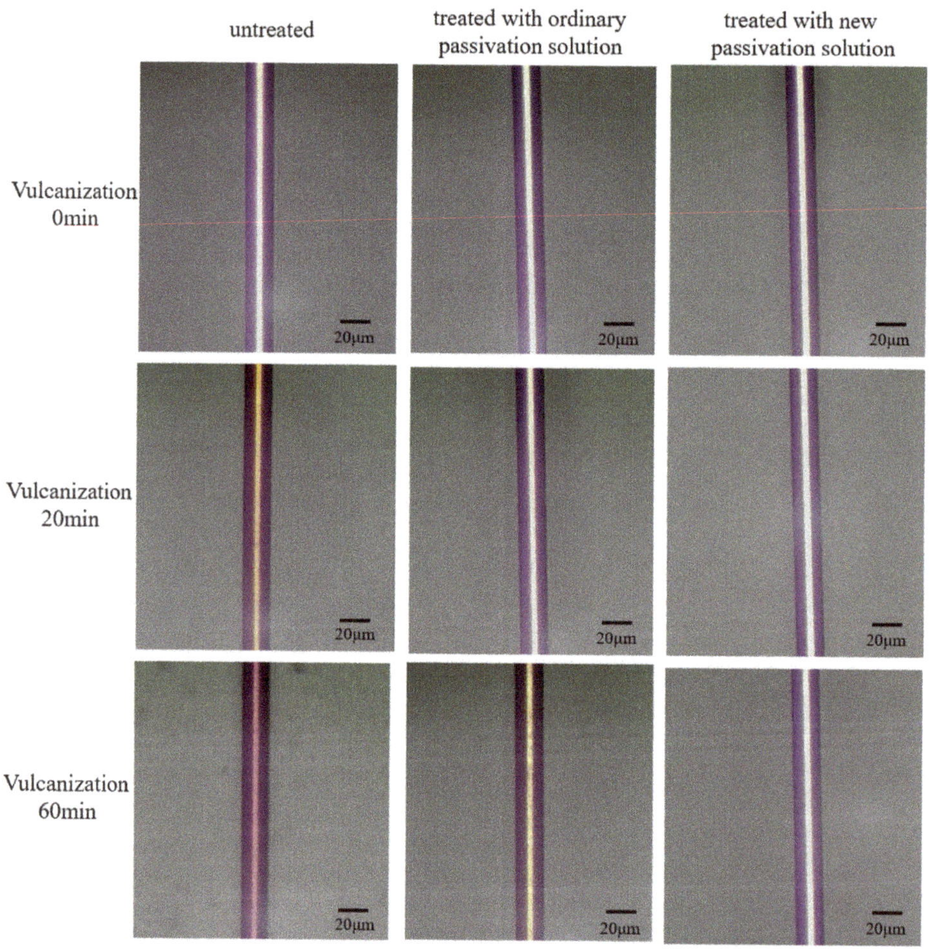

Figure 2. Comparison of the surface of the passivated silver alloy bonding wires in different solutions, before and after vulcanization for 20 and 60 min.

3.2. Surface Analysis

Figure 3 shows the XPS spectra of the passivated silver plates in the Pd-free solution after storage for 5 days. Peaks of Ag3d, O1s, Cr2p, and C1s electrons are apparent. The Ag3d fine spectrum (Figure 3b) shows 2 characteristic peaks at binding energies of 368.40 and 374.40 eV, corresponding to the $Ag3d_{5/2}$ and $Ag3d_{3/2}$ electrons, respectively. It has been reported that the peaks of Ag are at 368.2 eV [20] and 374.27 eV [21], and the peaks of Ag_2O are at 367.4 eV [22] and 373.90 eV [23], respectively. As the peaks of the Ag3d fine spectrum in Figure 3b show highly symmetrical shapes, it can be concluded that most of the Ag element is in the form of Ag, but not AgO. The $Cr2p_{3/2}$ fine spectrum (Figure 3c) shows 2 characteristic peaks at binding energies of 577.28 and 576.30 eV, corresponding to $Cr(OH)_3$ and Cr_2O_3, respectively [24–26], The peak of Cr is at 574.13 eV [27], which is not observed in the spectrum in Figure 3. The Cr ions are present because they are in the passivation solution. The O1s fine spectrum (Figure 4d) shows 3 characteristic peaks after peak fitting, with binding energies of 532.00, 531.28 and 529.44 eV, respectively. The peak at 529.44 eV corresponds to O^{2-} [28], which indicates that the surface contains metallic oxides. The peak at 531.28 eV corresponds to organic C–O and OH^- [29,30], and the peak at 532.00 eV corresponds to organic C=O and OH^- [29,31]. These two peaks and the existence of the C1s peak indicate that the surface of the passivated silver plate contains some residual organics, which are difficult to avoid in XPS measurements [32]. Based on the XPS spectra, it can be concluded the metallic species on the surface of the passivated silver plates in the Pd-free solution are Ag, $Cr(OH)_3$ and Cr_2O_3.

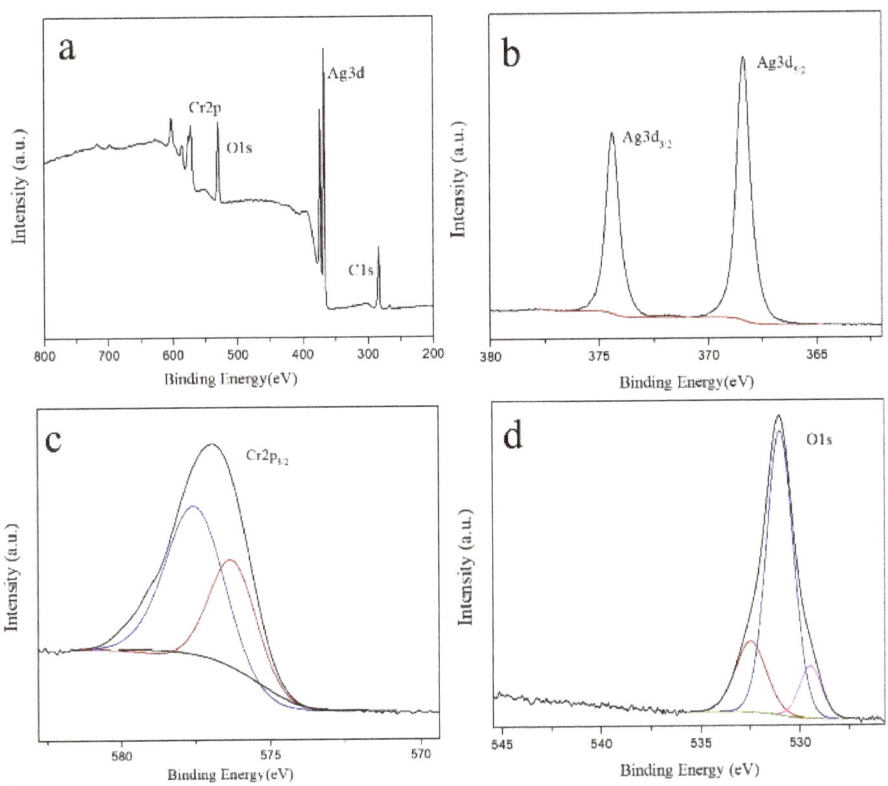

Figure 3. XPS spectra of the passivated silver plates in the Pd-free solution after storage for 5 days. (**a**) Survey spectra; (**b**) Ag3d spectra; (**c**) Cr2p spectra; (**d**) O1s spectra.

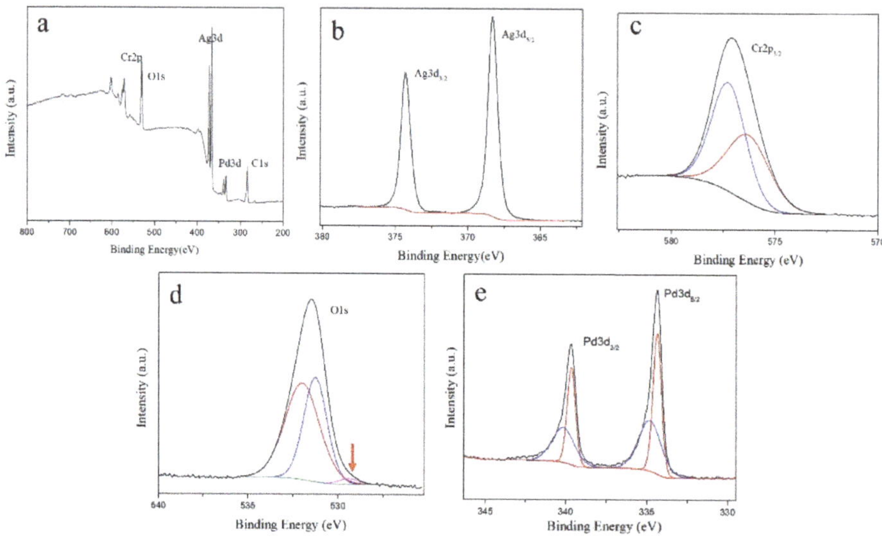

Figure 4. XPS spectra of the passivated silver plates in the Pd-containing solution after storage for 5 days. (**a**) Survey spectra; (**b**) Ag3d spectra; (**c**) Cr2p spectra; (**d**) O1s spectra; (**e**) Pd3d spectra.

Figure 4 shows the XPS spectra of the passivated silver plates in the Pd-containing solution after storage for 5 days. Peaks of Ag3d, O1s, Pd3d, Cr2p, and C1s electrons can be observed. The spectra of Ag3d (Figure 4b) and Cr2p (Figure 4c) are similar to that of the passivated silver plates in the Pd-free solution. In the O1s fine spectrum (Figure 4d), the O^{2-} peak at 529.44 eV is relatively smaller than that in Figure 3d, as marked by the red arrow. In Figure 4e, the $Pd3d_{3/2}$ fine spectrum shows 2 characteristic peaks at 340.18 and 339.68 eV, and the $Pd3d_{5/2}$ fine spectrum shows 2 characteristic peaks at 334.85 and 334.41 eV. These peaks correspond to Pd [33]. The peaks of PdO, PdO_2 and PdO_3 are at 336.90, 337.50 and 337.7 eV, respectively [34,35], which are not observed in the spectrum in Figure 4. Their absence indicates that the Pd on the surface of the passivated silver plate is in the form of Pd. Based on the XPS spectra, it can be concluded the metallic species on the surface of the passivated silver plates in the Pd-containing solution mainly contains Ag, $Cr(OH)_3$, Cr_2O_3 and Pd. Based on the XPS spectra, the relative atomic concentrations of different metallic species on the surfaces of the passivated Ag plates are calculated and shown in Table 3. In comparison, the surface of the silver plates in the Pd-containing solution contains a small amount of Pd, and the concentrations of Ag and Cr_2O_3 have slightly decreased.

Table 3. Relative atomic concentrations of different metallic species on the surfaces of the passivated Ag plates.

Treatment	Concentrations (in at. %)			
	Ag	$Cr(OH)_3$	Cr_2O_3	Pd
Passivated in Pd-free solution	68.8%	16.4%	9.2%	-
Passivated in Pd-containing solution	62.3%	13.5%	8.4%	15.8%

3.3. *Shear Tests of the Passivated Silver Alloy Bonding Wires in Different Solutions*

Figure 5 shows the cross-section images of the bonded wires on the pads before the shear force tests. As the wires are melted during the wire bonding process, the solder ball and the pad form a contact angle, which are summarized in Table 4. For the untreated wire (Figure 5a), the average contact angle is 104°, whereas for the wires passivated in the Pd-free solution (Figure 5b), the average contact angle is 136°, corresponding to a lower wettability. As a result, the contact interface between the solder ball and the pad tends to shrink, and eventually forms a sharp angle on the edge. Due to the stress concentration caused by the sharp angle and the relatively low contact area, a low bonding strength can be expected. For the wire passivated in the Pd-containing solution (Figure 5c), the average contact angle decreased to 108°, which is close to the contact angle of the untreated wires.

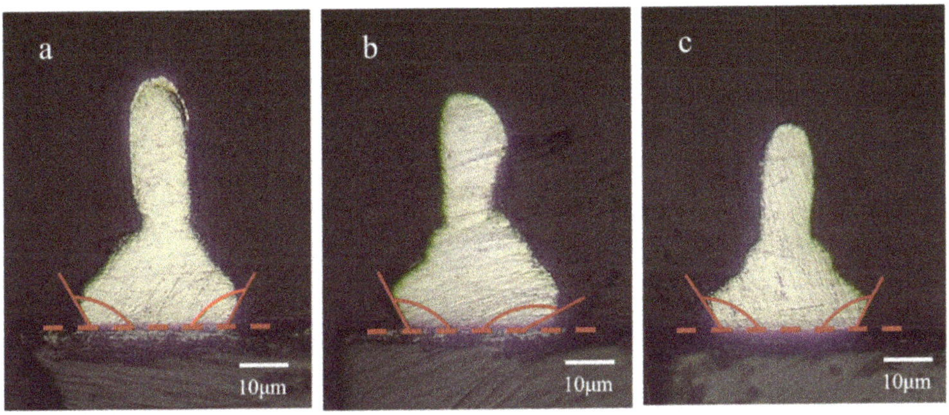

Figure 5. Cross-section images of the bonded wires on the pads before the shear force tests. (**a**) Untreated wire. (**b**) Wire passivated in the Pd-free solution. (**c**) Wire passivated in the Pd-containing solution.

Table 4. Measured contact angles of the bonded wires on the pads.

Wire Treatment Type	Contact Angle (°)			
	Maximum	Minimum	Mean	Standard Error
Untreated	113	93	104	8
Passivated in Pd-free solution	157	104	136	16
Passivated in Pd-containing solution	120	94	108	8

One possible reason for the different contact angles of the wires relates to the content of elements on the surface of the wires. For the wires passivated in the Pd-containing solution, there is a small amount of Pd on the surface. As revealed in previous studies [36–40], the existence of noble metal elements, such as Pd, is beneficial for the oxidation and vulcanization resistance of the Cu and Ag bonding wires. The presence of Pd in the passivated layer might hinder the oxidation of the wires during the bonding process, which decreases the surface energy and increases the wettability of the solder ball. Therefore, smaller contact angles are formed.

Figure 6 shows the measured bonding strength of the wires. The untreated wires, the wires passivated in the Pd-free solution, and the wires passivated in the Pd-containing solution. fractured at mean shear forces of 0.30, 0.20, and 0.27 N, respectively. Through the use of the Pd-containing solution for passivation, the strength of the treated bonding wire increased by 35%, compared to the wire treated in the standard Pd-free solution.

Figure 7 shows the fracture surface of the untreated silver alloy bonding wires and silver alloy bonding wires passivated in different solutions. For the untreated wires, the whole fracture surface is covered by striped fracture patterns, indicating a good interconnection between the bonding wire and the pad. The average bond force is relatively high (0.30 N). For the passivated wire in the Pd-free solution, only half of the fracture surface is covered with the fracture pattern, and the other half of the surface retains the original flat morphology. The presence of some of the original morphology indicates that the bonding wire has not made sufficient contact with the pad. The average bond force is relatively low

(0.20 N). For the wire passivated in the Pd-containing solution, the fracture surface is almost completely covered with the fracture pattern, indicating a good interconnection between the bonding wire and the pad. The average bond force is close to that of the untreated wires (0.27 N).

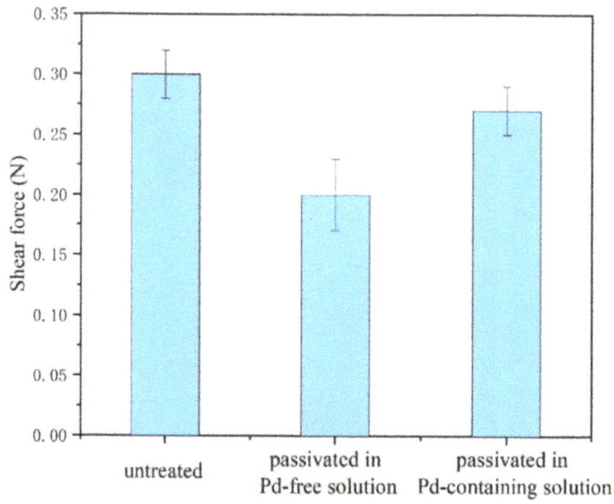

Figure 6. Bonding strength of the untreated silver alloy bonding wires and the silver alloy bonding wires passivated in different solutions.

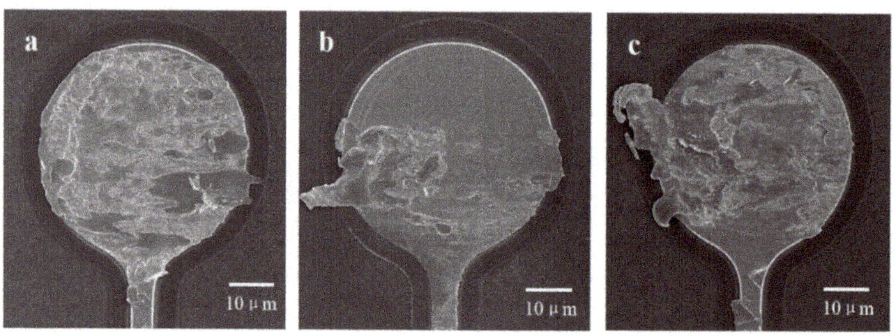

Figure 7. Fracture surface of the untreated and passivated silver alloy bonding wires in different solutions. (**a**) Untreated wire. (**b**) Wire passivated in the Pd-free solution. (**c**) Wire passivated in the Pd-containing solution.

Based on the measured contact angles and bonding strength, it can be found that a smaller contact angle corresponds to a higher strength. This is because the obtuse angle between the solder ball and the pad is beneficial to reducing the stress concentration on the interface. These results indicate that wettability modulation may be an effective way to improve the interface reliability of wire bonding.

Based on the above results and discussion, the improved vulcanization resistance and bonding performance of the silver alloy bonding wires passivated in the Pd-containing solution can be reasonably understood. The results show that Pd plays a key role in improving the performance of the bonding wire. If Pd is added to the bonding wire by alloying, the required amount of Pd is relatively high, which adds to product cost and may lead to problematic inhomogeneous compositions. In comparison, the cathodic passivation method has the advantages of simple processing, fast reaction times, reduced material consumption, and applicability for different material components.

4. Conclusions

A cathodic passivation solution with palladium nitrate has been developed to improve the resistance to vulcanization while retaining the bonding performance of silver alloy bonding wires. For the passivated wires treated with a Pd-free solution, the vulcanization resistance is improved, but the bonding strength of the wire significantly decreases from 0.30 to 0.20 N. With the addition of palladium nitrate to the passivation solution, the surface of the wire can maintain its original silver-white color after vulcanization in 0.3 $\mu g/m^3$ of ammonium sulfide for 60 min. The wettability of the wires is improved, as the contact angle between the solder balls and the Au pad decreases from 136° to 108°. The bonding strength of the wire is 0.27 N, 35% stronger than the wire treated with the Pd-free solution. The cathodic passivation treatment in the Pd-containing solution in this study provides an effective and facile method for the improvement of the properties of silver alloy bonding wires.

Author Contributions: G.Y.: Conceptualization, Data curation, Formal analysis, Investigation, Methodology, Software, Visualization, Writing—original draft. Z.Z.: Investigation, Visualization. H.Z.: Visualization. Y.Z.: Funding acquisition. Z.P.: Conceptualization, Supervision. P.G.: Methodology, Visualization. X.W.: Methodology, Visualization. C.C.: Supervision, Funding acquisition. All authors have read and agreed to the published version of the manuscript.

Funding: We acknowledge the support of the National Natural Science Foundation of China (Grant Nos. 62174039, 61874155), Guangdong Basic and Applied Basic Research (Grant Nos. 2021A1515110656, 2022A1515010141, 2021A1515011642), National Key R&D Program of China (Grant No. 2018YFE0204601), and the Key Research and Development Program of Guangdong Province (Grant Nos. 2021B0101290001, 2020B0101290001).

Institutional Review Board Statement: Not applicable.

Informed Consent Statement: Not applicable.

Data Availability Statement: The datasets generated during and/or analyzed during the current study are available from the corresponding author on reasonable request.

Conflicts of Interest: The authors declare no conflict of interest. The authors declare that they have no known competing financial interests or personal relationships that could have appeared to influence the work reported in this paper.

References

1. Yang, J. Wire Bonding in Microelectronics. *Assem. Autom.* **2011**, *31*, 395. [CrossRef]
2. Liu, H.; Chen, Q.; Zhao, Z.; Wang, Q.; Zeng, J.; Chae, J.; Lee, J. Reliability of Au-Ag Alloy Wire Bonding. In Proceedings of the 2010 Proceedings 60th Electronic Components and Technology Conference, Las Vegas, NV, USA, 1–4 June 2010; pp. 234–239.
3. Tseng, Y.-W.; Hung, F.-Y.; Lui, T.-S.; Chen, M.-Y.; Hsueh, H.-W. Effect of annealing on the microstructure and bonding interface properties of Ag–2Pd alloy wire. *Microelectron. Reliab.* **2015**, *55*, 1256–1261. [CrossRef]
4. Chuang, T.-H.; Tsai, C.-H.; Wang, H.-C.; Chang, C.-C.; Chuang, C.-H.; Lee, J.-D.; Tsai, H.-H. Effects of Annealing Twins on the Grain Growth and Mechanical Properties of Ag-8Au-3Pd Bonding Wires. *J. Electron. Mater.* **2012**, *41*, 3215–3222. [CrossRef]
5. Yurechko, M.; Grushko, B.; Velikanova, T.; Urban, K. Isothermal sections of the Al–Pd–Co alloy system for 50–100 at.% Al. *J. Alloys Compd.* **2002**, *337*, 172–181. [CrossRef]
6. Wu, J.; Rockey, T.; Yauw, O.; Shen, L.; Chylak, B. Bonding of Ag-alloy wire in LED packages. In Proceedings of the 2012 35th IEEE/CPMT International Electronics Manufacturing Technology Conference (IEMT), Ipoh, Malaysia, 6–8 November 2012; pp. 1–4.
7. Guo, R.; Gao, L.; Mao, D.; Li, M.; Wang, X.; Lv, Z.; Chiu, H. Study of free air ball formation in Ag–8Au–3Pd alloy wire bonding. *Microelectron. Reliab.* **2014**, *54*, 2550–2554. [CrossRef]
8. Fan, S.H.; Amp, A.S. Failure Mode and Failure Mechanism Analysis of Semiconductor Device Wire Bonding. *Environ. Technol.* **2018**, *36*, 54–61.
9. Ma, Z.-Q.; Hu, H.-N.; Wang, Q. Failure Analysis of a Model Audion Inner Lead Bonding. *Fail. Anal. Prev.* **2007**, *4*, 50–54.
10. Zhang, S.-U.; Lee, B.W. Fatigue life evaluation of wire bonds in LED packages using numerical analysis. *Microelectron. Reliab.* **2014**, *54*, 2853–2859. [CrossRef]
11. Wu, W.; Held, M.; Jacob, P.; Scacco, P.; Birolini, A. Investigation on the long term reliability of power IGBT modules. In Proceedings of the International Symposium on Power Semiconductor Devices and IC's: ISPSD'95, Yokohama, Japan, 23–25 May 1995; pp. 443–448.

12. Zheng, L.B.; Li, H.; Jin, P.Y.; Fang, H.C.; Wang, C.L.; Wang, Z.P. Investigation of the Temperature Character of IGBT Wire Bonding Lift-Off Based the 3-D Thermal-Electro Coupling FEM. *Adv. Mater. Res.* **2013**, *616–618*, 1689–1692. [CrossRef]
13. Cheng, C.H.; Hsiao, H.L.; Chu, S.I.; Shieh, Y.Y.; Sun, C.Y.; Peng, C. Low cost silver alloy wire bonding with excellent reliability performance. In Proceedings of the 2013 IEEE 63rd Electronic Components and Technology Conference (ECTC), Las Vegas, NV, USA, 28–31 May 2013; pp. 1569–1573.
14. Cho, J.-S.; Yoo, K.-A.; Moon, J.-T.; Son, S.-B.; Lee, S.-H.; Oh, K.H. Pd effect on reliability of Ag bonding wires in microelectronic devices in high-humidity environments. *Met. Mater. Int.* **2012**, *18*, 881–885. [CrossRef]
15. Tseng, Y.; Hung, F.; Lui, T. Microstructure, tensile and electrical properties of gold-coated silver bonding wire. *Microelectron. Reliab.* **2015**, *55*, 608–612. [CrossRef]
16. Hung, F.-Y.; Lui, T.-S.; Chu, K.-M.; Tseng, Y.-W. Aluminium wires have the free air balls (FABs): Electronic flame-off, fracture strength, electrical properties, and bonding characteristics of nano Zn Film Al–Si bonding wires. *Metals* **2017**, *7*, 152. [CrossRef]
17. Cho, J.-S.; Jeong, H.-S.; Moon, J.-T.; Yoo, S.-J.; Seo, J.-S.; Lee, S.-M.; Ha, S.-W.; Her, E.-K.; Kang, S.-H.; Oh, K.-H. Thermal reliability & IMC behavior of low cost alternative Au-Ag-Pd wire bonds to Al metallization. In Proceedings of the 2009 59th Electronic Components & Technology Conference, San Diego, CA, USA, 26–29 May 2009; pp. 1569–1573.
18. Xu, H.; Qin, I.; Clauberg, H.; Chylak, B.; L.Acoff, V. Behavior of palladium and its impact on intermetallic growth in palladium-coated Cu wire bonding. *Acta Mater.* **2013**, *61*, 79–88. [CrossRef]
19. Chuang, T.-H.; Lin, H.-J.; Chuang, C.-H.; Shiue, Y.-Y.; Shieu, F.-S.; Huang, Y.-L.; Hsu, P.-C.; Lee, J.-D.; Tsai, H.-H. Thermal stability of grain structure and material properties in an annealing twinned Ag–4Pd alloy wire. *J. Alloys Compd.* **2014**, *615*, 891–898. [CrossRef]
20. Johansson, G.; Hedman, J.; Berndtsson, A.; Klasson, M.; Nilsson, R. Calibration of electron spectra. *J. Electron Spectrosc. Relat. Phenom.* **1973**, *2*, 295–317. [CrossRef]
21. Yan, L.; Jordan, R.G.; Qiu, S.L. Electronic structures of ordered Ag-Mg alloys. *Phys. Rev. B Condens. Matter* **1994**, *49*, 4478.
22. Hammond, J.S.; Gaarenstroom, S.W.; Winograd, N. X-ray photoelectron spectroscopic studies of cadmium- and silver-oxygen surfaces. *Anal. Chem.* **2002**, *47*, 2193–2199. [CrossRef]
23. Tjeng, L.H.; Meinders, M.B.J.; Vanelp, J.; Ghijsen, J.; Sawatzky, G.A.; Johnson, R.L. Electronic structure of Ag_2O. *Phys. Rev. B Condens. Matter* **1990**, *41*, 3190. [CrossRef] [PubMed]
24. Shuttleworth, D. Preparation of metal-polymer dispersions by plasma techniques. An ESCA investigation. *J. Phys. Chem.* **1980**, *84*, 1629–1634. [CrossRef]
25. Capece, F.M.; Castro, V.D.; Furlani, C.; Mattogno, G.; Fragale, C.; Gargano, M.; Rossi, M. "Copper chromite" catalysts: XPS structure elucidation and correlation with catalytic activity. *J. Electron. Spectrosc. Relat. Phenom.* **1982**, *27*, 119–128. [CrossRef]
26. Howng, W.-Y.; Thorn, R.J. Investigation of the electronic structure of $La_{1-x}(M^{2+})_xCrO_3$, Cr_2O_3 and La_2O_3 by X-ray photoelectron spectroscopy. *J. Phys. Chem. Solids* **1980**, *41*, 75–81. [CrossRef]
27. Moffat, T.P.; Latanision, R.M.; Ruf, R.R. An X-ray photoelectron spectroscopy study of chromium-metalloid alloys—III. *Electrochim. Acta* **1995**, *40*, 1723–1734. [CrossRef]
28. Sugimoto, K.; Sawada, Y. The role of molybdenum additions to austenitic stainless steels in the inhibition of pitting in acid chloride solutions. *Corros. Sci.* **1977**, *17*, 425–445. [CrossRef]
29. Asami, K.; Hashimoto, K. The X-ray photo-electron spectra of several oxides of iron and chromium. *Corros. Sci.* **1977**, *17*, 559–570. [CrossRef]
30. López, G.; Castner, D.G.; Ratner, B.D. XPS O 1s binding energies for polymers containing hydroxyl, ether, ketone and ester groups. *Surf. Interface Anal.* **1991**, *17*, 267–272. [CrossRef]
31. Khan, M.T.; Srivastava, S. Some new ruthenium (III) schiff base complexes: A photoelectron spectroscopic study. *Polyhedron* **1988**, *7*, 1063–1065. [CrossRef]
32. Yang, G.; Zou, Q.; Wang, P.; Lai, H.; Lai, T.; Zeng, X.; Li, Z.; Luo, J.; Zhang, Y.; Cui, C. Towards understanding the facile synthesis of well-covered Cu-Ag core-shell nanoparticles from a complexing model. *J. Alloys Compd.* **2021**, *874*, 159900. [CrossRef]
33. Schleich, B.; Schmeisser, D.; Göpel, W. Structure and reactivity of the system $Si/SiO_2/Pd$: A combined XPS, UPS and HREELS study. *Surf. Sci.* **1987**, *191*, 367–384. [CrossRef]
34. Tressaud, A.; Khairoun, S.; Touhara, H.; Watanabe, N. X-ray Photoelectron Spectroscopy of Palladium Fluorides. *Z. Anorg. Allg. Chem.* **2010**, *540*, 291–299. [CrossRef]
35. Shafeev, G.A.; Themlin, J.M.; Bellard, L.; Marine, W.; Cros, A. enhanced adherence of areaselective electroless metal plating on insulators. *J. Vac. Sci. Technol. A* **1996**, *14*, 319–326. [CrossRef]
36. Uno, T. Enhancing bondability with coated copper bonding wire. *Microelectron. Reliab.* **2011**, *51*, 88–96. [CrossRef]
37. Zhang, B.; Qian, K.; Wang, T.; Cong, Y.; Wang, J. Behaviors of Palladium in Palladium coated copper wire bonding process. In Proceedings of the 2009 International Conference on Electronic Packaging Technology & High Density Packaging (ICEPT-HDP'09), Beijing, China, 10–13 August 2009.
38. Koh, W.; Lee, T.-K.; Ng, H.-S.; Goh, K.-S.; Ho, H.-M. Investigation of palladium coverage on bonded balls of palladium-coated copper wires. In Proceedings of the 2011 12th International Conference on Electronic Packaging Technology and High Density Packaging, Shanghai, China, 8–11 August 2011; pp. 1–7.
39. Rui, G.; Cheng, Y.; Mao, D.; Ming, L.; Chiu, H. Thermal reliability of Ag-8Au-3Pd alloy wire bonds. In Proceedings of the 2013 3rd IEEE CPMT Symposium Japan, Kyoto, Japan, 11–13 November 2013.
40. Tura, J.M.; Regull, P.; Victori, L.; Castellar, M. XPS and IR (ATR) analysis of Pd oxide films obtained by electrochemical methods. *Surf. Interface Anal.* **1988**, *11*, 447–449. [CrossRef]

materials

MDPI

Article

Experimental and Simulation Investigation of Nd Additions on As-Cast Microstructure and Precipitate Development in Mg–Nd System Alloys

Xuewei Yan [1], Bin Su [2,*], Xuemei Yang [3], Qingdong Xu [2], Xiaopeng Zhang [2], Jing Wang [2] and Zhenhua Wen [1,*]

1 School of Aero Engine, Zhengzhou University of Aeronautics, Zhengzhou 450046, China; yanxuewei@zua.edu.cn

2 Institute of Materials, China Academy of Engineering Physics, Jiangyou 621700, China; 138268104xu@163.com (Q.X.); saviorggg@163.com (X.Z.); wangjing7@caep.cn (J.W.)

3 School of Aerospace Engineering, Zhengzhou University of Aeronautics, Zhengzhou 450046, China; yangxuemei@zua.edu.cn

* Correspondence: sub703@126.com (B.S.); zhenhuawen@zua.edu.cn (Z.W.); Tel.: +86-166-5960-9901 (B.S.); +86-371-6191-0097 (Z.W.)

Abstract: The microstructure and precipitate evolution of as-cast Mg–Nd alloys with different contents of Nd was investigated via experimental and simulation methods. The research showed that the as-cast microstructure of Mg–Nd alloy consisted of α-Mg dendrites and the intermetallic phases. A metastable β phase precipitated, followed by α-Mg dendrites that could be confirmed as $Mg_{12}Nd$ by X-ray diffraction (XRD) analysis. The amount of β-$Mg_{12}Nd$ presented a rising trend with increasing Nd additions. In addition, the tertiary phase was also observed in as-cast Mg–Nd alloy when Nd content was greater than 3 wt.%, which precipitated from the oversaturated α-Mg matrix. The tertiary phase should be β_1-Mg_3Nd, which is also a metastable phase with a face-centered cubic lattice. However, it is a pity that the tertiary phase was not detected by the XRD technique. Moreover, an effective cellular automaton (CA) model was explored and applied to simulate the time-dependent α-Mg/β_1-Mg_3Nd eutectic growth. The simulated results of α-Mg/β_1-Mg_3Nd eutectic growth in Mg-3Nd presented that the growth of α-Mg dendrites was accompanied by the nucleation and growth of β_1-Mg_3Nd precipitates and eventually formed a eutectic structure. The eutectic morphologies for Mg–Nd system alloys with different Nd contents were also simulated using the proposed model, and the results revealed that α-Mg dendrite was a refinement, and the amount of α-Mg/β_1-Mg_3Nd eutectic was promoted, with increasing Nd content.

Keywords: Mg–Nd alloy; microstructure; precipitate; eutectic growth; modelling and simulation

Citation: Yan, X.; Su, B.; Yang, X.; Xu, Q.; Zhang, X.; Wang, J.; Wen, Z. Experimental and Simulation Investigation of Nd Additions on As-Cast Microstructure and Precipitate Development in Mg–Nd System Alloys. *Materials* **2022**, *15*, 2535. https://doi.org/10.3390/ma15072535

Academic Editors: Zine El Abiddine Fellah and Tony Spassov

Received: 26 February 2022
Accepted: 28 March 2022
Published: 30 March 2022

1. Introduction

Magnesium (Mg) and its alloys have received increasing attention in recent years due to their low-density, high specific strength and excellent creep resistance [1–3]. The abundance of magnesium in the earth's crust and the potential low-cost make it a viable alternative to steels or aluminum alloys for structural components in the automotive, aerospace and electronics industries [4,5]. However, Mg and Mg alloys suffer from poor mechanical properties, corrosion resistance and high-temperature strength because the hexagonal close-packed (HCP) structure leads to insufficient slip systems, strong basal texture and poor plasticity [6–8]. Improvements to these deficient properties of Mg alloys could be achieved through alloying and the consequent heat treatment, in particular through adding alloyed elements to fine grains and increasing the amount of precipitates and solid solution [9,10].

Recently, many studies have focused on the modification of the alloy elements, with a special emphasis on additions to improve the mechanical properties of Mg alloys. Alloys such as Mg–Al [11], Mg–Zn [12], Mg–Zr [13] and Mg–RE (rare earth) [14] system alloys have been developed towards the high creep resistance. Among them, Mg–RE alloys were attractive due to their remarkable age-hardening response, excellent creep resistance and good formability [15]. Specifically, Mg–Gd [16], Mg–Y [17] and Mg–Nd [18] are the most common heat-resistant Mg alloys. Over the past few decades, more and more research information concerning Mg–Nd alloy both experimentally and theoretically have been extensively reported in the literature. Yan et al. [19] investigated the creep behavior of Mg-2Nd alloy under different temperature and applied stress conditions and found that this alloy exhibited good creep resistance due to both solution hardening and, especially, precipitation hardening. Hantzsche et al. [20] systematically evaluated the effect of Ce, Nd and Y additions on the microstructure and texture development in Mg–RE alloys, and the authors concluded that the amount of the RE addition required for sufficient texture weakening was connected with the solid solubility of the respective element. Zhu et al. [21] investigated the relationship between the microstructure and creep resistance in Mg–RE alloys, and the authors reckoned that the strengthening of α-Mg matrix by solid solution and/or precipitation was more important than grain boundary reinforcement by intermetallic phases for the creep resistance of Mg–RE alloys. Liu et al. [22] studied the effect of Al content on the microstructure and mechanical properties of as-cast Mg–5Nd alloys, and the results indicated that Al additions could significantly lead to the grain refinement of Mg–5Nd alloy and ascribed the greatly improved mechanical properties to the refined grains and secondary phase strengthening. It is well known that microstructures are the strategic link between a material's process and performance, and, thus, further investigation of the detailed interaction between the microstructure and the properties of Mg–Nd alloys is of great importance to control the desirable microstructure and improve the performance.

In addition, some recent reports [23–25] displayed that the high strength and good creep resistance of Mg–Nd alloy resulted from the formation of fine precipitates of metastable phases. Thus, a deep understanding of precipitated behavior in Mg–Nd alloy may provide a capability for the rational design of composition and processes to improve the performance of Mg–Nd alloy. However, the precipitation sequence of Mg–Nd alloys remains controversial and have been the topic of a number of experimental and computational studies. Saito et al. [26] observed the microstructure of precipitates formed in an Mg–0.5Nd alloy by high-angle annular detector dark-field scanning transmission electron microscopy, and the results indicated that the precipitation sequence could be Mg-solution $\rightarrow$ GP-zone $\rightarrow \beta'$ (Mg_7Nd, orthorhombic) $\rightarrow \beta_1$ (Mg_3Nd, FCC). Liu et al. [27] developed a phase-field model to examine the heterogeneous nucleation of β_1 precipitates in Mg–0.5 at.% Nd alloy, and the authors found that β_1 precipitated as ultra-thin laths with abnormally large aspect ratios under the influence of the stress field of a screw dislocation. Zhu et al. [28] presented the formation of linear-chain distribution features of precipitates in Mg–Nd alloys, and the results showed that the configuration consists of bamboo trunks of β_1 variants and trunk connections of a hexagonal lattice β_2 phase. Natarajan et al. [29] reported a combined computational and experimental study of phase stability and precipitation in Mg–Nd alloys, and the results supported the following precipitation sequence for binary Mg–Nd alloys: SSSS (supersaturated solid solutions) $\rightarrow$ GP zones (Guinier–Preston zones, hexagons) $\rightarrow$ $\beta''' \rightarrow \beta_1$ (Mg_3Nd) $\rightarrow \beta$ ($Mg_{12}Nd$) $\rightarrow \beta e$ ($Mg_{41}Nd_5$). Considering the studies mentioned above, precipitation has long been considered one of the most effective approaches for manipulating the kinetics and homogeneity in Mg–Nd alloys, and the related research has made significant progress over the past few years. However, the microstructure and precipitate controls in Mg–Nd alloys are still challenges due to the complex precipitation behavior, especially in the as-cast alloys. Therefore, a further study is needed for the satisfied microstructures and accurate precipitates in Mg–Nd system alloys.

The aim of the present work was to investigate the microstructure and precipitate development in as-cast Mg–Nd alloys using combined experimental and computational methods during the solidification process. The influence of different Nd additions on the microstructure and precipitate morphologies was examined using an optical microscope

(OM) (ZEISS, Oberkochen, Baden Wurttemberg, Germany) and a scanning electron microscope (SEM) (FEI Company, Hillsboro, ORE, USA). Chemical composition and phase characterization were also performed by the energy-dispersive spectroscopy (EDS) and X-ray diffraction (XRD) techniques. In addition, an effective model for predicting eutectic evolution behavior in Mg–Nd system alloys was established, and the eutectic growth and the final morphologies were simulated to analyze the effects of the Nd additions.

2. Experimental Procedure and Simulation Method

2.1. Materials and Experimental Methods

The experimental materials employed in this work were Mg–x Nd (x = 1, 2, 3, 4, 5, 6 and 7, wt.%) system alloys, which were prepared in a 60 KW level electric resistance furnace (Henan Sante Furnace Technology Co., Ltd., Luoyang, Henan, China) under the protection of a mixed gas atmosphere of 1% SF_6 and 99% CO_2. The high-purity Mg (99.98 wt.%) (Luoyang Maige Magnesium Industry Co., Ltd., Luoyang, Henan, China) was preheated to 150 °C for several minutes to displace the moisture and then melted at 750 °C. When the pure Mg melted sufficiently, the temperature was elevated to 780 °C and moderate Mg-20 wt.% Nd master alloy was added. After being stirred to remove slag for 10 min, the melting alloy then stood for 30 min at 750 °C. Subsequently, the power was cut off, the melting alloy cooled to ~720 °C, and it was then poured into a mild steel mold (80 mm in diameter, preheated to ~300 °C), which was covered with a mold release agent (boron nitride) (Hefei Kejing Material Technology Co., Ltd., Hefei, Anhui, China). The filled mold then cooled in the air until the melting alloy was completely solidified. The as-cast ingots were machined and polished, and the chemical compositions were measured using energy-dispersive X-ray fluorescence (EDXRF). The results as well as the designed compositions are listed in Table 1.

Table 1. Chemical compositions of Mg–x Nd alloys.

Alloys	Designed Compositions		Analyzed Compositions	
	Nd (wt.%)	Mg	Nd (wt.%)	Mg
Mg–1 Nd	1	Bal.	0.95	Bal.
Mg–2 Nd	2	Bal.	1.98	Bal.
Mg–3 Nd	3	Bal.	3.10	Bal.
Mg–4 Nd	4	Bal.	3.92	Bal.
Mg–5 Nd	5	Bal.	4.85	Bal.
Mg–6 Nd	6	Bal.	5.89	Bal.
Mg–7 Nd	7	Bal.	6.92	Bal.

The samples for microstructural analysis were ground, mechanically polished and then etched with a mixture of 5% HNO_3 and 95% ethanol. The microstructures were characterized by an optical microscope equipped with a digital camera, and the precipitated phases were observed using a focused ion beam-scanning electron microscope (FIB-SEM, FEI Helios Nanolab 600i) (FEI Company, Hillsboro, ORE, USA) at 15 kV with a working distance of 4 mm. The precipitates were qualitatively analyzed using X-ray diffraction (XRD, TDF-3000, Dandong Tongda Science & Technology Co., Ltd., Dandong, China), and a microanalysis was performed on the different phases in the microstructure using energy-dispersive X-ray spectroscopy (EDS) equipped with an EDS detector (EDAX Inc., Philadelphia, PA, USA) for elemental analysis. The volume fraction and average size of the secondary phase were statistically measured using OM micrographs captured from different regions and calculated by Image-Pro Plus software (Version 6.0, Media Cybernetics, Inc., Los Angeles, CA, USA). Partial thermophysical parameters (thermal conductivity and specific heat) were measured at different temperatures, and the results provided access to the data needed in the simulations.

2.2. Model Description and Simulation Parameters

According to the references [19,30], the microstructure of as-cast Mg–Nd alloy consists of α-Mg matrix and divorced eutectic $Mg_{12}Nd$ (β phase). Other reports [31,32], however, have shown that the intermetallic phase formed in the eutectic was Mg_3Nd (β_1 phase) in as-cast Mg–Nd alloy, and the eutectic presented a lamellar structure. The β and β_1 are both metastable precipitates; the β phase has a body-centered tetragonal structure (a = 1.031 nm, c = 0.593 nm) and the β_1 phase has a face-centered cubic lattice (a = 0.74 nm). In this study, we mainly focused on the modelling and simulation for the eutectic formation of α-Mg matrix and β_1-Mg_3Nd precipitates. The detailed analytical model of lamellar and rod eutectic growth was first established by Jackson and Hunt (JH) in 1966 [33]. In the last decades, numerous researchers developed the JH model for investigating the eutectic growth mechanism. Although great progress has been made in the study of eutectic growth behavior, it must be pointed out that the knowledge and quantitative understanding of eutectic formation in the final solidification stage is still limited. Following the models established by JH and other researchers, a cellular automaton (CA) model was presented to investigate the evolution behavior of the α-Mg/β_1 eutectic structure in Mg–Nd system alloys. Figure 1 shows the schematic diagrams of the lattice correspondence of α-Mg with β_1 and the lamellar eutectic at the dendrite tip. In Figure 1b, S_α and $S_{\beta 1}$ represent half the widths of the α and β_1 phases, respectively, and v is the growth velocity of the lamellar eutectic.

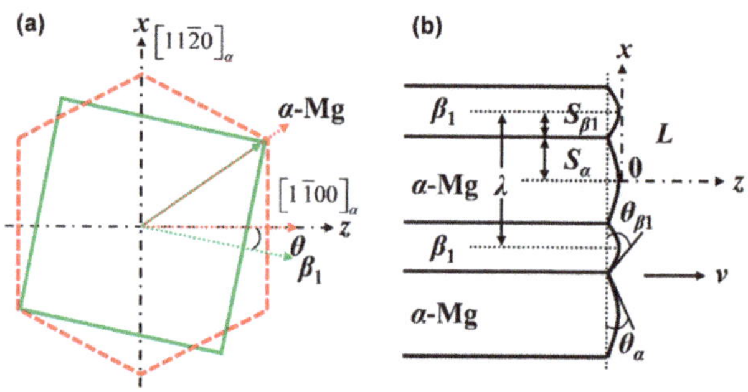

Figure 1. (**a**) Lattice vectors corresponding to the (011) β_1 and (0001) $_\alpha$ planes of β_1 and α-Mg [34]; (**b**) schematic diagram of the lamellar eutectic structure at the eutectic dendrite tip region [35].

Solute diffusion is very important in determining eutectic growth. In this work, solute diffusion along y direction was ignored, and the solute field in liquid is given by:

$$\frac{\partial^2 C}{\partial x^2} + \frac{\partial^2 C}{\partial z^2} + \frac{v}{D}\frac{\partial C}{\partial z} = 0, \quad (1)$$

where C is the solute concentration in the liquid. The general periodic solution of the diffusion equation can be characterized by the following equation:

$$C = C_\infty + \sum_{n=0}^{\infty} B_n e^{-\omega_e z} \cos(b_n x), \quad (2)$$

where C_∞ is the solute concentration in the liquid far from the interface, $b_n = n\pi/(S_\alpha + S_{\beta 1})$, and $\omega_e = (v/2D) + \left[(v/2D)^2 b_n^2\right]^{1/2}$. The coefficient B_n is obtained from the boundary conditions at the interface:

$$-\left(\frac{\partial C}{\partial z}\right)_{z=0} = \left(\frac{v}{D}\right) C(x,0)(1-k_\alpha)(\alpha\text{–Mg}), \quad (3)$$

$$-\left(\frac{\partial C}{\partial z}\right)_{z=0} = -\left(\frac{v}{D}\right)[1 - C(x,0)](1 - k_{\beta 1})(\beta_1 \text{ phase}), \quad (4)$$

where $C(x, 0)$ is the interface concentration along the α or β_1 phase, and k_α and $k_{\beta 1}$ are the solute distribution coefficients in the α and β_1 phases, respectively.

The growth kinetics are determined by the local undercooling (ΔT), which consists of thermal, constitutional and curvature effects that can be defined as:

$$\Delta T = \Delta T_e + m_i(\omega - \omega_e) - \Gamma_i \overline{K}_i, \quad (5)$$

where ΔT_e is the metastable condition eutectic undercooling, m_i is the liquidus slope (i denotes the α or β_1 phase), ω is the concentration of the interfacial cell, Γ_i is the Gibbs–Thomson coefficient of the i phase and $\overline{K}_i$ is the mean curvature of the interfacial cell, which can be approximately given by:

$$\overline{K}_i = \left[1 - 2\left(f_{s,i} + \sum_{j=1}^{N} f_{s,i}^j\right) / (N+1)\right] / \Delta x, \quad (6)$$

where $f_{s,i}$ and $f_{s,i}^j$ are the solid fraction of the i phase in the interfacial cell and its neighboring cells, respectively. N is the number of neighboring cells, which is equal to 8 in the actual calculations, Δx is the cell size and the growth velocity, v, can be analytically expressed as:

$$v = a_i \cdot (\Delta T)^2, \quad (7)$$

where a_i is the growth kinetic coefficient, and the value is 2.9 ($\mu m \cdot s^{-1} \cdot K^{-2}$) for the non-faceted α phase, while it is 5.8 ($\mu m \cdot s^{-1} \cdot K^{-2}$) for the faceted β_1 phase [36]. The solid fraction of the α or β_1 phase is calculated separately by the following equations:

$$\Delta f_{s,\alpha} = v \cdot \Delta t_n / \Delta x, \quad (8)$$

$$\Delta f_{s,\beta} = \cos\theta \cdot v \cdot \Delta t_n / \Delta x, \quad (9)$$

where Δt_n is the time step, and θ is the angle between the growth direction and the linking line, which is between the interfacial cell and the position of the nucleus of the β_1 phase.

Some thermophysical parameters of Mg–Nd system alloys were calculated by JmatPro software (CnTech Co., Ltd., Version 6.1, Shanghai, China), and they were compared with the experimental results, as shown in Figure 2. It can be seen that the density increases with the Nd additions, as the density of Nd is greater than that of Mg, and the values vary between 1.55 and 1.85 g/cm^3. Meanwhile, in Figure 2c, the thermal conductivity decreases with increases in Nd content. The diffusion coefficient is shown in Figure 2b; the order of magnitude is ~10^{-9} in liquid, and it is ~10^{-12} in solid. In addition, the experimental measured results of the thermal conductivity and specific heat (shown in Figure 2e,f) fit well with the calculated results (Figure 2c,d). However, due to the limitation of JmatPro software, an optimized table look-up technique was adopted to obtain the parameters that were needed in the simulations, as shown in Table 2.

Table 2. Thermophysical parameters of Mg–Nd alloy from Refs [31,36–38].

Definition and Units	Values
Eutectic temperature (°C)	552
Eutectic composition (wt.%)	33
Liquid slope m_α (°C/wt.%)	−5.1
Liquid slope $m_{\beta 1}$ (°C/wt.%)	13.2
Gibbs–Thomson coefficient of α-Mg (m·K)	6.2×10^{-7}
Gibbs–Thomson coefficient of the β_1 phase (m·K)	1.7×10^{-7}
Solute distribution coefficient of α-Mg	0.4
Solute distribution coefficient of the β_1 phase	0.113

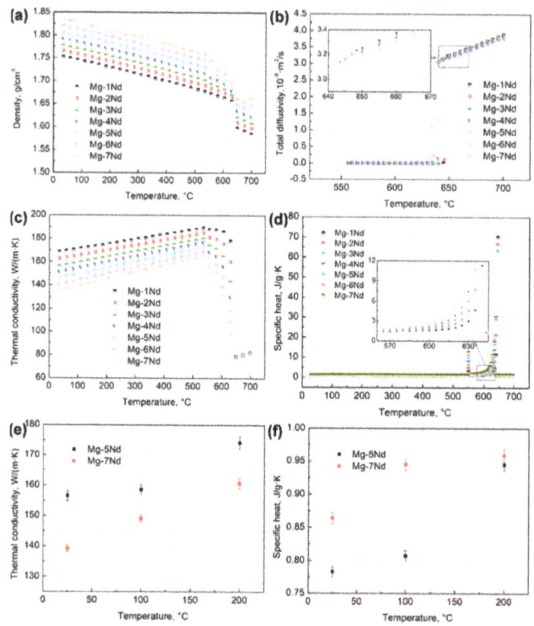

Figure 2. Thermophysical parameters of Mg–Nd system alloys: (**a**) density; (**b**) total diffusivity; (**c**,**e**) thermal conductivity, obtained by calculation and experiment; (**d**,**f**) specific heat, obtained by calculation and experiment.

3. Results and Discussion

3.1. Microstructure and Precipitate Morphologies

Low- and high-magnification optical micrographs of the microstructure and secondary phase morphologies in different Mg–Nd alloys are shown in Figure 3a–g, which presents the volume fractions and average sizes of the secondary phase variation in the Mg–Nd system alloys. It can be seen that the microstructure of Mg–Nd alloy consists of the dendritic α-Mg and a secondary phase (pointed by black arrows). The secondary phase is a $Mg_{12}Nd$ intermetallic compound, which is a metastable phase and always exists in the as-solidified alloys [19]. It is worth noting that the amount of secondary phase increases with Nd additions, and their morphologies change from blocky-shaped particles to a continuous network structure. According to previous reports [39–41] the maximum solubility of Nd in Mg is ~3.6 wt.% at the eutectic temperature (552 °C), while regarding the Mg–Nd binary phase diagram [42], Nd has the maximum solid solubility in Mg at ~548 ± 2 °C and almost zero at room temperature. Therefore, the solid solution of Nd in the Mg matrix is easy to saturate, and, as the Nd content increases, some of the Nd element does not go into the solid solution; it then exists in the form of a secondary phase, which always distributes discontinuously along the grain boundary. Due to the rapid growth of secondary phase along grain boundary, Nd-free bands are produced near the grain boundaries. In addition, it is observed that the microstructure of Mg–6Nd alloy consists of an obvious dendritic structure and network intermetallic phase (Figure 3f), which is consistent with those of a previous study [19]. The quantitative statistical results demonstrated that the volume fraction of the secondary phase increased from 0.95 % in Mg–1Nd to 20.04% in Mg–7Nd (Figure 3h). The average size of the secondary phase presented a relatively stable interval when Nd content was lower than 5 wt.%, and the value was approximately 4.91–10.41 μm, while the average size significantly increased to ~38.36 μm in Mg–6Nd alloy and decreased to ~22.03 μm when the content of Nd increased to 7 wt.% (Figure 3i). In addition, according to the study [19], the addition of Nd to Mg caused a significant improvement in the creep properties and the creep resistance increased with the increase in Nd addition, which is accounted for by the combination of precipitation and solid solution hardening.

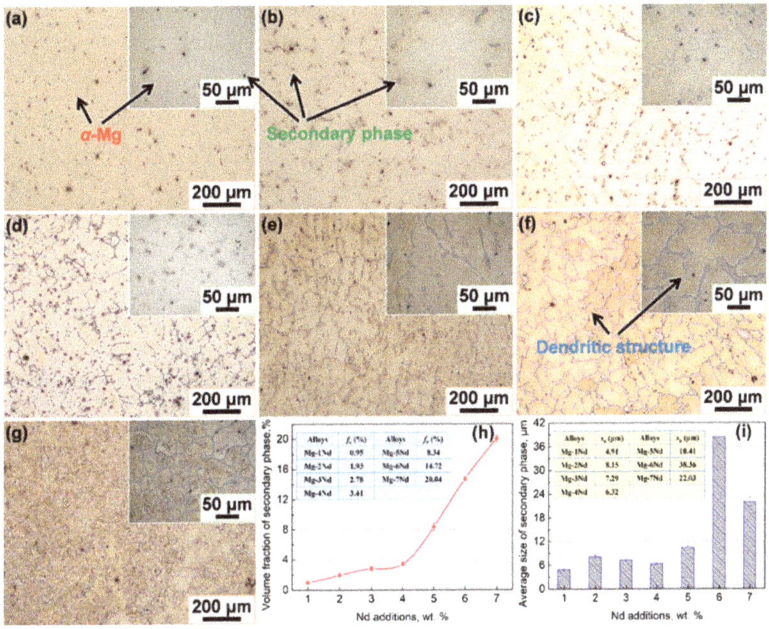

Figure 3. Low- and high-magnification optical micrographs of the microstructure and the secondary phase in different Mg–Nd system alloys: (**a**) Mg-1Nd; (**b**) Mg-2Nd; (**c**) Mg-3Nd; (**d**) Mg-4Nd; (**e**) Mg-5Nd; (**f**) Mg-6Nd; (**g**) Mg-7Nd; (**h**,**i**) the volume fraction and average size of secondary phase variation in Mg–Nd system alloys, respectively.

In order to observe the precipitate development in Mg–Nd system alloys more clearly, FIB-SEM examinations were performed, as shown in Figure 4. SEM images are able to provide a good contrast between intermetallic or eutectic phases and the α-Mg matrix. The contrast among different phases in the microstructure was exploited to obtain different grayscale images from which the intermetallic or eutectic phases and α-Mg matrix were determined. It can be seen that the lower Nd content alloys (Mg–1Nd and Mg–2Nd in Figure 4a,b) are mainly composed of two different phases, i.e., α-Mg and a secondary phase (β phase, $Mg_{12}Nd$), which is in agreement with the OM observations mentioned above. The formation of a secondary phase is due to the non-equilibrium solidification of Nd in α-Mg being oversaturated, and part of the Nd forms divorced eutectic β-$Mg_{12}Nd$ instead of precipitating in the as-cast alloys. With the increasing of Nd additions, however, a number of tiny precipitates form from the oversaturated α-Mg matrix, temporarily named the tertiary phase (marked by black arrows in Figure 4c,f). Based on the previous reports [28,31], the tertiary phase might be a β_1 phase (Mg_3Nd), and the microstructure of this alloy, in fact, contains a eutectic mixture of the intermetallic phase and α-Mg. It is noted that the result is quite different from previous studies [6,19,30], which have only found the divorced $Mg_{12}Nd$ intermetallic in the as-cast binary Mg–Nd alloys. In addition, according to the Mg–Nd binary phase diagram [42], a β_1-Mg_3Nd phase tends to form in the high-Nd-content alloys, for example, in the master alloy with the Nd content over 20 wt. %. Nevertheless, at the relatively low content of Nd, Mg3Nd could also form at a higher cooling rate [31], in an extruded state Mg–Nd alloy [32] or after high-pressure torsion [43]. The β_1 is also a metastable phase and presents a zigzag structure that seems to have precipitated homogeneously in the α-Mg matrix. The morphology and distribution of the β_1 phase changed with Nd contents, and the morphology of β_1 phase in the lower

Nd content alloys is obviously different from that in the Mg–7 Nd alloy (Figure 4g). The variation may be due to the different interaction mechanism between strain and dislocation in different Nd addition alloys.

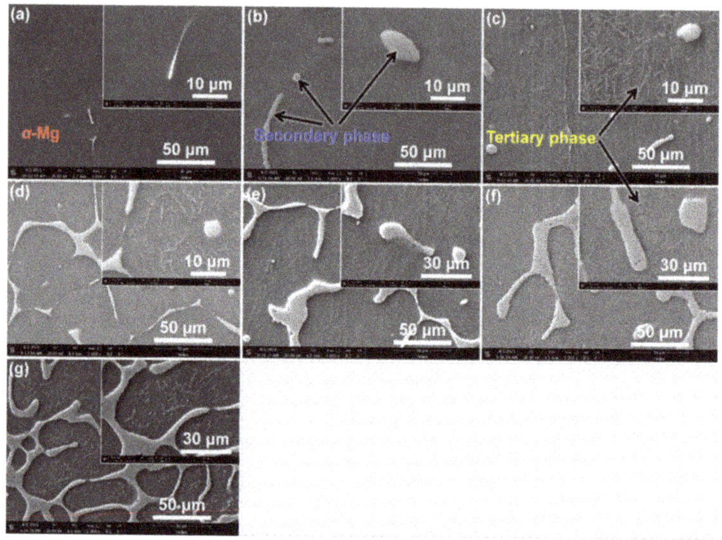

Figure 4. Low- and high-magnification SEM images of precipitate development in Mg–Nd system alloys: (**a**) Mg-1Nd; (**b**) Mg-2Nd; (**c**) Mg-3Nd; (**d**) Mg-4Nd; (**e**) Mg-5Nd; (**f**) Mg-6Nd; (**g**) Mg-7Nd.

3.2. *Chemical Composition and Phase Characterization*

Figures 5 and 6 present the micrographs and chemical composition analysis of Mg–3Nd and Mg–6Nd alloys, respectively. The elemental mappings in Figure 5b,c indicate that the secondary phase is rich in Nd, and the matrix is mainly composed of Mg element, where the lighter pixels represent a higher element content, and the darker ones illustrate a lower element content. Nevertheless, it is hard to identify the element distribution of the tertiary phase in elemental mappings, and a similar phenomenon could be found in Figure 6b,c. In addition, the impurity elements, such as Fe and P, were not detected, which was probably due to artifacts, such as a peak overlap in the EDS mappings. In addition, with the increasing Nd content, the enrichment of Nd presents semi-continuously intermixed with the Mg matrix. In order to obtain further insight into the distribution of chemical elements, EDS analysis was pursued. Figures 5d–f and 6d–f display the point EDS results in the secondary phase (light-grey region, Point 1 in the figures) and Mg matrix (dark-grey region, Point 2 in the figures). Since the point EDS actually measured the composition in a region larger than the spots in both Mg–3Nd and Mg–6Nd alloys, the analysis included a portion of the matrix and a number of particles. It can be seen that the elemental composition at Point 1 and Point 2 in Mg–3Nd alloy were approximately 78.29 wt.% Mg and 27.11 wt.% Nd and 97.48 wt.% Mg and2.52 wt.% Nd, respectively, while it was about 69.46 wt.% Mg and 30.54 wt.% Nd and 97.39 wt.% Mg and 2.61 wt.% Nd in Mg–6Nd alloy. The amount of Nd in the matrix presents a relatively small change between Mg–3Nd and Mg–6Nd alloys, and this is because the additional amount of Nd is more than its solubility limit in both alloys. It should be noted here that the point EDS analysis shows that Fe element was detected in both alloys, which most likely derived from the raw materials or the tools used for melting and casting, such as the steel mold and stirring tool.

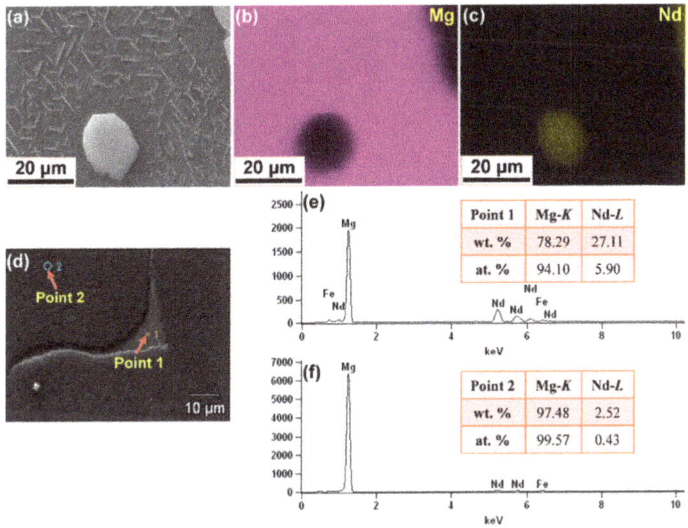

Figure 5. Micrographs and chemical composition analysis of Mg–3Nd alloy: (**a**–**c**) BSE image and corresponding elemental mappings; (**d**–**f**) SEM image and point EDS results of the secondary phase and Mg matrix.

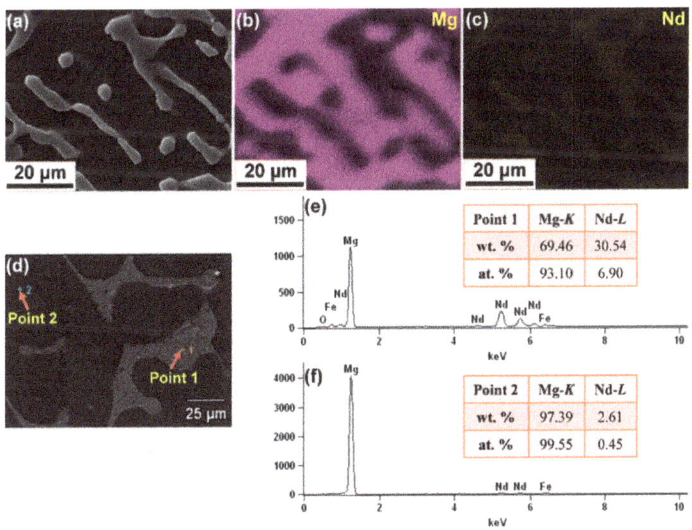

Figure 6. Micrographs and chemical composition analysis of Mg–6Nd alloy: (**a**–**c**) BSE image and corresponding elemental mappings; (**d**–**f**) SEM image and point EDS results of the secondary phase and Mg matrix.

Figure 7 displays the X-ray diffraction patterns of Mg–Nd system alloys. It can be seen that all the peaks are indexed as arising from two different phases, i.e., α-Mg and β-$Mg_{12}Nd$, which is consistent with the OM observations in Figure 2. The β-$Mg_{12}Nd$ is a metastable phase, and its amount increases with the Nd levels, based on the statistical results mentioned above. Other reports [18,44,45], however, indicated that the β_e-$Mg_{41}Nd_5$ phase could also be detected in the as-extruded Mg–Nd alloys and at high temperatures. Actually, the sequence and type of precipitation are affected by the alloy state, solidification condition, heat treatment process, etc. In the present work, the β-$Mg_{12}Nd$ phase easily formed in the as-cast Mg–Nd alloys under the determined solidification condition. Regrettably, the tertiary phase was not detected in all Mg–Nd system alloys; this may be because the amount of tertiary phase was less and the volume was small in the as-cast Mg–Nd system alloys. According to the analysis and deduction above in this work, the tertiary phase (β_1-Mg_3Nd) is a metastable eutectic phase that likely precipitates from the supersaturated α-Mg during cooling after casting. In addition, the metastable intermetallic phases (β or β_1) would transform to the equilibrium phase at a certain temperature during heating, such as under the subsequent solution and aging conditions.

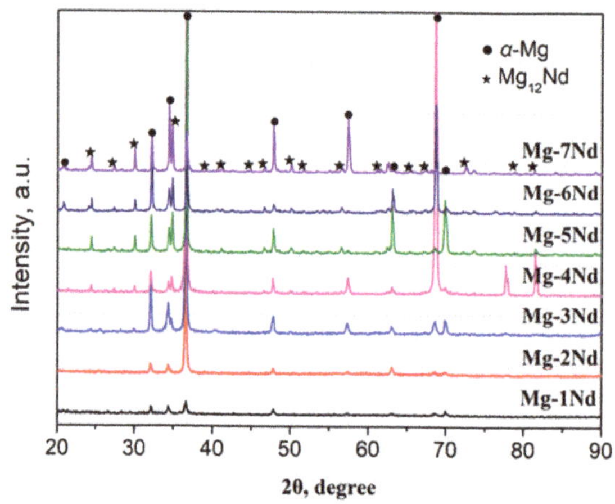

Figure 7. X-ray diffraction patterns of Mg–Nd system alloys.

3.3. *Simulation of Eutectic Growth and Development*

In this section, to provide further investigation of the evolution behavior of the eutectic in Mg–Nd system alloys, the proposed CA model was applied to simulate the α-Mg/β_1-Mg_3Nd eutectic growth and morphologies. The corresponding parameters used in the simulation cases are shown in Figure 2 and Table 2, some of which vary with the solidification conditions. As a representative, Mg-3Nd alloy was selected to illustrate the growth process of the eutectic structure, and the simulation results are shown in Figure 8. A square-shaped domain was used for the simulation cases consisting of 400×400 cells. No mass flux conditions were imposed at the calculation boundaries, and the temperature was constant. The total eutectic nuclei number was equal to 50 in the simulation case, with randomly assigned locations and preferential growth orientations. The red color in Figure 8 represents the α-Mg dendrites, the yellow color represents the β_1-Mg_3Nd precipitates and the blue color represents the liquid phase. Initially, the nucleated grains are randomly distributed in the calculation domain, and the α-Mg dendrites start to develop along the preferential growth orientations that present a petaliform morphology. It can be seen that the growth of α-Mg dendrites is accompanied by the nucleation and growth of β_1-Mg_3Nd precipitates and eventually forms the eutectic structure. The β_1-Mg_3Nd precipitates tend to be suppressed by the faster-growing α-Mg dendrites, resulting in the discontinuous growth of β_1. Actually, the eutectic growth mainly depends on the solute distribution. With the continuous solidification of the liquid alloy, excess solute is rejected into the melt, and the interaction effects of the solute field between the adjacent α-Mg dendrites are intensified (Figure 8b,c). The β_1-Mg_3Nd will absorb amounts of solute from the surrounding liquid and nucleate at grain boundaries. It can be seen that β_1-Mg_3Nd precipitates at grain boundaries present the zigzag shape (Figure 8d,e), which is consistent with the experimental observations in Section 3.1. Figure 8f shows the final morphology of the α-Mg/β_1-Mg_3Nd eutectic structure. It can be found that most β_1-Mg_3Nd precipitates distribute along the grain boundaries and form interconnected networks. In addition, it is also noticeable that the β_1-Mg_3Nd precipitates formed at the α-Mg dendritic arms are eliminated or decreased; this is probably because β_1 is a metastable phase, and it will be dissolved as the temperature drops lower. The divorced eutectic β-$Mg_{12}Nd$ also has an important influence on the formation of β_1-Mg_3Nd at grain boundaries, which are limited by the proposed CA model; however, the microstructure in the simulation results do not contain the divorced eutectic β-$Mg_{12}Nd$.

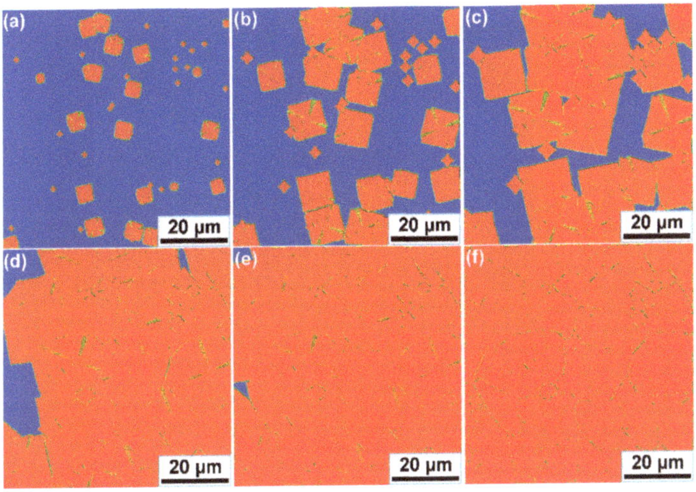

Figure 8. Simulation results of α-Mg/β_1-Mg_3Nd eutectic growth for Mg-3Nd alloy at various times: (**a**) 0.27 s; (**b**) 0.53 s; (**c**) 0.82 s; (**d**) 1.15 s; (**e**) 1.53 s; (**f**) 2.21s.

In order to investigate the effect of Nd additions on the α-Mg/β_1-Mg_3Nd eutectic, simulation cases with various Nd contents were performed, and the simulation results are shown in Figure 9. In the Mg–Nd system alloy, α-Mg is the faster growing phase, and the volume fraction of β_1-Mg_3Nd in the eutectic structure is very small; thus, it is difficult to describe this intermetallic phase when the content of Nd is lower. Therefore, Mg–Nd system alloys containing 4 wt.%, 5 wt.%, 6 wt.% and 7 wt.% were selected for the simulations. It can be observed that the eutectic morphologies with different Nd contents have a similar characteristic, consisting of α-Mg dendrites, zigzag-shaped β_1-Mg_3Nd at grain boundaries and rod-like β_1-Mg_3Nd in the dendritic regions. As the Nd content increases, the α-Mg dendrite is refined, and the amount of the eutectic structure is clearly promoted. In addition, as the eutectic fraction increases with the Nd content, the morphologies undergo a prominent transition from isolated zigzag or rod-like shapes to interconnected networks. The proposed model incorporated several aspects, including growth algorithms, kinetics and solute diffusion, to achieve the eutectic growth mechanism for Mg–Nd system alloys. It can be seen that the simulation and experimental results presented a good agreement, indicating that this model can successfully reproduce the eutectic growth of Mg–Nd system alloys. As known to us, in addition to the influence of alloying elements, other factors such as the cooling rate and undercooling also have significant effects on the eutectic growth. However, due to the limitation of the established model and technical algorithm, this study only focused on the α-Mg/β_1-Mg_3Nd eutectic development with the Nd additions. In our following work, we will make an effort to investigate the precipitation mechanism of β_1-Mg_3Nd to further improve the simulation precision.

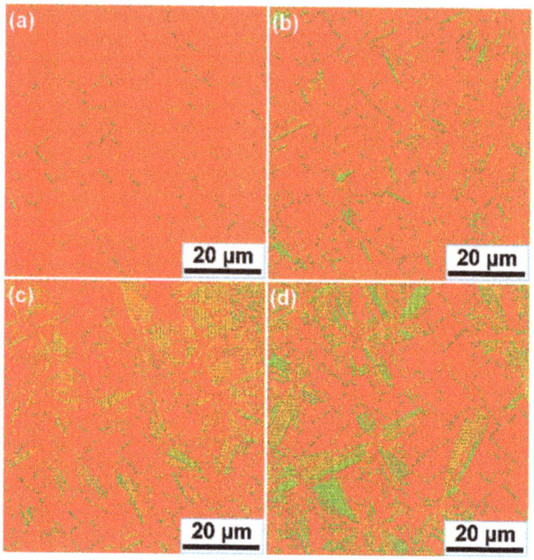

Figure 9. Simulation results of α-Mg/β_1-Mg_3Nd eutectic morphologies for Mg–Nd system alloys: (**a**) Mg-4Nd; (**b**) Mg-5Nd (**c**) Mg-6Nd; (**d**) Mg-7Nd.

4. Conclusions

In this work, experimental and computational methods were used to investigate the microstructure and precipitate development in as-cast Mg–Nd system alloys. OM, SEM and XRD analyses were performed, and a eutectic growth model was established. The main conclusions are shown as follows:

(1) The Mg–Nd alloys with different Nd contents (1, 2, 3, 4, 5, 6 and 7 wt.%) were prepared, and the microstructure and precipitate development in Mg–Nd system alloys was observed by OM and SEM techniques. According to the OM micrographs, the microstructure of Mg–Nd alloy consisted of α-Mg dendrites and the secondary phase, and the amount of secondary phase presented an increasing trend with the increasing Nd additions. SEM images indicated that a tertiary phase precipitated from the oversaturated α-Mg matrix in the higher Nd content alloys, and this is different from the previous studies, which only found a divorced eutectic $Mg_{12}Nd$ in as-cast Mg–Nd alloys.

(2) Chemical composition and phase characterizations were performed by EDS and XRD methods, respectively. The point EDS results indicated that the Nd content in the α-Mg matrix presented a relatively small change between Mg–3Nd and Mg–6Nd alloys, as the additional amount of Nd was more than the solubility limit in both alloys. The secondary phase was further confirmed to be $Mg_{12}Nd$ by XRD analysis. Regrettably, the tertiary phase was not detected in all Mg–Nd system alloys; this may be because the amount of tertiary phase was less and the volume was small in the as-cast alloys.

(3) An effective CA model was explored, with the advantage of describing the time-dependent α-Mg/β_1-Mg_3Nd eutectic growth. The α-Mg/β_1-Mg_3Nd eutectic growth in Mg–3Nd alloy was simulated using the proposed CA model, and the simulated results revealed that the growth of α-Mg dendrites was accompanied by the nucleation and growth of β_1-Mg_3Nd precipitates and eventually formed the eutectic structure. In addition, eutectic morphologies for Mg–Nd system alloys with different Nd contents were also simulated, and the results indicated that the α-Mg dendrite was refined and the amount of α-Mg/β_1-Mg_3Nd eutectic was promoted with an increase in the Nd content.

Author Contributions: Conceptualization, B.S., Experiment, Q.X., X.Z. and J.W.; Methodology, B.S. and Z.W.; Modelling and Simulation, X.Y. (Xuewei Yan) and X.Y. (Xuemei Yang); Investigation, Resources and Data Curation, X.Y. (Xuemei Yang), Q.X., X.Z. and J.W.; Writing—Original Draft Preparation, X.Y. (Xuewei Yan); Writing—Review and Editing, B.S. and Z.W. All authors have read and agreed to the published version of the manuscript.

Funding: This research was funded by the National Key R&D Program of China (No. 2016YFB0701201), the National Natural Science Foundation of China (Nos. 51904276, 51975539), the Program for Science and Technology Innovation Talents in Universities of Henan Province (No. 22HASTIT031) and the Key Scientific Research Project of Colleges and Universities in Henan Province (No. 20A430032).

Institutional Review Board Statement: Not applicable.

Informed Consent Statement: Not applicable.

Data Availability Statement: The data presented in this work are available on request from the corresponding authors.

Conflicts of Interest: The authors declare no conflict of interest.

References

1. Polmear, I.J. Magnesium alloys and applications. *Mater. Sci. Technol.* **1994**, *10*, 1–16. [CrossRef]
2. Merson, D.; Brilevsky, A.; Myagkikh, P.; Tarkova, A.; Prokhorikhin, A.; Kretov, E.; Frolova, T.; Vinogradov, A. The Functional Properties of Mg–Zn–X Biodegradable Magnesium Alloys. *Materials* **2020**, *13*, 544. [CrossRef] [PubMed]
3. Wu, G.; Wang, C.; Sun, M.; Ding, W. Recent developments and applications on high-performance cast magnesium rare-earth alloys. *J. Magnes. Alloy.* **2020**, *9*, 1–20. [CrossRef]
4. Hisa, M.; Barry, J.C.; Dunlop, G.L. New type of precipitate in Mg-rare-earth alloys. *Philos. Mag. A* **2002**, *82*, 497–510. [CrossRef]
5. Liu, H.; Zhu, Y.M.; Wilson, N.C.; Nie, J.F. On the structure and role of βF′ in β1 precipitation in Mg–Nd alloys. *Acta Mater.* **2017**, *133*, 408–426. [CrossRef]
6. Yan, J.; Sun, Y.; Xue, F.; Xue, S.; Xiao, Y.; Tao, W. Creep behavior of Mg-2 wt. % Nd binary alloy. *Mater. Sci. Eng. A* **2009**, *524*, 102–107. [CrossRef]
7. Chen, L.; Sheng, Y.; Wang, X.; Zhao, X.; Liu, H.; Li, W. Effect of the Microstructure and Distribution of the Second Phase on the Stress Corrosion Cracking of Biomedical Mg-Zn-Zr-xSr Alloys. *Materials* **2018**, *11*, 551. [CrossRef] [PubMed]
8. Huang, Z.; Yang, C.; Allison, J.E.; Qi, L.; Misra, A. Dislocation cross-slip in precipitation hardened Mg–Nd alloys. *J. Alloy. Compd.* **2020**, *859*, 157858. [CrossRef]
9. Wei, L.Y.; Dunlop, G.L.; Westengen, H. Age hardening and precipitation in a cast magnesium-rare-earth alloy. *J. Mater. Sci.* **1996**, *31*, 387–397. [CrossRef]
10. Tang, J.; Huo, Q.; Zhang, Z.; Zhang, Y.; Zhao, S.; Hashimoto, A.; Yang, X. Enhancing the creep resistance of a dilute Mg-1.5 wt%Nd alloy plate via pre-compression and subsequent peak-aging. *J. Alloys Compd.* **2021**, *861*, 158590. [CrossRef]
11. Li, S.; Zheng, W.; Tang, B.; Zeng, D.; Guo, X. Grain Coarsening Behavior of Mg-Al Alloys with Mischmetal Addition. *J. Rare Earths* **2007**, *25*, 227–232. [CrossRef]
12. Kojima, Y.; Aizawa, T.; Kamado, S. Mechanical Properties of Rapidly Solidified Mg-Zn Alloys. *Mater. Sci. Forum* **2000**, *350–351*, 105–110. [CrossRef]
13. Niu, R.-L.; Yan, F.-J.; Wang, Y.-S.; Duan, D.-P.; Yang, X.-M. Effect of Zr content on damping property of Mg–Zr binary alloys. *Mater. Sci. Eng. A* **2018**, *718*, 418–426. [CrossRef]
14. Guo, H.; Liu, S.; Huang, L.; Wang, D.; Du, Y.; Chu, M. Thermal Conductivity of As-Cast and Annealed Mg-RE Binary Alloys. *Metals* **2021**, *11*, 554. [CrossRef]
15. Liu, D.; Yang, D.; Li, X.; Hu, S. Mechanical properties, corrosion resistance and biocompatibilities of degradable Mg-RE alloys: A review. *J. Mater. Res. Technol.* **2018**, *8*, 1538–1549. [CrossRef]
16. Yang, L.; Yuan, Y.; Chen, T.; Dai, X.; Zhang, L.; Li, D.; Tang, A.; Yi, W.; Zhang, L.; Pan, F. Diffusion behaviour and mechanical properties of binary Mg–Gd system. *Intermetallics* **2021**, *133*, 107171. [CrossRef]
17. Sudholz, A.; Gusieva, K.; Chen, X.-B.; Muddle, B.; Gibson, M.; Birbilis, N. Electrochemical behaviour and corrosion of Mg–Y alloys. *Corros. Sci.* **2011**, *53*, 2277–2282. [CrossRef]
18. Zhang, Y.; Huang, Y.; Feyerabend, F.; Blawert, C.; Gan, W.; Maawad, E.; You, S.; Gavras, S.; Scharnagl, N.; Bode, J.; et al. Influence of the amount of intermetallics on the degradation of Mg-Nd alloys under physiological conditions. *Acta Biomater.* **2021**, *121*, 695–712. [CrossRef] [PubMed]
19. Yan, J.; Sun, Y.; Xue, F.; Xue, S.; Tao, W. Microstructure and mechanical properties in cast magnesium–neodymium binary alloys. *Mater. Sci. Eng. A* **2008**, *476*, 366–371. [CrossRef]
20. Hantzsche, K.; Bohlen, J.; Wendt, J.; Kainer, K.U.; Yi, S.B.; Letzig, D. Effect of rare earth additions on microstructure and texture development of magnesium alloy sheets. *Scr. Mater.* **2010**, *63*, 725–730. [CrossRef]

21. Zhu, S.; Gibson, M.; Easton, M.; Nie, J. The relationship between microstructure and creep resistance in die-cast magnesium–rare earth alloys. *Scr. Mater.* **2010**, *63*, 698–703. [CrossRef]
22. Liu, D.; Song, J.; Jiang, B.; Zeng, Y.; Wang, Q.; Jiang, Z.; Liu, B.; Huang, G.; Pan, F. Effect of Al content on microstructure and mechanical properties of as-cast Mg-5Nd alloys. *J. Alloy. Compd.* **2017**, *737*, 263–270. [CrossRef]
23. Xu, Z.; Weyland, M.; Nie, J. On the strain accommodation of β1 precipitates in magnesium alloy WE54. *Acta Mater.* **2014**, *75*, 122–133. [CrossRef]
24. DeWitt, S.; Solomon, E.L.S.; Natarajan, A.R.; Araullo-Peters, V.; Rudraraju, S.; Aagesen, L.K.; Puchala, B.; Marquis, E.A.; Ven, A.V.D.; Thornton, K.; et al. Misfit-driven β''' precipitate composition and morphology in Mg-Nd alloys. *Acta Mater.* **2017**, *136*, 378–389. [CrossRef]
25. Sun, B.; Tan, J.; Zhang, H.; Sun, Y. Atomic scale investigation of a novel metastable structure in aged Mg–Nd alloys. *Scr. Mater.* **2018**, *161*, 6–12. [CrossRef]
26. Saito, K.; Hiraga, K. The Structures of Precipitates in an Mg-0.5 at%Nd Age-Hardened Alloy Studied by HAADF-STEM Technique. *Mater. Trans.* **2011**, *52*, 1860–1867. [CrossRef]
27. Liu, H.; Gao, Y.; Zhu, Y.; Wang, Y.; Nie, J. A simulation study of β 1 precipitation on dislocations in an Mg–rare earth alloy. *Acta Mater.* **2014**, *77*, 133–150. [CrossRef]
28. Zhu, Y.; Liu, H.; Xu, Z.; Wang, Y.; Nie, J. Linear-chain configuration of precipitates in Mg–Nd alloys. *Acta Mater.* **2015**, *83*, 239–247. [CrossRef]
29. Natarajan, A.R.; Solomon, E.L.; Puchala, B.; Marquis, E.; Van der Ven, A. On the early stages of precipitation in dilute Mg–Nd alloys. *Acta Mater.* **2016**, *108*, 367–379. [CrossRef]
30. Liu, Z.J.; Wu, G.H.; Pang, S.; Ding, W.J. Mg-Nd alloy and application of Nd in magnesium alloys. *Foundry Technol.* **2012**, *33*, 151–154, CNKI:SUN:ZZJS.0.2012-02-005.
31. Chia, T.; Easton, M.; Zhu, S.; Gibson, M.; Birbilis, N.; Nie, J. The effect of alloy composition on the microstructure and tensile properties of binary Mg-rare earth alloys. *Intermetallics* **2009**, *17*, 481–490. [CrossRef]
32. Zhou, B.; Wang, L.; Zhu, G.; Wang, J.; Wen, W.; Zeng, X. Understanding the Strengthening Effect of β1 Precipitates in Mg-Nd Using In Situ Synchrotron X-ray Diffraction. *JOM* **2018**, *70*, 2315–2320. [CrossRef]
33. Jackson, K.; Hunt, J. Lamellar and Rod Eutectic Growth. In *Dynamics of Curved Fronts*; Academic Press: Cambridge, MA, USA, 1988; pp. 363–376. [CrossRef]
34. Choudhuri, D.; Dendge, N.; Nag, S.; Gibson, M.A.; Banerjee, R. Role of applied uniaxial stress during creep testing on precipitation in Mg–Nd alloys. *Mater. Sci. Eng. A* **2014**, *612*, 140–152. [CrossRef]
35. Li, J.; Zhou, Y. Eutectic growth in bulk undercooled melts. *Acta Mater.* **2005**, *53*, 2351–2359. [CrossRef]
36. Chen, R.; Xu, Q.-Y.; Liu, B.-C. Modeling of aluminum-silicon irregular eutectic growth by cellular automaton model. *China Foundry* **2016**, *13*, 114–122. [CrossRef]
37. Mezbahul-Islam, M.; Mostafa, A.; Medraj, M. Essential Magnesium Alloys Binary Phase Diagrams and Their Thermochemical Data. *J. Mater.* **2014**, *2014*, 1–33. [CrossRef]
38. Yang, M.H.; Guo, Z.P.; Xiong, S.M. Numerical simulation of dendritic growth of magnesium alloy with convection. *Chin. J. Nonferrous Met.* **2015**, *25*, 835–843.
39. Gorsse, S.; Hutchinson, C.; Chevalier, B.; Nie, J.-F. A thermodynamic assessment of the Mg–Nd binary system using random solution and associate models for the liquid phase. *J. Alloy. Compd.* **2005**, *392*, 253–262. [CrossRef]
40. Sun, M.; Hu, X.; Peng, L.; Fu, P.; Ding, W.; Peng, Y. On the production of Mg-Nd master alloy from NdFeB magnet scraps. *J. Mater. Process. Technol.* **2015**, *218*, 57–61. [CrossRef]
41. Kubásek, J.; Vojtěch, D. Structural and corrosion characterization of biodegradable Mg–RE (RE=Gd, Y, Nd) alloys. *Trans. Nonferrous Met. Soc. China* **2013**, *23*, 1215–1225. [CrossRef]
42. Wang, X. Investigation on High-Performance Mg-Nd Alloys. Master Thesis, Southeast University, Nanjing, China, 2007.
43. Bourezg, Y.I.; Azzeddine, H.; Hennet, L.; Thiaudière, D.; Huang, Y.; Bradai, D.; Langdon, T.G. The sequence and kinetics of pre-precipitation in Mg-Nd alloys after HPT processing: A synchrotron and DSC study. *J. Alloys Compd.* **2017**, *719*, 236–241. [CrossRef]
44. Zhang, Y.; Huang, Y.; Feyerabend, F.; Gavras, S.; Xu, Y.; Willumeit-Römer, R.; Kainer, K.U.; Hort, N. Effects of Intermetallic Microstructure on Degradation of Mg-5Nd Alloy. *Met. Mater. Trans. A* **2020**, *51*, 5498–5515. [CrossRef]
45. Nie, J.-F. Precipitation and Hardening in Magnesium Alloys. *MTA* **2012**, *43*, 3891–3939. [CrossRef]

materials

MDPI

Article

Alloying Elements Effect on the Recrystallization Process in Magnesium-Rich Aluminum Alloy

Vladimir Aryshenskii [1], Fedor Grechnikov [1], Evgenii Aryshenskii [1], Yaroslav Erisov [1], Sergey Konovalov [2,*], Maksim Tepterev [1] and Alexander Kuzin [1]

1 Institute of Aerospace Engineering, Samara National Research University, Moskovskoye Shosse 34, 443086 Samara, Russia
2 Departament of Mechanics and Engineering, Siberian State Industrial University, Kirova 42, 654007 Novokuznetsk, Russia
* Correspondence: konovalov@sibsiu.ru; Tel.: +7-3843-742-016

Abstract: This paper addresses the study of the complex effect of alloying elements (magnesium, manganese, copper and zirconium) on changes in magnesium-rich aluminum alloy composition, fine and coarse particle size and number, recrystallization characteristics and mechanical properties. The data obtained made it possible to analyze change in the chemical composition, sizes of intermetallic compounds and dispersoids depending on alloying elements content. The effect of the chemical composition on the driving force and the number of recrystallization nuclei was studied. It was established that the addition of alloying elements leads to grain refinement, including through the activation of a particle-stimulated nucleation mechanism. As a result, with Mg increase from 4 to 5%, addition of 0.5% Mn and 0.5% Cu, the grain size decreased from 72 to 15 μm. Grain refinement occurred due to an increase in the number of particle-stimulated nuclei, the number of which at minimal alloying rose from 3.47×10^{11} to 81.2×10^{11} with the maximum concentration of Mg, Mn, Cu additives. The retarding force of recrystallization, which in the original alloy was 1.57×10^{-3} N/m^2, increased to 5.49×10^{-3} N/m^2 at maximum alloying. The influence of copper was especially noticeable, the introduction of 0.5% increasing the retarding force of recrystallization by 2.39×10^{-3} N/m^2. This is due to the fact that copper has the most significant effect on the size and number of intermetallic particles. It was established that strength increase without ductility change occurs when magnesium, manganese and copper content increases.

Keywords: aluminum; recrystallization; microstructure-property characterization simulation and modeling; heat treatment and surface treatment

Citation: Aryshenskii, V.; Grechnikov, F.; Aryshenskii, E.; Erisov, Y.; Konovalov, S.; Tepterev, M.; Kuzin, A. Alloying Elements Effect on the Recrystallization Process in Magnesium-Rich Aluminum Alloy. *Materials* **2022**, *15*, 7062. https://doi.org/10.3390/ma15207062

Academic Editors: Pan Gong, Xin Wang, Maojun Li and Guangchao Han

Received: 29 August 2022
Accepted: 7 October 2022
Published: 11 October 2022

1. Introduction

Aluminum is one of the most commonly used metals in modern industry [1–12]. Aluminum alloys with a high magnesium content are very popular multifunctional alloys, and are used in the automotive industry, shipbuilding, aerospace engineering and the packaging industry [13–18]. The advantages of these alloys include strength, corrosion resistance, weldability and ductility, and the absence of yield plateau [19,20]. Attainment of these characteristics is possible only with the correct choice of parameters of thermomechanical treatment [21–23]. Recrystallization is the most important process that must be controlled by the correct choice of modes during thermomechanical treatment of these alloys, since in many respects the features of recrystallization behavior (or its absence) determine the properties of the aluminum alloys listed above [22,24,25]. In addition to parameters of the technological process, the mechanisms of recrystallization are strongly affected by the size and number of intermetallic particles. Finely dispersed particles (less than 1 μm in diameter) inhibit the recrystallization process, sometimes making it impossible due to their large amount [26,27]. Large particles play the role of nuclei in the recrystallized structure according to the mechanism described in the literature as Particle Stimulated Nucleation PSN [28,29].

A number of works have been devoted to the study of size and number of intermetallic particles and their effect on recrystallization in alloys containing magnesium. For example, in [30], alloys AA5182, AA5052 and AA5005 were compared. These alloys were alloyed in various combinations of Mg; Mn; Si; Fe; Cr; Cu and Ti. The number of PSN nuclei for each of the alloys was given, and it was found that the larger the PSN, the smaller the cubic texture, but the research did not show how specific alloying elements affected the number and size of intermetallic particles. In [31], recrystallization was studied on the basis of the 5083 alloy. This alloy contained Mn, Zr, Sc, Cr, Fe, Si and Mg, and the content of the first element remained constant. The effect of other elements on large intermetallic and finely dispersed particles was studied. The influence of the chemical composition on the amount and size of intermetallic compounds, as well as the decelerating force of recrystallization, was not evaluated. In [32], using AA5251 alloy as an example, the relationship between A1(Fe,Mn)Si, Mg_xSi intermetallics and the recrystallization process, as well as the reduced volume fraction occupied by particles of both phases, were shown. The relationship between intermetallic compounds and the number of nuclei and the retarding force of recrystallization was not described. In [33] a comparison of particles in alloys 5754 and 5182 containing Si, Fe, Cu, Mn, Mg, Cr, Zn and Ti was made, and the number of both large intermetallic and fine particles was described, dependent on microchemistry. However, the research considered the homogenized state and the relationship of these particles, but the recrystallization process was not studied. In [34], the amount number of intermetallic particles was also studied, but their effect on nucleation was not investigated. In [35], the effect of homogenization on the number of fine particles and on the inhibition of recrystallization was studied. Thus, despite a large number of studies devoted to this issue, they have been mainly devoted either to a homogenized state and the connection between particles paying little attention to recrystallization, or to the comparison of two different alloys with several different elements that change at the same time and may not completely coincide. This makes it difficult to analyze their effect on the number and size of intermetallic particles and on nucleation during recrystallization. Therefore, a study of how a gradual increase in the most popular elements in alloying of high-magnesium aluminum alloys affects the size and number of particles, and, consequently, the patterns and mechanisms of the recrystallization process, is an urgent and not fully explored issue. The purpose of this work was to study the influence of magnesium in the range of 4–5%, manganese in the range of 0.2–0.5%, and copper in the range of 0.1–0.5% on intermetallic particles and their influence on nucleation and inhibition of the recrystallization process during annealing of a hot-rolled aluminum alloy. These element concentrations were chosen because they are typical for high magnesium alloys.

2. Materials and Methods

An aluminum-magnesium alloy with 4% magnesium content was chosen as the base alloy. During the research, a gradual increase in magnesium (up to 5%), manganese (from 0.2 to 0.5%), copper (from 0.25 to 0.5%) and zirconium (up to 0.05%) was performed. The studied alloys' chemical composition is presented in Table 1. Ingots sized 35 × 200 × 300 mm^3 were cast in a metal mold. Experimental melting was performed in a graphite crucible in a medium frequency induction furnace; the molten metal weight was 4–5 kg and the cast ingot weight was 3 kg. The casting temperature was 720–740 °C, the crystallization rate was 2 °C per second, and the rate of cooling after casting was 1 °C per second. Prior to molten metal casting in the casting mold, the alloy was refined by carnallite flux, dosed in proportion of 5 g of flux per 1 kg of charge. After that the segregation layer was skimmed from metal surface. The solidified ingot was removed from the mold and water cooled. The following materials were used to produce the studied alloys: A85 grade prime aluminum, MG90 grade prime magnesium, M1 grade copper, Mn90Al10 grade alloying pellets.

Table 1. Chemical compositions of the investigated Al-Mg alloys.

	Content of Elements, %				
Alloy No.	Mg	Mn	Cu	Si	Fe
1	4	-	-	<0.05	<0.05
2	4.25	-	-	<0.05	<0.05
3	4.5	-	-	<0.05	<0.05
4	4.75	-	-	<0.05	<0.05
5	5	-	-	<0.05	<0.05
6	5	0.2	-	<0.05	<0.05
7	5	0.35	-	<0.05	<0.05
8	5	0.5	-	<0.05	<0.05
9	5	0.5	0.1	<0.05	<0.05
10	5	0.5	0.25	<0.05	<0.05
11	5	0.5	0.5	<0.05	<0.05

After casting, the ingots were milled on four sides to 30 × 180 × 250 mm^3 size (to prepare for rolling). After machining the ingots, 8-hour homogenization annealing was performed at 460–480 °C.

Thermo-mechanical modes were calculated for ingot hot rolling. The optimal ingot pre-heat temperature for rolling (450–500 °C) and reduction per pass, depending on the previous pass thickness, were also determined. The hot rolling schedule, enabling us to obtain the required structure, is presented in Table 2. It should be noted that impurities are were the level typical for aluminum alloys with a high magnesium content used in industry. The procedure for the thermomechanical treatment made in the study was: 1. Casting; 2. Homogenization and heating in laboratory furnace; 3. Hot rolling on the laboratory mill; 4. Annealing.

Table 2. Ingot hot rolling schedule.

Pass No.	Initial Thickness, mm	Final Thickness, mm	Thickness Reduction, %
1	30	27	10
2	27	24	11.1
3	24	21	12.5
4	21	19	9.5
5	19	17	10.5
6	17	15	11.8
7	15	13	13.3
8	13	12	7.7
9	12	11	8.3
10	11	10	9.1
11	10	8	20
12	8	6	25
13	6	4	33.3
14	4	3	25

Rolling was performed using a 300 (Dima Maschinen, Esslingen am Neckar, Germany) laboratory mill. A 30 mm thick workpiece was hot rolled to attain a 3 mm thickness (90% total deformation). The maximum force during hot rolling did not exceed 650 kN. NOASAR 8109 rolling oil was used to reduce force and friction during hot rolling.

After hot rolling, the samples were edge-trimmed. The samples were cut to a maximum 300 mm length. They were also marked and annealed at 360 °C for 3 h. The samples were annealed and cooled in a THERMCONCEPT KM 70/06/A laboratory furnace (Thermconcept, Bremen, Germany). After annealing, all samples were cold rolled from 3 mm to 0.3 mm thickness (with equal reductions per pass), which corresponds to 90% total deformation. Cold rolling was performed to obtain the required mechanical properties and target thickness. Rolling forces did not exceed 550 kN during cold rolling.

Mechanical properties of the cold-rolled samples were studied in the longitudinal direction after coating, curing, and pasteurization simulation. The samples were tested using a Zwick/Roell Z050 stretching machine (ZwickRoell, Denmark, Germany) in accordance with EN 541-2006 and EN 10002-1. Five samples per each chemical composition were taken in the longitudinal direction. The samples, ruptured outside the working area during the test, were not accounted for in the final result. The averaged test results are presented below. Mechanical tests results of ultimate yield strength were applied as the criterion for establishing the specific chemical element content during other component additions (Table 1).

Samples for microstructure, electrical conductivity and micro-hardness studies were taken at the end of each processing stage. Micro-hardness and electrical conductivity measurements were carried out because they are very sensitive to changes of the dispersoid fraction, as well as the concentration of the alloying elements in a supersaturated solid solution. It should be noted that electrical conductivity is often used in studies related to aluminum alloys microalloying [36–38].

The cross-sectional microstructure of samples was studied in polarized light after section electropolishing in a fluoroboric electrolyte (boric acid—11 g; hydrofluoric acid—30 mL; distilled water—2200 mL). The structure was studied after etching using an Axiovert 40 MAT microscope (Carl Zeiss AG, Oberkochen, Germany). Electrical conductivity was measured using a portable VE-17NTs (LLC NPP "Sigma", Nalchik, Russia). The excitation current frequency was 100 kHz.

Based on optical microscopy data for samples annealed after hot rolling, the average grain size was determined by the secant method. This was determined only for fully recrystallized samples when magnified to capture 80–200 grains. To determine the average grain size, at least eight secants were chosen in eight areas. An eyepiece-micrometer ruler was used as a secant. Positioning the ruler across the direction of deformation and at an angle of 45°, the number of grains (n) intersected by this secant was counted. The average grain size was determined by:

$$D_{cp} = \frac{L \cdot k}{\sum n}, \tag{1}$$

where: L is the length of the eyepiece-micrometer ruler, mm; k is the number of secants, and n is the sum of grains intersected by all secants.

In addition to optical microscopy, some samples were studied using automatic analysis of electron backscattered diffraction (EBSD) patterns using a TESCAN VEGA LMH scanning electron microscope with a LaB6 cathode (SEM) equipped with an Oxford Instruments Advanced AZtecEnergy X-ray energy-dispersive microanalysis system and an Oxford Instruments NordLysMax2 EBSD attachment, using AZtec version 2.2 software. For analysis of grain (sub-grain) boundary misorientation, maps were built from areas 690 × 400 μm in size with a scanning step of 0.5 μm. For data collection, a 2 × 2 binning mode was used, providing a high level of structural detail.

Data for analysis were obtained with an indexation coefficient of 95% or more. The error in determining the orientation of the crystal lattice (the average angular deviation between the detected and simulated Kikuchi bands) was no more than ±0.5°. Boundaries with misorientations less 15° were taken as low-angle boundaries. Boundaries with misorientations of more than 15° were considered as high-angle boundaries. When high-angle boundaries were determined, individual grains were defined. For each of them, the misorientation between neighboring point was calculated. If the average misorientation between points was less than 2°, the grain was considered recrystallized and marked as blue. If the average misorientation between points was less than 2°, but subgrain misorentation was also less than 2°, this grain colored yellow and considered as substructed. If the average misorientation between points was more than 2°, the grain was considered deformed and marked in red.

Studies of alloys with additions of 0.5% manganese and 0.5% copper were carried out using a Tecnai G2 F20 S-TWIN TMP transmission electron microscope with a thermal field cathode at an accelerating voltage of 200 kV. A study of the chemical composition of structural components was carried out by energy dispersive spectroscopy (EDS) using an X-Max 80T detector in the energy range 0–10 keV. The energy resolution of the detector was 122 eV.

Micro-hardness was measured to indirectly assess the alloying effect on strength properties. Micro-hardness tests were carried out with a Wolpert 402MVD hardness tester (Wolpert, Bretzfeld, Germany) using the Micro-Vickers method in accordance with GOST 9450-76. The measurements were taken only for hot-rolled samples and the samples after recrystallization annealing since it was not possible to measure the micro-hardness of thin cold-rolled samples. Each sample was measured at five points spaced 0.01 mm (10 measurements at each point) with a 25 gf load (values deviating from the mean by more than five units were not accounted for). Arithmetic means are used in the graphs and in the discussion.

To assess the effect of the alloying element content on the change in the size and number of large and fine particles, a study was carried out using a JEOL 6390A SEM scanning electron microscope (Akishima, Tokyo, Japan). Hot-rolled samples were used for the study. High-purity samples were preliminarily polished. The samples were examined at 300 (intermetallic compounds) and 10,000 magnifications (finely dispersed intermetallic compounds) after polishing. The content of chemical elements in the solid solution and large intermetallic compounds was determined using an EDAX energy-dispersive X-ray microanalyzer attachment.

Although the detected particles had different morphologies, for convenience, their average radius was determined. For this, the area of all objects in the study area was found. Then, the assumption was made that all objects had the shape of a circle. After that, the radius of each of them was found and then the average radius estimated. Taking into account the average radius, the total volume of these particles was found. The volume fraction of particles (F_v) was calculated as the ratio of the volume of particles (equal to the number of particles on each side of the cube, multiplied by the average volume of these particles) to the volume in which they were measured.

The different types of recrystallization nuclei were estimated using the well-established method for aluminum alloys described in [9]. Data on the number of intermetallic particles were taken directly from the results of this study. Subgrain size data and their dependence on the Hollomon-Zener parameter were taken from [7] for 5182 alloys (4.8% Mg; 0.37% Mn; 0.15% Zn; 0.2% Ti; 0.060% Cu; 0.4% Si; 0.01% Fe), their chemical compositions being close to the investigated alloys. These data could be used because chemical composition change does not have a significant effect on magnesium-rich aluminum alloy subgrain size [39].

The retarding force of recrystallization P_z was estimated using Formula (2) according to [40].

$$P_Z = \frac{3\gamma_B F_V}{r_D} \tag{2}$$

where F_V is the volume fraction of particles of the second phase, and r_D is the average particle size.

The number of recrystallization nuclei depending on the Holomon-Zener parameter, including those formed by the PSN mechanism, was found according to the models proposed in [41,42].

3. Results and Discussion

3.1. Grain Structure

The grain size after recrystallization annealing decreased with increased magnesium content (Figure 1). This occurred because of two factors. The first was the development of an additional number of nuclei representing large Mg_2Si-type intermetallic particles [28]. In addition, an increased number of nuclei was observed due to size decrease and, consequently, subgrain number increase, occurring inevitably with an aluminum alloying level increase [39]. Second, additional structure refinement occurred during crystallization. It should be also noted that a hot-rolled magnesium-rich alloy has a more elongated grain structure. In general, the grain size decreased with the addition of magnesium from 72 to 32 μm.

Manganese also contributes to grain refinement. During recrystallization, modification occurs due to increase of $Al_6(FeMn)$ nuclei number. Addition of up to 0.5% manganese increased the total amount of intermetallic particles by more than 20% and reduced the size of the grain structure to 25 μm.

In the case of copper addition, grain refinement during recrystallization is facilitated by development of $CuMg_4Al6$-type particles [43]. Addition of 0.5% copper led to an increase of large intermetallic particles. Addition of 0.5% copper increased the total number of particles by seven times, and the grain size, in its turn, decreased to 15 μm. At the same time, a significant increase in the amount of finely dispersed intermetallic particles resulted in partial blocking of the recrystallization process.

It should be noted that in all the studied alloys after hot rolling, structural deformations were observed, as in [44–47]. However, in alloy No. 1, a recrystallized structure was observed. This can be explained by recrystallization, which occurs during cooling of a hot rolled workpiece. Recrystallization occurred in alloys No. 1 because it did not contain particles that inhibit the process. Grain growth during next annealing process can be explained by secondary recrystallization, which is also caused by the absence of fine particles [48].

The results of EBSD in Figure 2 show that in the hot-rolled state, a structure consisting of deformed grains elongated in the direction of rolling was observed. It should be noted that, according to optical microscopy (Figure), the grain structure for this alloying was stretched in the rolling direction up to 500 μm, so the grain could not be fully represented. The grains consisted mainly of sections, and subgrains having a high dislocation density and a polygonized structure with an average size of 2 μm were formed within them, which is in good agreement with the data presented in [39]. The presence of small, well-formed subgrains, compared to low-alloyed aluminum alloys, can be explained by a decrease of stacking fault energies in high-magnesium alloys [49], which makes dislocations less mobile, and dislocations annihilate less quickly. In addition to areas with well-formed subgrains, there were areas with recrystallization nuclei. Their presence is explained by the fact that, despite the inhibition of recrystallization by finely dispersed particles and the fairly rapid cooling of the laboratory ingot in air, recrystallization nuclei had time for formation. Unfortunately, the method of obtaining an EBSD image did not make it possible to identify whether these nuclei were formed from subgrains or from particles. However, the presence of recrystallized areas indicated precisely their formation in the latter. The pole figure and the inverse pole (Figure 3c,d) showed classical rolling texture patterns [50]. At the same time, the set of crystallites for analysis in the hot-rolled state was very limited in the EBSD method. As result this could not provide the general pattern of texture distribution. Therefore, for a more detailed study of the texture, distribution-ray diffraction analysis is required, but was beyond the scope of this work.

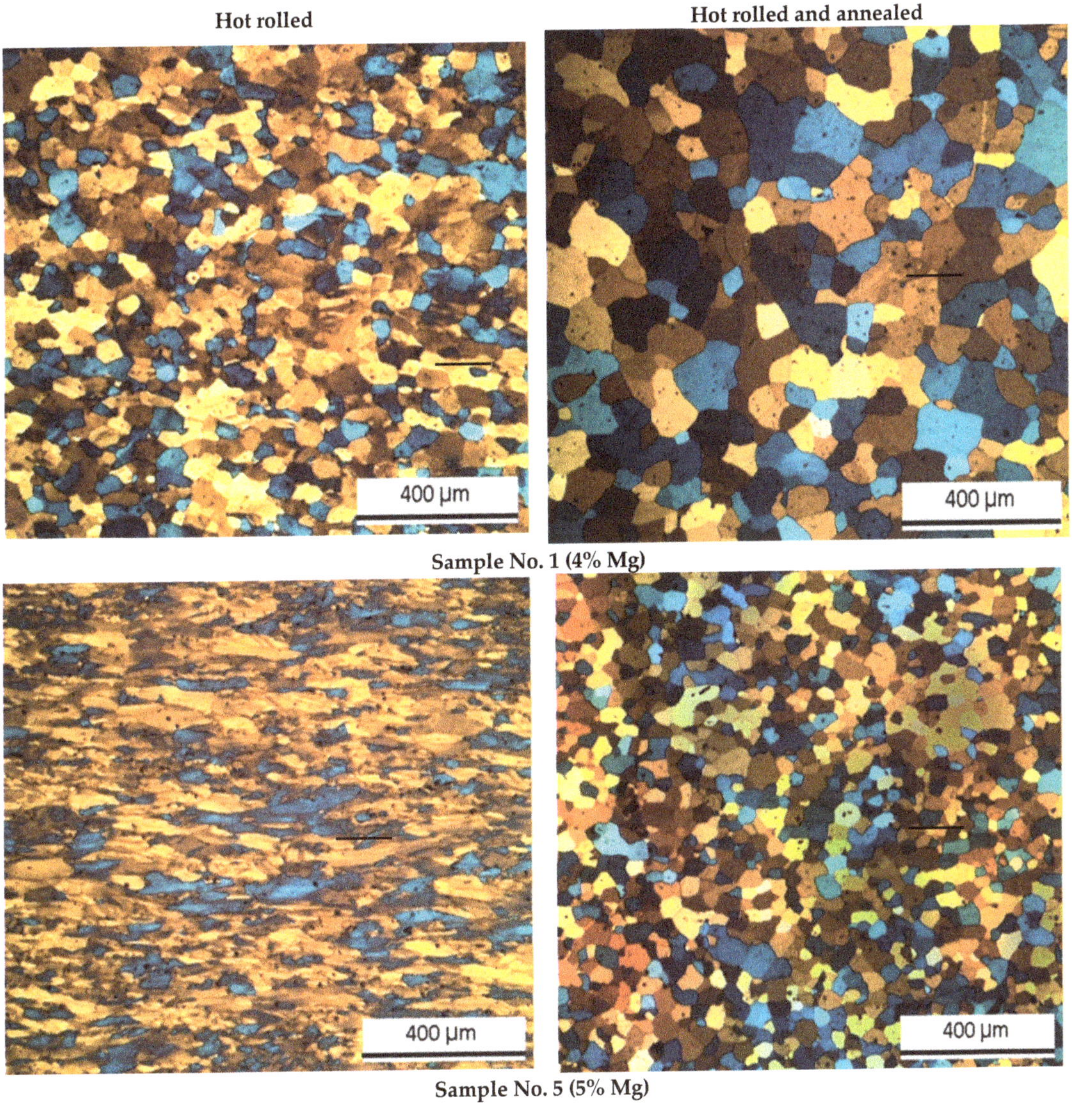

Figure 1. *Cont.*

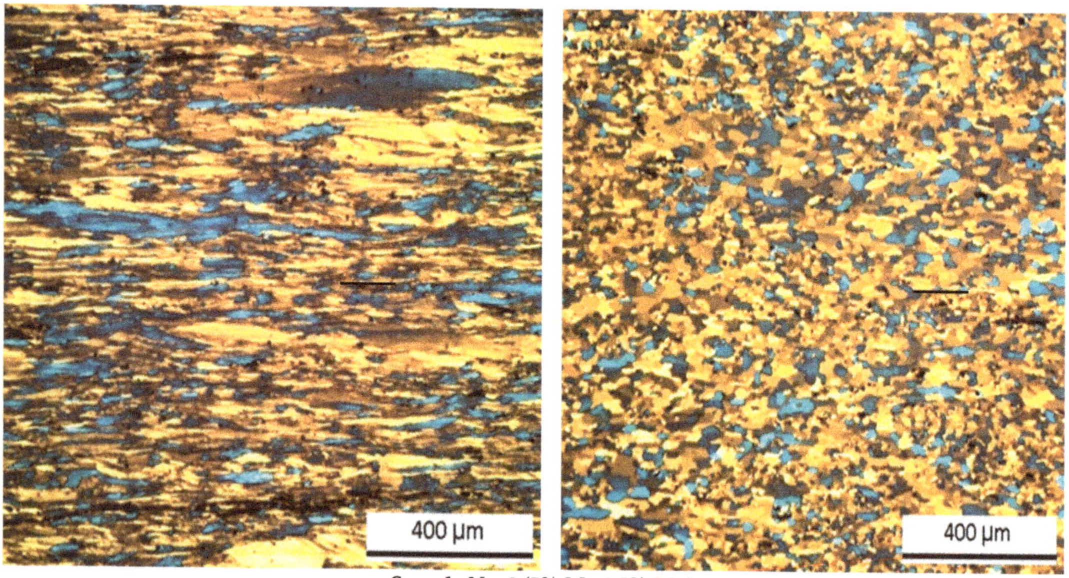

Sample No. 8 (5% Mg 0.5% Mn)

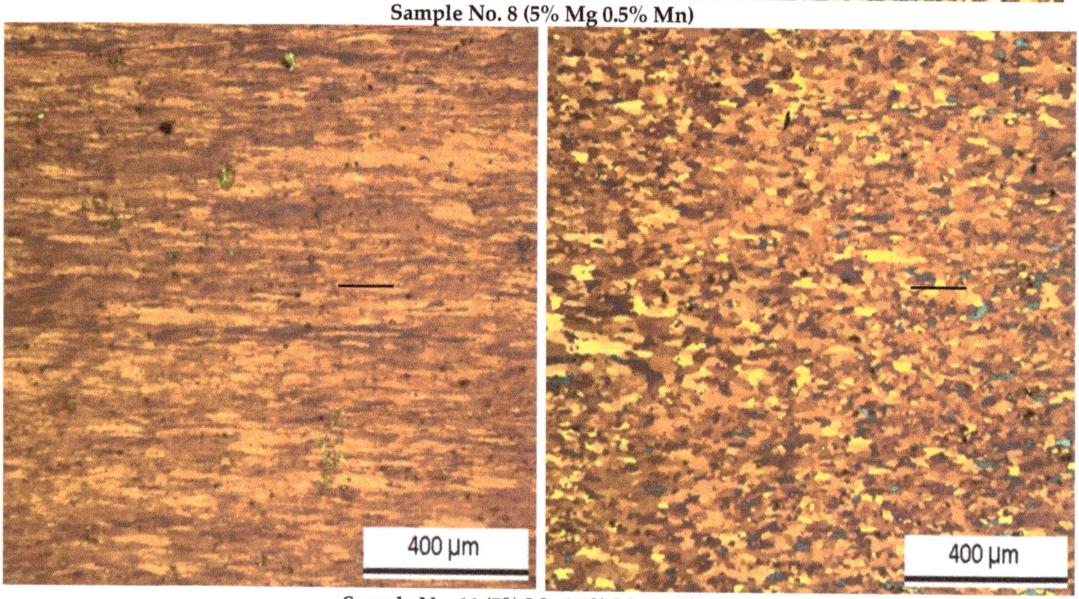

Sample No. 11 (5% Mg 0,5% Mn 0.5% Cu)

Figure 1. Grain structure of hot-rolled and annealed samples related to alloy chemical composition.

The EBSD data for the 5Mg0.5Mn0.5Cu alloy in the annealed state in Figure 3 shows the presence of small-sized grains of 30 µm in which low-angle boundaries were not found. This, as well as the fact that the grains changed their shape and size, indicates that recrystallization was completed. Small inclusions of a non-recrystallized structure were observed. This suggests that recrystallization in these volumes was blocked by the action of finely dispersed intermetallic particles. In general, the EBSD data were in good agreement with the optical microscopy data. Based on analysis of inverse pole figures and pole figures, pronounced texture components were absent. This suggests that PNS is the dominant nucleation mechanism, since grains which grow during recrystallization from this nuclei type do not have of any texture components [28].

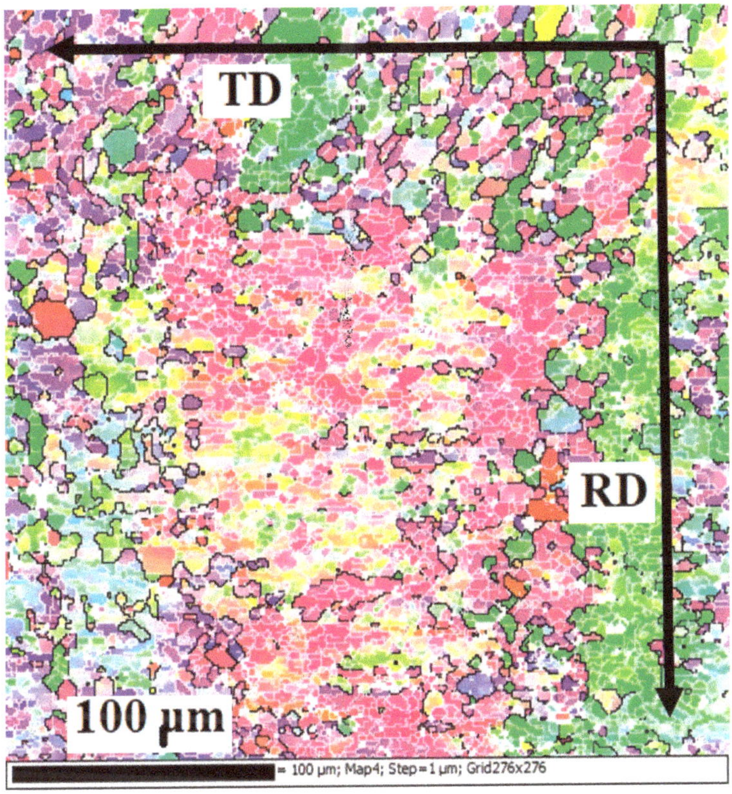

(**a**)

Figure 2. *Cont.*

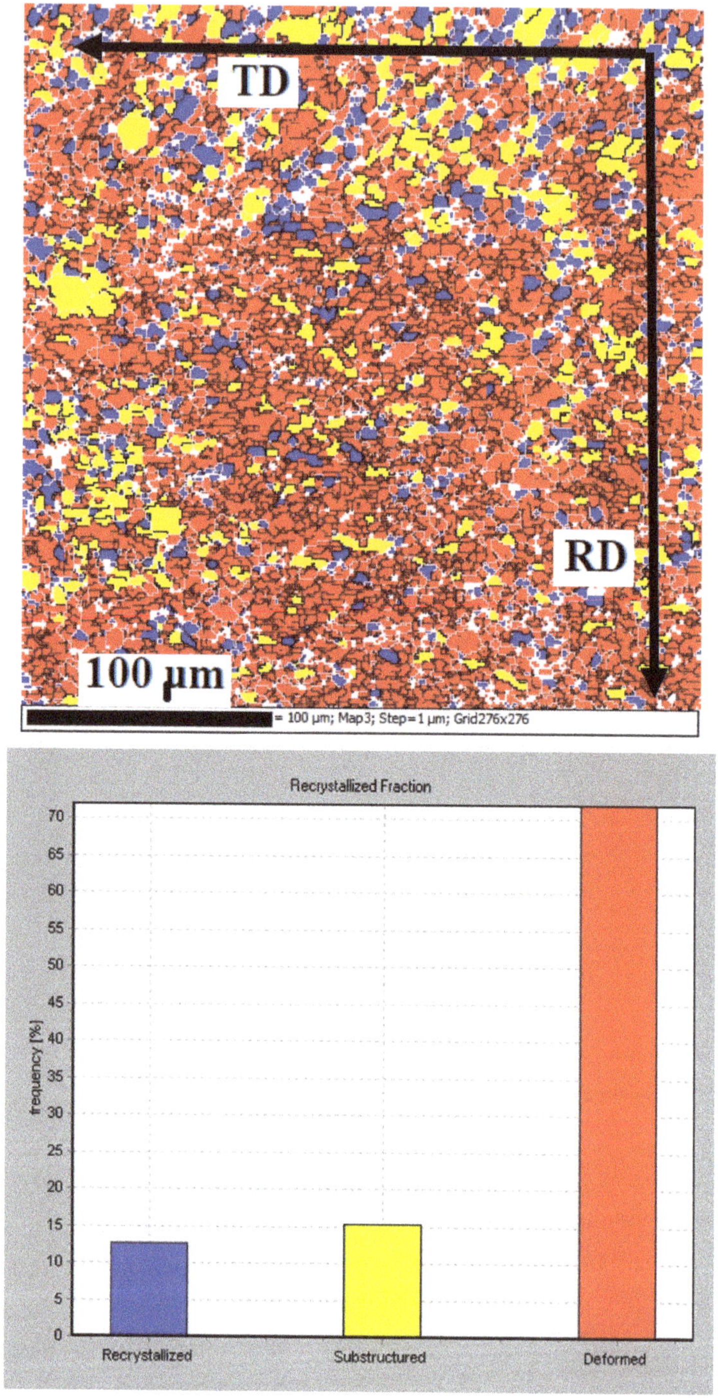

(b)

Figure 2. *Cont.*

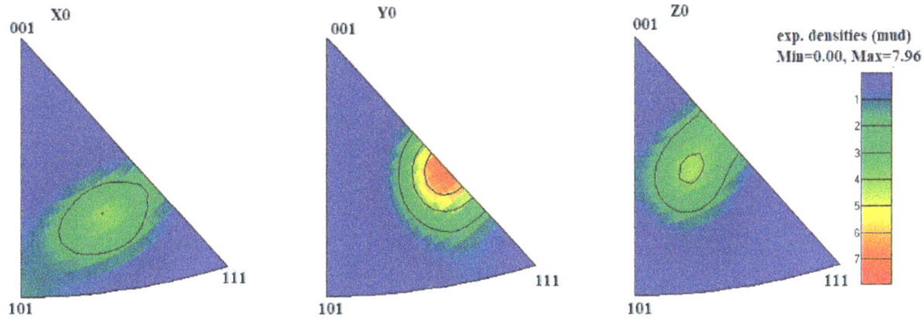

(**c**)

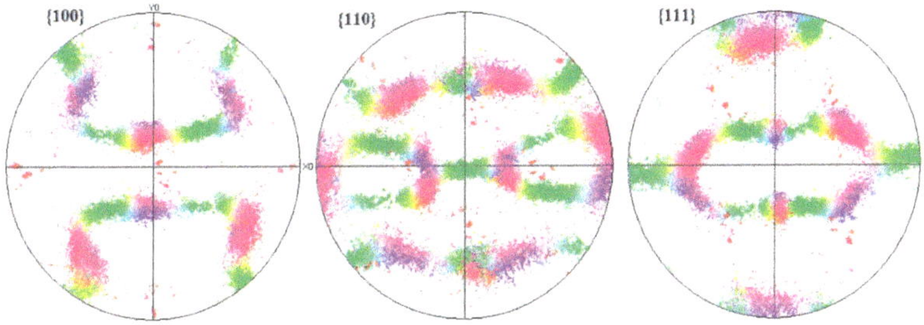

(**d**)

Figure 2. EBSD results for 5Mg0.5Mn0.5Cu alloy after hot-rolling: (**a**) subgrain structure; (**b**) fraction mapping; (**c**) inverse pole figure (**d**) pole figures.

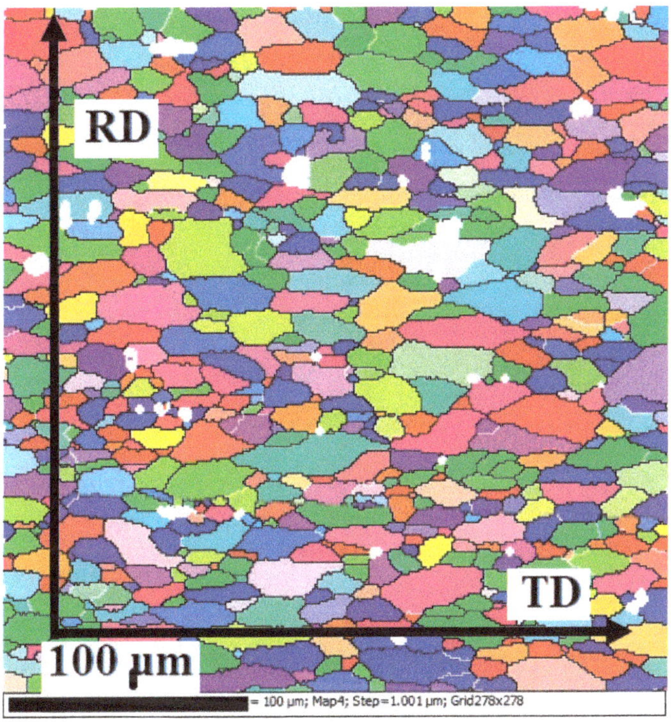

(**a**)

Figure 3. *Cont.*

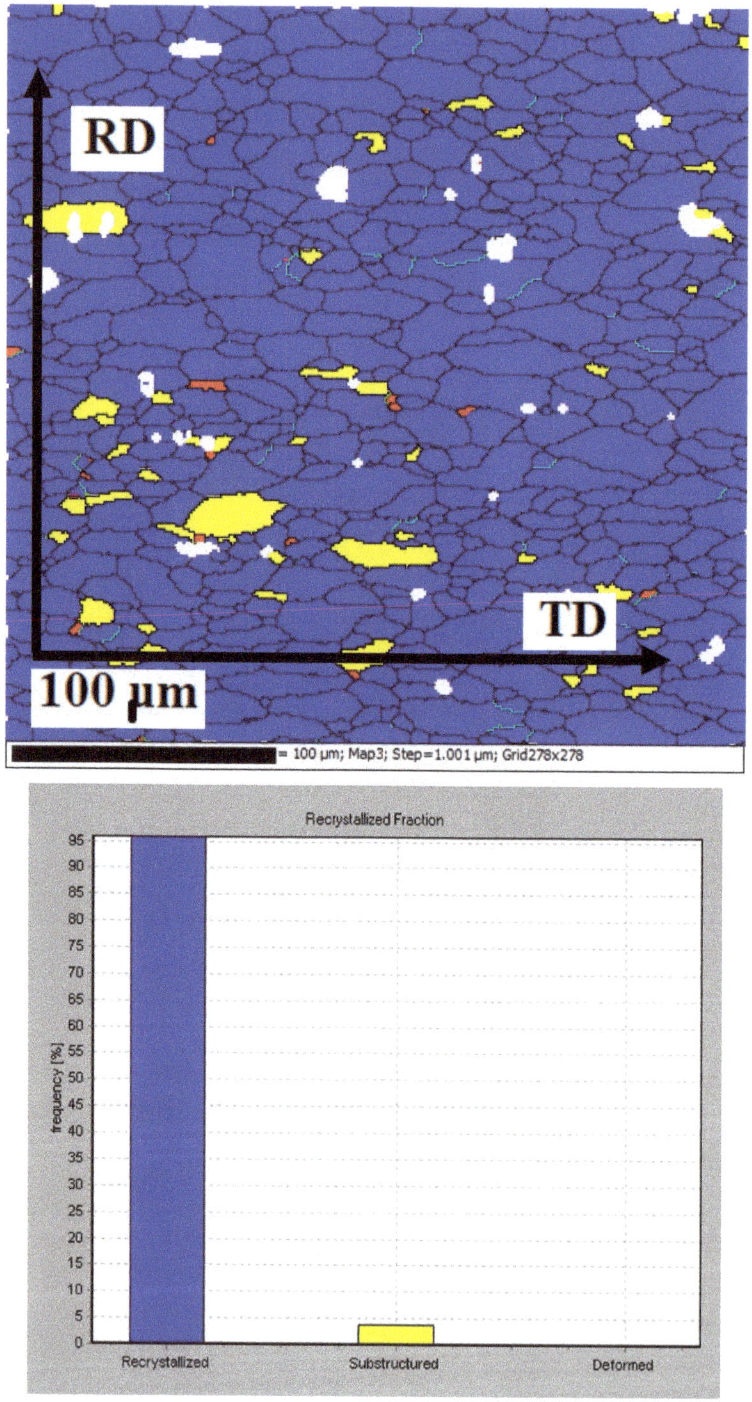

(**b**)

Figure 3. *Cont.*

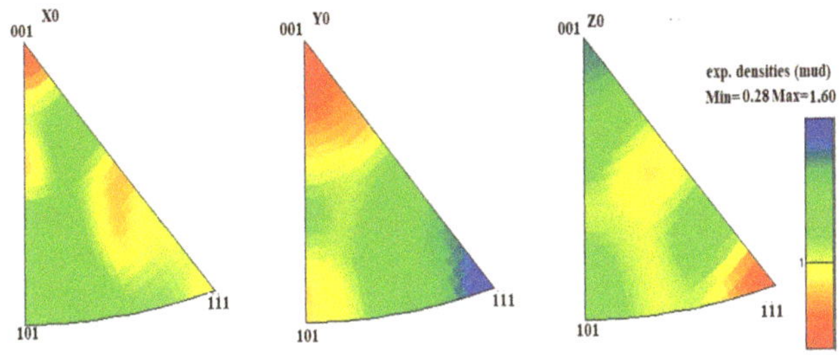

(c)

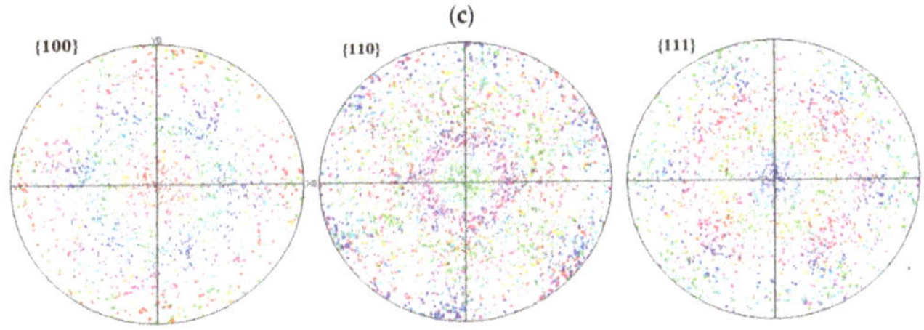

(d)

Figure 3. EBSD results for5Mg0.5Mn0.5Cu alloy after annealing: (**a**) subgrain structure; (**b**) fraction mapping; (**c**) inverse pole figure; (**d**) pole figure.

3.2. *Micro-Hardness and Electrical Conductivity*

Measured micro-hardness and electrical conductivity results are presented in Figures 4 and 5. A magnesium content increase from 4 to 5% resulted in micro-hardness increases from 58.72 to 86.52 HV and 52.92 from to 59.22 HV in the hot-rolled state and after annealing, respectively. Half a percent of manganese led to micro-hardness increases from 86.52 to 97.52 HV and from 59.22 to 63.72 HV in the hot-rolled state and after annealing, respectively. Copper led to micro-hardness growth from 97.52 to 114.42 HV, and after annealing and in the condition of hot-rolling it remained almost the same.

The greatest increase in hardness and, consequently, solid solution hardening, occurred when magnesium content increased. Manganese and copper did not have such a strong effect on this factor. The main mechanism of increase in micro-hardness with the addition of these elements was the presence of fine particles.

The micro-hardness difference in the hot-rolled and annealed states can be explained by the subgrain structure that was observed in the hot-rolled state. Subgrain structure in the annealed state would lead to a large number of dislocations, increasing micro-hardness [51]. Moreover, an increased amount of alloying elements, especially magnesium, always facilitates subgrain structure refinement [39]. Dislocation density, observed after annealing, is much lower [52] and reduces micro-hardness.

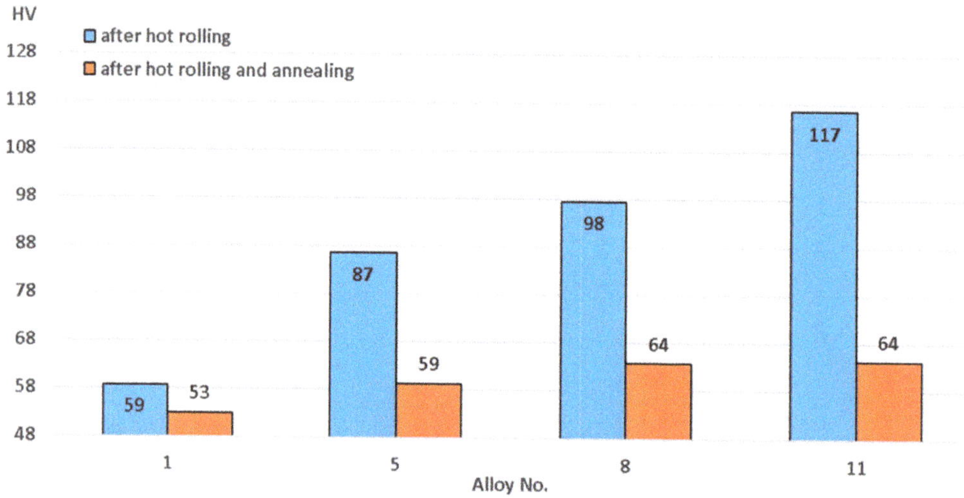

Figure 4. Changes in micro-hardness related to alloy chemical composition.

Electrical conductivity decreased in samples when the content of alloying components increased, regardless of the state (Figure 5). This occurred due to a large amount of alloying components melting in the supersaturated solid solution. In addition, during hot rolling, fine particles precipitate from the supersaturated solid solution. Therefore, subsequent annealing does not make much difference. An increase of magnesium content from 4 to 5% led to an electrical conductivity decrease from 20.4 to 18.3 μS. Adding 0.5% manganese reduced the electrical conductivity from 18.3 to 15.4 μS. Introduction of 0.5% copper led to an electrical conductivity decrease from 15.4 to14.8 μS.

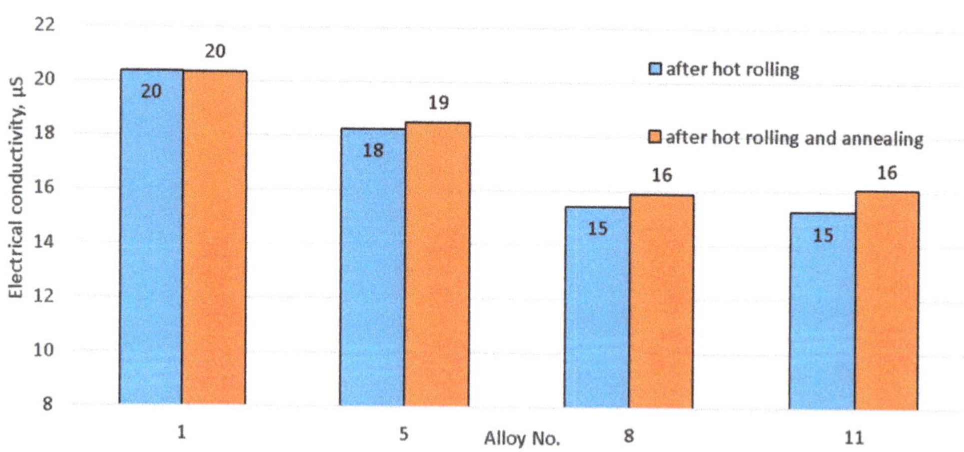

Figure 5. Change of electrical conductivity related to alloy chemical composition.

3.3. *Coarse Intermetallic Particles Study Using Scanning Microscopy*

Coarse particle (intermetallic compounds) parameters measured using electron microscopy at the stage of hot rolling are shown in Figure 6. Magnesium and manganese did not have a large impact on intermetallic compound size and quantity. With the addition of 0.5% copper, the number of large intermetallic particles increased sevenfold, and their average radius decreased from 0.76 to 0.46 μm. The abovementioned changes occurred after copper was added and, as is shown below, this was caused by the development of a large amount of smaller $Al_6Cu_2Mn_3$ intermetallic compounds. These particles formed due to copper poor solubility in aluminum matrix at the temperatures below 350 °C [53].

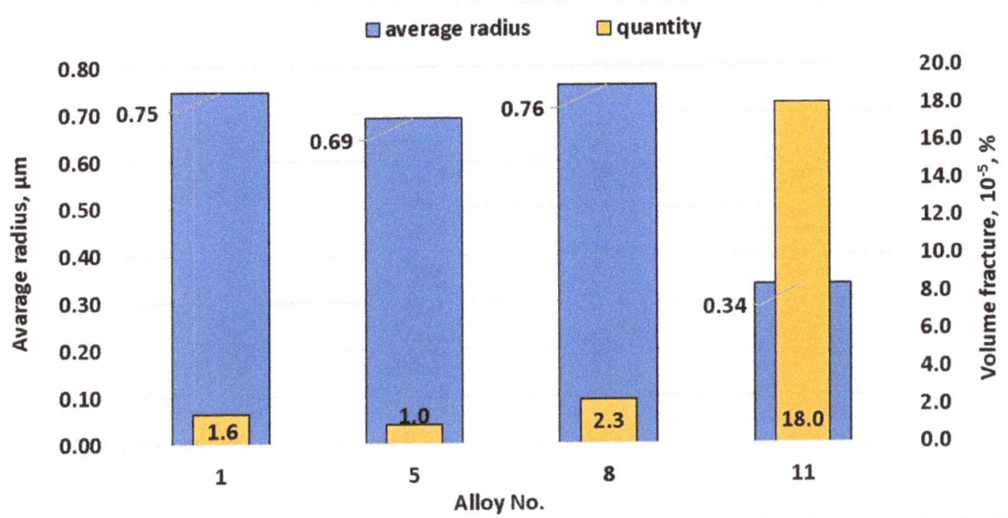

Figure 6. Changes in intermetallic compound size and number related to alloy chemical composition.

The chemical compositions of intermetallic compounds are shown in Figure 7. The percentage of elements in each spectrum are shown in Table 3. Elements such as Fe and Si were identified in all samples. These elements were not added to the alloy and were present as impurities contained in the aluminum ligature.

The main particles in magnesium alloys consisted of Mg_2Si. They had an elongated shape and average size of about 5–10 µm (Figure 8). Based on the presence of magnesium and silicon in them, it can be assumed with a high degree of probability that these particles were, indeed, Mg_2Si [53,54], and they often appear in high-magnesium alloys [33,55]. These particles are primary intermetallic compounds that "survived" the process of homogenization. EDX analysis showed the presence of aluminum in these alloys, which, however, was a consequence of the influence of the solid solution involved in this research method, as, for example, in [33,54,56]. The supersaturated solid solution contained magnesium, which explains strengthening of the solid solution (Figure 4) and the decrease of electrical conductivity (Figure 5).

When manganese was added, particles appeared which, in terms of the content of chemical elements, appeared to be Al_6FeMn, which is common in these types of alloys [34,57,58]. They had a 10–15 microns size and elongated shape. Al_8Mg_5 type particles, typical for magnesium-rich aluminum alloys, were found in [33,35,59]. The absence of such phases in alloys 1–5 can be explained by the limited area where large intermetallic particles chemical analysis was performed. In addition, Mg_2Si and Al_3Si particles are similar in color, making it difficult to select zones where both types of particles are present.

$Al_6(FeMn)$ type particles appeared in alloys 6–8 due to manganese addition. As can be seen in Figure 7, the solid solution also contained manganese.

Intermetallic compounds can be identified by chemical composition as phases such as Mg_2Si and $Al_6(FeMn)$. There was a phase close to $Al_{20}Cu_2Mn_3$ observed in alloys with joint manganese-copper alloying [60,61]. Average sizes of intermetallic compounds were 3–5, 5–8 and 10–15 µm, respectively. Mg_2Si particles were represented by elongated dark inclusions, $Al_6(FeMn)$ particles were light and had elongated shape, and $Al_{20}Cu_2Mn_3$ particles were light square-shaped particles. Manganese, copper and magnesium were also observed in a supersaturated solid solution.

Note that Fe and Si impurities, despite their rather low content, play an important role by participating in the formation of such particles as Mg_2Si and $Al_6(FeMn)$. Therefore, our data are relevant for alloys with the usual level of impurity elements. Additional research on the relationship between intermetallic particles and the recrystallization process in alloys made from high purity master alloys may be needed to account for the effect of Si and Fe reduction.

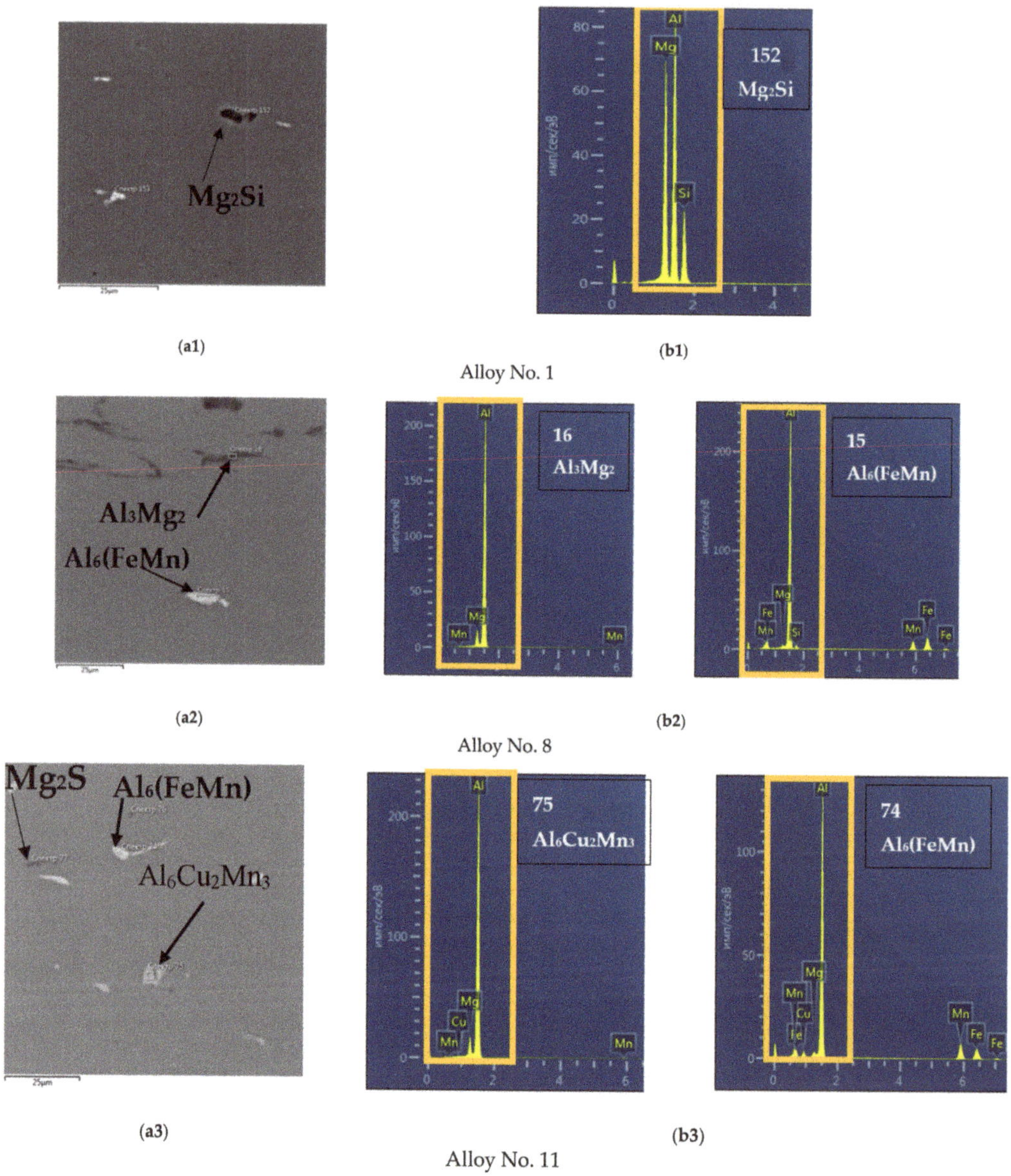

Figure 7. The types of intermetallic compounds (**a1–a3**); result of the EDS analyses (**b1–b3**).

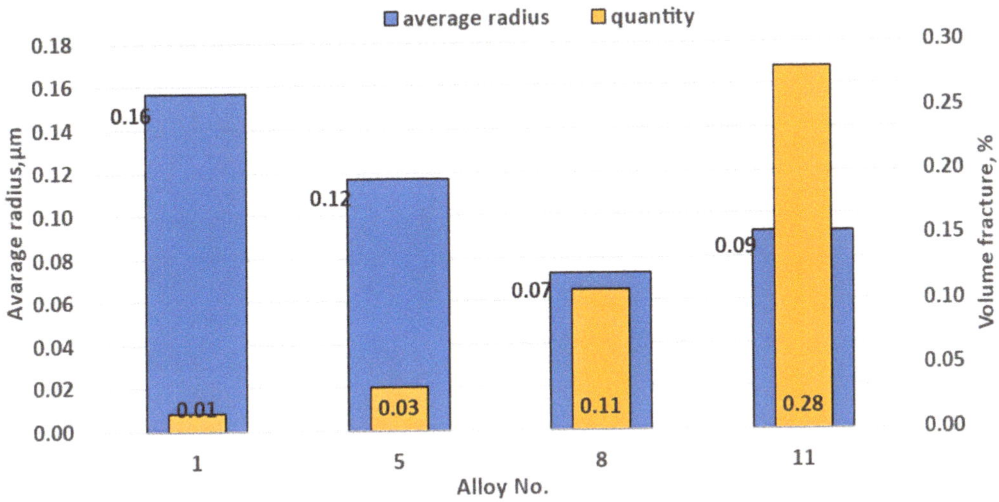

Figure 8. Change of dispersoid size and number depending on alloy chemical composition.

Table 3. Content of elements in particles.

Spectrum	Particle	Elements Content, %					
		Al	Mg	Mn	Cu	Fe	Si
15	$Al_6(FeMn)$	69.23	0.64	10.6	-	17.55	1.98
16	Al_3Mg_2	67.50	32.29	-	-	-	0.22
74	$Al_6(FeMn)$	68.09	1.08	15.73	3.93	11.17	-
75	$Al_6Cu_2Mn_3$	75.10	0.65	12.01	0.52	11.72	-
152	Mg_2Si	51.73	25.94	-	-	-	22.33

3.4. Fine Particle Studies using Scanning and Transmission Microscopy

Result of fine particle electron microscopy investigations are shown in Figure 8. After magnesium content reached 5%, the size of the dispersoids decreased from 0.16 to 0.12 µm, and their number increased three times. Addition of 0.5% manganese led to reduction of their radius from 0.12 to 0.07 µm, and their quantity increased more than 3.66 times. Addition of 0.5% copper increased manganese dispersoid radius from 0.07 to 0.1 µm, the number grew more than three times. As a result, copper had most prominent impact on recrystallization during hot rolling.

It should be noted that very few dispersoids are present in alloys containing magnesium only. They were detected by SEM (Figure 9), where it was possible to study larger areas, but this method does not allow their chemical composition to be analyzed. However, it should be noted that such a small number of particles has practically no effect on the recrystallization process, so there is no need for the laborious process of detection using transmission microscopy. As the proportion of manganese increased, the number of finely dispersed particles increased also (Figure 9). At the same time, manganese was the main element found in these particles. In addition, there were particles containing both manganese and iron, which from the chemical composition consisted of Al_6Mn and Al_6FeMn [62,63] (Figure 10). The presence of magnesium in the EDS results can be explained by solid solution impact. Although the particle number increased, and the size decreased, these particles did not have a decisive influence on the course of the recrystallization process.

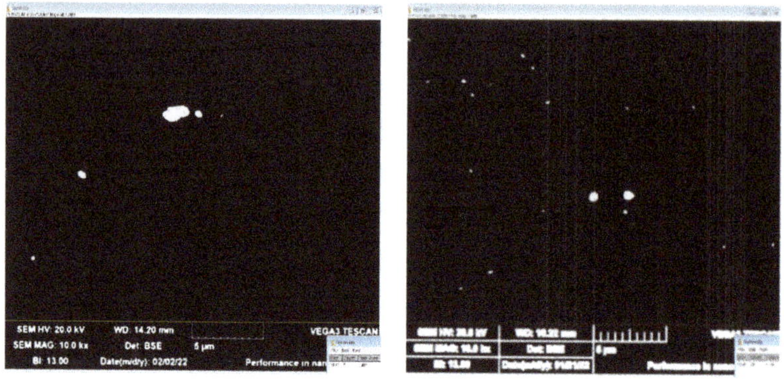

(**a**) (**b**)

Figure 9. Scanning electronic microscopy of (**a**) 5Mg and (**b**) 5Mg0.5Mn alloys.

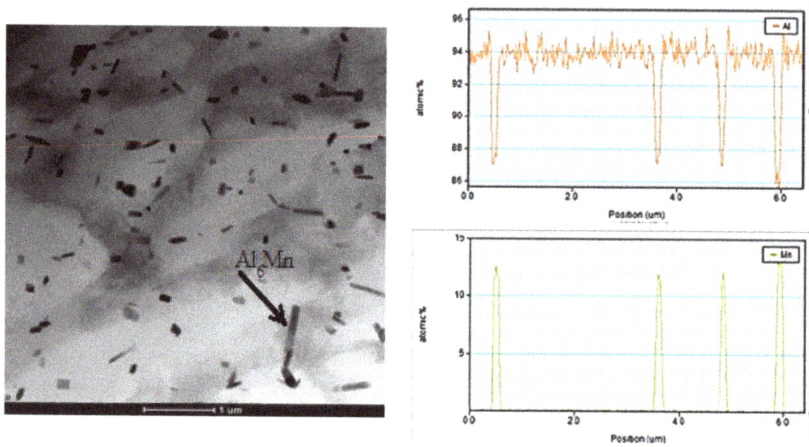

(**a**) (**b**)

Figure 10. *Cont.*

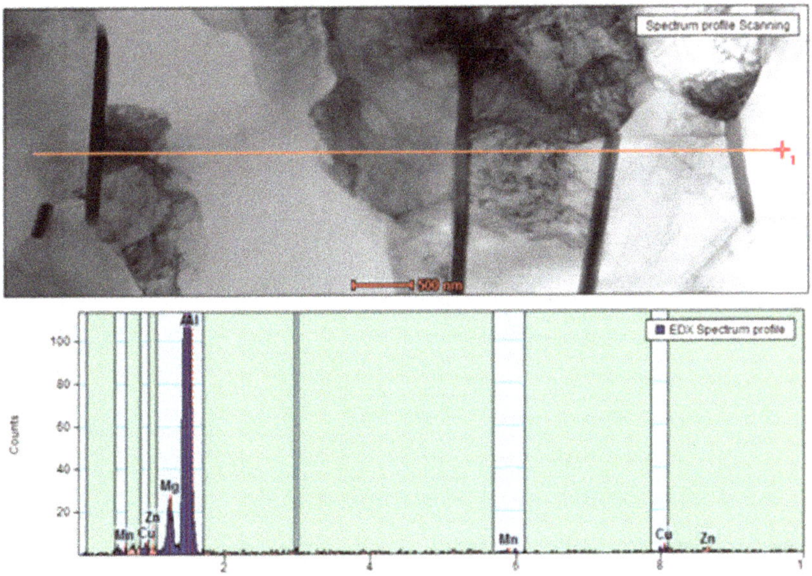

(c)

Figure 10. Transmission electronic microscopy of the 5Mg0.5Mn alloy (**a**); EDS results for the 5Mg0.5Mn alloy (**b**,**c**).

TEM analysis made it possible to reveal particles in the form of plates close in composition to the S Al_2CuMg phase, since the Cu:Mg ratio in them was close to 1:1 [62,63], and the T of which did not exceed 300 nm. In addition, there were plates that could be described as the T($Al_{20}Cu_2Mn_3$) phase, since the Mn:Cu ratio was close to 3:2 [64,65] (Figure 11). The size of this phase was not more than 100 nm. It should be noted that these phases are the main blockers of the recrystallization process. At the same time, as copper content grew, the number of fine particles increased, and the retarding force of recrystallization also increased. It should be noted that although the S phase in the form of S″ can be strengthening, in this case, as in the T phase, strengthening occurred only due to the inhibition of recrystallization. It is necessary to carry out hardening and subsequent artificial aging at lower temperatures to obtain S″. The particles close to S and T observed in this case were most likely formed during the decomposition of a supersaturated solid solution, which occurred during hot rolling and annealing at temperatures of 360–500 °C; therefore, by their nature, they should be close to equilibrium. In this way, with copper addition, new particles formed, so that the total number significantly increased as confirmed by the SEM result (Figure 12).

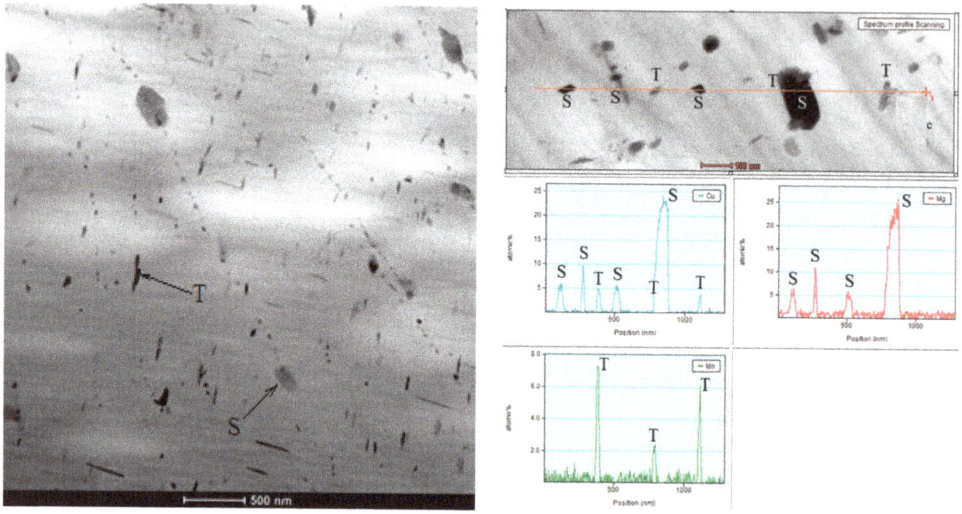

(**a**) (**b**)

Figure 11. Transmission electronic microscopy of the 5Mg0.5Mn0.5Cu alloy (**a**). EDS results for the 5Mg0.5Mn0.5Cu alloy (**b**).

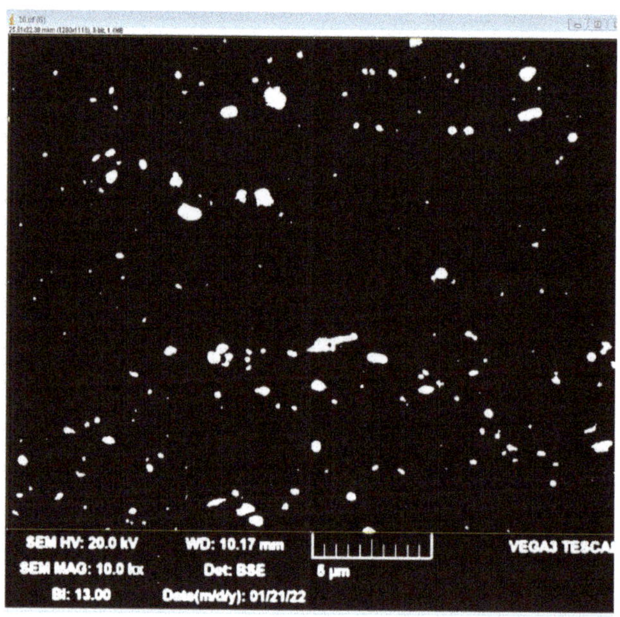

Figure 12. Scanning electronic microscopy of the 5Mg0.5Mn0.5Cu alloy.

3.5. Calculations of the Effect of the Size and Number of Intermetallic Particles on the Inhibition Force and Nucleation during Recrystallization

Figure 13 shows the effect of chemical composition on the number of nuclei for recrystallization which proceeds based on the particle stimulated nucleation (PSN) mechanism. When the amount of alloying elements was increased, a slight increase in PSN nuclei number (until copper was added to the alloy) was observed. At the same time, the proportion of PSN nuclei of the total nuclei number did not change.

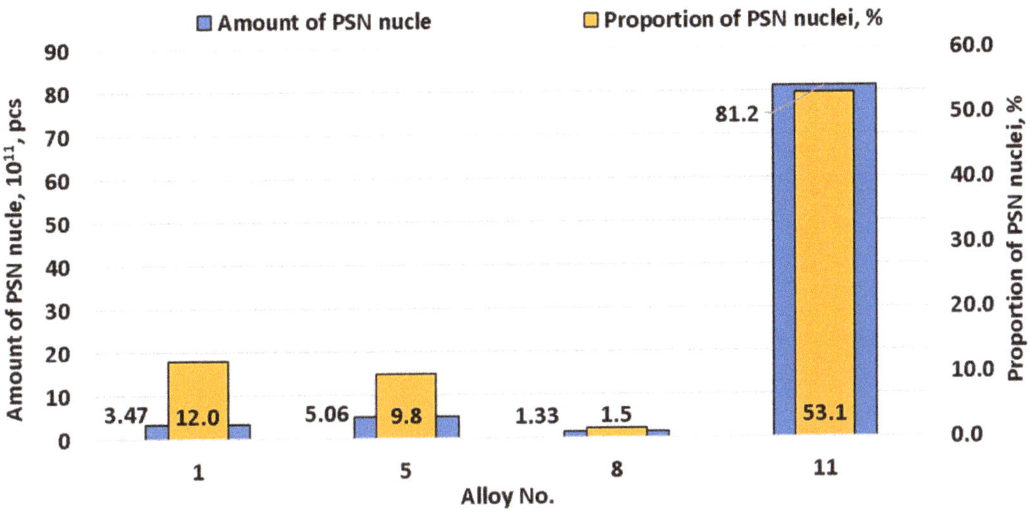

Figure 13. Change of PSN nuclei number depending on alloy chemical composition.

Figure 14 shows the effect of chemical composition and the Zener-Hollomon parameter on the number of nuclei formed from PSN or from subgrains. Subgrains are the predominant source of nucleation at low Zener-Hollomon parameter values. However, as Zener-Hollomon parameter increased, the subgrain size rapidly decreased and particle-based nucleation becomes predominant. Particle-stimulated nucleation prevailing over subgrain-based nucleation is generally typical for magnesium-rich aluminum alloys. This is caused, primarily, not by the large size and number of second-phase particles, but by small-sized subgrains in this type of alloy [30,66]. This results in intense crystallographic hardening and lower stacking fault energy compared to other alloys [67]. The ratio of subgrain-based and second-phase particles-based nuclei numbers remained almost unchanged before the addition of copper to the alloy. This was associated with an insignificant increase of intermetallic compound size and number. However, their number spiked after copper was added. As a result of this chemical composition, second-phase particles became the main nuclei sources during recrystallization. Thus, in terms of recrystallization, the studied alloy nuclei development mechanism was close to that of the 1565ch alloy, where intermetallic particle-based nucleation prevails over subgrain-based nucleation [28]. It must be mentioned that magnesium and manganese did not have a pronounced impact on PSN nuclei, but the number of nuclei from intermetallic compounds grew 4.4 times after copper was added.

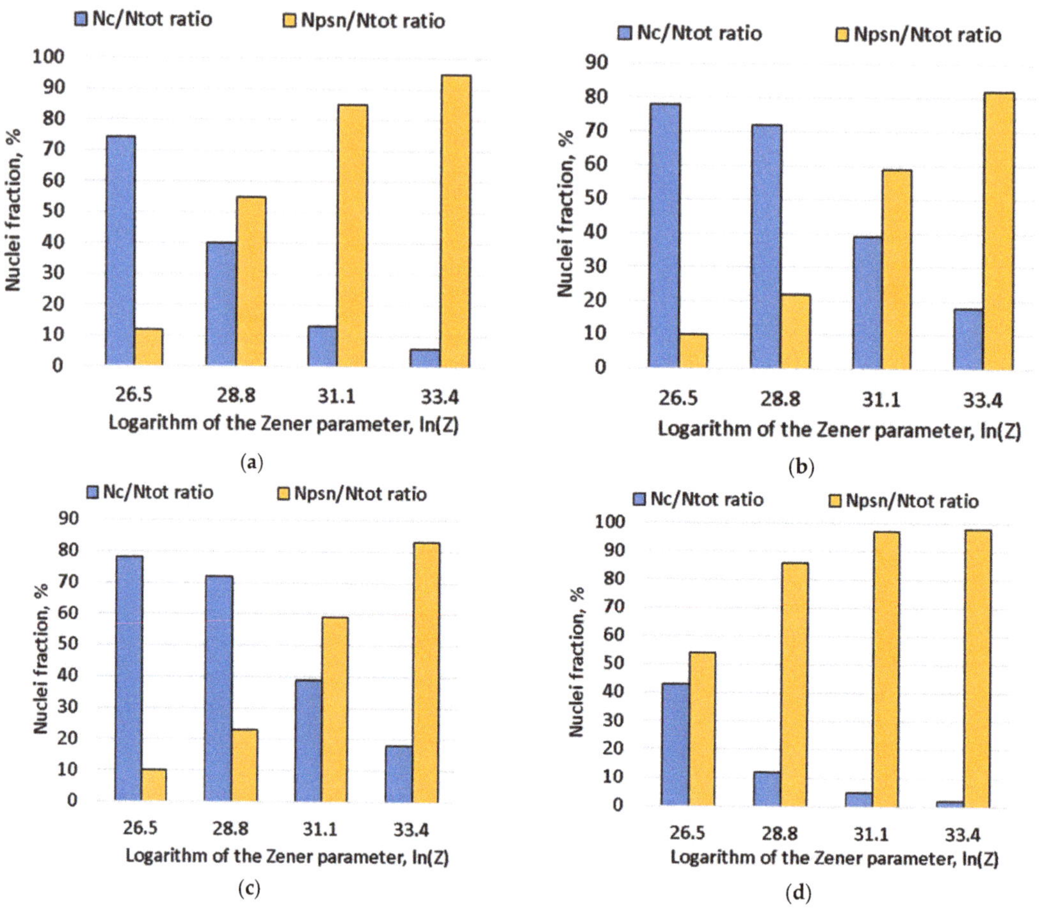

Figure 14. Changing t of N_c subgrains-based nuclei and N_{PSN} second-phase particles-based nuclei fraction depending on Zener-Hollomon parameter and alloy chemical composition: (**a**) alloy No. 1, (**b**) alloy No. 5, (**c**) alloy No. 8, (**d**) alloy No. 11.

Figure 15 shows the recrystallization retarding force. The retarding force of recrystallization did not change significantly when the content of magnesium and manganese increased. This was because the number of fine particles did not grow significantly with increasing magnesium and manganese content (Figure 7). However, with 0.5% copper addition, dispersoids increased dramatically, leading to an increase of the recrystallization retarding force from 1.53×10^{-3} N/m^2 to 3.92 N/m^2 (Figure 8). Zirconium produced an even greater significant effect on the recrystallization retarding force. The results, presented in Figure 8 correlate with the optical microscopy data (Figure 1), demonstrating partial blocking of the recrystallization process due to copper addition.

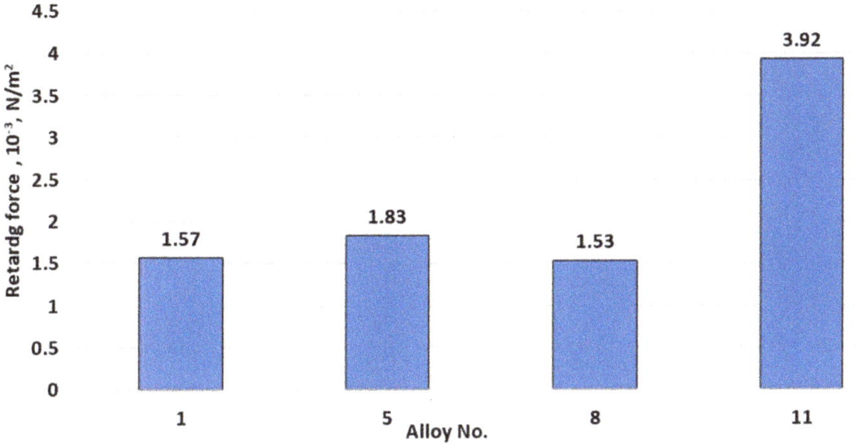

Figure 15. Changes in the recrystallization retarding force depending on alloy chemical composition.

4. Conclusions

Increasing magnesium, manganese and copper content affected grain size due to an increase in nuclei number as a result of large intermetallic particle development. The main large intermetallic particles were Mg_2Si, $Al_6(FeMn)$ and $Al_{20}Cu_2Mn_3$.

In the alloys containing no other alloying components, apart from magnesium, the level of finely dispersed components was very low. When manganese was added, Al_6Mn and $Al_6(FeMn)$ types dispersoids appeared, and when copper was added, fine particles close to $T(Al_{20}Cu_2Mn_3)$ and $S(Al_2CuMg)$ appeared. Copper addition significantly increased the total number of intermetallic particles and dispersoids, leading to recrystallization retarding force intensification. Therefore, copper addition partially blocked this process.

Nucleation modeling during recrystallization showed that this alloy (like other magnesium alloys) is prone to PSN mechanism activation. However, subgrain-based recrystallization nuclei development prevailed at low values of the Zener-Hollomon parameter without copper addition. The PSN mechanism began to prevail when copper was added, even at low Zener-Hollomon parameter values.

Author Contributions: Conceptualization, V.A., F.G. and E.A.; methodology, Y.E. and M.T.; software, E.A.; validation, S.K., Y.E. and F.G.; formal analysis, V.A. and E.A.; investigation, M.T. and A.K.; resources, V.A. and F.G.; data curation, V.A. and S.K.; writing—original draft preparation, E.A., M.T. and A.K.; writing—review and editing, V.A., F.G. and S.K.; visualization, Y.E. and A.K.; supervision, V.A. and F.G.; project administration, V.A.; funding acquisition, V.A. All authors have read and agreed to the published version of the manuscript.

Funding: The study was supported by a grant from the Russian Science Foundation, project No. 22-19-00810, https//rscf.ru/project/22-19-00810/ (accessed on 10 November 2021).

Institutional Review Board Statement: Study did not require ethical approval.

Informed Consent Statement: Written informed consent has been obtained from the patient(s) to publish this paper.

Data Availability Statement: The data presented in this study are available on request from the corresponding author.

Acknowledgments: The study was supported by a grant from the Russian Science Foundation No. 22-19-00810, https//rscf.ru/project/22-19-00810/ (accessed on 10 November 2021).

Conflicts of Interest: The authors declare no conflict of interest.

References

1. Du, H.; Zhang, S.; Zhang, B.; Tao, X.; Yao, Z.; Belov, N.; Liu, Z. Ca-modified Al–Mg–Sc alloy with high strength at elevated temperatures due to a hierarchical microstructure. *J. Mater. Sci.* **2021**, *56*, 16145–16157. [CrossRef]
2. Savchenkov, S.; Kosov, Y.; Bazhin, V.; Krylov, K.; Kawalla, R. Microstructural Master Alloys Features of Aluminum–Erbium System. *Crystals* **2021**, *11*, 1353. [CrossRef]
3. Belov, N.A.; Shurkin, P.K.; Korotkova, N.O.; Cherkasov, S.O. The effect of heat treatment on the structure and mechanical properties of cold-rolled sheets made of Al-Cu-Mn alloys with varying copper to manganese ratios. *Tsetnye Met.* **2021**, *9*, 80–86. [CrossRef]
4. Belov, N.A.; Cherkasov, S.O.; Korotkova, N.O.; Yakovleva, A.O.; Tsydenov, K.A. Effect of Iron and Silicon on the Phase Composition and Microstructure of the Al–2% Cu–2% Mn (wt%) Cold Rolled Alloy. *Phys. Met. Metallogr.* **2021**, *122*, 1095–1102. [CrossRef]
5. Safyari, M.; Moshtaghi, M.; Hojo, T.; Akiyama, E. Mechanisms of hydrogen embrittlement in high-strength aluminum alloys containing coherent or incoherent dispersoids. *Corros. Sci.* **2022**, *194*, 109895. [CrossRef]
6. Safyari, M.; Hojo, T.; Moshtaghi, M. Effect of environmental relative humidity on hydrogen-induced mechanical degradation in an Al–Zn–Mg–Cu alloy. *Vacuum* **2021**, *192*, 110489. [CrossRef]
7. Bazhin, V.Y.; Alattar, A.L.; Danilov, I.V. Development of technologies for the production of multi-component ligatures Al-Cu-BC with high thermal characteristics. *IOP Conf. Ser. Mater. Sci. Eng.* **2019**, *537*, 022037. [CrossRef]
8. Deev, V.B.; Ri, E.H.; Prusov, E.S.; Ermakov, M.A.; Goncharov, A.V. Grain Refinement of Casting Aluminum Alloys of the Al–Mg–Si System by Processing the Liquid Phase Using Nanosecond Electromagnetic Pulses. *Russ. J. Non-Ferr. Met.* **2021**, *65*, 522–530. [CrossRef]
9. Prusov, E.S.; Deev, V.B.; Aborkin, A.V.; Ri, E.K.; Rakhuba, E.M. Structural and Morphological Characteristics of the Friction Surfaces of In-Situ Cast Aluminum Matrix Composites. *J. Surf. Investig. X-ray Synchrotron Neutron Tech.* **2021**, *15*, 1332–1337. [CrossRef]
10. Annamalai, A.R.; Srikanth, M.; Muthuchamy, A.; Acharya, S.; Khisti, A.; Agrawal, D.K.; Jen, C.P. Spark Plasma Sintering and Characterization of Al-TiB2 Composites. *Metals* **2020**, *10*, 1110. [CrossRef]
11. Somasundaram, M.; Uttamchand, N.K.; Annamalai, A.R.; Jen, C. Insights on Spark Plasma Sintering of Magnesium Composites: —A Review. *Nanomaterials* **2022**, *12*, 2178. [CrossRef]
12. Muthuchamy, A.; Annamalai, A.R.; Acharyya, S.G.; Nagaraju, N.; Agrawal, D.K. Microstructural and electrochemical behaviour of aluminium alloy composites produced using different sintering techniques. *Mater. Res.* **2018**, *21*. [CrossRef]
13. Witkin, D.; Lee, Z.; Rodriguez, R.; Nutt, S.; Lavernia, E. Al–Mg alloy engineered with bimodal grain size for high strength and increased ductility. *Scr. Mater.* **2003**, *49*, 297–302. [CrossRef]
14. Patlan, V.; Vinogradov, A.; Higashi, K.; Kitagawa, K. Overview of fatigue properties of fine grain 5056 Al-Mg alloy processed by equal-channel angular pressing. *Mater. Sci. Eng. A* **2001**, *300*, 171–182. [CrossRef]
15. Papadopoulou, S.; Kontopoulou, A.; Gavalas, E.; Papaefthymiou, S. The effects of reduction and thermal treatment on the recrystallization and crystallographic texture evolution of 5182 aluminum alloy. *Metals* **2020**, *10*, 1380. [CrossRef]
16. Zhou, D.; Zhang, X.; Wang, H.; Li, Y.; Sun, B.; Zhang, D. Influence of Mg on tensile deformation behavior of high Mg-content Al-Mg alloys. *Int. J. Plast.* **2022**, *157*, 103405. [CrossRef]
17. Romhanji, E.; Popović, M.; Glišić, D.; Stefanović, M.; Milovanović, M. On the Al-Mg alloy sheets for automotive application: Problems and solutions. *Metalurgija* **2004**, *10*, 205–216.
18. Zhang, S.; Du, H.; Yao, Z.; Liu, Z.; Zhu, Y.; Shuai, L.; Belov, N. Superior high temperature creep resistance of a cast Al–Mg–Ca-Sc alloy with multi-scale hierarchical microstructures. *Mater. Sci. Eng. A* **2022**, *850*, 143533. [CrossRef]
19. Aryshenskii, E.V.; Guk, S.V.; Galiev, E.E.; Drits, A.M.; Kavalla, R. Possibility of application of a 1565ch alloy in the automotive industry. *Russ. Metall.* **2018**, *10*, 995–1001. [CrossRef]
20. Ovchinnikov, V.V.; Drits, A.M.; Gureeva, M.A.; Malov, D.V.; Rastopchin, R.N. Mechanical properties of the welded joints of 1565ch alloy sheets at low temperatures. *Russ. Metall.* **2017**, *6*, 483. [CrossRef]
21. Hirsch, J. Recent development in aluminium for automotive applications. *Trans. Nonferrous Met. Soc. China* **2014**, *24*, 1995–2002. [CrossRef]
22. Engler, O.; Löchte, L.; Hirsch, J. Through-process simulation of texture and properties during the thermomechanical processing of aluminium sheets. *Acta Mater.* **2007**, *55*, 5449–5463. [CrossRef]
23. Hirsch, J. Aluminium alloys for automotive application. *Mater. Sci. Forum* **1997**, *242*, 33–50. [CrossRef]
24. Wells, M.A.; Samarasekera, I.V.; Brimacombe, J.K.; Hawbolt, E.B.; Lloyd, D.J. Modeling the microstructural changes during hot tandem rolling of AA5XXX aluminum alloys: Part, I. Microstructural evolution. *Metall. Mater. Trans. B* **1998**, *29*, 611–620. [CrossRef]
25. Hirsch, J. Thermomechanical control in aluminium sheet production. *Mater. Sci. Forum* **2003**, *426*, 185–194. [CrossRef]
26. Králík, R.; Křivská, B.; Bajtošová, L.; Cieslar, M. The influence of Zener drag on recrystallization behaviour of twin-roll cast AA8079 alloy after homogenization. *IOP Conf. Ser. Mater. Sci. Eng.* **2021**, *1178*, 012034. [CrossRef]
27. Nes, E.; Ryum, N.; Hunderi, O. On the Zener Drag. *Acta Metall.* **1985**, *33*, 11–22. [CrossRef]
28. Chan, H.M.; Humphreys, F.J. Effect of particle stimulated nucleation on orientation of recrystallized grains. *Met. Sci.* **1984**, *18*, 527–530. [CrossRef]

29. Humphreys, F.J. The nucleation of recrystallization at second phase particles in deformed aluminium. *Acta Metall.* **1977**, *25*, 1323–1344. [CrossRef]
30. Vatne, H.E.; Wells, M.A. Modelling of the recrystallization behaviour of AA5XXX aluminum alloys after hot deformation. *Can. Metall. Q.* **2003**, *42*, 79–88. [CrossRef]
31. Vetrano, J.S.; Bruemmer, S.M.; Pawlowski, L.M.; Robertson, I.M. Influence of the particle size on recrystallization and grain growth in Al-Mg-X alloys. *Mater. Sci. Eng. A* **1997**, *238*, 101–107. [CrossRef]
32. Attallah, M.M.; Strangwood, M.; Davis, C.L. Influence of the heating rate on the initiation of primary recrystallization in a deformed Al–Mg alloy. *Scr. Mater.* **2010**, *63*, 371–374. [CrossRef]
33. Engler, O.; Kuhnke, K.; Hasenclever, J. Development of intermetallic particles during solidification and homogenization of two AA 5xxx series Al-Mg alloys with different Mg contents. *J. Alloys Compd.* **2017**, *728*, 669–681. [CrossRef]
34. Radetić, T.; Popović, M.; Romhanji, E. Microstructure evolution of a modified AA5083 aluminum alloy during a multistage homogenization treatment. *Mater. Charact.* **2012**, *65*, 16–27. [CrossRef]
35. Engler, O.; Miller-Jupp, S. Control of second-phase particles in the Al-Mg-Mn alloy AA 5083. *J. Alloys Compd.* **2016**, *689*, 998–1010. [CrossRef]
36. Dorin, T.; Ramajayam, M.; Babaniaris, S.; Jiang, L.; Langan, T.J. Precipitation sequence in Al–Mg–Si–Sc–Zr alloys during isochronal aging. *Materialia* **2019**, *8*, 100437. [CrossRef]
37. Babaniaris, S.; Ramajayam, M.; Jiang, L.; Langan, T.; Dorin, T. Developing an optimized homogenization process for Sc and Zr containing Al-Mg-Si alloys. In *Light Metals*; Springer: Cham, Switzerland, 2019; pp. 1445–1453. [CrossRef]
38. Vo, N.Q.; Dunand, D.C.; Seidman, D.N. Improving aging and creep resistance in a dilute Al–Sc alloy by microalloying with Si, Zr and Er. *Acta Mater.* **2014**, *63*, 73–85. [CrossRef]
39. Aryshenskii, E.V.; Aryshenskii, V.Y.; Beglov, E.D.; Chitnaeva, E.S.; Konovalov, S.V. Investigation of subgrain and fine intermetallic participles size impact on grain boundary mobility in aluminum alloys with transitional metal addition. *Mater. Today Proc.* **2019**, *19*, 2183–2188. [CrossRef]
40. Vatne, H.E.; Furu, T.; Ørsund, R.; Nes, E. Modelling recrystallization after hot deformation of aluminium. *Acta Mater.* **1996**, *44*, 4463–4473. [CrossRef]
41. Aryshenskii, E.; Beglov, E.; Konovalov, S.; Aryshenskii, V.; Khalimova, A. Approach to oriented grain growth accounting during aluminum alloys recrystallization simulation. *Mater. Today Proc.* **2021**, *46*, 957–960. [CrossRef]
42. Aryshenskii, E.; Kawalla, R.; Hirsch, J. Development of new fast algorithms for calculation of texture evolution during hot continuous rolling of Al–Fe alloys. *Steel Res. Int.* **2017**, *88*, 1700053. [CrossRef]
43. Mondolfo, L.F. *Aluminum Alloys: Structure and Properties*; Elsevier: Amsterdam, The Netherlands, 2013.
44. Kumar, R.; Gupta, A.; Kumar, A.; Chouhan, R.N.; Khatirkar, R.K. Microstructure and texture development during deformation and recrystallisation in strip cast AA8011 aluminum alloy. *J. Alloys Compd.* **2018**, *742*, 369–382. [CrossRef]
45. Sidor, J.J.; Petrov, R.H.; Kestens, L.A. Modeling the crystallographic texture changes in aluminum alloys during recrystallization. *Acta Mater.* **2011**, *59*, 5735–5748. [CrossRef]
46. Slámová, M.; Očenášek, V.; Vander Voort, G. Polarized light microscopy: Utilization in the investigation of the recrystallization of aluminum alloys. *Mater. Charact.* **2004**, *52*, 165–177. [CrossRef]
47. Lee, K.J.; Woo, K.D. Effect of the hot-rolling microstructure on texture and surface roughening of Al-Mg-Si series aluminum alloy sheets. *Met. Mater. Int.* **2011**, *17*, 689–695. [CrossRef]
48. Chang, J.K.; Takata, K.; Ichitani, K.; Taleff, E.M. Abnormal grain growth and recrystallization in Al-Mg alloy AA5182 following hot deformation. *Metall. Mater. Trans. A* **2010**, *41*, 1942–1953. [CrossRef]
49. Zhao, D.; Løvvik, O.M.; Marthinsen, K.; Li, Y. Impurity effect of Mg on the generalized planar fault energy of Al. *J. Mater. Sci.* **2016**, *51*, 6552–6568. [CrossRef]
50. Engler, O.; Hirsch, J.; Lücke, K. Texture development in Al 1.8 wt% Cu depending on the precipitation state—I. Rolling textures. *Acta Metall.* **1989**, *37*, 2743–2753. [CrossRef]
51. Ameri, A.A.; Elewa, N.N.; Ashraf, M.; Escobedo-Diaz, J.P. General methodology to estimate the dislocation density from microhardness measurements. *Mater. Charact.* **2017**, *131*, 324–330. [CrossRef]
52. Humphreys, F.J.; Hatherly, M. *Recrystallization and Related Annealing Phenomena*; Elsevier: Amsterdam, The Netherlands, 2012.
53. Kumar, S.; Babu, N.H.; Scamans, G.M.; Eskin, D.G.; Fan, Z. Solidification behaviour of an AA5754 Al alloy ingot cast with high impurity content. *Int. J. Mater. Res.* **2012**, *103*, 1228–1234. [CrossRef]
54. Engler, O.; Liu, Z.; Kuhnke, K. Impact of homogenization on particles in the Al–Mg–Mn alloy AA 5454–Experiment and simulation. *J. Alloys Compd.* **2013**, *560*, 111–122. [CrossRef]
55. Kang, J.; Wilkinson, D.S.; Malakhov, D.V.; Halim, H.; Jain, M.; Embury, J.D.; Mishra, R.K. Effect of processing route on the spatial distributions of constituent particles and their role in the fracture process in AA5754 alloy sheet materials. *Mater. Sci. Eng. A* **2007**, *456*, 85–92. [CrossRef]
56. Engler, O.; Kuhnke, K.; Westphal, K.; Hasenclever, J. Impact of chromium on the microchemistry evolution during solidification and homogenization of the Al-Mg alloy AA 5052. *J. Alloys Compd.* **2018**, *744*, 561–573. [CrossRef]
57. Dons, A.L. Precipitation of Primary Particles in Al-Mn-Fe Alloys-Experiments and Simulation/Bildung von Primärteilchen in Al-Mn-Fe Legierungen-Experimente und Simulation. *Int. J. Mater. Res.* **1990**, *81*, 484–489. [CrossRef]

58. Ratchev, P.; Verlinden, B.; Van Houtte, P. Effect of preheat temperature on the orientation relationship of (Mn,Fe)Al6 precipitates in an AA 5182 aluminiummagnesium alloy. *Acta Metall Mater.* **1995**, *43*, 621–629.
59. Simensen, C.J.; Södervall, U. Investigation of trace elements in an Al–4.8 wt.% Mg–0.3 wt.% Mn alloy. Surface and Interface Analysis: An International Journal devoted to the development and application of techniques for the analysis of surfaces. *Interfaces Thin Film.* **2000**, *30*, 309–314.
60. Shuncai, W.; Chunzhi, L.; Minggao, Y. Determination of structure of Al20Cu2Mn3 phase in Al Cu Mn alloys. *Mater. Res. Bull.* **1989**, *24*, 1267–1270. [CrossRef]
61. Shen, Z.; Liu, C.; Ding, Q.; Wang, S.; Wei, X.; Chen, L.; Zhang, Z. The structure determination of Al20Cu2Mn3 by near atomic resolution chemical mapping. *J. Alloys Compd.* **2014**, *601*, 25–30. [CrossRef]
62. Yan, J.; Hodge, A.M. Study of β precipitation and layer structure formation in Al 5083: The role of dispersoids and grain boundaries. *J. Alloys Compd.* **2017**, *703*, 242–250. [CrossRef]
63. Liu, Z.; Mohles, V.; Engler, O.; Gottstein, G. Characterization and statistical modeling of the precipitation kinetics in the commercial aluminum alloy AA5182. *MRS Online Proc. Libr.* **2011**, *1369*, 1–6. [CrossRef]
64. Hashimoto, T.; Zhang, X.; Zhou, X.; Skeldon, P.; Haigh, S.J.; Thompson, G.E. Investigation of dealloying of S phase (Al2CuMg) in AA 2024-T3 aluminium alloy using high resolution 2D and 3D electron imaging. *Corros. Sci.* **2016**, *103*, 157–164. [CrossRef]
65. Pang, X.; Yang, W.; Yang, J.; Pang, M.; Zhan, Y. Atomic structure, stability and electronic properties of S (Al2CuMg)/Al interface: A first-principles study. *Intermetallics* **2018**, *93*, 329–337. [CrossRef]
66. Aryshenskii, E.V.; Hirsch, J.; Konovalov, S.V.; Prahl, U. Specific features of microstructural evolution during hot rolling of the as-cast magnesium-rich aluminum alloys with added transition metal elements. *Metall. Mater. Trans. A* **2019**, *50*, 5782–5799. [CrossRef]
67. Aryshenskii, E.; Hirsch, J.; Drits, A.; Konovalov, S.; Aryshenskii, V.; Tepterev, M. Texture Development in Aluminum Alloys with High Magnesium Content. *Metals* **2022**, *12*, 723. [CrossRef]

materials

MDPI

Article

Effect of Aging Treatment on Microstructural Evolution and Mechanical Properties of the Electron Beam Cold Hearth Melting Ti-6Al-4V Alloy

Jiaxin Yu [1,2], Zhengpei Yin [3], Zhirong Huang [1,2], Shuai Zhao [1,2], Haiguang Huang [1], Kun Yu [3], Rongfeng Zhou [1,2] and Han Xiao [1,2,*]

1 Faculty of Materials Science and Engineering, Kunming University of Science and Technology, Kunming 650093, China
2 National-Local Joint Engineering Laboratory for Technology of Advanced Metallic Solidification Forming and Equipment, Kunming University of Science and Technology, Kunming 650093, China
3 Yunnan Titanium Industry Co., Ltd., Chuxiong 651209, China
* Correspondence: kmxh@kust.edu.cn

Abstract: Ti-6Al-4V (Ti64 or TC4) alloy is widely used in the industrial field. However, there have been few studies of the TC4 alloy melted by electron beam cold hearth melting (EBCHM) technology. Aging treatment has a considerable influence on the secondary α-phase in titanium alloys. Therefore, TC4 alloy melted by EBCHM technology was investigated in this study. The effect of different aging times on the microstructural evolution and mechanical properties of titanium alloy sheets was evaluated. The results showed that, with increase in aging time, the primary α-phase enlarged and grain globularization occurred. In addition, some transformed β-phases disappeared. The strength and Vickers hardness of the heat-treated sheets decreased, while the plasticity increased with increase in aging time, indicating that the mechanical properties developed with evolution of the microstructure. After aging at 560 °C for 2 h, the properties overall were optimal. The type of fracture of the samples was ductile fracture; the dimples became larger with increase in aging time. After heat treatment, the recrystallized nucleus, substructures and HAGBs increased, while the deformed structure and LAGBs decreased. Some grains had rotated following heat treatment, indicating that anisotropy was greatly reduced.

Keywords: TC4 alloy; EBCHM; aging treatment; microstructure; mechanical properties

Citation: Yu, J.; Yin, Z.; Huang, Z.; Zhao, S.; Huang, H.; Yu, K.; Zhou, R.; Xiao, H. Effect of Aging Treatment on Microstructural Evolution and Mechanical Properties of the Electron Beam Cold Hearth Melting Ti-6Al-4V Alloy. *Materials* **2022**, *15*, 7122. https://doi.org/10.3390/ma15207122

Academic Editor: Yong-Cheng Lin

Received: 16 August 2022
Accepted: 4 October 2022
Published: 13 October 2022

1. Introduction

Ti-6Al-4V (Ti64 or TC4) alloy is widely used in the aerospace and chemical industries, in ships and other fields due to its low density, high specific strength, good corrosion resistance and biocompatibility [1–3]. TC4 titanium alloy is a representative α + β titanium alloy, which has α- and β-phases [4]. It is one of the most used titanium alloy materials [5].

Traditional titanium alloy production mostly obtains plates by forging, machining and rolling following a triple-vacuum arc remelting (VAR) process [6]. However, this smelting method produces many inclusions. VAR melting is a slow process as the liquid flow condition is unsuitable for acceleration of the dissolution of inclusions. High-density inclusions sink to the bottom of the pool, where the temperature is lower than at other positions. Compared to the triple VAR melting process, electron beam cold hearth melting (EBCHM) technology has significant advantages for the production of very clean metal. During the melting process, volatile impurities evaporate due to exposure in a vacuum atmosphere (10^{-2}~10^{-3} Pa). The melt-flow conditions lead to a long inclusion residence time, which promotes the dissolution of inclusions [7]. If EBCHM technology is used, the content of inclusions can be effectively reduced [8]. Moreover, EBCHM technology has the advantages of high energy density and high vacuum, which can improve the purity of titanium alloys [9]. The ingots obtained by EBCHM technology can be rolled directly after melting, which simplifies the production process and reduces the cost. Sheets rolled from TC4 alloy ingots melted by EBCHM technology can be used in aviation, weapon manufacture and other fields. Many studies have shown that stress can be reduced and the microstructure of the alloy can be improved by appropriate heat treatment [10,11]. The microstructure plays an important role in the mechanical properties of the material, and can be adjusted to improve mechanical properties of the material [12–14]. The strength and hardness of titanium alloys can be greatly improved using a supersaturated solid solution formed by solution treatment [15,16]. The dispersed phase can be precipitated by subsequent aging treatment to improve strengthening, and the plasticity and toughness can be improved at the same time [17,18]. The microstructure of titanium alloys is greatly affected by the process [19,20]. To enable titanium alloys to be better used in industrial production, many investigations have been carried out into the structure and properties of titanium alloys. Fan et al. [21] investigated the effect of solution aging at 950 °C/AC/1 h + 540 °C/AC/4 h on the microstructure and properties of a Ti-6Al-4V alloy produced by selective laser melting. The results showed that a basket-weave structure can be obtained after solution with aging treatment, resulting in excellent mechanical properties. Lei et al. [22] investigated the effects of different structures on the tensile properties and impact toughness of Ti-6Al-4V alloy. The results showed that a lamellar structure had higher impact toughness and strength, but plasticity was lower than for a globular structure. Chen et al. [23] studied the effect of aging heat treatment on the microstructure and tensile properties of a high strength titanium alloy. The results showed that the β grain size can be restrained by a primary α-phase during aging treatment, and the morphology of the secondary α is sensitive to time. Xu et al. [24] studied the microstructural evolution and mechanical properties of a titanium alloy during solution-plus-aging treatment. The results showed that an α+β solution-plus-aging treatment resulted in an excellent combination of strength and plasticity.

The TC4 alloy is a very widely used alloy. Hence, it is important to investigate for TC4 alloy low-cost production technology. If EBCHM technology is used, it can not only reduce the content of inclusions, but can also provide technical methods enabling low-cost production. To date few studies have investigated TC4 alloy ingots melted by EBCHM technology, so, results are not available for direct industrial application. Aging treatment has a significant influence on the secondary α-phase precipitated in titanium alloys, which results in their different properties. Therefore, it is crucial to study the microstructure evolution of titanium alloys because of its potential applications. In this study, TC4 alloy melted by EBCHM technology was used and the effect of aging treatment for different times on the microstructure and mechanical properties of TC4 alloy sheets was investigated. This provided not only a theoretical basis for the production of TC4 alloy sheets with excellent performance and low cost, but also a protocol for the subsequent heat treatment process.

2. Experimental Procedures

The experimental material was TC4 alloy. A billet was cut in the middle part of the ingot with a size of 80 × 70 × 50 mm^3, which was produced by an electron beam cold hearth melting process at an average melting rate of 500 kg/h. A melting power of 200~210 kW was used to melt the alloy, and a power of 180~185 kW was used to ensure a constant temperature in the surface. The vacuum pressure was 0.1~1.0 Pa during the melting process. The dimension of the crystallizer was 1270 × 220 × 200 mm^3 and the size of the cold hearth was 1450 × 400 × 250 mm^3. The hot-rolled sheet was obtained by first rolling along the length direction and then rolling along the width direction. The chemical composition of the titanium alloy sheet was measured by chemical analysis, as shown in Table 1. The β-phase transformation temperature was obtained by differential scanning calorimetry (DSC) of the alloy at about 958.6 °C, as shown in Figure 1. The samples were further heat-treated with solution-plus-aging treatment. First, the sheet was heated to 930 °C and kept for 15 min, then cooled in water for solution treatment, kept at 560 °C for

2 h, 3 h and 4 h, and then furnace cooled (FC) for aging treatment. The standard tensile specimens built in a horizontal plane for tensile property evaluation, as shown in Figure 2, were cut in the rolling direction (RD) and transverse direction (TD).

Table 1. Chemical composition of TC4 alloy (wt, %).

Ti	Al	V	Fe	C	N	H	O
Bal.	5.93	4.27	0.07	0.05	0.009	0.004	0.14

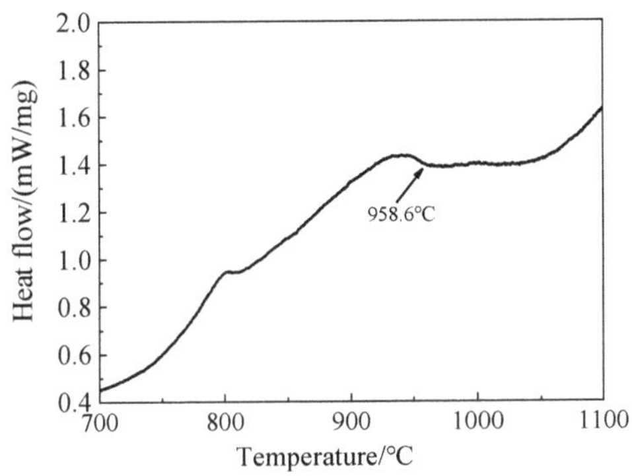

Figure 1. DSC curves of TC4 alloy.

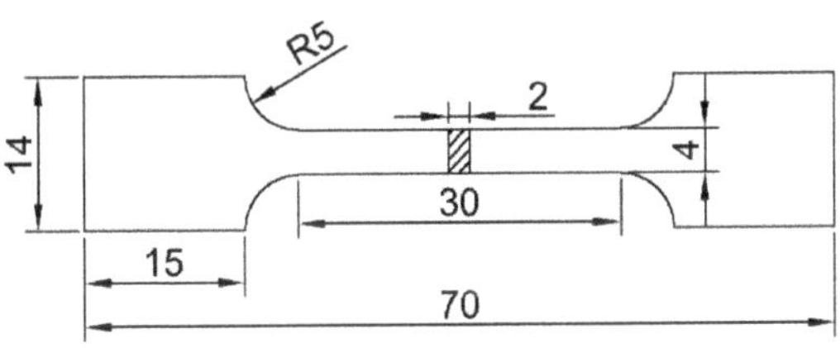

Figure 2. The size of tensile specimen (mm).

A small part of the geometry of 10 × 10 × 10 mm was used for microstructure analysis and electron backscattered diffraction (EBSD) tests. Standard Kroll's reagent (HF: HNO_3: H_2O = 3:5:12) was used to identify the various phases. The microstructure and fracture morphology of specimens were observed using an ECLIPSE MA200 microscope and a ZEISS EVO18 (Zeiss, Oberkochen, Germany) scanning electron microscopy (SEM). Image-Pro-Plus software was used to calculate the phase volume fraction. X-ray diffraction was used to measure the Cu target Kα radiation; the scanning rate was 10°/min. Tensile tests were performed using a Zwick/Roll-Z150 machine (Zwick/Roll, Ulm, Germany); the tensile rate was 1.8 mm/min. Each group of specimens was repeated three times and the average value taken. Vickers hardness tests were performed using an HMV-G21S machine (Shimadzu, Kyoto, Japan) with a load of 200 g and a loading time of 15 s; each group of specimens was repeated seven times and the average value taken.

3. Results and Discussion

3.1. Microstructural Characterization of TC4 Aged Sheets

The purpose of aging treatment is to decompose the metastable phase formed after heat treatment. Solution treatment and then quenching in water to effect aging is performed to precipitate the secondary phase with a dispersed distribution to enable strengthening [15,17,18]. The microstructures of the TC4 alloy which were obtained by solution treatment and aging treatment for different times (2 h, 3 h, and 4 h) are shown in Figure 3. It can be seen from the figure that the structure was composed of a large number of α-phase colonies and a small number of β-phase colonies, with a small amount of recrystallized equiaxed α-phase dispersed in it. From Figure 3a, b, it can be seen that the thick α-phase colonies in the longitudinal section and the cross-section of the hot-rolled sheet were distributed, and that the β-phase was distributed among them with the same degree of distribution. The average grain sizes of the equiaxed α-phase in the longitudinal section and cross-section were about 1.52 μm and 1.33 μm, respectively. The microstructures after aging for 2 h in the longitudinal section and cross-section are shown in Figure 3d, e. The microstructure consisted of a few equiaxed α-phases and some β phases. The average grain sizes of the equiaxed α-phase in the longitudinal section and cross-section were about 2.32 μm and 2.21 μm, respectively. For the transverse cross-section, the SEM images of the alloy are shown in Figure 4. It can be seen from Figure 4b that many secondary α-phases were precipitated on the transformed β matrix; the volume fraction of the primary equiaxed α-phase was about 31.2%. As shown in Figure 3g,h, the average grain sizes of the equiaxed α-phase in the longitudinal section and cross-section were about 2.83 μm and 2.69 μm, respectively. It can be seen from Figure 3g,h that the microstructure was composed of many primary α-phases (equiaxed and lamellar) and transformed β phases, and thick α laths were distributed along the rolling direction. It can be seen from Figure 4c that the microstructure was composed of many primary equiaxed α-phases and transformed β phases; the volume fraction of the primary equiaxed α-phase was about 31.0%. Compared with the microstructure after aging for 2 h, the primary α phase, lamellar α-phase and secondary α-phase on the β matrix had merged and developed. As shown in Figure 3j,k, the average grain sizes of the equiaxed α-phase in the longitudinal section and cross-section were about 2.98 μm and 2.77 μm, respectively. It can be seen from Figure 3j,k that, with prolongation of the aging time, the microstructure was still composed of some equiaxed primary α-phases and a few transformed β phases. The volume fraction of the equiaxed α-phase was calculated according to the area fraction of the phase in the figure. It can be seen from Figure 4d,e that the average grain size of the equiaxed primary α-phase increased. This phenomenon reflected the grain growth of the α phase; however, the volume fraction of the primary α-phase did not change too much and its value was about 30.2%. The α-phase tended to be equiaxed, indicating that grain globularization had occurred, and the content of the transformed β-phase decreased gradually. To sum up, the structural evolution of the sheet under different aging times is shown in Figure 5. With increase in aging time, the average grain size of the primary equiaxed α-phase increased gradually and grain globularization occurred; however, the volume fraction of the primary α-phase was not obviously changed. The morphology of the transformed β-phase was obviously changed with increase in aging time; the secondary α-phase precipitated on the β matrix appeared to be merged and matured, and the content of the transformed β-phase gradually decreased. This phenomenon was also observed in other titanium alloys [24,25].

Figure 6 shows the energy spectrum of the sheet obtained after aging for 2 h. It can be seen from Figure 6b that the element Ti in the hot-rolled sheet after solution-plus-aging treatment was uniformly distributed in the structure, and there was no segregation. As shown in Figure 6c,d, the elements Al and V were also uniformly distributed in the structure, and there was basically no aggregation of Al or V elements at the grain boundaries of the equiaxed α phase.

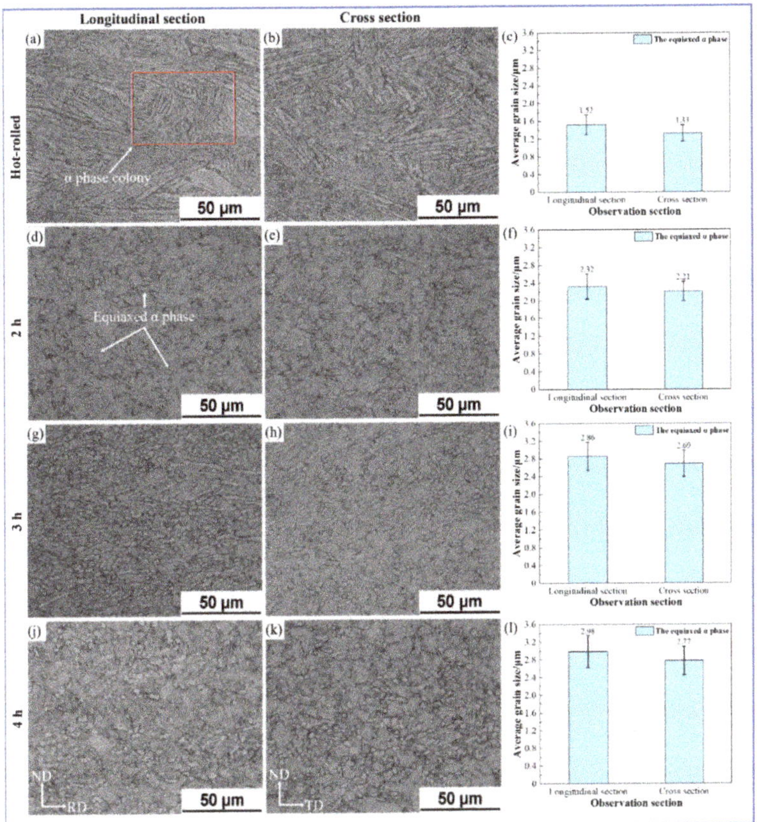

Figure 3. Microstructure and average grain size of TC4 alloy aged sheets.

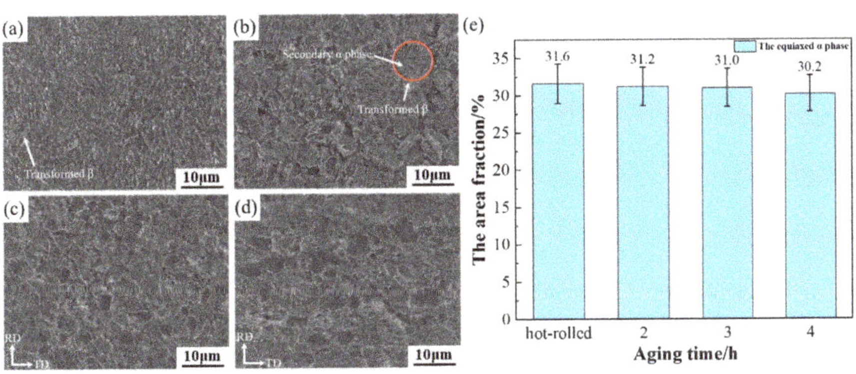

Figure 4. SEM images of TC4 alloy aged sheets: (**a**) Hot-rolled (**b**) 2 h (**c**) 3 h (**d**) 4 h (**e**) the area fraction of equiaxed α phase.

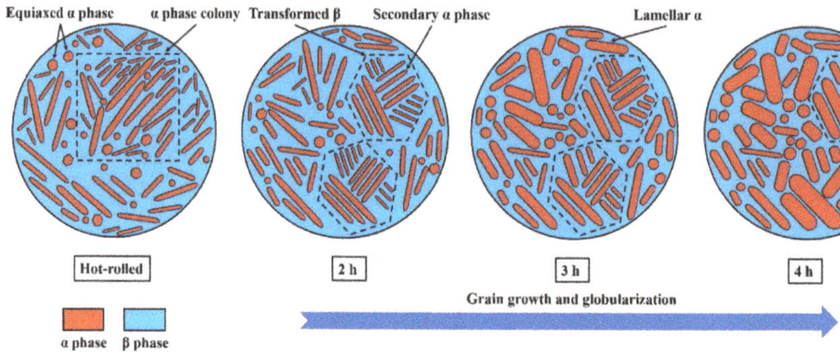

Figure 5. Microstructure evolution of TC4 alloy aged sheets.

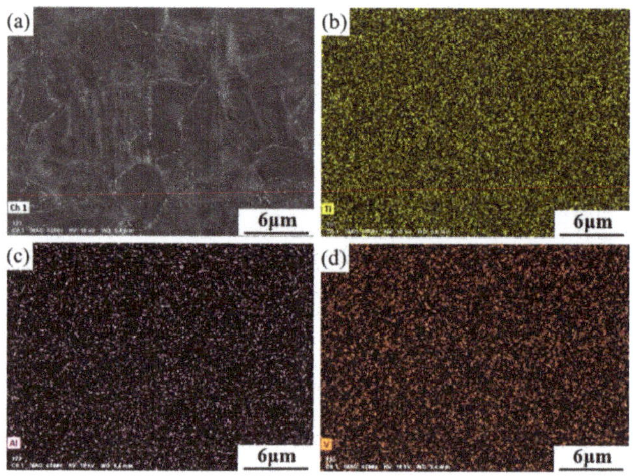

Figure 6. EDS of the sheet aging for 2 h: (**a**) Scanning surface (**b**) Ti (**c**) Al (**d**) V.

The XRD patterns of the sheets are shown in Figure 7a. It can be seen from the figure that the phase compositions of the hot-rolled and heat-treated sheets were mainly α/α'-Ti and β-Ti [26] and exhibited the highest diffraction peak at 40.5°. The peak intensities of each phase of the heat-treated sheet were significantly higher than that of the unheated sheet. With increase in aging time, the peak intensities of $(10\bar{1}1)$ increased gradually, indicating that the volume fraction of α/α'-Ti increased after solution-plus-aging treatment. As can be seen from the structural evolution, the structure became coarse and matured with increase in aging time, so the intensities of each peak were enhanced to different degrees. In order to investigate the effect of the aging time on the grain size, the Scheler equation was used to calculate the crystallite size of $(10\bar{1}1)$ [27]:

$$D = \frac{K\lambda}{\beta_{hkl} \cos\theta} \tag{1}$$

where D is the crystallite size (nm), K is a constant (0.89 is used in this case), λ is 0.154 nm, β_{hkl} is the FWHM of diffraction peak, θ is the peak position, and all angles are in radians. It can be seen from Figure 7b that the FWHM decreased, and the calculated crystallite size perpendicular to $(10\bar{1}1)$ was increased with increase in aging time. The results showed that the crystallite size of $(10\bar{1}1)$ increased with increased aging time, showing the same trend as for grain size.

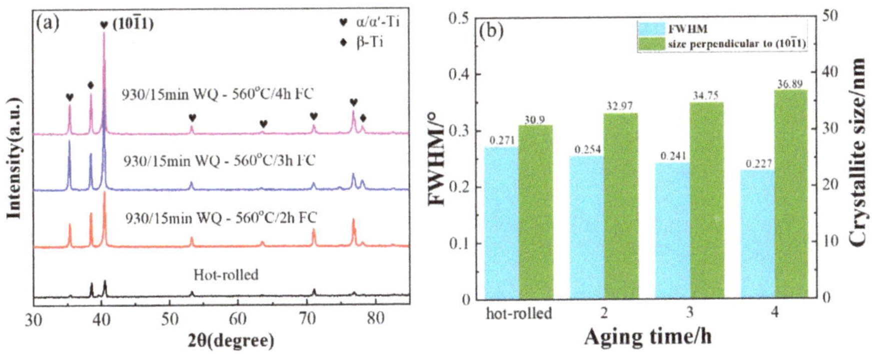

Figure 7. (**a**) XRD patterns of TC4 alloy aged sheets (**b**) the FWHM of (10$\bar{1}$1) and the calculated size.

3.2. Mechanical Properties of TC4 Aged Sheets

The TC4 alloy sheets obtained by heat treatment were subjected to tensile tests unidirectionally at room temperature in the RD and TD directions. The mechanical properties are shown in Figure 8. It can be seen from the figure that the tensile strength (UTS), yield strength (YS) and elongation (EL) of the hot-rolled sheets in the RD and TD directions were 912.7 and 941.5 MPa, 792.1 and 884.3 MPa, and 11.2% and 9.4%, respectively. The UTS, YS and EL of the sheets after aging for 2 h were 1073.6 and 1077.4 MPa, 955.0 and 991.0 MPa, and 10.4% and 10.5%, respectively. After aging for 3 h, they were 1038.1 and 1048.7 MPa, 922.3 and 962.4 MPa, and 10.5% and 10.7%, respectively. They were 1036.2 and 1041.8 MPa, 917.0 and 920.0 MPa, and 10.7% and 11.3% after aging for 4 h, respectively. With increase in aging time, the strength of the sheet decreased and the plasticity slightly increased This corresponded to the evolution of the microstructure [15,17]. The differences in UTS, YS and EL of the hot-rolled and heat-treated sheets in RD and TD directions were: 28.8 MPa, 92.2 MPa and 1.8% (hot-rolled); 3.8 MPa, 36.0 MPa and 0.1% (2 h); 10.6 MPa, 40.1 MPa and 0.2% (3 h); 5.6 MPa, 3.0 MPa and 0.6% (4 h). The difference between the anisotropy of the sheets obtained after different aging times was small, and the comprehensive properties of the sheet after aging for 2 h were the best. The effect on the anisotropy of the sheets was similar to other titanium alloys [13,28]. Compared with the hot-rolled sheet, the UTS and YS of the sheet obtained after aging for 2 h in the RD and TD directions increased by 17.63% and 14.43%, 20.57% and 12.07%, respectively. The EL decreased by 7.14% in the RD direction and increased by 11.70% in the TD direction. In summary, as the aging time increased, the strength of the heat-treated sheets decreased and the plasticity increased slightly. After aging for 2 h, the sheet had the best comprehensive properties, the UTS and YS of the sheet increased, and the EL decreased compared with the hot-rolled sheet.

The Vickers hardness values for the TC4 alloy sheets are shown in Figure 9. It can be seen that the Vickers hardness value decreased gradually with increase in aging time. The sheet obtained after aging for 2 h had the largest Vickers hardness which was 308.8 HV. The reason for the decrease in Vickers hardness may be that, as the aging time increased, its structure and secondary α-phase on the β matrix had merged and grown up, and the transformed β had basically disappeared [3,29]. As a result, its strength and hardness were reduced.

The fracture morphologies of the samples are shown in Figure 10. The macroscopic fracture morphologies in the RD and TD directions were relatively uneven and showed a dark gray fiber shape, which had a certain degree of necking. The fractures of the hot-rolled and heat-treated samples were composed of dimples with different sizes; they were deep and there were many small dimples in the large dimples. Moreover, the fracture dimples became larger with increase in aging time and all the samples were ductile fractures. The small dimples were formed by the secondary α, flaky α phase, and β phase, and the large dimples were formed by the equiaxed α-phase [17]. With increasing aging time, the volume fraction of the secondary α-phase and transformed β-phase declined, and the volume fraction of the equiaxed α-phase increased. Therefore, the fracture dimple became larger with increase in aging time. This phenomenon was also observed in other studies [3,4]. In summary, the fracture dimples were ductile fractures in both the RD and TD directions and the plasticity was improved due to the dimples becoming larger with increase in aging time.

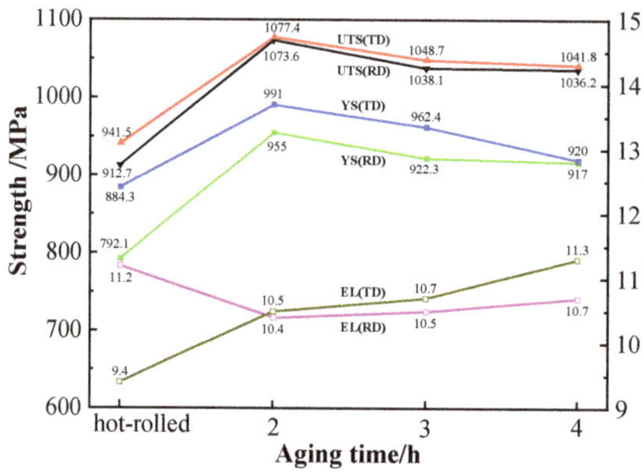

Figure 8. The mechanical properties of TC4 alloy aged sheets.

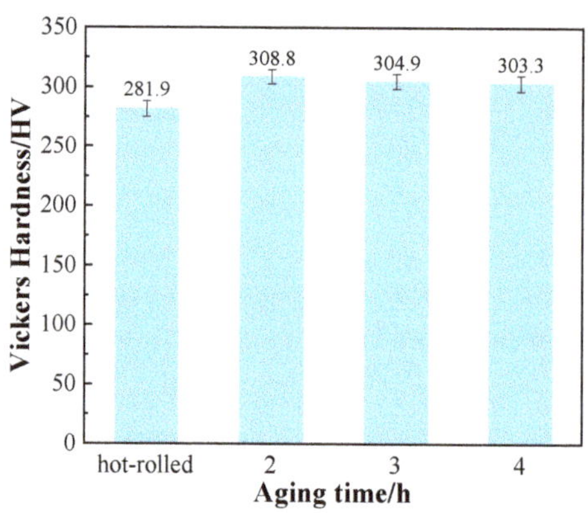

Figure 9. Vickers hardness of TC4 alloy aged sheets.

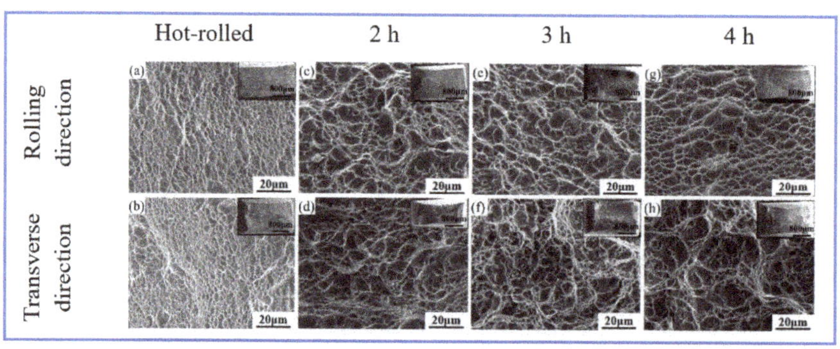

Figure 10. Fracture morphologies of TC4 alloy aged sheets.

3.3. Evolution of Grain and Orientation Distribution after Heat Treatment

To explore the recrystallization, grain boundaries and texture changes after heat treatment, the hot-rolled sheet and the sample obtained after aging for 2 h were characterized by EBSD testing. The internal average misorientation angle (IAMA) map and the proportion of hot-rolled sample are shown in Figure 11a,c; the figures for the heat treatment sample are shown in Figure 11b,d. In the IAMA maps, the angles which are above 1° are classified as deformed grain (red), the angles which are under 1° and the angles from subgrain to subgrain which are above 1° are classified as substructure grains (yellow). The remaining grains are classified as recrystallized grains (blue) [30]. It can be seen from the figure that, after solution-plus-aging treatment, the recrystallized nucleus and substructures were significantly increased, and the deformed matrix structure was significantly reduced. The deformed matrix structure of the sheet was reduced from 91% to 18.24%; the recrystallized nuclei and substructure increased from 7.57% to 50.63% and 1.43% to 31.13%, respectively. The increase in recrystallization nuclei was due to recrystallization during the heat treatment process. The formation mechanism may be that, during the heat treatment process, a certain section of the original grain boundary penetrated the larger residual strain grains and the deformation storage energy disappeared [10,29]. Moreover, many strain-free recrystallized nuclei were formed through the merger of subgrain or grain boundaries [31,32].

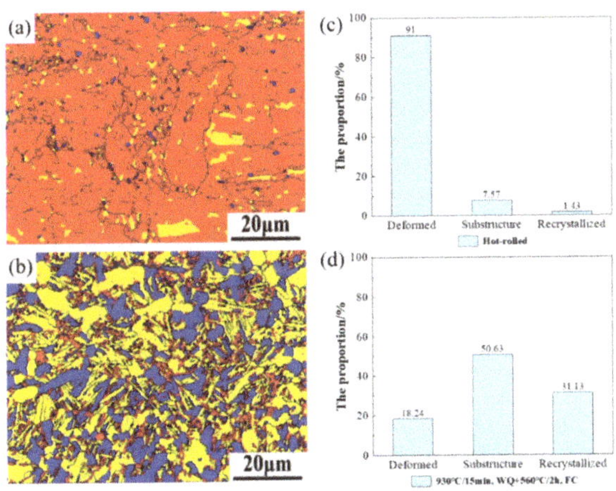

Figure 11. Recrystallized (blue), substructure (yellow), and deformed (red) grains in TC4 alloy: (**a**,**c**) Hot-rolled; (**b**,**d**) 930 °C/15 min, WQ + 560 °C/2 h, FC.

Figure 12 shows the misorientation angle distributions of TC4 alloy sheet after heat treatment. The misorientation distribution map represents the proportion of orientation angles of grain boundaries; the random distribution of polycrystalline orientation is represented by the black line, indicating that it has no preferred orientation. The grain boundaries significantly changed after heat treatment, the low-angle grain boundaries (LAGBs, <5°) and high-angle grain boundaries (HAGBs, >15°) of the hot-rolled sheet accounted for 90.19% and 9.81%, and the LAGBs and HAGBs of the heat-treated sheet accounted for 38.83% and 61.17%, respectively. Moreover, a large number of LAGBs in the hot-rolled sheet were distributed in the range 0° to 15°, showing a characteristic "single peak", and there were some obvious peaks in the heat-treated sheet around 4°, 64° and 90°. There were a number of grain boundary orientations around 60°, which was related to the different variants of the β-phase to the α phase. The α-phase precipitated in a Burgers relationship to the β-phase with twelve crystallographic variants; the misorientations could be divided into six types according to the axis-angle pairs, as shown in Table 2 [33]. The distribution of the sheet after heat treatment was similar to random distribution, but there was a certain deviation. Compared with the hot-rolled sheet, the HAGBs of the heat-treated sheet increased and the degree of preferred orientation decreased. In summary, the HAGBs increased and the LAGBs decreased after solution-plus-aging treatment. The reason was that recrystallization occurred during the heat treatment process, which caused the original lamellar structure to be spheroidized [25,34]. In addition, the HAGBs of the sheet after solution-plus-aging treatment accounted for the largest proportion, indicating that its degree of recrystallization was the largest; the degree of preferred orientation was reduced, showing the characteristic of random orientation.

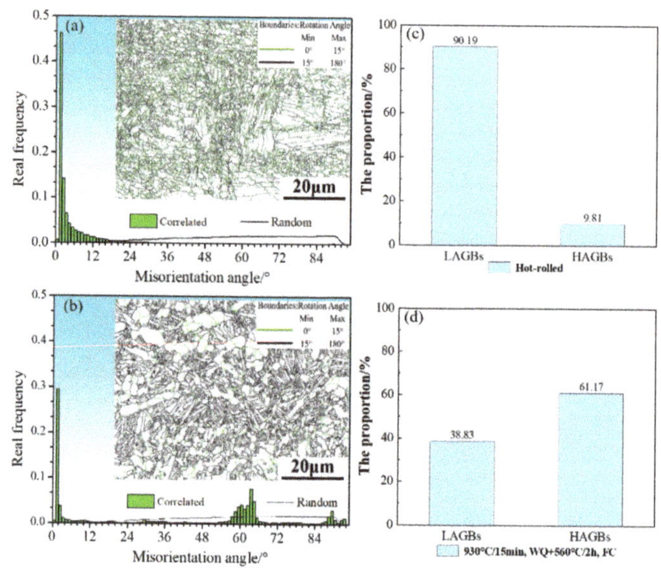

Figure 12. Misorientation angle distributions of TC4 alloy sheets: (**a**,**c**) hot-rolled; (**b**,**d**) 930 °C/15 min, WQ + 560 °C/2 h, FC.

Table 2. The six types of pair-wise misorientation between α and β phase.

Type	Angle Pair	Axis	Probability Occurrence
A	0		15.38%
B	10.53°	[0 0 0 1]	7.69%
C	60°	[1 1 −2 0]	15.38%
D	60.83°	[−1.377 −1 2.377 0.359]	30.77%
E	63.26°	[−10 5 5 3]	15.38%
F	90°	[1 −2.38 1.38 0]	15.38%

The pole figures (PFs) are shown in Figure 13. The hot-rolled sheet had a strong basal texture in the {0001} pole figure, and the c-axis of the grains was deflected away from the ND direction to the RD direction; the angle between the c-axis and ND direction was about 40°. There was variant distribution and coarsening behavior of the α-phase [33]. In this study, the intensity of the heat-treated sheet was lower than that of the hot-rolled sheet, and the c-axis of the α-phase grains was almost parallel to the TD, which showed typical transverse texture characteristics [34]. In summary, the hot-rolled and heat-treated sheets both had basal texture in the {0001} pole figure, and the intensity was obviously weakened after heat treatment. In addition, the grains rotated and the transverse texture appeared [35].

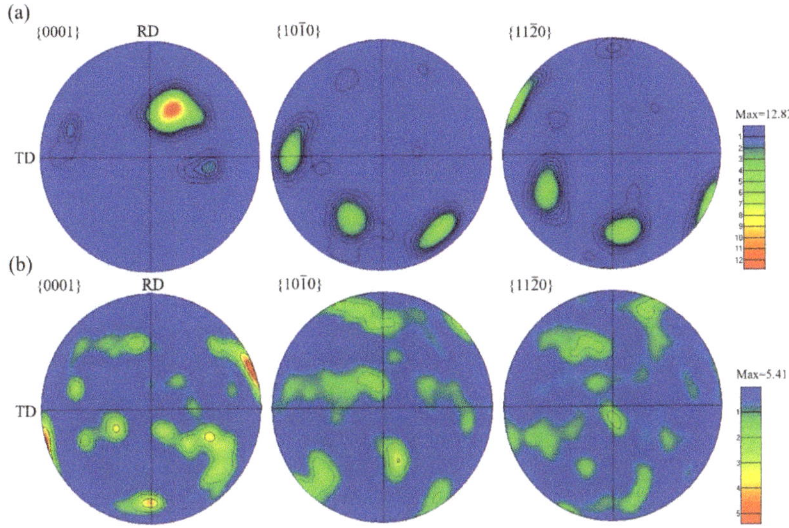

Figure 13. PFs of TC4 alloy sheets: (**a**) Hot-rolled (**b**) 930 °C/15 min, WQ + 560 °C/2 h, FC.

The inverse pole figures (IPFs) are shown in Figure 14. It can be seen from Figure 14a that the grain orientation in the RD direction of the hot-rolled sheet was between <0001> and <$\bar{1}2\bar{1}0$>, and the grain orientations in the TD and ND direction were between <0001> and <$01\bar{1}0$> . It can be seen from Figure 14b that most of the grains in the RD direction of the heat-treated sheet were between <$01\bar{1}0$> and <$\bar{1}2\bar{1}0$>, and some grains were between <0001> and <$01\bar{1}0$> in the TD direction. The grains in the ND direction were oriented away from <0001>, and their intensity decreased after heat treatment, indicating that some grains had rotated. This phenomenon represents the weakening and randomization of the texture [36].

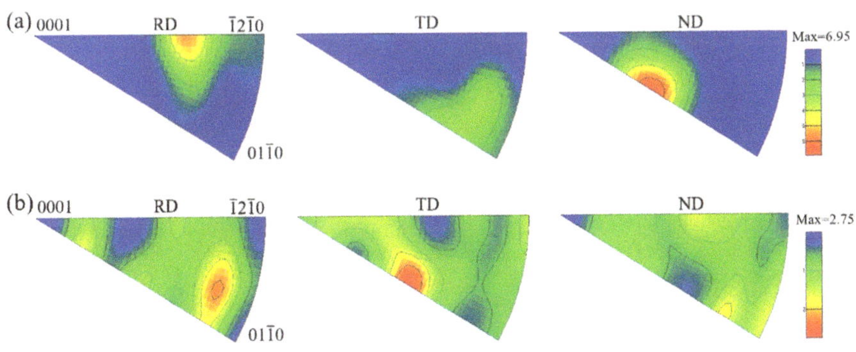

Figure 14. IPFs of TC4 alloy sheets: (**a**) Hot-rolled (**b**) 930 °C/15 min, WQ + 560 °C/2 h, FC.

4. Conclusions

(1) After solution-plus-aging treatment, the phase of the TC4 hot-rolled and aged sheets was mainly composed of α/α'-Ti and β-Ti, and the peak intensities of each phase were higher than those of the unheated sheet, indicating that the grains had grown up. With increase in aging time, the equiaxed primary α-phase grew up gradually and grain globularization occurred, but the volume fraction of the primary α-phase did not change significantly. The morphology of the transformed β-phase was obviously

changed; many secondary α-phases precipitated on the β matrix appeared to be merged and grown up, and some transformed β-phases disappeared.

(2) With increase in aging time, the strength of the heat-treated sheets decreased and the plasticity slightly increased; the difference between the anisotropy of the sheets was small. After aging for 2 h, the comprehensive properties were the best. The Vickers hardness of the sheet showed a downward trend; the value was proportional to its strength, and inversely proportional to its elongation. The fracture dimples, which were ductile fracture, in the RD and TD directions of the samples were similar, and the dimples became larger with increase in aging time.

(3) After heat treatment, the recrystallized nucleus and substructures were increased, and the deformed matrix structure was reduced. The LAGBs decreased and the HAGBs increased after heat treatment. There was a basal texture in the {0001} pole figure, the intensity was obviously weakened after heat treatment, and some grains had rotated, indicating that the anisotropy was greatly improved.

Author Contributions: Conceptualization, H.X.; methodology, J.Y. and Z.H.; software, S.Z.; validation, H.H. and K.Y.; formal analysis, J.Y.; investigation, J.Y. and Z.Y.; resources, H.H. and K.Y.; data curation, K.Y.; writing—original draft preparation, J.Y.; writing—review and editing, H.X. and R.Z.; supervision, H.X.; project administration, H.X.; funding acquisition, R.Z. All authors have read and agreed to the published version of the manuscript.

Funding: This research was funded by the Yunnan Provincial Major Science and Technology Special Plan of China (No. 202002AB080001-3).

Institutional Review Board Statement: Not applicable.

Informed Consent Statement: Not applicable.

Data Availability Statement: The data presented in this study are available upon reasonable request from the corresponding.

Acknowledgments: The author (Jiaxin Yu) greatly wish to thank Shuai Zhao for experimental guide and valuable discussions. Also thank to Han Xiao and Rongfeng Zhou from Kunming University of Science and Technology for their help in experiments.

Conflicts of Interest: The authors declare no conflict of interest.

References

1. Banerjee, D.; Williams, J.C. Perspectives on titanium science and technology. *Acta Mater.* **2013**, *61*, 844–879. [CrossRef]
2. Bai, Y.; Gai, X.; Li, S.J.; Zhang, L.C.; Liu, Y.J.; Hao, Y.L.; Zhang, X.; Yang, R.; Gao, Y.B. Improved corrosion behaviour of electron beam melted Ti-6Al-4V alloy in phosphate buffered saline. *Corros. Sci.* **2017**, *123*, 289–296. [CrossRef]
3. Costa, B.C.; Tokuhara, C.K.; Rocha, L.A.; Oliveira, R.C.; Lisboa-Filho, P.N.; Pessoa, J.C. Vanadium ionic species from degradation of Ti-6Al-4V metallic implants: In vitro cytotoxicity and speciation evaluation. *Mater. Sci. Eng. C* **2019**, *96*, 730–739. [CrossRef]
4. Liu, Z.; Zhao, Z.B.; Liu, J.R.; Wang, Q.J.; Guo, Z.H.; Liu, Z.M.; Zeng, Y.D.; Yang, G.; Gong, S.L. Effects of solution-aging treatments on microstructure features, mechanical properties and damage behaviors of additive manufactured Ti-6Al-4V alloy. *Mater. Sci. Eng. A* **2021**, *800*, 140380. [CrossRef]
5. Knobbe, H.; Köster, P.; Christ, H.-J.; Fritzen, C.-P.; Riedler, M. Initiation and propagation of short fatigue cracks in forged Ti-6Al-4V. *Procedia Eng.* **2010**, *2*, 931–940. [CrossRef]
6. Mitchell, A. The electron beam melting and refining of titanium alloys. *Mater. Sci. Eng. A* **1999**, *263*, 217–223. [CrossRef]
7. Zhang, Y.M.; Zhou, L.; Sun, J.; Han, M.C.; Georg, R.; Jochen, F.; Yang, J.C.; Zhao, Y.Q. An Investigation on Electron Beam Cold Hearth Melting of Ti64 Alloy. *Rare Met. Mater. Eng.* **2008**, *37*, 1973–1977. [CrossRef]
8. Liu, H.P.; Shi, S.; You, Q.F.; Zhao, L.H.; Tan, Y. Analysis on elemental volatilization behavior of titanium alloys during electron beam smelting. *Vacuum* **2018**, *157*, 395–401. [CrossRef]
9. Koleva, E.; Vutova, K.; Mladenov, G. The role of ingot-crucible thermal contact in mathematical modelling of the heat transfer during electron beam melting. *Vacuum* **2001**, *62*, 189–196. [CrossRef]
10. Su, B.X.; Luo, L.S.; Wang, B.B.; Su, Y.Q.; Wang, L.; Ritchie, R.O.; Guo, E.Y.; Li, T.; Yang, H.M.; Huang, H.G.; et al. Annealed microstructure dependent corrosion behavior of Ti-6Al-3Nb-2Zr-1Mo alloy. *J. Mater. Sci. Technol.* **2021**, *62*, 234–248. [CrossRef]
11. Gu, B.; Chekhonin, P.; Xin, S.W.; Liu, G.Q.; Ma, C.L.; Zhou, L.; Skrotzki, W. Microstructure and texture development during hot-compression of Ti5321. *Mater. Charact.* **2021**, *179*, 111297. [CrossRef]

12. Yang, Z.J.; Li, J.X.; Zhang, B.B.; Li, J.H. Microstructures and mechanical properties of a titanium alloy thick plate joint after electron beam welding plus solution-aging. *J. Mater. Res. Technol.* **2022**, *19*, 913–922. [CrossRef]
13. Nasiri-Abarbekoh, H.; Ekrami, A.; Ziaei-Moayyed, A.A.; Shohani, M. Effects of rolling reduction on mechanical properties anisotropy of commercially pure titanium. *Mater. Des.* **2012**, *34*, 268–274. [CrossRef]
14. Chen, W.; Lin, Y.C.; Zhang, X.Y.; Zhou, K.C. Balancing strength and ductility by controllable heat-treatment twinning in a near β-Ti alloy. *J. Mater. Res. Technol.* **2020**, *9*, 6962–6968. [CrossRef]
15. Wang, Z.Y.; Liu, L.B.; Wu, D.; Zhang, L.G.; Wang, W.L.; Zhou, K.C. α″ phase-assisted nucleation to obtain ultrafine α precipitates for designing high-strength near-β titanium alloys. *Trans. Nonferrous Met. Soc. China* **2020**, *30*, 2681–2696. [CrossRef]
16. Xiao, J.F.; Nie, Z.H.; Tan, C.W.; Zhou, G.; Chen, R.; Li, M.R.; Yu, X.D.; Zhao, X.C.; Hui, S.X.; Ye, W.J.; et al. Effect of reverse β-to-ω transformation on twinning and martensitic transformation in a metastable β titanium alloy. *Mater. Sci. Eng. A* **2019**, *759*, 680–687. [CrossRef]
17. Zhao, Q.Y.; Sun, Q.Y.; Xin, S.W.; Chen, Y.N.; Wu, C.; Wang, H.; Xu, J.W.; Wan, M.P.; Zeng, W.D.; Zhao, Y.Q. High-strength titanium alloys for aerospace engineering applications: A review on melting-forging process. *Mater. Sci. Eng. A* **2022**, *845*, 143260. [CrossRef]
18. Li, T.; Kent, D.; Sha, G.; Stephenson, L.T.; Ceguerra, A.V.; Ringer, S.P.; Dargusch, M.S.; Cairney, J.M. New insights into the phase transformations to isothermal ω and ω-assisted α in near β-Ti alloys. *Acta Mater.* **2016**, *106*, 353–366. [CrossRef]
19. Tian, Y.X.; Li, S.J.; Hao, Y.L.; Yang, R. High temperature deformation behavior and microstructure evolution mechanism transformation in Ti2448 alloy. *Acta Metall. Sin.* **2012**, *48*, 837–844. Available online: https://www.ams.org.cn/EN/10.3724/SP.J.1037.2012.00007 (accessed on 1 July 2012). [CrossRef]
20. Zhang, Z.Q.; Dong, L.M.; Yang, Y.; Guan, S.X.; Liu, Y.Y.; Yang, R. Influences of quenching temperature on the microstructure and deformation behaviors of TC16 titanium alloy. *Acta Metall. Sin.* **2011**, *47*, 1257–1262.
21. Fan, Z.C.; Feng, H.W. Study on selective laser melting and heat treatment of Ti-6Al-4V alloy. *Results Phys.* **2018**, *10*, 660–664. [CrossRef]
22. Lei, L.; Zhao, Y.Q.; Zhao, Q.Y.; Wu, C.; Huang, S.X.; Jia, W.J.; Zeng, W.D. Impact toughness and deformation modes of Ti-6Al-4V alloy with different microstructures. *Mater. Sci. Eng. A* **2021**, *801*, 140411. [CrossRef]
23. Chen, Y.Y.; Du, Z.X.; Xiao, S.L.; Xu, L.J.; Tian, J. Effect of aging heat treatment on microstructure and tensile properties of a new β high strength titanium alloy. *J. Alloys Compd.* **2014**, *586*, 588–592. [CrossRef]
24. Xu, S.H.; Liu, Y.; Liu, B.; Wang, X.; Chen, Z.X. Microstructural evolution and mechanical properties of Ti-5Al-5Mo-5V-3Cr alloy by heat treatment with continuous temperature gradient. *Trans. Nonferrous Met. Soc. China* **2018**, *28*, 273–281. [CrossRef]
25. Qi, P.; Li, B.L.; Wang, T.B.; Zhou, L.; Nie, Z.R. Microstructural evolution of near-β Ti-6Zr-5Fe alloy fabricated by selective laser melting before and after solution treatment. *J. Alloys Compd.* **2021**, *862*, 158496. [CrossRef]
26. Matsumoto, H.; Yoneda, H.; Sato, K.; Kurosu, S.; Maire, E.; Fabregue, D.; Konno, T.J.; Chiba, A. Room-temperature ductility of Ti-6Al-4V alloy with α′ martensite microstructure. *Mater. Sci. Eng. A* **2011**, *528*, 1512–1520. [CrossRef]
27. Su, J.L.; Ji, X.K.; Liu, J.; Teng, J.; Jiang, F.L.; Fu, D.F.; Zhang, H. Revealing the decomposition mechanisms of dislocations and metastable α′ phase and their effects on mechanical properties in a Ti-6Al-4V alloy. *J. Mater. Sci. Technol.* **2022**, *107*, 136–148. [CrossRef]
28. Sun, L.B.; Jiang, F.C.; Huang, R.S.; Yuan, D.; Guo, C.H.; Wang, J.D. Anisotropic mechanical properties and deformation behavior of low-carbon high-strength steel component fabricated by wire and arc additive manufacturing. *Mater. Sci. Eng. A* **2020**, *787*, 139514. [CrossRef]
29. Shahmir, H.; Langdon, T.G. Using heat treatments, high-pressure torsion and post-deformation annealing to optimize the properties of Ti-6Al-4V alloys. *Acta Mater.* **2017**, *141*, 419–426. [CrossRef]
30. Wang, K.; Wu, M.Y.; Yan, Z.B.; Li, D.R.; Xin, R.L.; Liu, Q. Microstructure evolution and static recrystallization during hot rolling and annealing of an equiaxed-structure TC21 titanium alloy. *J. Alloys Compd.* **2018**, *752*, 14–22. [CrossRef]
31. Sakai, T.; Belyakov, A.; Kaibyshev, R.; Miura, H.; Jonas, J.J. Dynamic and post-dynamic recrystallization under hot, cold and severe plastic deformation conditions. *Prog. Mater. Sci.* **2014**, *60*, 130–207. [CrossRef]
32. Pradhan, S.K.; Tripathy, S.; Singh, R.; Murugaiyan, P.; Roy, D.; Humane, M.M.; Chowdhury, S.G. On the grain boundary character evolution in non equiatomic high entropy alloy during hot rolling induced dynamic recrystallization. *J. Alloys Compd.* **2022**, *922*, 166126. [CrossRef]
33. Balachandran, S.; Kashiwar, A.; Choudhury, A.; Banerjee, D.; Shi, R.P.; Wang, Y.Z. On variant distribution and coarsening behavior of the α phase in a metastable β titanium alloy. *Acta Mater.* **2016**, *106*, 347–387. [CrossRef]
34. Liu, Z.G.; Li, P.J.; Xiong, L.T.; Liu, T.Y.; He, L.J. High-temperature tensile deformation behavior and microstructure evolution of Ti55 titanium alloy. *Mater. Sci. Eng. A* **2017**, *680*, 259–269. [CrossRef]
35. Ma, X.L.; Huang, C.X.; Moering, J.; Ruppert, M.; Höppel, H.W.; Göken, M.; Narayan, J.; Zhu, Y.T. Mechanical properties of copper/bronze laminates: Role of interfaces. *Acta Mater.* **2016**, *116*, 43–52. [CrossRef]
36. Gu, J.L.; Wang, X.S.; Bai, J.; Ding, J.L.; Williams, S.; Zhai, Y.C.; Liu, K. Deformation microstructures and strengthening mechanisms for the wire+arc additively manufactured Al-Mg4.5Mn alloy with inter-layer rolling. *Mater. Sci. Eng. A* **2018**, *712*, 292–301. [CrossRef]

materials MDPI

Article

Glass-Forming Ability and Corrosion Behavior of Ti-Based Amorphous Alloy Ti-Zr-Si-Fe

Ling Bai [1,*], Ziyang Ding [1], Haiying Zhang [1] and Chunxiang Cui [2]

1 School of New Materials Engineering, Zhengzhou Technical College, Zhengzhou 450121, China
2 School of Materials Science and Engineering, Hebei University of Technology, Tianjin 300130, China
* Correspondence: bling3322@126.com; Tel.: +86-150-3831-0831

Abstract: Ti-based alloy Ti75Zr11Si9Fe5 (At %) and Ti66Zr11Si15Fe5Mo3 (At %) ribbons are fabricated by a single roller spun-melt technique, according to the three empirical rules. Both alloys are found to have a large, supercooled liquid region (ΔT_x) before crystallization that reaches 80–90 K. The results show that both alloys possess excellent glass-forming abilities. The electrochemical measurement proves both amorphous alloys possess relatively high corrosion resistance in 3 mass% NaCl solution.

Keywords: glass-forming ability; corrosion behavior; amorphous alloy

Citation: Bai, L.; Ding, Z.; Zhang, H.; Cui, C. Glass-Forming Ability and Corrosion Behavior of Ti-Based Amorphous Alloy Ti-Zr-Si-Fe. *Materials* **2022**, *15*, 7229. https://doi.org/10.3390/ma15207229

Academic Editor: Bruno Chrcanovic

Received: 15 September 2022
Accepted: 11 October 2022
Published: 17 October 2022

1. Introduction

Ti-based alloys have the advantages of high specific strength and strong corrosion resistance and are widely used as engineering materials [1–3]. At the same time, it has been found that amorphous alloys have superior mechanical properties, ductility and corrosion resistance compared to crystalline alloys [4–7]. Therefore, in recent decades, people have shown more and more interest in Ti-based amorphous alloys, and a lot of scientific research has been carried out [8,9]. Compared with crystalline Ti-based alloys, amorphous titanium alloys exhibit higher mechanical properties [10,11], low elastic modulus, excellent biocompatibility and high corrosion resistance. With the development of titanium-based amorphous alloys, Ti-rich bulk glassy alloys were found in rapidly solidified Ti-Cu-Al [12,13], Ti-Cu-Ni (Co) [14–16], Ti-Cu-Zr-Ni [17,18], Ti-Cu-Co-Al-Zr [19] and Ti-Cu-Ni-Sn (Si) [20,21] systems to date. Since 1988, many alloys with high glass-forming ability and capable of being formed into bulk metallic glasses (BMG) have been discovered more and more, for example, in multicomponent Mg-, La-, Zr-, Fe- and Pd-based metals in the alloy system. Later, Takeuchi and Inoue [22] classified BMGs into seven groups by studying the chemical species, internal atomic size differences and mixing heat of the constituent elements. This classification helps to grasp the various characteristics of BMG. In G-IV and G-VI, the Ti-based BMGs are generalized. However, the constituent elements are Ti-Cu-Ni and Ti-Zr-Cu systems, respectively. In this paper, Ti75Zr11Si9Fe5 (At %) and Ti66Zr11Si15Fe5Mo3 (At %) are fabricated successfully, whose constituent elements are Ti-Zr-Si. The main reason for choosing Ti75Zr11Si9Fe5 (At %) and Ti66Zr11Si15Fe5Mo3 (At %) in this research study is that the addition of alloying elements has a great influence on the phase composition and glass-forming ability of amorphous alloys. When selecting the added elements, it is necessary to consider not only the good glass-forming ability after addition but also the compatibility and harmfulness of the implants with the human body. For example, the metals Cu and Ni will cause cytotoxicity and neurotoxicity to the human body, while Al may cause Alzheimer's and other symptoms. The novelty of this work includes three main aspects. In the first aspect, new ti-based amorphous alloys Ti75Zr11Si9Fe5 (At %) and Ti66Zr11Si15Fe5Mo3 (At %) are designed and developed for the first time. Secondly, Ti75Zr11Si9Fe5 (At %) and Ti66Zr11Si15Fe5Mo3 (At %) amorphous alloys have been prepared for the first time, which have good glass-forming ability. Thirdly, the corrosion resistance of the two new titanium-based alloys is significantly improved compared with pure titanium and Ti6Al4V alloys. The results show that both alloys possess excellent glass-forming ability. In addition, this paper also presents the thermal stability and corrosion behavior in NaCl solutions of both alloys.

2. Experimental Procedures

There are two main hypotheses in this study. Hypothesis 1, Ti75Zr11Si9Fe5 (At %) and Ti66Zr11Si15Fe5Mo3 (At %) have good glass-forming ability. Hypothesis 2, Ti75Zr11Si9Fe5 (At %) and Ti66Zr11Si15Fe5Mo3 (At %) amorphous alloys have good corrosion resistance. The alloy ribbons with a width of 5–8 mm and thickness of about 80 um were fabricated from the buttony Ti-based alloy ingots with nominal compositions of Ti75Zr11Si9Fe5 (At %) and Ti66Zr11Si15Fe5Mo3 (At %) by a single-roller spun-melt technique under an argon atmosphere. The phase composition and microstructure of the as-quenched alloy ribbons were examined by X-ray diffraction (XRD) using Cu K_α radiation and transmission electron microscopy (TEM), respectively. The thermal stability associated with glass transition temperature (T_g) and crystallization temperature (T_x) was investigated by differential scanning calorimetry (DSC) at a heating rate of 10 K/s. The corrosion behavior of both Ti75Zr11Si9Fe5 (At %) and Ti66Zr11Si15Fe5Mo3 (At %) was evaluated by electrochemical measurement. The electrolyte of 3% NaCl solution open to air was used at room temperature (about 298 k). The electrochemical measurements were conducted in a three-electrode cell using a saturated calomel (SCE) and graphite counter electrodes. The potentiodynamic cathodic and anodic polarization curves were measured separately at a potential sweep rate of 50 mV min^{-1}. Hypothesis 1 can be supported by the XRD experiment and DSC experiment. Hypothesis 2 can be supported by electrochemical measurements.

3. Results and Discussion

3.1. *Amorphous Structure of the Melt-Spun*

Figure 1 below shows the melt-spun amorphous structure found in the patterns obtained from XRD experiments. From Figure 1, a broad peak can be clearly observed in the 2θ range of 31–48°, and binding experiments can analyze that this broad peak is produced by the amorphous phase. This indicates that both alloys designed have very good amorphous-forming ability. However, it can be seen from the figure that the glass-forming ability of the two alloys is not the same, and the alloy Ti66Zr11Si15Fe5Mo3 (At %) is better than that of Ti75Zr11Si9Fe5 (At %). The high glass-forming ability (GFA) of these two alloys by stabilizing the supercooled liquid region is mainly due to the formation of a new type of glassy structure; the structure has some special features, such as high-density packing, a new local atomic configuration and long-range homogeneity with attractive interactions [23–25].

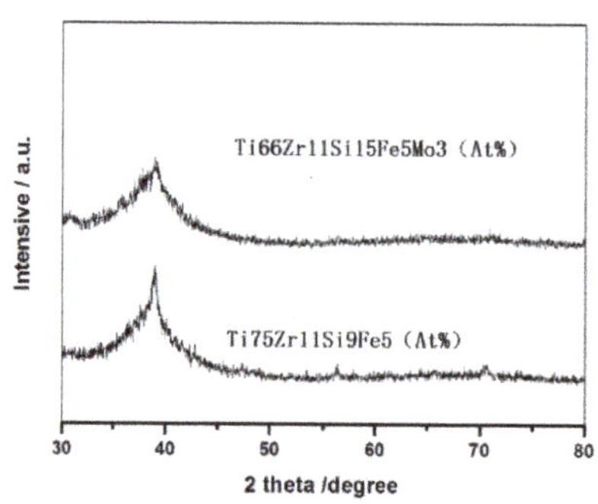

Figure 1. X-ray diffraction patterns of the as-quenched Ti66Zr11Si15Fe5Mo3 (At %) and Ti75Zr11Si9Fe5 (At %).

In this paper, guided by the three empirical theories of Japanese scholar Inoue, combined with the theory of supercooled region before the crystallization of molten liquid in thermodynamics, as well as the most basic principle of a phase diagram, etc., the glass-forming ability of the two alloys was studied. Except for Pd-based alloys, most of the currently reported bulk glass alloys have those simple empirical composition rules. On the basis of a comprehensive study of important quantities such as alloying element additions, the different atomic size ratios and the interaction between constituent elements, Inoue outlined the three empirical rules [22]. First, the Ti75Zr11Si9Fe5 (At %) and Ti66Zr11Si15Fe5Mo3 (At %) alloys used in the experiment contain at least four types of elements, both of which exceed three. In the Ti66Zr11Si15Fe5Mo3 (At %) alloy, compared with the Ti75Zr11Si9Fe5 (At %) alloy, Mo is added, and the addition of this element makes the alloy more in line with the three empirical rules [22], which may be the reason the amorphous phase formation ability of the alloy is better than that of Ti75Zr11Si9Fe5 (At %). Second, the atomic size ratios between the main components of the two alloys both exceed 12% and are significantly different. The rearrangement of constituent elements of different atomic sizes may lead to higher packing densities. During the formation of titanium-based amorphous alloys, constituent elements of different atomic sizes are rearranged, and this process results in alloys with higher stacking densities. The atomic dimensions of the elements are derived from the atomic radii listed in the data book [26], which are calculated as half the distance between atoms in the crystalline state.

Table 1 lists the atomic radius of different elements. The atomic radii of both alloys exhibit the following rules: Si < Fe < Ti < Zr. The heat of mixing between the components of the alloy is negative. Since the crystallization of amorphous alloys requires a large amount of activation energy, the existence of atomic pairs with different negative mixing heats in amorphous alloys can greatly improve the thermal stability of supercooled liquids. In addition, in the amorphous alloy Ti66Zr11Si15Fe5Mo3 (At %), due to the addition of Mo, the degree of negative mixing heat between atomic pairs in the alloy is greatly deepened, such as the existence of atomic pairs such as Ti-Mo and Si-Mo, and this can increase the value of the negative heat of mixing in the alloy system. In addition, in the amorphous alloy Ti66Zr11Si15Fe5Mo3 (At %), due to the addition of Mo, the number of atomic pairs with negative heat of mixing, such as Ti-Mo and Si-Mo, can be effectively increased. Here, we express the heat of mixing between pairs of alloy atoms by the enthalpy of mixing between pairs of atoms containing equal atomic contents. The values of heat of mixing were quoted as the enthalpy of mixing (ΔH_{AB}^{mix}) [27,28] of the binary liquid in an A–B systems at an equi-atomic composition.

Table 1. Atomic radius of the elements/nm.

Element	Ti	Zr	Fe	Si	Mo
Atomic radii	0.417	0.162	0.124	0.117	0.136

Table 2 lists the values of ΔH_{AB}^{mix} (KJ/mol), for the atomic pairs between the two alloying elements were calculated according to the Miedema model. Generally, the mixing enthalpy of the solid alloy solution is affected by structural factors of the material itself, the chemical and elastic [29]. However, since, in a liquid (glassy) melt, there are no elastic and structural factors, in the research, the mixing enthalpy only considers chemical factors, and the mixing enthalpy can be determined based on the regular melt model [29,30], where $\Omega_{ij}(=4\Delta H_{AB}^{mix})$ is the regular melt interaction parameter between the ith and jth elements, c_i is the atomic percentage of the ith component, and ΔH_{AB}^{mix} is the mixing enthalpy of binary liquid alloys.

$$\Delta H^{mix} = \sum_{i=1, i\neq j}^{n} \Omega_{ij} c_i c_j$$

The calibration value of ΔH_{AB}^{mix} is used as $\Delta H_{AB}^{mix(cali)} = \Delta H_{AB}^{mix} - \Delta H^{trans}/2$ for containing one nontransition metal and $\Delta H_{AB}^{mix(cali)} = (\Delta H_i^{trans} + \Delta H_j^{trans})/2$ for containing two nontransition metals. ΔH^{trans} are 100, 30, 180, 310, 17, 34 and 25 KJ mol^{-1} for containing H, B, C, N, Si, P and Ge [29], respectively. In summary, the enthalpies of mixing of the both alloys are −159.348 KJ/mol and −117.8748 KJ/mol, respectively. It is worth noting that the negative effect of the former alloy is greater than that of the latter alloy, which is part of the reason why the amorphous forming ability of Ti66Zr11Si15Fe5Mo3 (At %) is better than Ti75Zr11Si9Fe5 (At %). To sum up, the alloy system contains more than three alloys, the atomic size difference between the main components is greater than 12% and the large negative mixing enthalpy can lead to an alloy with a lower atomic dispersibility and glassy structure with a higher atomic stacking density. In general, the higher the stacking density of atoms, the better the thermal stability of the amorphous alloy and the greater the resistance of the supercooled liquid to the crystal transformation—that is, the higher the glass-forming ability of the amorphous alloy [31].

Table 2. The values of ΔH_{AB}^{mix}/KJ/mol calculated by Miedema's model for atomic pairs between elements.

	Ti	**Zr**	**Fe**	**Si**	**Mo**
Ti	0	0	−17	−66	−4
Zr	0	0	−25	−84	−6
Fe	−17	−25	0	−35	−2
Si	−66	−84	-35	0	−35
Mo	−4	−6	−2	−35	0

3.2. Microstructure Morphologies

The SEM microstructure of Ti75Zr11Si9Fe5 (At %) and Ti66Zr11Si15Fe5Mo3 (At %) is shown in Figure 2. It can be seen from Figure 2 that the SEM microstructures of the two alloys have relatively simple surface morphologies, showing a disordered glass structure. Compared with Ti75Zr11Si9Fe5 (At %), Ti66Zr11Si15Fe5Mo3 (At %) has smaller grains in the structure. The main reason is that the atomic content of metalloid element Si is increased in the alloy Ti66Zr11Si15Fe5Mo3 (At %), and there is a large atomic size difference between Si and the main metal element Ti and a large negative mixed heat; this greatly increases the topologically disordered arrangement of atoms or molecules in the three-dimensional space of the alloy system, which hinders the progress of crystallization.

The TEM morphology of Ti75Zr11Si9Fe5 (At %) and Ti66Zr11Si15Fe5Mo3 (At %) is shown in Figure 3. It can be seen from the figure that the grain boundaries of the matrix phases of the two alloys are blurred, and even no grain boundaries can be seen in some areas, which is very similar to the amorphous structure of metallic glass. At the same time, a large number of nanoparticles cause precipitation along the fuzzy grain boundaries of the alloy, especially at the intersection of multiple grain boundaries, and the particle size of the precipitates is significantly larger than the other grain boundaries. The main reason is that the second phase has a greater probability of nucleation in the high-energy region at the defect, and the grain boundaries, dislocations and vacancies belong to such high-energy regions, and, relatively speaking, the grain boundary is more conducive to the second phase: nucleation. In addition, a sufficient amount of elements in the alloy are beneficial to the formation of the β-phase structure, so that a sufficient amount of β-phase structure has been formed when the alloy is in the molten state. The lower phase can also appear in the higher energy grain boundary region in the form of a second phase.

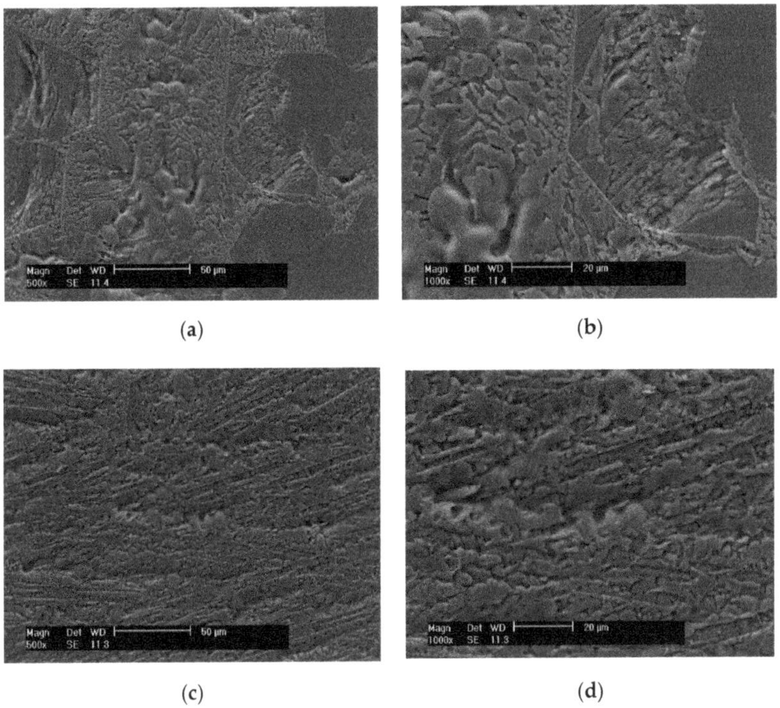

Figure 2. The SEM microstructure of Ti75Zr11Si9Fe5 (At %) and Ti66Zr11Si15Fe5Mo3 (At %): (**a**) Ti75Zr11Si9Fe5 (At %) (500×), (**b**) Ti75Zr11Si9Fe5 (At %) (1000×), (**c**) Ti66Zr11Si15Fe5Mo3 (At %) (500×) and (**d**) Ti66Zr11Si15Fe5Mo3 (At %) (1000×).

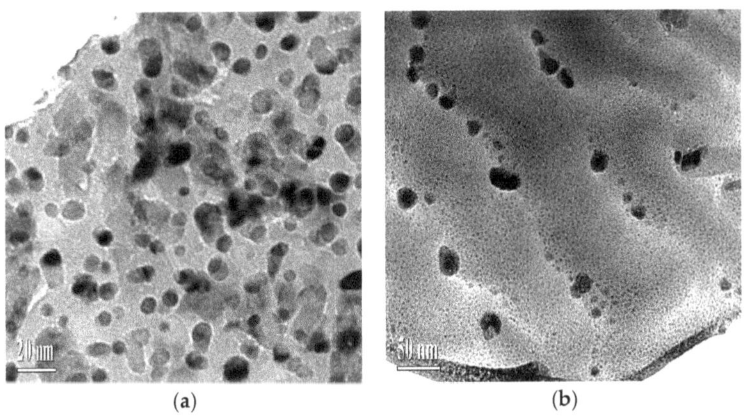

Figure 3. TEM morphology of Ti75Zr11Si9Fe5 (At %) and Ti66Zr11Si15Fe5Mo3 (At %): (**a**) Ti75Zr11Si9Fe5 (At %) and (**b**) Ti66Zr11Si15Fe5Mo3 (At %).

3.3. Analysis of DSC

Figure 4 shows the DSC thermograms of amorphous alloys of both Ti75Zr11Si9Fe5 (At %) and Ti66Zr11Si15Fe5Mo3 (At %). From the figure, it can be found that there is an obvious and broad glass transition region. Due to the large atomic size difference and negative mixing heat among the multi-components of the bulk amorphous alloy, the nucleation and growth rate of the crystalline phase of the alloy are inhibited. As a result, a stable supercooled liquid phase region appears in the alloy before crystallization. It can be seen from the figure that there are differences in the supercooled liquid phase (ΔT_x) of the two alloys before crystallization, which are about 80 K and 90 K, respectively, and this is not common in the Ti-based amorphous alloy system. The single exothermic peak is generated by the interaction of several crystal phases. The crystalline mode implies that the atomic rearrangements of the constituent elements on a long-range scale are necessary for the progress of the crystallization reaction. This inevitably leads to a delay in the crystallization reaction, resulting in a high thermal stability of the supercooled liquid. In such supercooled liquids, the topological and chemical short-range order will be greatly enhanced, and the structure will also change, which is conducive to the formation of high-density random packing structures, which are typically characterized by low atomic diffusivity. In general, the higher the stacking density of atoms, the better the thermal stability of the amorphous alloy and the greater the resistance of the supercooled liquid to the crystal transformation—that is, the higher the glass-forming ability of the amorphous alloy [31]. All these further improve that both Ti75Zr11Si9Fe5 (At %) and Ti66Zr11Si15Fe5Mo3 (At %) alloy ribbons with a large, supercooled liquid region possesses a high glass-forming ability and can be consolidated into a bulk form by taking advantage of the large viscous flow in the supercooled liquid region.

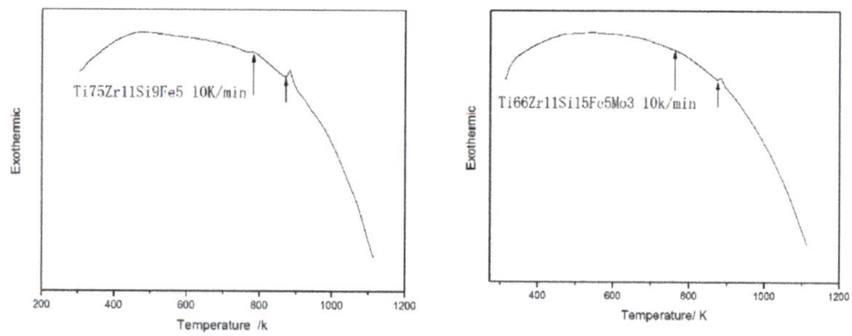

Figure 4. DSC thermograms of amorphous alloys of both Ti75Zr11Si9Fe5 (At %) and Ti66Zr11Si15Fe5Mo3 (At %).

3.4. The Potentiodynamic Polarization Curve

In order to gain a deeper and more comprehensive understanding of the corrosion behavior of these two amorphous alloys, electrochemical measurement was performed in 3 mass% NaCl solution. Before the start of the corrosion test, the alloy specimens were mechanically polished in cyclohexane with silicon carbide paper up to grit 2000, degreased in acetone, washed in distilled water, dried in air and further exposed to air for 24 h for good reproducibility. Figure 5 shows the potentiodynamic polarization curve of both Ti75Zr11Si9Fe5 (At %) and Ti66Zr11Si15Fe5Mo3 (At %) amorphous alloys in 3 mass% NaCl solution open to air at 298 K. As seen in Figure 5, pure Ti and Ti6Al4V alloys are passivated with different current densities. However, both amorphous alloys kept steady current densities at about 0 A.m^{-2}, which suggests that both were hardly eroded by the 3 mass% NaCl solution. Based on the data, it is concluded that the both amorphous alloys relatively possess excellent corrosion resistance compared with pure Ti and Ti6Al4V alloys in the Cl^--containing solutions or brine. Ti and Zr are known as strong passive valve metals in aggressive environments, and the enrichment of Ti and Zr in the surface film was found in Ti- and Zr-containing Ni-based glassy alloys [31]. This indicates that the improved corrosion resistance of the two amorphous alloys is mainly due to the formation of a protective film rich in Ti and Zr on the surface of the alloys in the NaCl solution. In addition, the improvement of corrosion resistance is also related to the structural and chemical uniformity of amorphous alloys.

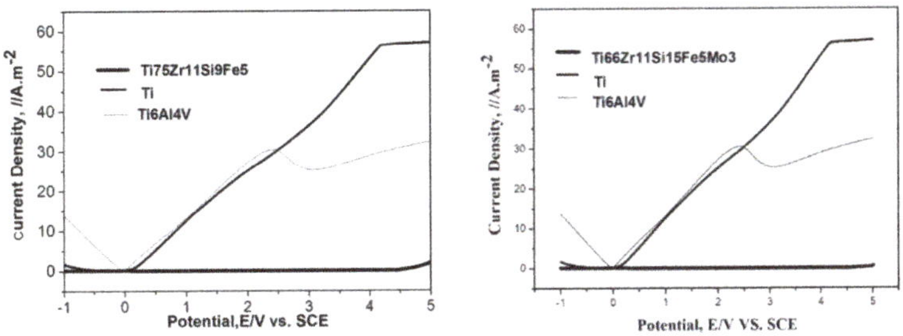

Figure 5. The potentiodynamic polarization curve of both Ti75Zr11Si9Fe5 (At %) and Ti66Zr11Si15Fe5Mo3 (At %) amorphous alloys in 3 mass% NaCl solution open to air at 298 K.

4. Conclusions

(1) Ti−based alloys with nominal compositions of Ti75Zr11Si9Fe5 (At %) and Ti66Zr11Si15Fe5Mo3 (At %) have excellent amorphous-forming ability, which both satisfies the three empirical rules.
(2) Both 75Zr11Si9Fe5 (At %) and Ti66Zr11Si15Fe5Mo3 (At %) amorphous alloys have a large, supercooled liquid region before crystallization, which is important to the amorphous-forming ability of the alloys.
(3) The addition of Si and Mo improve the amorphous-forming ability of Ti66Zr11Si15Fe5Mo3 (At %) alloy.
(4) Both 75Zr11Si9Fe5 (At %) and Ti66Zr11Si15Fe5Mo3 (At %) possess relatively excellent corrosion resistance, compared with pure Ti and Ti6Al4V alloys.

In this study, in order to ensure the accuracy of the two alloy compositions and the reliability of the results, it is suggested to test the composition of the two alloys before the experiment. Titanium-based alloys are widely used in biomedical fields because of their excellent biocompatibility and biological activity. Especially as a biological implant material, it plays an important role. The elastic modulus, excellent biocompatibility and excellent corrosion resistance make it develop rapidly. The results of this study about Ti75Zr11Si9Fe5 (At %) and Ti66Zr11Si15Fe5Mo3 (At %) will be beneficial to promote the development and application of titanium alloy implant materials.

Author Contributions: Conceptualization, L.B.; Experiment, L.B. and C.C.; Methodology, H.Z.; Resources and Data Curation, H.Z. and Z.D.; Writing—Original Draft Preparation, L.B. and Writing—Review and Editing, L.B. and Z.D. All authors have read and agreed to the published version of the manuscript.

Funding: This research was funded by the Program of 2021 Henan Province Key R&D and Promotion Project (Science and Technology) (212102210551).

Institutional Review Board Statement: Not applicable.

Informed Consent Statement: Not applicable.

Data Availability Statement: The data presented in this work are available on request from the corresponding authors.

Acknowledgments: Many thanks to the support of the research group members.

Conflicts of Interest: The authors declare no conflict of interest.

References

1. Straumal, B.B.; Gornakova, A.S.; Kilmametov, A.R.; Rabkin, E.; Anisimova, N.Y.; Kiselevskiy, M.V. β-Ti-Based Alloys for Medical Applications. *Russ. J. Non-Ferr. Met.* **2021**, *62*, 54–63. [CrossRef]
2. Jung, H.D. Titanium and Its alloys for biomedical applications. *Metals* **2021**, *11*, 1945. [CrossRef]
3. Zarka, M.; Dikici, B.; Niinomi, M.; Ezirmik, K.V.; Nakai, M.; Yilmazer, H. A systematic study of β-type Ti-based PVD coatings on magnesium for biomedical application. *Vacuum* **2021**, *183*, 109850. [CrossRef]

4. Jiang, H.; Shang, T.; Xian, H.; Sun, B.; Zhang, Q.; Yu, Q.; Bai, H.; Gu, L.; Wang, W. Structures and Functional Properties of Amorphous Alloys. *Small Struct.* **2020**, *2*, 2000057. [CrossRef]
5. Li, X.; Cai, W.; Li, D.-S.; Xu, J.; Tao, H.; Liu, B. Amorphous alloys for electrocatalysis: The significant role of the amorphous alloy structure. *Nano Res.* **2021**, 1–12. [CrossRef]
6. Wu, Y.Q.; Zhang, Y.; Zhang, T. Application of 3D Balanced Growth Theory to the Formation of Bulk Amorphous Alloys. *Front. Mater.* **2021**, *8*, 694920. [CrossRef]
7. Li, X.; Zhang, L.; Shi, T.; Yu, W.; Shao, J.; Zhou, X.; He, Y.; Zhang, B. Preparation and electrochemical characterization of Ti-based amorphous metallic glass with high strength. *Mater. Corros.* **2020**, *72*, 951–959. [CrossRef]
8. Chen, M.; Wei, D.; Li, C.; Liu, L.; Zhou, Q.; Wu, F. Influence of laser pulse on solidification of molten pool during laser welding of dissimilar Ti-based amorphous alloys. *Sci. Technol. Weld. Join.* **2021**, *26*, 264–272. [CrossRef]
9. Li, K.; Jiao, Z.; Li, Z.; Wang, Z. Strain-rate effect on yielding behavior of an in-situ Ti-based metallic glass matrix composite upon dynamic compression. *Mater. Sci. Eng. A* **2021**, *815*, 141267. [CrossRef]
10. Jia, Q.; Zhou, Q.; Ren, Y.; Du, Y.; Zhao, X.; Wang, X.-Z.; Wang, H.; Beake, B.D.; Zhou, F. Tribological characteristics of Ti-based bulk metallic glass via deep cryogenic-cycling treatment. *Mater. Charact.* **2021**, *179*, 111356. [CrossRef]
11. Wang, F.; Yin, D.; Lv, J.; Zhang, S.; Ma, M.; Zhang, X.; Liu, R. Effect on microstructure and plastic deformation behavior of a Zr-based amorphous alloy by cooling rate control. *J. Mater. Sci. Technol.* **2021**, *82*, 1–9. [CrossRef]
12. Cui, C.; Bai, L.; Liu, S.; Qi, Y.; Zhao, L. Ti 5 Si 3/β-Ti nano-core–shell structure toughened glassy Ti alloy matrix biocomposites. *RSC Adv.* **2015**, *5*, 8355–8361. [CrossRef]
13. Jin, Z.S.; Yang, Y.J.; Zhang, Z.P.; Ma, X.Z.; Lv, J.W.; Wang, F.L.; Ma, M.Z.; Zhang, X.Y.; Liu, R.P. Effect of Hf substitution Cu on glass-forming ability, mechanical properties and corrosion resistance of Ni-free Zr–Ti–Cu–Al bulk metallic glasses. *J. Alloy. Compd.* **2019**, *806*, 668–675. [CrossRef]
14. Shan, F.; Sun, T.; Song, W.; Peng, C.; Sun, H.; Gong, J.; Song, K. A bridge from metallic glasses to medium-entropy alloys in Ti-Cu-Zr-Pd-Co system: Design, microstructure, and deformation-induced-martensitic transformation. *J. Non-Crystalline Solids* **2022**, *587*, 121608. [CrossRef]
15. Lee, S.Y.; Lee, H.J.; Baek, J.H.; Park, S.S.; Lee, J.G. Microstructural and Corrosion Properties of Ti-to-Zr Dissimilar Alloy Joints Brazed with a Zr-Ti-Cu-Ni Amorphous Filler Alloy. *Metals* **2021**, *11*, 192. [CrossRef]
16. Yue, X.; Hu, R.; Qi, J.; He, Y.; Meng, Q.; Wei, F.; Ren, Y.; Sui, Y. Fabrication and Degradation Properties of Nanoporous Copper with Tunable Pores by Dealloying Amorphous Ti-Cu Alloys with Minor Co Addition. *J. Mater. Eng. Perform.* **2021**, *30*, 1759–1767. [CrossRef]
17. Chen, F.C.; Dai, F.P.; Yang, X.Y.; Ruan, Y.; Wei, B.B. Effect of Sn and Al additions on the microstructure and mechanical properties of amorphous Ti–Cu–Zr–Ni alloys. *Chin. Phys. B* **2020**, *29*, 066401. [CrossRef]
18. Kang, Y.; Feng, K.; Zhang, W.; Mao, Y. Microstructural and mechanical properties of CFC composite/Ti6Al4V joints brazed with Ag–Cu–Ti and refractory metal foils. *Arch. Civ. Mech. Eng.* **2021**, *21*, 1–10. [CrossRef]
19. Prabhu, Y.; Vincent, S.; Bhatt, J. Thermodynamic modelling to optimize glass forming composition in multicomponent Zr-Cu-Co-Al system. *Mater. Today Proc.* **2020**, *28*, 1239–1244. [CrossRef]
20. Louzguine-Luzgin, D.V. High-strength Ti-based alloys containing Fe as one of the main alloying elements. *Mater. Trans.* **2018**, M2018114. [CrossRef]
21. Park, H.J.; Lee, H.J.; Kim, T.K.; Hong, S.H.; Wang, W.M.; Choi, T.J.; Kim, K.B. Formation of photo-reactive heterostructure from a multicomponent amorphous alloy with atomically random distribution. *J. Mater. Sci. Technol.* **2022**, *109*, 245–253. [CrossRef]
22. Takeuchi, A.; Inoue, A. Classificaion of Bulk Metallic Glasses by Atomic Size Difference, Heat of Mixing and Period of Constituent Elements and Its Application to Characterization of the Main Alloying Element. *Mater. Trans.* **2005**, *46*, 2817–2829. [CrossRef]
23. Kim, W.C.; Kim, Y.J.; Kim, Y.S.; Hyun, J.I.; Hong, S.H.; Kim, W.T.; Kim, D.H. Enhancement of superelastic property in Ti–Zr–Ni–Cu alloy by using glass alloy precursor with high glass forming ability. *Acta Mater.* **2019**, *173*, 130–141. [CrossRef]
24. Maiorova, A.V.; Kulikova, T.V.; Bykov, A.S. New criteria for predicting compositions with high glass-forming ability in Zr–Co–Al alloys. *J. Non-Cryst. Solids* **2022**, *594*, 121812. [CrossRef]
25. Saini, J.S.; Miska, J.P.; Lei, F.; AuYeung, N.; Xu, D. Hafnium based metallic glasses with high density and high glass-forming ability. *J. Alloys Compd.* **2021**, *882*, 160896. [CrossRef]
26. Ma, E. Unusual dislocation behavior in high-entropy alloys. *Scr. Mater.* **2020**, *181*, 127–133. [CrossRef]
27. Boer, F.R.; Perrifor, D.G. *Cohesion in Metals*; Elsevier Science Publishers B. V.: Amsterdam, The Netherlands, 1988.
28. Zeng, X.; Liu, Z.; Wu, G.; Tong, X.; Xiong, Y.; Cheng, X.; Wang, X.; Yamaguchi, T. Microstructure and high-temperature properties of laser cladded AlCoCrFeNiTi0.5 high-entropy coating on Ti 6Al-4V alloy. *Surf. Coat. Technol.* **2021**, *418*, 127243. [CrossRef]
29. LaRosa, C.R.; Shih, M.; Varvenne, C.; Ghazisaeidi, M. Solid solution strengthening theories of high-entropy alloys. *Mater. Charact.* **2019**, *151*, 310–317. [CrossRef]
30. Luan, H.; Zhang, X.; Ding, H.; Zhang, F.; Luan, J.H.; Jiao, Z.B.; Yang, Y.-C.; Bu, H.; Wang, R.; Gu, J.; et al. High-entropy induced a glass-to-glass transition in a metallic glass. *Nat. Commun.* **2022**, *13*, 1–11. [CrossRef]
31. Bocchetta, P.; Chen, L.Y.; Tardelli, J.D.C.; Reis, A.C.D.; Almeraya-Calderón, F.; Leo, P. Passive Layers and Corrosion Resistance of Biomedical Ti-6Al-4V and β-Ti Alloys. *Coatings* **2021**, *11*, 487. [CrossRef]

materials

MDPI

Article

Versatile Medium Entropy Ti-Based Bulk Metallic Glass Composites

Tianzeng Liu [1,2], Yanchun Zhao [1,*], Li Feng [1] and Pan Gong [3,*]

1 State Key Laboratory of Advanced Processing and Recycling of Non-Ferrous Metals, Lanzhou University of Technology, Lanzhou 730050, China
2 Iron and Steel Research Institute, Jiuquan Iron and Steel Corporation, Jiayuguan 735100, China
3 State Key Laboratory of Materials Processing and Die & Mould Technology, School of Materials Science and Engineering, Huazhong University of Science and Technology, Wuhan 430074, China
* Correspondence: yanchun_zhao@163.com (Y.Z.); pangong@mail.hust.edu.cn (P.G.)

Abstract: An ultra-strong Ti-based bulk metallic glass composite was developed via the transformation-induced plasticity (TRIP) effect to enhance both the ductility and work-hardening capability of the amorphous matrix. The functionally graded composites with a continuous gradient microstructure were obtained. It was found that the austenitic center possesses good plasticity and toughness. Furthermore, the amorphous surface exhibited high strength and hardness, as well as excellent wear corrosion resistance. Compared with the Ti-6Al-4V alloy, bulk metallic glass composites (BMGCs) exhibit better spontaneous passivation behavior during the potential dynamic polarization. No crystallization was observed on the friction surface, indicating their good friction-reduction and anti-wear properties.

Keywords: bulk metallic glass; phase transformation; mechanical behavior; corrosion wear resistance

Citation: Liu, T.; Zhao, Y.; Feng, L.; Gong, P. Versatile Medium Entropy Ti-Based Bulk Metallic Glass Composites. *Materials* **2022**, *15*, 7304. https://doi.org/10.3390/ma15207304

Academic Editor: Mihai Stoica

Received: 1 October 2022
Accepted: 17 October 2022
Published: 19 October 2022

Publisher's Note: MDPI stays neutral with regard to jurisdictional claims in published maps and institutional affiliations.

1. Introduction

Because of the excellent integrated properties, bulk metallic glasses (BMGs) have become a shining rookie of the material world. They can be used as the functional structural materials with excellent physical, chemical and mechanical properties [1,2]. Ti-based BMGs have drawn special interest due to their high specific strength, good corrosive wear resistance and excellent biocompatibility, and are widely used in medicine, environmental protection, energy conservation, aerospace and other military industries [3]. As known, among the traditional biomedical materials, titanium-based alloys have become the first choice of orthopedic and dental implant products due to their good biocompatibility and mechanical properties, high corrosion resistance and relatively low Young's modulus [4,5]. Biomedical titanium based amorphous alloys possess higher strength and hardness, which allows them to exhibit better wear resistance under dry and wet friction [6].

However, a significant drawback of BMGs includes the limited strain softening with loading, which restricts their application as an advanced structural material. Furthermore, BMGs are significantly less ductile as compared to crystalline alloys. The plastic deformation of crystalline materials typically occurs by dislocation motion. Due to the periodic arrangement of atoms and the long-range translational symmetry, the propagation of dislocations can be carried out at lower energy or stress states. Although dislocations do not exist in BMGs, the plastic deformation of BMGs occur via the rearrangement of localized atoms. This rearrangement requires higher energy or stress than dislocations and is not easy to slip, thus forming localized softening shear bands. The shear band is the main carrier of the plastic deformation of BMGs and typically forms during the percolation of the stress-induced shear transition region (the active atomic cluster motion region) in the nanoscale layer.

The appearance of in-situ BMGCs is predestined. As is well known, transformation-induced Plasticity (TRIP) is an effective way to enhance the work-hardening ability of conventional materials. For instance, TRIP has been widely used in steel and ceramics to enhance their ductility, toughness and work-hardening behavior. Recently, a great amount of remarkable works utilized the TRIP method to develop some novel advanced structural alloys. For instance, this method was used to improve both the strength and ductility in BMGs and high entropy alloys (HEAs) [7–9]. Furthermore, some alloy systems have the possibility of producing in situ composites which consist of shape memory crystals and glassy matrix [10,11].

To beat the strength and ductility trade-off, in-situ BMGCs intervene in the proliferation and expansion of the multiple main plastic deformation carrier, shear bands (SBs). However, SBs form during percolation of the localized stress-induced shear transition region in the nanoscale layer and thus usually induce the strain softening as loading. To improve the stain hardening, TRIP-BMGCs introduce the crystals which can realize stress-induced martensite phase transition and twinning to guarantee a strong work-hardening rate and toughening degree. Based on current understanding, for crystalline alloys the increasing stress-induced martensite transformation via dislocations pile-up in the unstable austenite phase, which delays the plastic deformation period and thereby enhances both the strength and plasticity. However, the strong work hardening behavior mechanism, how the amorphous matrix interacts with the crystals, is not fully explained. In this paper, a novel system of the low elastic modulus, ultra-strong Ti-based BMGCs were developed, through TRIP effect of the B2/B19′ dual shape-memory crystals (DSMCs) to enhance both the ductility and work-hardening capability. The functionally graded composites with a continuous gradient microstructure were obtained during the casting solidification. The amorphous surface of the BMGC was found to exhibit high strength and hardness, as well as excellent wear resistance and corrosion resistance. Importantly, since the Ti-based BMGCs exhibit elastic modulus values that are comparable to those for biological organisms, they have potential for use as implant materials. Meanwhile, the austenitic center of the Ti BMGC possesses good plasticity and toughness. Thus, the DSMC toughened Ti-based BMGCs are expected to have a wide range of application as functional and structural materials [12–14].

For this work, to better understand the strong work hardening behavior mechanism, as well as the interaction between amorphous matrix and the crystals, a systematic investigation was conducted to fundamentally understand the effect of DSMC on the macro and micro-mechanical behavior, work-hardening mechanism and functional properties of Ti BMGCs. To accomplish this endeavor, the microstructure, compressive properties and micro-mechanical, nano fretting and creep behaviors at the ambient temperature were examined. Thus, the connection between the macro-mechanical behavior and the microstructure of the alloy, the nanofretting behavior and creep strain in different regions of the sample, which are rarely reported, will be studied. Additionally, the work-hardening response and wear corrosion behavior in different mediums were analyzed, and will expand the application of Ti-based BMGCs as functional and structural materials

2. Materials and Methods

The ingots of the $Ti_{40}Ni_{40}Cu_{20}$ (atomic percent) alloys were prepared by mixing the constituent elements with purities >99.99 (weight percent) in a high-frequency induction vacuum furnace with a water-cooled Cu crucible under an argon atmosphere. Each ingot was re-melted at least six times to ensure a homogeneous microstructure. Ti-based cylinder samples with Φ3 mm were prepared by suction casting into copper mold. The microstructure of the alloys was studied, using the X-ray diffractometer (XRD, D/max-2400, Rigaku, Tokyo, Japan) with Cu-Kα radiation (40kV-30mA), thermal field-emission scanning electron microscopy (SEM, JSM-5600LV, JEOL, Akishima, Japan) and transmission electron microscopy (TEM, JEM-2010, JEOL, Akishima, Japan) coupled with the selected area diffraction pattern (SADP). The thermal response of the alloy was investigated using a differential scanning calorimeter (DSC, Nietzsche STA449F3, Munich, Germany) under flowing purified argon at a heating rate of 20 K/min.

For compression tests, Φ3 mm cylinder samples with a height-diameter ratio of 2:1 were fabricated. The two compression faces of each sample were then carefully polished such that they were parallel to one other. For the mechanical testing experiments, an Instron 3382 electronic-testing machine was employed. Here, samples were subjected to a strain rate of 5×10^{-4} s^{-1}. The nanoindentation measurements were carried out at room temperature using a Hysitron TI-950 in-situ nanomechanical testing system equipped with a diamond Berkovich tip (Minneapolis, MN, USA). The load resolution and background noise were less than 1 and 30 nN, respectively. Furthermore, the displacement resolution and background noise were less than 0.02 and 0.2 nm, respectively. The specimens were mechanically polished to a mirror finish before nanoindentations. The displacement–load curves of the edge, transition and center zones were obtained by loading and unloading at a constant loading rate of 0.2 mN/s, loading up to the maximum 10 mN for 10 s, then unloading to 0 mN at the same loading rate.

Electrochemical measurements were conducted using the Potentiostat Workstation μAutolab Type III. Tests were carried out using a three-electrode cell equipped with a platinum counter electrode (CE) and an Ag/AgCl reference electrode (RE). The samples were immersed in solutions, where they were used as working electrodes (WE). The tests were performed in 310 K Phosphate-buffered Saline (PBS) solution (8 g/L NaCl, 0.2 g/L KCl, 0.14 g/L NaH_2PO_4, 0.2 g/L KH_2PO_4, prepared from analytic reagents and distilled water, and diluted with 1mol/L NaOH up to PH 7.3), and 298 K Artificial Sea Water (ASW) (24.53 g/L NaCl, 5.20 g/L $MgCl_2$, 4.09 g/L Na_2SO_4, 1.16 g/L $CaCl_2$, 0.695 g/L KCl, 0.201 g/L $NaHCO_3$, 0.101 g/L KBr, 0.027 g/L H_3BO_3, 0.025 g/L $SrCl_2$, 0.003 g/L NaF, prepared from analytic reagents and distilled water, and diluted with 1mol/L NaOH up to PH 8.2), respectively. The potentiodynamic polarization curves were recorded using a potential sweep rate of 5 mV/s in which the potential ranged from −0.8 to +0.8 V/E. After immersing the specimens for 30 min, the experiments were performed in simulated seawater and simulated body solutions that was exposed to air. Furthermore, the experiment was run until the open-circuit potential (OCP) reached steady-state conditions. All electrochemical measurements were repeated at least three times to verify the reproducibility of the results. The surface linear reciprocating sliding friction and wear tests were carried out on an MFT-R4000 high-speed friction and wear testing apparatus using GCr15 balls with a diameter of 9.6 mm under a normal load of 20N. From the results, the morphology and corrosion products of wear corrosion were determined via a SEM JSM-5600LV with energy dispersive spectrometry (EDS) mapping.

3. Results and Discussion

3.1. Microstructure

The X-ray diffraction patterns of the as-cast and fracture sample after compressive loading are shown in Figure 1a. As can be seen, the XRD pattern of the as-cast alloy exhibits a diffuse hump for angles ranging from $2\theta = 35°$ to $2\theta = 50°$. Moreover, sharp crystal diffraction peaks which represent the B2-Ti (Ni, Cu) and B19′-Ti (Ni, Cu) phases are superimposed over the diffuse scattering peak, indicating that the structure contains both a metallic glass matrix and crystal composite structures. The radius of the Ti, Ni and Cu atoms were also determined to be 0.086, 0.069 and 0.073 nm, respectively. As Cu is larger than Ni, Cu atoms will increase the interatomic spacing when they replace Ni atoms in the B2-TiNi and B19′-TiNi structures, which leads to the formation of the Ti (Ni, Cu) solid solutions. Consequently, the XRD diffraction peaks will shift to lower angles. The TEM selected area electron diffraction (SAED) patterns and HRTEM image of the as-cast $Ti_{40}Ni_{40}Cu_{20}$ are shown in Figure 1b,c. As mentioned in the X-ray diffraction patterns, the corresponding dispersion halo ring and the diffraction patterns confirm the amorphous and DSMC mixed structure. The high-resolution transmission electron microscopy (HRTEM) and the fast Fourier transform (FFT), further certified the microstructure of DSMC BMGCs. The mixing enthalpies,

ΔH_{mix} Ti-Ni = −35 KJ/mol, ΔH_{mix} Ti-Cu = −9 KJ/mol and ΔH_{mix} Ni-Cu = 4 KJ/mol, indicate the stronger bonding between Ti and Ni. Additionally, as the cooling rates significantly increase from inner region of the material to the surface, the thermally-induced B19′-Ti (Ni, Cu) martensite transformations can be observed (Figure 1d,e). In this scenario, the Cu atoms are discharged from the dendritic gap, which results in an increase in the glass forming ability (GFA) of the residual liquid. Furthermore, the martensite interface loses coherence and undergoes amorphization (Figure 1c,d). Finally, liquid atoms that are located in the edge zone arrange into an amorphous configuration directly before nucleation and diffusion (Figure 1d). It can be observed that the Ni content in the inner B2-Ti (Ni, Cu) dendrite is significantly greater than that of the surface. Since Ni can effectively reduce the martensite phase transformation temperature in the BMGC, the inner B2 dendrites are stable (Figure 1f). Thus, the functionally graded composites with a continuously graded microstructure can be obtained when there is a temperature gradient in the material during solidification. In this condition, the amorphous surface may exhibit high strength and hardness, excellent wear resistance, corrosion resistance and biological compatibility, while the austenitic center will possess good plasticity.

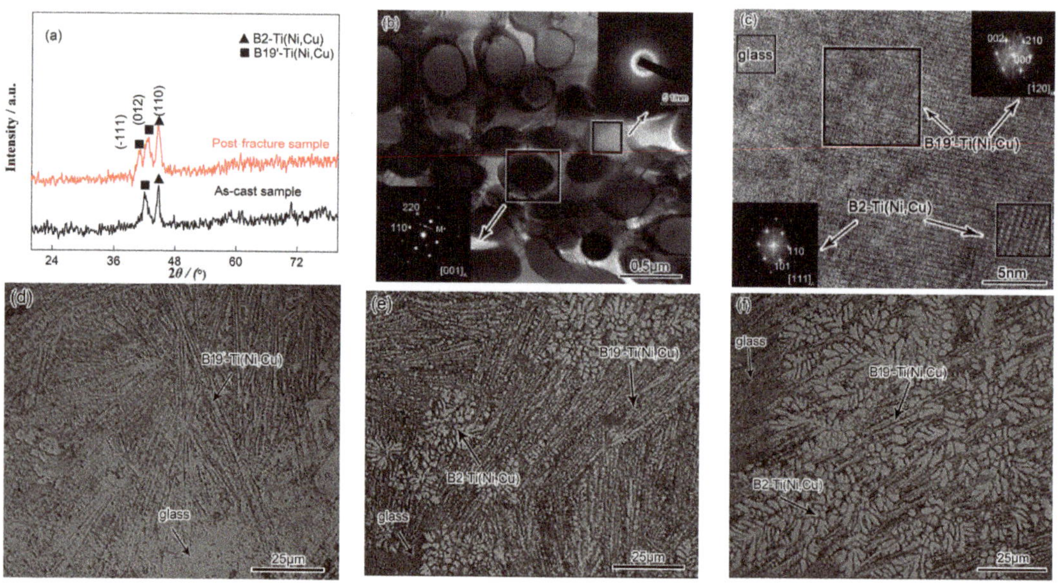

Figure 1. Microstructure of Φ3 mm $Ti_{40}Ni_{40}Cu_{20}$ alloy. (**a**) XRD patterns of as-cast and post-fracture samples. (**b**) TEM and SADP of as-cast samples, dispersion halo ring and diffraction patterns confirm the amorphous and DSMC mixed structure. (**c**) High resolution transmission electron microscopy (HRTEM) and Fast Fourier transform (FFT) further certified the microstructure of DSMC BMGCs. (**d–f**) Optical microscope (OM) of the edge, transition and inner zones; (**d**) contains a more amorphous featureless phase, fine lath martensite and some small austenite plates; (**e**) contains more coarse lath martensite and some fine austenite dendrites, and a small amount of amorphous phase; (**f**) mostly contains the developed austenite dendrites, and little lath martensite and amorphous featureless phase.

3.2. Macro-Mechanical Properties

From the XRD pattern of the post-fracture sample displayed in Figure 1a, the stress-induced preferentially-oriented-(−111) martensitic transformation from the (012) partial austenite forms during loading. Figure 2a shows the work hardening rate, $\theta = d\sigma/d\varepsilon$, and the true stress versus true strain curves of the alloy. The relationship between the work hardening rate and the true strain can be divided into three stages. The initial work hardening stage, shows that the work hardening rate sharply decreases with the increase of strain. This stage can be regarded as the preparation condition for the work hardening, and the size of this stage indicates the degree of difficulty in which the material work hardening is. Furthermore, in this stage, the alloy undergoes a phase transition in which dislocations are formed that promote the work hardening in the alloy. The second stage consists of the main stage for the strengthening and toughening of the material, where the work hardening rate decreases with the increase of the strain. The longer the second stage is, the larger the work hardening rate will be. The third stage refers to the linear hardening stage, which is characterized by a work hardening rate that monotonically decreases with the increase of strain. Compared to the other stages, the change of the work hardening rate during this stage is the slowest. In this stage, the strengthening process will end, and the longer the stage is, the larger the work hardening rate will be. That is to say, the higher the strength of material as well as the better plasticity. Figure 2b gives the true compressive stress versus strain data of Ti-based dendrite-composite alloys [15,16], and some other advanced metastable alloys [17–19]. It is remarkable that the DSMC BMGCs deal better with the strength–ductility trade-off.

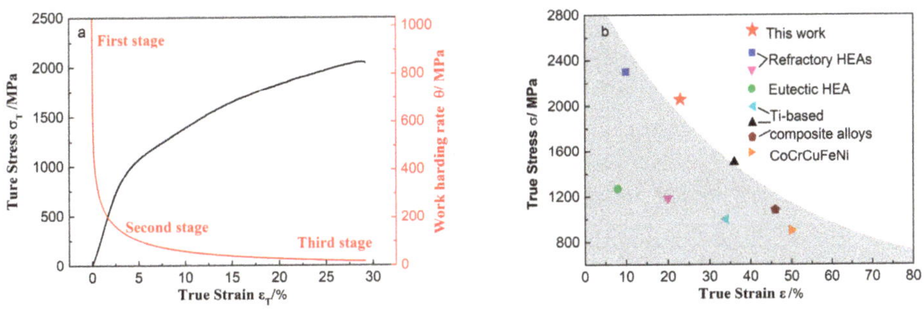

Figure 2. (**a**) The work hardening rate and true stress versus true strain curves engineering stress-strain curve of Φ3 mm $Ti_{40}Ni_{40}Cu_{20}$ alloy; (**b**) compressive properties compared to some other advanced alloys.

3.3. *Micro-Mechanical Properties*

To better understand the connection between the macro-mechanical behavior and the microstructure of the alloy, the nanofretting behavior and creep strain in different regions of the sample were examined. Additionally, the energy dissipation factor (EDF) was calculated. Nanofretting is the special friction mode in which the relative reciprocating motion amplitude is in the nanometer scale. The area of a hysteresis loop in the load–displacement curve of radial nanofretting before closure reflects the process of the energy dissipation.

The maximum elastic strain energy during loading is defined as the total deformation energy (W_L), which is the area under the load–displacement curve from Figure 3a (1):

$$W_I = \int_0^{h'} P_I(h)dh \quad (1)$$

where $P_L(h)$ and h' are the loading curve function and the maximum indention depth, respectively. Then the elastic recovery energy during unloading, W_U, can be calculated by the area of enclosed zone of unloading curve from Figure 3a (2):

$$W_U = \int_{h_f}^{h'} P_U(h)dh \quad (2)$$

where $P_U(h)$ is the unloading curve function. Therefore, the energy dissipated in each reaction is [20]:

$$EDF = \frac{\int_0^{h'} P_L(h)dh - \int_{h_f}^{h'} P_U(h)dh}{\int_0^{h'} P_L(h)dh} \quad (3)$$

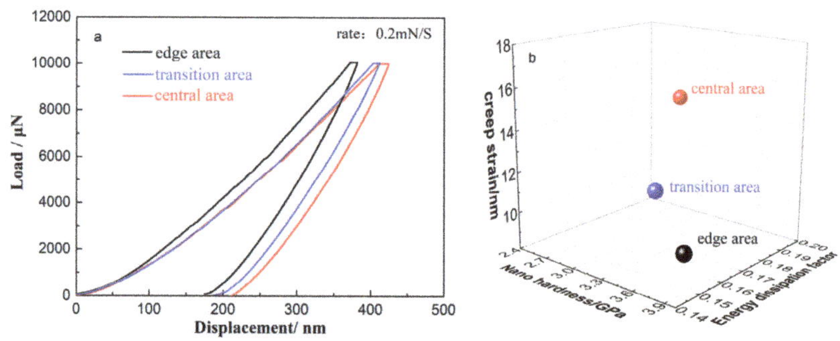

Figure 3. (**a**) P-H curves of nanofretting; (**b**) nano hardness, energy dissipation factor and creep strain in different areas.

According to the load–displacement (L–D) curve results and the Oliver–Pharr method, the calculated nano-hardness in the edge, transition and central regions increases, whereas the plastic deformation, energy consumption factor and creep show the opposite trend.

3.4. Corrosion and Friction Behavior

Figure 4a gives the potentiodynamic polarization curves for the ASW and PBS of the Ti–6Al–4V (TC4) and $Ti_{40}Ni_{40}Cu_{20}$ BMGC. Both alloys show a spontaneous passivation behavior. The free corrosion potential E_{corr} values of the TC4 and BMGC were −0.490 and −0.346 V in the ASW at 298 K. In the PBS, these values were found to be −0.3608 and −0.287 at 310 K. As compared with the TC4, the BMGCs exhibit higher E_{corr} values, such that it is not easy to lose electrons, demonstrating a lower tendency to corrode. The self-corrosion current density (I_{corr}) is given by the Tafel extrapolation method, where the higher polarization resistance (R_p) can be calculated using the Stern–Geary equation [21]:

$$R_P = \frac{\beta_a \beta_c}{2.3 I_{corr} (\beta_a + \beta_c)} \quad (4)$$

where β_a is anodic Tafel slope and β_c is cathodic Tafel slope. I_{corr} and the R_p of the TC4 and the BMGC are 2.53 $\mu A \cdot cm^{-2}$, 1.5×10^5 $\Omega \cdot cm^2$ and 1.601 $\mu A \cdot cm^{-2}$, 1.9×10^6 $\Omega \cdot cm^2$ in the ASW, while they were 0.488 $\mu A \cdot cm^{-2}$, 4.6×10^5 $\Omega \cdot cm^2$ and 0.343 $\mu A \cdot cm^{-2}$, 3.2×10^6 $\Omega \cdot cm^2$ in the PBS. Lower I_{corr} and higher R_p values indicate that the alloy has lower thermodynamic tendency as well as slower kinetic rates of corrosion.

The standard electrode potential for the Ti, Ni and Cu elements are −1.63, −0.257 and −0.342 V, respectively [22]. Thus, it is easier for the Ti atom to lose an outer electron which results in the rapid formation of the TiO_2 oxide film, the n-type semiconductor, hindering the environmental and body hazardous element Ni from resolving-out [23]. The oxide film is generated via the following chemical process:

$$Ti + 2H_2O = TiO_2 + 4H^+ + 4e \quad (5)$$

$$TiO_2 + 2H_2O = Ti^{4+} + 4OH^- \quad (6)$$

When the generation–resolution rate is equal, the reactions reach equilibrium and a stable oxide film is formed. A higher R_p value indicates the formation of a more compact passivation film with higher resistance. As the homogeneous amorphous surface seldom contains defects, non-uniform corrosion does not occur easily in the BMGC, and the corrosion-resistance is greater than that of the crystalline alloy [24].

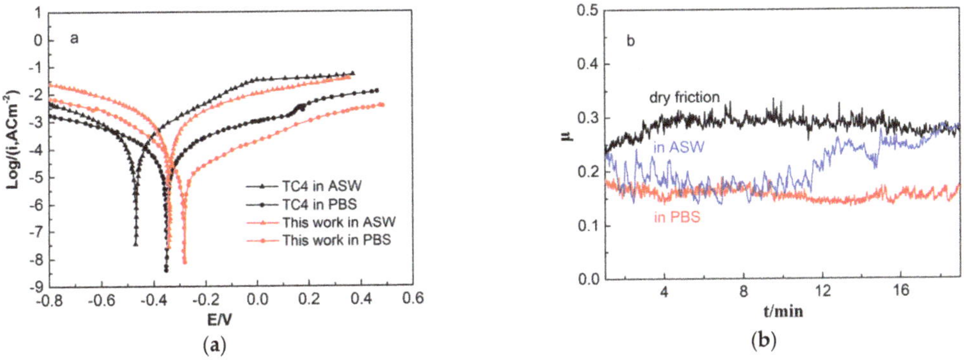

Figure 4. (**a**) Potentiodynamic polarization curves of Ti–6Al–4V and $Ti_{40}Ni_{40}Cu_{20}$ BMGC in ASW at 298 K and PBS at 310 K; (**b**) friction coefficient of $Ti_{40}Ni_{40}Cu_{20}$/GCr15 under dry friction and in ASW, PBS solutions.

However, the TiO_2 film is fragile and can easily be locally eroded if defects exist in the microstructure below the passive film. Defects such as grain boundaries, inclusions, second phase precipitates, micro pores and cracks will provide a condition for the initiation of pitting corrosion in the medium with activated anions. In Figure 4a, the potentiodynamic polarization curve for the BMGC in the ASW fluctuates around 0.2 V. Also, in Figure 4b, the friction coefficient curve in the ASW also shows obvious fluctuations. In the scratched oxide film, defects are absorbed by the anions, leading to anodic dissolution on the bare alloy surface. In the protective area of the oxide film, the cathode reaction occurs, and can be written as [25].

$$2H_2O + O_2 + 4e = 4OH^- \ (pH \geq 7) \tag{7}$$

Thus, the active-passive cell forms, and the Ti ions in located in the etching pits continually increase in number. Furthermore, more Cl ions migrate into the pits. Consequently, the $TiCl_4$ phase forms as the corrosion product that subsequently causes hydrolyzation. The hydrolyzation reaction is defined as:

$$TiCl_4 + (x + 2)H_2O = TiO_2 \cdot xH_2O + 4HCl \tag{8}$$

The pH value of the etching pits decreases as the reaction proceeds, leading to accelerated pitting. As illustrated in Figure 4a, the corrosion current density increases after pitting occurs in the artificial seawater. Meanwhile, the concentration of the newly formed TiO_2 increases to a certain extent, which hinders further development of the pitting. Thus, the polarization curve presents repassivation after the pitting occurs. In addition, the corrosion pit and its surrounding corrosion structure can be observed in Figure 5. As can be seen, no obvious corrosion products are found in the rest area. EDS mappings show that the O elements distribute on the surface, concentrated in the corrosion area. The Ti content is greater and with a more uniform distribution on the surface. The Cl is mainly concentrated on the boundary of the uncorroded region and the defects-caused corroded region. The Ni and Cu elements were observed to uniformly distribute outside the corrosion area.

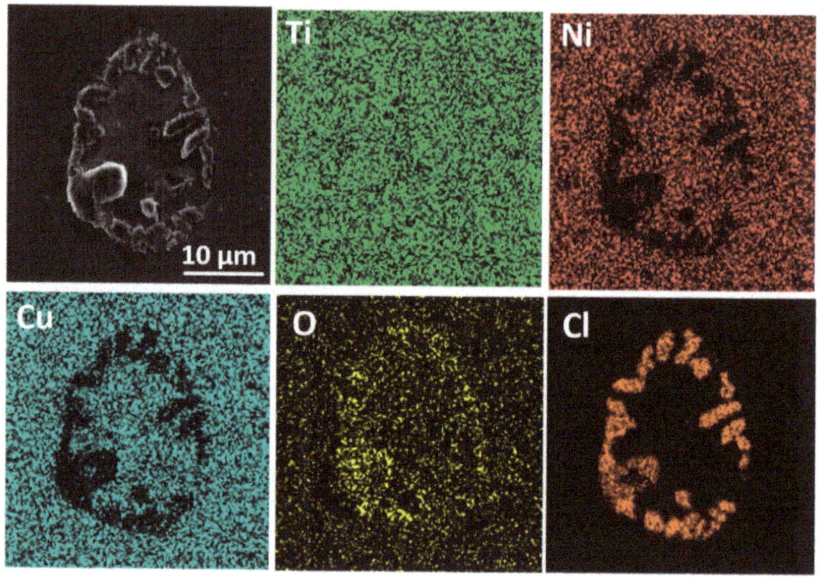

Figure 5. SEM and EDS mappings of $Ti_{40}Ni_{40}Cu_{20}$ BMGC wear surface in ASW.

That is, the uncorroded region is covered with a uniform oxide film. In the corroded region and its surroundings, more O and Cl elements are enriched due to the newly formed TiO_2 and migrating Cl^-. Ni and Cu have not participated in the corrosion process owing to the protection of the passive film.

From Figure 4b, the friction coefficients μ of the ASW and PBS are only 0.261, 0.183 and 0.160 in air, and show a decreasing trend. The main wear behavior of the dry friction are fatigue and abrasive damage. As for the artificial seawater, the wear mechanisms are attributed to the fatigue and pitting wear. With respect to the PBS, the same mechanism is associated with the fatigue wear. No crystallization was found to occur on the friction surface of each sample, indicating that the BMGCs exhibits good friction-reduction and anti-wear properties.

4. Conclusions

In the present work, Ti-based BMGCs were designed via TRIP effect of DSMCs to enhance both the ductility and work-hardening capability. Based on the macro- and micro-mechanical behavior, as well as three stages of work hardening, the TRIP effect depends on the consistency of diffusion and moving velocity of solute atoms and movable dislocations. Moreover, the excellent wear and corrosion resistance were attributed to the amorphous surface. The versatile DSMC BMGCs can serve as advanced metals with excellent mechanical and functional properties. However, the development and application of DSMC BMGCs are limited by size restrictions and their processing and forming abilities, which need further research in the future.

Author Contributions: Conceptualization, Y.Z.; methodology, Y.Z. and P.G.; validation, Y.Z. and and P.G.; formal analysis, Y.Z., T.L. and L.F.; investigation, Y.Z., T.L., L.F. and P.G.; resources, Y.Z. and P.G.; data curation, Y.Z., T.L., L.F. and P.G.; writing—original draft preparation, Y.Z. and T.L.; writing—review and editing, L.F. and P.G.; supervision, P.G.; project administration, Y.Z. and P.G.; funding acquisition, Y.Z. and P.G. All authors have read and agreed to the published version of the manuscript.

Funding: This research was funded by the National Natural Science Foundation of China, grant number 52061027; and the Hongliu Outstanding Talents Program of Lanzhou University of Technology, grant number 2018.

Institutional Review Board Statement: Not applicable.

Informed Consent Statement: Not applicable.

Acknowledgments: We acknowledge the support given by the Key Laboratory of Cloud Computing of Gansu Province (Gansu Computing Center).

Conflicts of Interest: The authors declare no conflict of interest. The funders had no role in the design of the study; in the collection, analyses, or interpretation of data; in the writing of the manuscript; or in the decision to publish the results.

References

1. Zhang, Y. Mechanical Properties and Structures of High Entropy Alloys and Bulk Metallic Glasses Composites. *Mater. Sci. Forum* **2010**, *654–656*, 1058–1061. [CrossRef]
2. Qiao, J.; Jia, H.; Zhang, A.; Liaw, P.; Li, L. Multi-step shear banding for bulk metallic glasses at ambient and cryogenic temperatures. *Mater. Chem. Phys.* **2012**, *136*, 75–79. [CrossRef]
3. Qiao, J.; Jia, H.; Liaw, P.K. Metallic glass matrix composites. *Mater. Sci. Eng. R Rep.* **2016**, *100*, 1–69. [CrossRef]
4. Spataru, M.C.; Butnaru, M.; Sandu, A.V.; Vulpe, V.; Vlad, M.D.; Baltatu, M.S.; Geanta, V.; Voiculescu, I.; Vizureanu, P.; Carmen Solcan, C. In-depth assessment of new Ti-based biocompatible materials. *Mater. Chem. Phys.* **2021**, *258*, 123959. [CrossRef]
5. Baltatu, I.; Sandu, A.V.; Vlad, M.D.; Spataru, M.C.; Vizureanu, P.; Baltatu, M.S. Mechanical characterization and in vitro assay of biocompatible titanium alloys. *Micromachines* **2022**, *13*, 430. [CrossRef]
6. Hua, N.; Wang, W.; Wang, Q.; Ye, Y.; Lin, S.; Zhang, L.; Guo, Q.; Brechtl, J.; Liaw, K.P. Mechanical, corrosion, and wear properties of biomedical Ti–Zr–Nb–Ta–Mo high entropy alloys. *J. Alloys Compd.* **2021**, *861*, 157997. [CrossRef]
7. Yeh, J.-W.; Lin, S.-J. Breakthrough applications of high-entropy materials. *J. Mater. Res.* **2018**, *33*, 3129–3137. [CrossRef]
8. Li, Z.; Pradeep, K.G.; Deng, Y.; Raabe, D.; Tasan, C.C. Metastable high-entropy dual-phase alloys overcome the strength–ductility trade-off. *Nature* **2016**, *534*, 227–230. [CrossRef]
9. Medasani, B.; Gamst, A.; Ding, H.; Chen, W.; Persson, K.A.; Asta, M.; Canning, A.; Haranczyk, M. Predicting defect behavior in B2 intermetallics by merging ab initio modeling and machine learning. *NPJ Comput. Mater.* **2016**, *1*, 1–10. [CrossRef]
10. Pauly, S.; Das, J.; Bednarcik, J.; Mattern, N.; Kim, K.B.; Kim, D.H.; Eckert, J. Deformation- induced martensitic transformation in Cu-Zr-(Al, Ti) bulk metallic glass composites. *Scr. Mater.* **2009**, *60*, 431–434. [CrossRef]
11. Wu, Y.; Ma, D.; Li, Q.; Stoica, A.; Song, W.; Wang, H.; Liu, X.; Stoica, G.; Wang, G.; An, K.; et al. Transformation-induced plasticity in bulk metallic glass composites evidenced by in-situ neutron diffraction. *Acta Mater.* **2017**, *124*, 478–488. [CrossRef]
12. Nieh, T.G.; Wadsworth, J.; Liu, C.T.; Ohkubo, T.; Hirotsu, Y. Plasticity and structural instability in a bulk metallic glass deformed in the supercooled liquid region. *Acta Mater.* **2001**, *49*, 2887–2896. [CrossRef]
13. Zhang, F.; Zhang, C.; Lv, D.; Zhu, J.; Cao, W.; Chen, S.; Schmid-Fetzer, R. Prediction of Glass Forming Ability Through High Throughput Calculation. *J. Phase Equilibria Diffus.* **2018**, *39*, 562–570. [CrossRef]
14. Schroers, J. Processing of Bulk Metallic Glass. *Adv. Mater.* **2010**, *22*, 1566–1597. [CrossRef]
15. Okulov, I.V.; Kühn, U.; Marr, T.; Freudenberger, J.; Schultz, L.; Oertel, C.-G.; Skrotzki, W.; Eckert, J. Deformation and fracture behavior of composite structured Ti-Nb-Al-Co(-Ni) alloys. *Appl. Phys. Lett.* **2014**, *104*, 071905. [CrossRef]
16. Okulov, I.; Sarmanova, M.; Volegov, A.; Okulov, A.; Kühn, U.; Skrotzki, W.; Eckert, J. Effect of boron on microstructure and mechanical properties of multicomponent titanium alloys. *Mater. Lett.* **2015**, *158*, 111–114. [CrossRef]
17. Fan, J.; Zhang, L.; Yu, P.; Zhang, M.; Li, G.; Liaw, P.; Liu, R. A novel high-entropy alloy with a dendrite-composite microstructure and remarkable compression performance. *Scr. Mater.* **2019**, *159*, 18–23. [CrossRef]
18. Couzinie, J.-P.; Lilensten, L.; Champion, Y.; Dirras, G.; Perrière, L.; Guillot, I. On the room temperature deformation mechanisms of a TiZrHfNbTa refractory high-entropy alloy. *Mater. Sci. Eng. A* **2015**, *645*, 255–263. [CrossRef]
19. Wang, X.; Zhang, Y.; Qiao, Y.; Chen, G. Novel microstructure and properties of multicomponent CoCrCuFeNiTix alloys. *Intermetallics* **2007**, *15*, 357–362. [CrossRef]
20. Oyen, M.L. Spherical Indentation Creep Following Ramp Loading. *J. Mater. Res.* **2006**, *20*, 2094–2100. [CrossRef]
21. Stern, M.; Geary, A.L. Electrochemical polarization: I. A theoretical analysis of the shape of polarization curves. *J. Electron. Mater.* **1957**, *104*, 56–63.
22. Ma, X.; Zhen, N.; Guo, J.J.; Li, Q.; Chang, C.T.; Sun, Y.F. Preparation of Ni-based bulk metallic glasses with high corrosion-resistance. *J. Non-Cryst. Solids* **2016**, *443*, 91–96. [CrossRef]
23. Liu, Y.; Pang, S.J.; Li, H.F.; Chen, B.; Zhang, T. Formation and properties of Ti-based Ti-Zr-Cu-Fe-Sn-Si bulk metallic glasses with different (Ti+Zr)/Cu ratios for biomedical application. *Intermetallics* **2016**, *72*, 36–43. [CrossRef]

24. Guo, S.F.; Chan, K.C.; Xie, S.H.; Yu, P.; Huang, Y.J.; Zhang, H.J. Novel centimeter-sized Fe-based bulk metallic glass with high corrosion resistance in simulated acid rain and seawater. *J. Non-Cryst. Solids* **2013**, *369*, 29–33. [CrossRef]
25. Tauseef, A.; Tariq, N.H.; Akhter, J.I.; Hasan, B.A.; Mehmood, M. Corrosion behavior of Zr-Cu-Ni-Al bulk metallic glasses in chloride medium. *J. Alloys Compd.* **2010**, *489*, 596–599. [CrossRef]

materials

Article

Grinding Temperature and Surface Integrity of Quenched Automotive Transmission Gear during the Form Grinding Process

Xiaoyang Jiang [1], Ke Liu [2], Yong Yan [3], Maojun Li [1,*], Pan Gong [4,*] and Hong He [1]

1 State Key Laboratory of Advanced Design and Manufacture for Vehicle Body, Hunan University, Changsha 410082, China
2 Jianglu Machinery & Electronics Group Co., Ltd., Xiangtan 411100, China
3 China North Advanced Technology Generalization Institute, Beijing 100089, China
4 State Key Laboratory of Materials Processing and Die & Mold Technology, School of Materials Science and Engineering, Huazhong University of Science and Technology, Wuhan 430074, China
* Correspondence: maojunli@hnu.edu.cn (M.L.); pangong@hust.edu.cn (P.G.)

Abstract: Grinding burn is an undesired defect in gear machining, and a white layer is an indication of severe burn that is detrimental to gear surface performance. In this work, the influence of grinding parameters on the thickness of the white layer during form grinding of quenched transmission gear was investigated, and the microstructure evolution and mechanism of severe burn formation were analyzed. The grinding temperature increased with the grinding depth and grinding speed, with the highest level of ~290 °C. The thickness of the white layer exceeded 100 μm when the grinding depth was 0.03 mm, and the top layer was a plastic deformation layer followed by a fine-grained martensite layer. Coarse-grained acicular martensite was found at the interface between the white layer and softened dark layer. The mechanical effect and thermal softening mainly contributed to the formation of white layer stratification. The ground surface topography showed several scratches and typical grooves; when grinding depth increased to 0.03 mm, the grinding surface roughness Sa was relatively high and reached up to ~0.60 μm, mainly owing to severe plastic deformation under grinding wheel extrusion and the thermal effect.

Keywords: gear grinding; white layer; grinding temperature; surface integrity

Citation: Jiang, X.; Liu, K.; Yan, Y.; Li, M.; Gong, P.; He, H. Grinding Temperature and Surface Integrity of Quenched Automotive Transmission Gear during the Form Grinding Process. *Materials* **2022**, *15*, 7723. https://doi.org/10.3390/ma15217723

Academic Editor: Adam Grajcar

Received: 2 October 2022
Accepted: 29 October 2022
Published: 2 November 2022

1. Introduction

With the fast development of the vehicle industry and the increasing demand for transmission gears, it is important to manufacture gears with high quality and efficiency. For heavy-duty gears, the gear blank is usually subjected to heat treatment such as carburizing, nitriding, and quenching to meet the high hardness and toughness requirements. However, the deformation caused by heat treatment reduces the dimensional accuracy of gears. Grinding is normally applied as the final process in gear manufacturing to eliminate geometric deformation generated during heat treatment and to produce components with good geometric accuracy and surface quality [1–4]. Owing to the friction/shearing between abrasive particles and workpiece material, the grinding temperature is relatively high. Grinding burn occurs easily when the grinding parameters are unreasonable, which can be defined as changes caused by heat release out of the grinding process in the surface and subsurface layer of the workpiece [5]. Grinding burn originates from tempering and thermal softening when the grinding temperature exceeds the tempering temperature during grinding, and its essence is the transformation of the metallographic phase in the surface layer [3,6]. Grinding burn is usually divided into three levels, including predominantly dark blue oxide layers, thermal softening when the grinding temperature exceeds the austenitizing temperature, and a white layer and dark tempered layer. Grinding burn can induce machining defects such as microcracks on the ground surface [7].

As grinding burn is highly related to the final quality of machined products, researchers carried out an analysis on the formation mechanisms of grinding burn and how to reduce the probability of it occurring. Guerrini et al. [1] found that, when the grinding temperature exceeded 450 °C, a softened dark layer caused by carbon coming out of the solution was observed below the gear surface, and the thickness of the tempered layer increased with the grinding temperature. When the temperature exceeded the austenitization temperature, a severe bright martensite layer was formed on the outermost surface. It was found that the increase in cutting depth and feed speed led to higher level of grinding burn [3]. Sun et al. [8] reported that a white layer structure was generated on the surface layer after grinding hardening and its composition was analyzed, while the formation mechanism of the white layer and its connection with grinding parameters have not been further discussed. As the grinding depth increased, greater grinding energy was required to remove materials [9]. This indicated that more energy was converted into heat and significantly affected the workpiece surface. Su et al. [10] reported that the increase in grinding depth increased the thickness of the ablated layer, while an appropriate increase in feed speed and decrease in grinding speed could reduce the grinding temperature. Zhou et al. [11] found that, when the abrasive particles cut or ploughed across the rail, the grinding temperature decreased with the increasing forward speed, resulting in a small thickness of the white layer. Jamshidi et al. [5] defined surface burn as a chemical reaction between oxygen and workpiece material and proposed a burn model that comprehensively considered maximum temperature and exposure time to predict oxide layer thickness. Jermolajev et al. [12] established a time–temperature diagram showing surface layer modification for profile gear grinding based on local contact zone temperature measurements and used the tempering time to explain workpiece surface burn rather than contact time, as the grinding temperature could not drop below the tempering temperature instantaneously when the grinding wheel moved away.

Based on the comprehensive literature review, it was found that previous studies mainly worked on the theoretical models of grinding burn during surface grinding and the prediction of temperature thresholds for generating grinding burn. Several studies focused on the relationship between burn formation during gear grinding and grinding parameters. In this paper, according to the variation in grinding temperature and thickness of the ablation layer under different grinding parameters, the influence of grinding speed and depth on grinding burn was analyzed. The change in microhardness within subsurface layers, the microstructure and morphology, as well as the surface roughness were comprehensively investigated. The result of this study can mostly likely provide benchmarks for improving both the quality and efficiency during gear grinding in vehicle industries.

2. Experimental Setup and Procedure

2.1. Workpiece Material

A gear made of 20Cr2Ni4A steel was selected as the workpiece in this research. It underwent a series of heat treatments including normalizing, carburizing, quenching, and tempering. The corresponding chemical composition is shown in Table 1. Good toughness and high strength are required when 20Cr2Ni4A steel is used to manufacture heavy-duty gears, and it needs to be carburized and quenched to improve the hardness of its surface layer. The initial workpiece was cut into several segments for grinding trials with different grinding parameters, and the size of the samples was designed to be clamped and ground easily.

Table 1. Chemical composition of 20Cr2Ni4A steel (mass fraction, %) [13].

C	Si	Mn	Cr	Ni	Fe	S
0.17–0.23	0.17–0.37	0.30–0.60	1.25–1.65	3.25–3.65	≥95.00	≤0.03

2.2. Experimental Setup

A three-axis automatic hydraulic high-precision grinding machine (DY-510ASM) was used for gear grinding experimental trials, which was driven by an O05-63B0 three-phase asynchronous motor. The maximum rotation speed of the grinder was 4000 r/min with a rated power of 7.5 kW. An alumina forming grinding wheel with #80 abrasive mesh and a diameter of 350 mm was used, and the grinding wheel was dressed by a diamond wheel dresser before each set of experiments. The wheel topography was observed with a high-definition charge-coupled device (HD CCD) microscope, as shown in Figure 1; white corundum abrasive grains were evenly bonded to the surface and sharp cutting edges of abrasives could be observed.

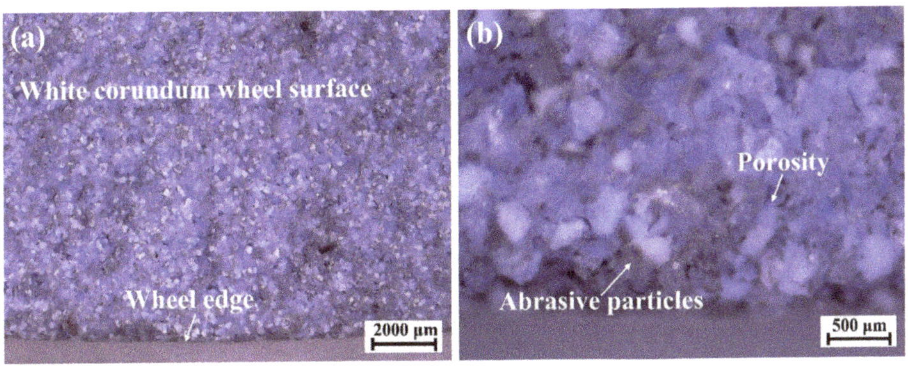

Figure 1. Topography of the grinding wheel surface observed by microscope with (**a**) low and (**b**) high magnification.

Before grinding, the runout on both sides of the tooth surface along the feed direction was limited within 0.001 mm by a dial indicator to ensure machining accuracy. The gear sample was clamped by the fixture on the perpendicular direction of feed speed. Considering the setting of gear machining parameters in industrial application, the total grinding thickness of one side was set at 0.24 mm, the feed speed and the grinding speed were selected with three levels, and the grinding depth for each pass was set to 0.02 mm and 0.03 mm. The specific grinding parameters are shown in Table 2.

Table 2. Parameters set in the gear forming grinding experiment.

Feed Speed (m/s)	Grinding Speed (m/s)	Grinding Depth (mm)	Number of Passes
0.19	26	0.02	12
	30	0.02	12
		0.03	8
	34	0.02	12
		0.03	8
0.26	26	0.02	12
	30	0.02	12
		0.03	8
	34	0.02	12
		0.03	8
0.35	26	0.02	12
	30	0.02	12
	34	0.02	12

2.3. Measuring Equipment

The temperature signals were captured by a GG-K-30-1000-CZ type thermocouple produced by KAIPUSEN when the grinding wheel passed the workpiece material surface.

For each gear sample, a narrow slot with a width of 1.0 mm was cut from the tooth tip to the tooth root along the middle of tooth width to accommodate the thermocouple sensing tip. As shown in Figure 2a, a thermocouple was inserted into the narrow slot on the workpiece to measure the grinding temperature. This method was generally used to record temperature variation with good accuracy [14,15]. The thermocouple sensing tips were located close to the grinding surface, and the narrow slots were sealed with waterproof and heat-resistant glue, which could prevent the cooling liquid from affecting the measurement results. Figure 2b shows the experimental setup of grinding temperature measurement; the gear sample selected for the grinding temperature measurement is shown in Figure 2c.

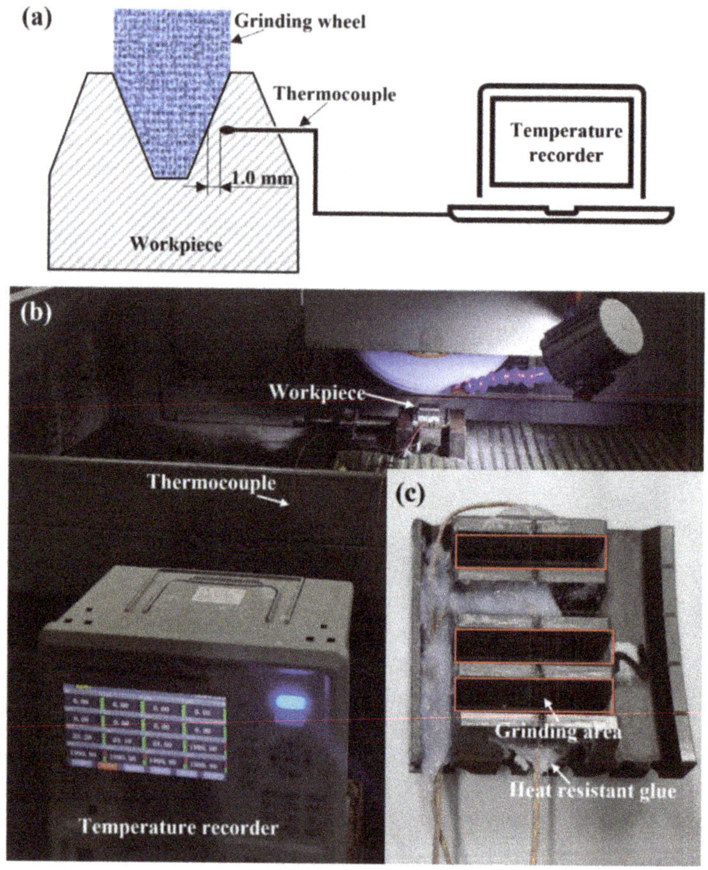

Figure 2. (**a**) Configuration of embedded thermocouples, (**b**) experimental setup of grinding temperature measurement, and (**c**) corresponding gear sample.

The gear teeth ground by different parameters were cut using a wire electrical discharge machine and embedded in resin for polishing to observe the microstructure of the ground surface layer. A high-resolution scanning electron microscope (SEM, TESCAN MIRA4 LMH) was used to observe the ground surface morphology. Microhardness at the subsurface was measured using a Vickers hardness tester with a load of 1000 g and intent time of 10 s, and each measurement was set with 100 μm depth intervals. An Atometrics-NA500 white light interferometer was also applied to measure the surface roughness of the ground surface.

3. Experimental Results and Discussion

3.1. Grinding Temperature

Most of the mechanical work occurring in the grinding process is converted into grinding heat, which increases the temperature of the gear tooth surface. If the thermal stress caused by the grinding process exceeds the yield strength of the workpiece material, residual tensile stress will be generated within the ground surface and it is possible to form dominant or recessive grinding cracks in the surface layer, which severely affect the gear strength and fatigue life [16]. In addition, as the contact arc surface of gear grinding is complex, the grinding temperature changes with the normal grinding depth and presents as a non-uniform form [17]. Thus, the grinding temperature is an important indicator of the workpiece heating state during the grinding process. The highest temperature during grinding under each set of parameters was recorded. Figure 3 shows the variation in grinding temperature under different grinding parameters. It is obvious that a high level of grinding parameters generally led to a relatively higher grinding temperature, with the maximum value of ~290 °C.

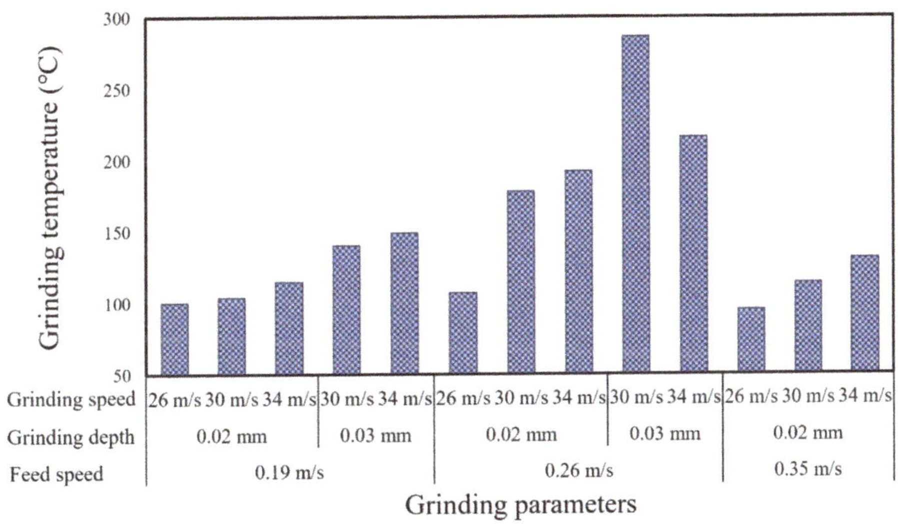

Figure 3. Temperature variation under different grinding parameters.

Statistical ANOVA method was applied to further analyze significant factors and the related percentage contribution ratio (PCR) for ground temperature. Figure 4 shows the detailed information concerning the main effects plots of grinding temperature related to different grinding parameters. Data from the ANOVA table confirmed that feed speed was statistically significant, affecting grinding temperature with a corresponding percentage contribution ratio of 33.7%, followed by grinding depth, with the details presented in Table 3.

Table 3. ANOVA of grinding temperature at different parameters.

Source	DOE	Adj SS	Adj MS	F	PCR
Feed speed (m/s)	2	14,512	7255.8	8.22 *	33.7%
Grinding speed (m/s)	2	2975	1487.5	1.69	3.2%
Grinding depth (mm)	1	5794	5793.6	6.56 *	13.0%
Error	7	6178	882.5		50.1%
Total	12	37,832			100%

* Significant at the 5% level, $F_{0.05,2,7} = 4.74$, $F_{0.05,1,7} = 5.59$.

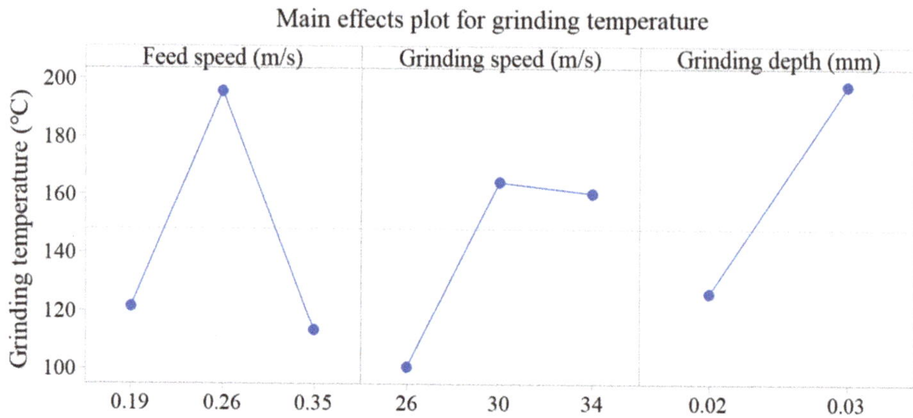

Figure 4. Main effects plot for grinding temperature at different parameters.

Generally, the increase in feed speed and grinding depth led to the increase in grinding temperature. Firstly, at a constant grinding speed, the grinding force increased with the increase in grinding depth and feed speed. It led to an increase in the total energy consumed by grinding, which indicated that more energy was converted into grinding heat [2]. The increase in grinding force also caused severe friction, which eventually led to the increase in grinding temperature. Secondly, as shown in Figure 5a, when the grinding depth increased, the actual cutting thickness increased, which not only enlarged the contact area between grinding wheel and workpiece, but also hindered the heat dissipation within the grinding area, and finally contributed to the increase in grinding temperature [18]. Figure 5b shows the influence of different feed speeds on grinding temperature. When the feed speed increased, the path traveled by the grinding wheel under per unit time was extended and more heat was generated, resulting in higher heat and grinding temperature. However, increasing the feed speed might also shorten the heat transfer time, and thus lead to a lower grinding temperature [3,19]. According to the recorded data, this kind of effect was more obvious when the feed speed increased to 0.35 m/s. With the same level of feed speed and grinding depth, the grinding temperature increased when the grinding speed increased from 30 m/s to 34 m/s, and the effect of the grinding speed was more significant when the grinding depth was set at 0.02 mm. The grinding temperature was relatively high when the grinding speed was selected as 34 m/s and the feed speed was 0.26 m/s. This was mainly because of the fact that the number of cuts per unit time increased at a higher grinding speed and, subsequently, more heat was generated [20]. The reduction in grinding temperature when increasing the grinding speed with the depth of 0.03 mm and feed of 0.26 m/s could be attributed to two aspects. One possible reason was the dimensional deformation of the gear blank during carburization and quenching processes, which might lead to uneven thermal field distribution in the grinding process. The other was that, at the depth of 0.03 mm and feed of 0.26 m/s, the scratch and ploughing effects were less significant when the grinding speed increased to 34 m/s, resulting in a decrease in the energy converted into grinding heat.

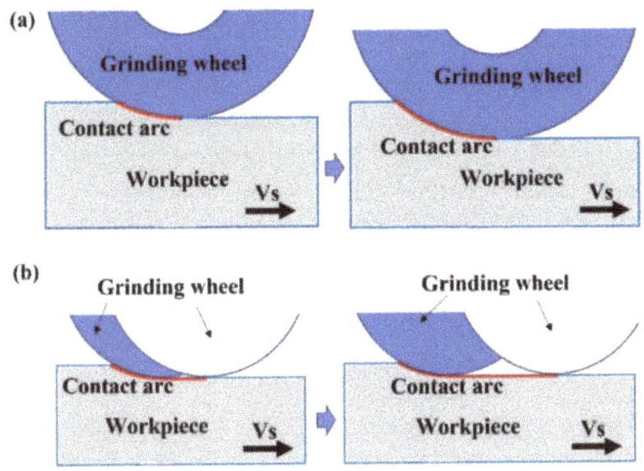

Figure 5. The influence of different (**a**) grinding depths and (**b**) feed speeds on contact arc length per unit time.

3.2. Characteristics of Grinding Burn

Figure 6 shows the morphology of the subsurface layer after grinding with different parameters. It was found that the bright white ablation layer generally occurred within the subsurface layer after grinding in the depth direction. The white layer was the reinforced structure of the surface material after hardening during abrasive grinding, and it was mainly composed of martensite, cementite, and residual austenite [8]. This phenomenon can be explained by the following reasons. The high temperature generated during grinding promoted the reaching of the austenitic field followed by rapid cooling, and the cooling fluid contributed to the transformation from the austenite to martensite phase [21]. White layer formation was fundamentally dependent on the thermal and mechanical effects of grinding on the workpiece material. These effects were mainly caused by the interaction of grinding tool and workpiece material and were significantly affected by grinding parameters [22].

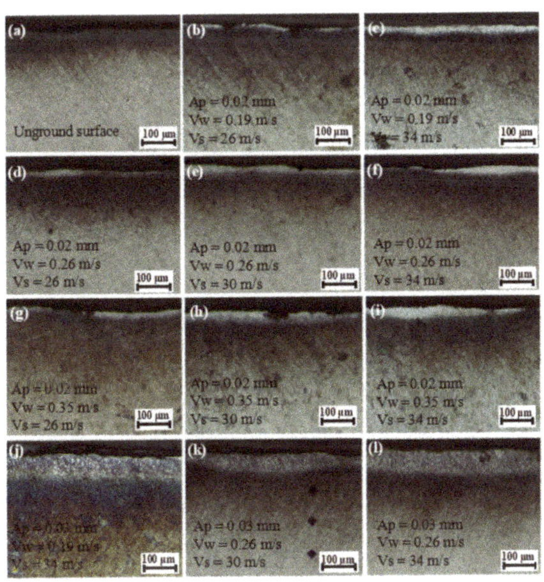

Figure 6. Morphology of (**a**) the unground surface layer and (**b–l**) the ground surface layer at different grinding parameters.

With the grinding depth of 0.02 mm, feed speed of 0.19 m/s, and grinding speed of 26 m/s, a relatively thin white layer with a thickness of less than 5 μm was generated on the ground surface. When the grinding speed increased from 26 m/s to 34 m/s, the thickness of the white layer increased slightly. When the feed speed was 0.35 m/s and the grinding speed was over 30 m/s, the thickness of the white layer exceeded ~50 μm. This was mainly because of the fact that the exposure time was the same when the feed speed was constant, while the temperature increased with a higher grinding speed, resulting in more obvious grinding burn and a thicker white layer. The grinding burn often indicated the changes in mechanical properties caused by microstructure transformation, which ultimately led to the scrapping of the machined part [3]. At the grinding depth of 0.02 mm and the same grinding speed, the variation in white layer thickness with different feed speeds was almost consistent. Under the condition of the same grinding depth and grinding speed, the increase in feed speed led to a higher heating rate and strain rate [23]. When the grinding depth increased to 0.03 mm, a white layer with a thickness of ~100 μm was generated. A higher grinding depth resulted in a great increase in grinding area and slow dissipation of heat, which further promoted grinding burns [18].

Figure 7 shows the SEM morphology of the ground surface layer under different grinding parameters. The effect of grinding parameters on the thickness of white layer and plastic deformation, which were most likely related to dynamic recrystallization caused by a high temperature and thermal stress during the grinding process [8]. This phenomenon caused higher hardness of the surface layer, which easily caused microcracks. Coarse-grained acicular martensite was observed at the interface between the white layer and original workpiece material, which might be the transition zone formed by the gradual weakening of thermal influence in this zone. The thickness of the white layer in Figure 7c reached ~100 μm, which could be divided into three levels. The outmost layer (layer I) was presumed to be a plastic deformation layer, and the microstructure was difficult to observe and was probably produced by abrasive particles of high-speed cutting at a high temperature. Layer II was composed of fine-grained martensite and possibly carbide particles precipitated by a high temperature. Owing to thermal and mechanical effects, the grains generated a series of dislocations and fragmentation, and finally formed the special layer after cooling [8]. In layer III, the martensite grains became coarse and retained austenite was observed, indicating that the quenching effect might be caused by a high level of the grinding temperature. In fact, when the temperature was above the austenitic transformation point and cooled rapidly, the martensitic microstructure would be generated [24].

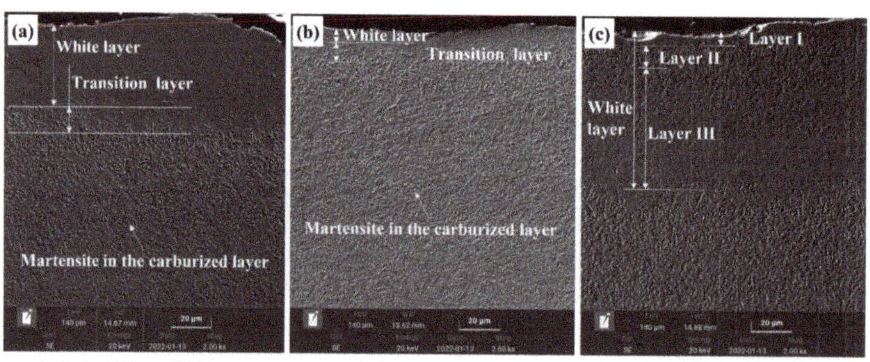

0.19 m/s; 0.02 mm; 34 m/s | 0.35 m/s; 0.02 mm; 26 m/s | 0.19 m/s; 0.03 mm; 34 m/s

Figure 7. SEM morphology of the ground surface layer at the feed speed, grinding depth, and grinding speed of (**a**) 0.19 m/s, 0.02 mm, and 34 m/s, respectively; (**b**) 0.35 m/s, 0.02 mm, and 26 m/s, respectively; and (**c**) 0.19 m/s, 0.03 mm, and 34 m/s, respectively.

There were two main reasons for the generation of the white layer during grinding quenched transmission gear; one was phase transformation caused by a high temperature and rapid cooling and the other one was plastic deformation caused by grinding abrasives [22]. Figure 8 shows the schematic of the formation mechanism of layer I, layer II, and layer III. When the temperature did not reach the austenite transformation point, the white layer could be formed by plastic deformation [24]. In the surface layer, owing to the strong mechanical extrusion caused by the increase in grinding depth, the material was plastically deformed and the refined grains formed layer I. Below this layer, the plastic deformation gradually decreased and, with the influence of grinding heat, a relatively obvious fine-grained martensite structure was generated, and it became the main composition of layer II. When grinding heat continued to be conducted to the subsurface, the influence of mechanical extrusion could be ignored and resulted in phase transformation during the rapid cooling process; subsequently, part of austenite transformed to martensite, forming layer III. The final properties of the surface layer were determined by the combination of work hardening caused by abrasive extrusion and heat softening caused by grinding temperature [20]. It was found that, with aggressive parameters, especially large grinding depths, undesirable grinding burn would occur on surface layer owing to the dominance of mechanical extrusion and thermal effects. The micro-structural transformation of surface layer material led to the deterioration of material properties, which ultimately caused the decrease in the surface rolling relief fatigue strength [22].

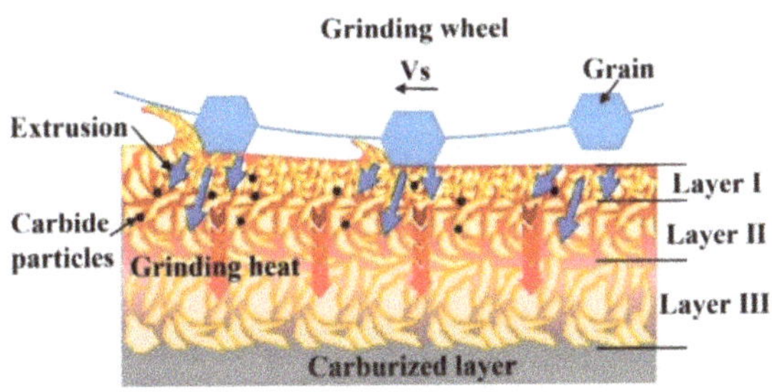

Figure 8. Schematic showing the forming mechanisms of the white layer during grinding of quenched gear samples.

3.3. Microhardness and Surface Roughness Variation

Figure 9 shows the microhardness variation of ground surface with different grinding parameters. Before grinding, the workpiece material had a very high carbon content (≥0.35 wt.%) owing to the existence of the carburized layer, which resulted in the high hardness of surface layer, and quenching treatment led to compressive stress on the surface. The microhardness gradually decreased with the depth below surface, and finally reached the value of base material (450 HV to 460 HV) [25]. The difference in the hardness gradient after grinding could be easily observed through the variation in the microhardness curves. The hardness decreased within a 200 μm thickness owing to the influence of grinding temperature, which resulted in heat softening in this depth range [26]. The softening effect was more pronounced with a grinding depth of 0.03 mm and grinding speed of 30 m/s and 34 m/s. This was in agreement with the theory of thermal softening during the grinding process, as high temperatures were more likely to be generated at a higher grinding depth and speed [27,28]. When the grinding depth was reduced to 0.02 mm, owing to the relatively low grinding temperature, the effect of thermal softening was difficult to observe and the work hardening could be found from the hardness gradient curve. With a relatively low level of grinding parameters, the highest microhardness value reached 680 HV at the depth of 200 μm, which normally formed at the place in which the most drastic plastic deformation took place because of the removal of materials under a high grinding pressure. The occurrence of work hardening and thermal softening was related to the grinding parameters. According to the findings from Ding et al. [29], a competing process between thermal softening and work hardening occurred when the workpiece material was subjected to a high grinding temperature and large grinding pressure. In this work, aggressive grinding parameters produced a higher grinding temperature and the effect of thermal softening on the hardness gradient was obvious, while a low level of grinding parameters weakened the thermal effect and contributed to grain refinement under mechanical extrusion [30].

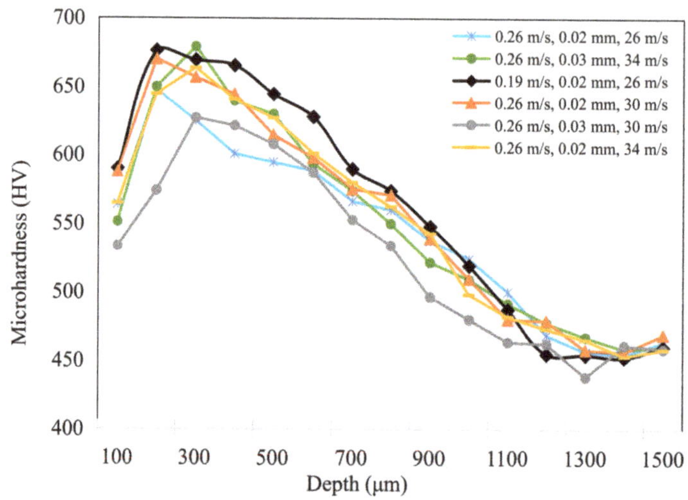

Figure 9. Microhardness variation within the subsurface layer when grinding quenched gear samples with different parameters.

Figure 10 shows the ground surface profile under different grinding parameters observed via white light interference. Grooves featured by the cutting marks of a large number of abrasive grains were observed on the ground surface regardless of grinding parameters. The inherent result was determined by characteristics such as the abrasion, size, and bond type of the grinding wheel during the grinding process [20]. The corresponding surface profile was featured with peaks and valleys, especially at higher feed speeds or grinding depths. With the increased feed speed, the number of abrasive grains involved in the cutting process per unit time decreased. Therefore, the depth of grooves generally increased. With the result of a high temperature and crushing of abrasive particles, deep grooves and chip adhesion were observed, mainly owing to the fact that residual materials accumulated under the plowing effect during the grinding process [31].

The material accumulation and deep grooves might be attributed to the factors including grain wear flat, the fractures of grains and bond bridges, together with workpiece material adhesion at high grinding forces and temperatures [26]. Figure 11 shows the topography of the ground surface with different grinding parameters. A large number of small scratches and grooves were observed. Obvious grooves were generated when the depth of scratches increased and the cross section of grooves was similar to the shapes of abrasive grains [11]. A large groove depth and small radius of curvature at the bottom of the groove could increase the coefficient of the effective stress concentration on the ground surface and increase the risk of fatigue failure [18]. It was found that particles penetrated the workpiece surface, which might be caused by the detachment and fracture of abrasive grains during the extrusion process of the grinding wheel. This defect added potential damage to the ground surface, as it not only affected surface integrity, but also probably led to surface corrosion and microcracking.

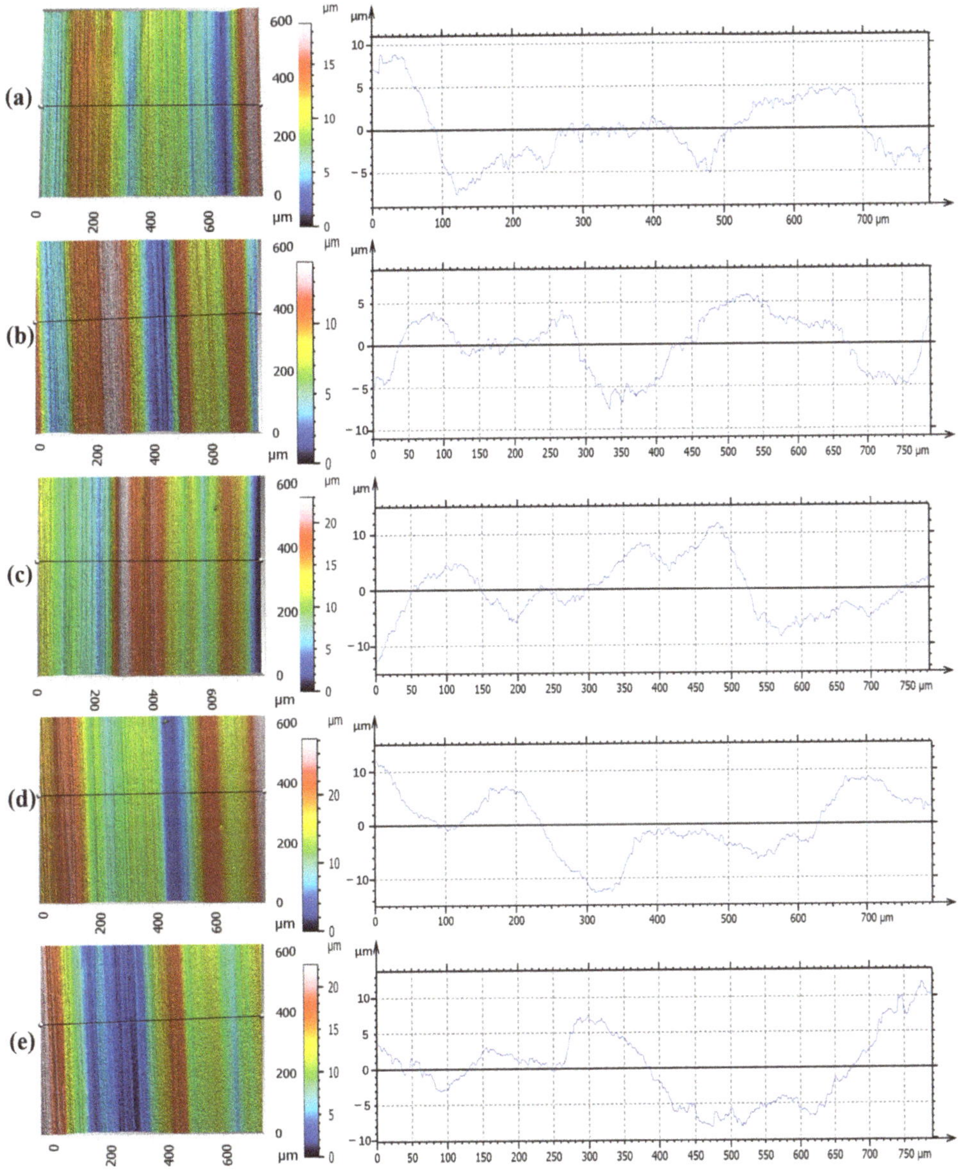

Figure 10. Ground surface profile under different grinding parameters (feed speed, grinding depth, and grinding speed): (**a**) 0.19 m/s, 0.02 mm, and 26 m/s, respectively; (**b**) 0.19 m/s, 0.02 mm, and 34 m/s, respectively; (**c**) 0.19 m/s, 0.03 mm, and 30 m/s, respectively; (**d**) 0.26 m/s, 0.02 mm, and 26 m/s, respectively; and (**e**) 0.35 m/s, 0.02 mm, and 34 m/s, respectively.

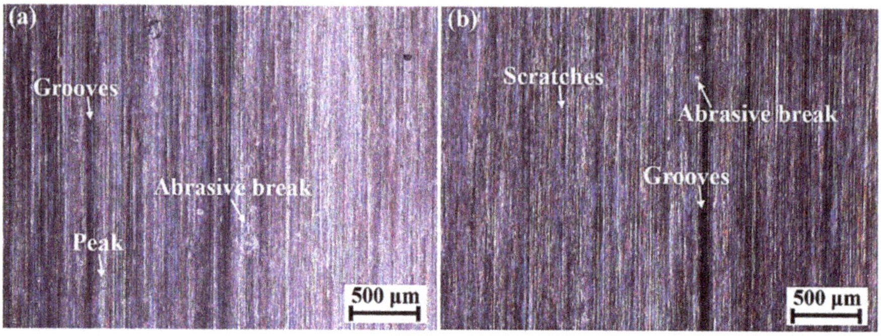

Figure 11. Ground surface morphology with the feed speeds, grinding depths, and grinding speeds of (**a**) 0.19 m/s, 0.02 mm, and 30 m/s, respectively, and (**b**) 0.19 m/s, 0.03 mm, and 34 m/s, respectively.

Figure 12 shows the variation in surface roughness (Sa) values of the ground surface with different grinding parameters. ANOVA was utilized to further analyze the significant factors and corresponding PCR for surface roughness (Sa). Compared with the data from ANOVA shown in Table 4, it was found that the grinding depth was statistically significant, affecting the surface roughness. Figure 13 shows the detailed information concerning the main effects plots of surface roughness related to different grinding parameters. When the grinding depth was 0.02 mm, the surface roughness gradually increased with the grinding speed and the maximum value reached 0.560 μm, which was obtained with the feed speed of 0.26 m/s and grinding speed of 34 m/s. Generally, the increase in grinding speed led to the decrease in surface roughness because it increased the number of abrasive cuts in the same period, which reduced the grinding force. However, a different trend of Sa values was observed with the grinding depth of 0.02 mm. It could be explained that the grinding temperature increased with a higher grinding speed, and the plastic deformation of workpiece material at higher temperature resulted in the increase in surface roughness [32]. In addition, with the grinding depth of 0.02 mm, a higher grinding speed produced smaller-diameter chips, and each area of the grinding wheel surface participated in the machining process more frequently, which accelerated the clogging of grinding wheel, resulting in increased grinding power and heat generation [33]. When the grinding depth increased to 0.03 mm, the surface roughness was deteriorated and increased to ~0.60 μm. Obviously, the increase in grinding depth produced a higher level of grinding force and grinding temperature and the material was squeezed by abrasive particles and flowed plastically to both sides, resulting in the increase in surface roughness [34]. At a grinding depth of 0.02 mm when the feed speed increased from 0.26 m/s to 0.35 m/s, and at the grinding depth of 0.03 mm when the feed speed increased from 0.19 m/s to 0.26 m/s, the decrease in Sa was less than 0.034 μm. However, at the feed speed of 0.19 m/s, the increase in Sa exceeded 0.097 μm when the grinding depth increased from 0.02 mm to 0.03 mm. The Sa decreased slightly when it increased from 0.26 m/s to 0.35 m/s, which might be caused by the squeezing effect; the influence on Sa was limited when compared with the effect of grinding depth. This is also consistent with the results of ANOVA in Table 4.

Table 4. ANOVA for surface roughness (Sa).

Source	DOE	Adj SS	Adj MS	F	PCR
Feed speed (m/s)	2	0.002239	0.001120	2.06	5.1%
Grinding speed (m/s)	2	0.000652	0.000326	0.60	0%
Grinding depth (mm)	1	0.012133	0.012133	22.35 *	51.0%
Error	7	0.003801	0.000543		43.9%
Total	12	0.022745			100%

* Significant at the 5% level, $F_{0.05,2,7} = 4.74$, $F_{0.05,1,7} = 5.59$.

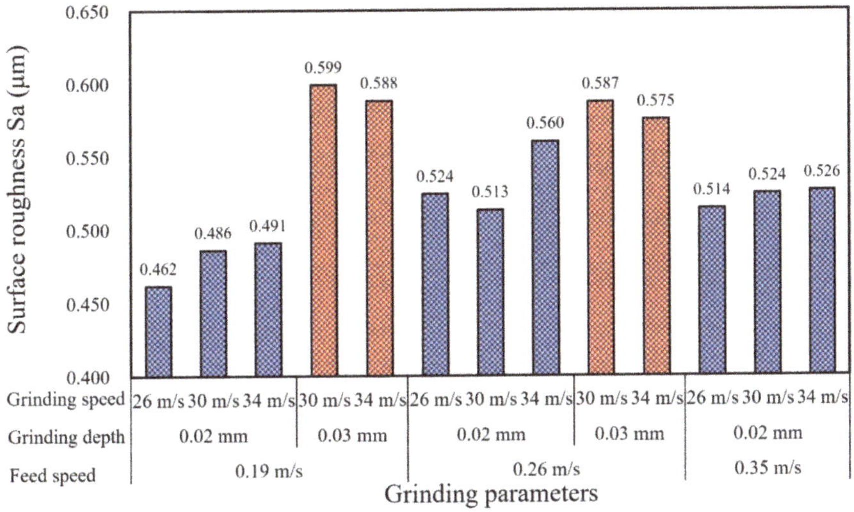

Figure 12. Surface roughness (Sa) variation under different grinding parameters.

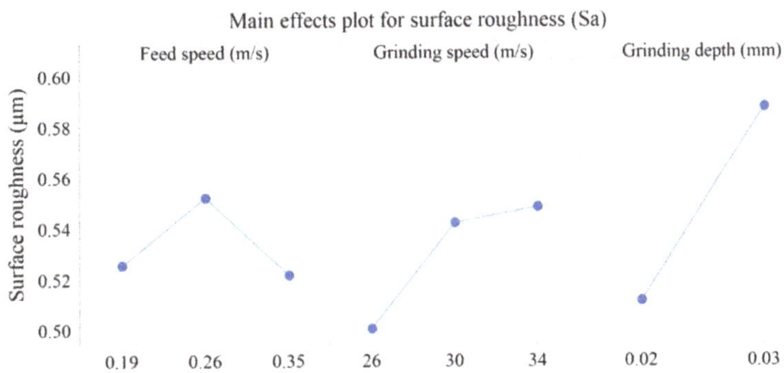

Figure 13. Main effects plot for surface roughness (Sa) with different parameters.

4. Conclusions

(1) The grinding temperature increased with the grinding depth and grinding speed, with the highest level of ~290 °C. The feed speed presented the most significant effect on grinding temperature with the corresponding percentage contribution ratio of 33.7%. Both work hardening and thermal softening were observed during grinding of quenched gear samples, and aggressive grinding parameters intensified the thermal softening effect, resulting in a reduction in microhardness within the subsurface layer.

(2) The thickness of the white layer exceeded 100 μm when the grinding depth increased to 0.03 mm and it could be subdivided into three different layers owing to the contribution of mechanical extrusion and the thermal effect. The top layer mainly featured plastic deformation and the second one was composed of fine-grained martensite. Coarse-grained acicular martensite was found at the interface between the white layer and the softened dark layer.

(3) The ground surface topography showed several scratches and typical grooves and some particles were embedded into the ground surface, which was probably caused by the shedding and breaking of abrasive particles. When the grinding depth increased to 0.03 mm, the grinding surface roughness (Sa) was relatively high and reached up to ~0.60 μm. The percentage contribution ratio of the grinding depth on surface

roughness was higher than 50%, mainly owing to severe plastic deformation under grinding wheel extrusion and the thermal effect.

Author Contributions: Conceptualization, M.L.; Data curation, X.J., K.L. and Y.Y.; Investigation, X.J. and P.G.; Methodology, M.L.; Writing—original draft, X.J.; Writing—review & editing, M.L., P.G. and H.H. All authors have read and agreed to the published version of the manuscript.

Funding: This research received no external funding.

Institutional Review Board Statement: Not applicable.

Informed Consent Statement: Not applicable.

Data Availability Statement: The data that support the findings of this study are available from the corresponding author upon reasonable request.

Conflicts of Interest: The authors declare no financial or commercial conflict of interest.

References

1. Guerrini, G.; Landi, E.; Peiffer, K.; Fortunato, A. Dry Grinding of Gears for Sustainable Automotive Transmission Production. *J. Clean. Prod.* **2018**, *176*, 76–88. [CrossRef]
2. Yao, C.F.; Wang, T.; Xiao, W.; Huang, X.C.; Ren, J.X. Experimental Study on Grinding Force and Grinding Temperature of Aermet 100 Steel in Surface Grinding. *J. Mater. Process. Technol.* **2014**, *214*, 2191–2199. [CrossRef]
3. Guerrini, G.; Lerra, F.; Fortunato, A. The Effect of Radial Infeed on Surface Integrity in Dry Generating Gear Grinding for Industrial Production of Automotive Transmission Gears. *J. Manuf. Process.* **2019**, *45*, 234–241. [CrossRef]
4. Manikandan, M.; Prabagaran, S. Investigations on Producing Optimal Grinding Spindle using 36CrNiMo4 Alloy Steel. *Mater. Today Proc.* **2021**, *37*, 1622–1628. [CrossRef]
5. Jamshidi, H.; Budak, E. On the Prediction of Surface Burn and its Thickness in Grinding Processes. *CIRP Ann.* **2021**, *70*, 285–288. [CrossRef]
6. Malkin, S.; Guo, C. Thermal Analysis of Grinding. *CIRP Ann.* **2007**, *56*, 760–782. [CrossRef]
7. Rasmussen, C.J.; Fæster, S.; Dhar, S.; Quaade, J.V.; Bini, M.; Danielsen, H.K. Surface Crack Formation on Rails at Grinding Induced Martensite White Etching Layers. *Wear* **2017**, *384–385*, 8–14. [CrossRef]
8. Sun, C.; Hong, Y.; Xiu, S.C.; Yao, Y.L. Grain Refinement Mechanism of Metamorphic Layers by Abrasive Grinding Hardening. *J. Manuf. Process.* **2021**, *69*, 125–141. [CrossRef]
9. Wu, C.J.; Guo, W.C.; Li, R.; Zhao, Y.D.; Zhou, Q.Z. Thermal Effect on Oxidation Layer Evolution and Phase Transformation in Grinding of Fe-Ni Super Alloy. *Mater. Lett.* **2020**, *275*, 128072. [CrossRef]
10. Su, J.X.; Zhang, Y.Z.; Deng, X.Z. Analysis and Experimental Study of Cycloid Gear Form Grinding Temperature Field. *Int. J. Adv. Manuf. Technol.* **2020**, *110*, 949–965. [CrossRef]
11. Zhou, K.; Ding, H.H.; Wang, R.X.; Yang, J.Y.; Guo, J.; Liu, Q.Y.; Wang, W.J. Experimental Investigation on Material Removal Mechanism during Rail Grinding at Different Forward Speeds. *Tribol. Int.* **2020**, *143*, 106040. [CrossRef]
12. Jermolajev, S.; Brinksmeier, E.; Heinzel, C. Surface Layer Modification Charts for Gear Grinding. *CIRP Ann.* **2018**, *67*, 333–336. [CrossRef]
13. Lu, H.; Ren, Y.P.; Chen, Y.; Zhu, H.; Xin, Z.D.; Wan, H.Y.; Li, Z.Y.; Tu, X.C.; Cheng, L.; He, K.; et al. Wear Resistance of 20Cr2Ni4A Alloy Steel Treated by Laser Shock Peening and Implantation of Diamond Nanoparticles. *Surf. Coat. Technol.* **2021**, *412*, 127070. [CrossRef]
14. Liu, D.H.; Wang, G.; Nie, Z.G.; Rong, Y.M. An in-situ Infrared Temperature-measurement Method with Back Focusing on Surface for Creep-feed Grinding. *Measurement* **2016**, *94*, 645–652. [CrossRef]
15. Li, H.N.; Axinte, D. On the Inverse Design of Discontinuous Abrasive Surface to Lower Friction-induced Temperature in Grinding: An Example of Engineered Abrasive Tools. *Int. J. Mach. Tools Manuf.* **2018**, *132*, 50–63. [CrossRef]
16. Saravanakumar, A.; Dhanabal, S.; Jayanand, E.; Logeshwaran, P. Analysis of Process Parameters in Surface Grinding Process. *Mater. Today Proc.* **2018**, *5*, 8131–8137. [CrossRef]
17. Zhao, B.; Guo, X.C.; Bie, W.B.; Chang, B.Q.; Zhao, C.Y. Thermo-mechanical Coupling Effect on Surface Residual Stress during Ultrasonic Vibration-assisted Forming Grinding gear. *J. Manuf. Process.* **2020**, *59*, 19–32. [CrossRef]
18. Zhang, S.Q.; Yang, Z.X.; Jiang, R.S.; Jin, Q.C.; Zhang, Q.; Wang, W.H. Effect of Creep Feed Grinding on Surface Integrity and Fatigue Life of Ni3Al based Superalloy IC10. *Chin. J. Aeronaut.* **2021**, *34*, 438–448. [CrossRef]
19. Sallem, H.; Hamdi, H. Analysis of Measured and Predicted Residual Stresses Induced by Finish Cylindrical Grinding of High Speed Steel with CBN Wheel. *Procedia CIRP* **2015**, *31*, 381–386. [CrossRef]
20. Dang, J.Q.; Zhang, H.; An, Q.L.; Ming, W.W.; Chen, M. Feasibility Study of Creep Feed Grinding of 300M Steel with Zirconium Corundum Wheel. *Chin. J. Aeronaut.* **2022**, *35*, 565–578. [CrossRef]
21. Ruchert, C.O.F.T.; Maciel, C.I.S.; Chemin, A.E.A. Sub Case Origin Fatigue in Teeth of Helical Gear of a TA 67n Turbo Reducer. *Eng. Failure Anal.* **2020**, *108*, 104286. [CrossRef]

22. Brown, M.; Wright, D.; M'Saoubi, R.; McGourlay, J.; Wallis, M.; Mantle, A.; Crawforth, P.; Ghadbeigi, H. Destructive and Non-destructive Testing Methods for Characterization and Detection of Machining-induced White Layer: A Review Paper. *CIRP J. Manuf. Sci. Technol.* **2018**, *23*, 39–53. [CrossRef]
23. Ding, Z.S.; Guo, F.; Guo, W.C.; Sun, G.X.; Lin, J.J.; Wu, C.J.; Liang, S.Y. Investigations on Grain Size Characteristics in Microstructure during Grinding of Maraging Steel 3J33. *J. Manuf. Process.* **2021**, *69*, 434–450. [CrossRef]
24. Poulachon, G.; Albert, A.; Schluraff, M.; Jawahir, I.S. An Experimental Investigation of Work Material Microstructure Effects on White Layer Formation in PCBN hard turning. *Int. J. Mach. Tools Manuf.* **2005**, *45*, 211–218. [CrossRef]
25. Dong, M.L.; Cui, X.F.; Zhang, Y.H.; Jin, G.; Yue, C.W.; Zhao, X.; Cai, Z.B.; Xu, B.S. Vacuum Carburization of 12Cr2Ni4A Low Carbon Alloy Steel with Lanthanum and Cerium Ion Implantation. *J. Rare Earths* **2017**, *35*, 1164–1170. [CrossRef]
26. Miao, Q.; Ding, W.J.; Kuang, W.F.; Yang, C.Y. Grinding Force and Surface Quality in Creep Feed Profile Grinding of Turbine Blade Root of Nickel-based Superalloy with Microcrystalline Alumina Abrasive Wheels. *Chin. J. Aeronaut.* **2021**, *34*, 576–585. [CrossRef]
27. Fathallah, B.B.; Fredj, N.B.; Sidhom, H.; Braham, C.; Ichida, Y. Effects of Abrasive Type Cooling Mode and Peripheral Grinding Wheel Speed on the AISI D2 Steel Ground Surface Integrity. *Int. J. Mach. Tools Manuf.* **2009**, *49*, 261–272. [CrossRef]
28. Barbacki, A.; Kawalec, M.; Hamrol, A. Turning and Grinding as a Source of Microstructural Changes in the Surface Layer of Hardened Steel. *J. Mater. Process. Technol.* **2003**, *133*, 21–25. [CrossRef]
29. Ding, W.F.; Xu, J.H.; Chen, Z.Z.; Su, H.H.; Fu, Y.C. Grindability and Surface Integrity of Cast Nickel-based Superalloy in Creep Feed Grinding with Brazed CBN Abrasive Wheels. *Chin. J. Aeronaut.* **2010**, *23*, 501–510.
30. Ren, X.P.; Liu, Z.Q. Influence of Cutting Parameters on Work Hardening Behavior of Surface Layer during Turning Superalloy Inconel 718. *Int. J. Adv. Manuf. Technol.* **2016**, *86*, 2319–2327. [CrossRef]
31. Zhou, W.H.; Tang, J.Y.; Chen, H.F.; Zhu, C.C.; Shao, W. A Comprehensive Investigation of Plowing and Grain-workpiece Micro Interactions on 3D Ground Surface Topography. *Int. J. Mech. Sci.* **2018**, *144*, 639–653. [CrossRef]
32. Ruzzi, R.S.; Silva, R.B.; Silva, L.R.R.; Machado, Á.R.; Jackson, M.J.; Hassui, A. Influence of Grinding Parameters on Inconel 625 Surface Grinding. *J. Manuf. Process.* **2020**, *55*, 174–185. [CrossRef]
33. Guba, N.; Hüsemann, T.; Karpuschewski, B. Influence of Gear Hobbing Feed Marks on the Resulting Gear Quality after Discontinuous Profile Grinding. *CIRP J. Manuf. Sci. Technol.* **2020**, *31*, 314–321. [CrossRef]
34. Kang, B.; Ma, H.R.; Li, J.; Xu, B.Q. Effect of grinding parameters on surface quality, microstructure and rolling contact fatigue behaviors of gear steel for vacuum pump. *Vacuum* **2020**, *180*, 109637. [CrossRef]

materials

Article

Grain Structure Formation and Texture Modification through Multi-Pass Friction Stir Processing in AlSi10Mg Alloy Produced by Laser Powder Bed Fusion

Akbar Heidarzadeh [1,*], Mousa Javidani [2], Mohammadreza Mofarrehi [2], Pouyan Motalleb-nejad [3], Roghayeh Mohammadzadeh [1], Hamidreza Jafarian [4] and X.-Grant Chen [2,*]

1 Department of Materials Engineering, Azarbaijan Shahid Madani University, Tabriz P.O. Box 53714-161, Iran
2 Department of Applied Science, University of Quebec at Chicoutimi, Saguenay, QC G7H 2B1, Canada
3 Faculty of Materials Engineering, Sahand University of Technology, Tabriz P.O. Box 51335-1996, Iran
4 School of Metallurgy and Materials Engineering, Iran University of Science and Technology, Narmak, Tehran P.O. Box 16846-13114, Iran
* Correspondence: xgrant_chen@uqac.ca (X.-G.C.); ac.heydarzadeh@azaruniv.ac.ir (A.H.)

Citation: Heidarzadeh, A.; Javidani, M.; Mofarrehi, M.; Motalleb-nejad, P.; Mohammadzadeh, R.; Jafarian, H.; Chen, X.-G. Grain Structure Formation and Texture Modification through Multi-Pass Friction Stir Processing in AlSi10Mg Alloy Produced by Laser Powder Bed Fusion. *Materials* **2023**, *16*, 944. https://doi.org/10.3390/ma16030944

Academic Editors: Pan Gong, Xin Wang, Maojun Li and Guangchao Han

Received: 24 November 2022
Revised: 14 January 2023
Accepted: 16 January 2023
Published: 19 January 2023

Abstract: A new strategy is proposed to modify the grain structure and crystallographic texture of laser-powder bed fusion AlSi10Mg alloy using multi-pass friction stir processing (FSP). Accordingly, 1–3 passes of FSP with 100% overlap were performed. Scanning electron microscopy and electron backscattered diffraction were used for microstructural characterization. Continuous dynamic recrystallization and geometric dynamic recrystallization are the governing mechanisms of grain refinement during FSP. The stir zones have bimodal grain structures containing large and fine grains. The multi-pass FSP caused a considerable increase in the volume fraction of the large-grained area in the stir zone, which contained higher values of low-angle boundaries and sharp shear texture components of $B(1\bar{1}2)[110]$ and $\bar{B}(\bar{1}1\bar{2})[\bar{1}\bar{1}0]$. The formation of low-energy grain boundaries in the stir zone and alignment of the low-energy crystallographic planes with the surface of the sample made the strategy of using multi-pass FSP a promising candidate for corrosion resistance enhancement in future studies. Moreover, the detailed evolution of the grains, texture components, grain boundaries, and Si particles is discussed.

Keywords: AlSi10Mg alloy; friction stir processing; grain boundary; texture; dynamic recrystallization

1. Introduction

Aluminum alloys are important materials in the aerospace and automotive industries and are appropriate candidates for production using laser-based additive manufacturing owing to their good printability [1]. Among aluminum alloys, AlSi10Mg has attracted significant attention owing to its high strength/weight ratio, low cost, and excellent printability [2]. Currently, the AlSi10Mg alloy is produced and employed in different industrial sectors by laser-powder bed fusion (L-PBF), which is a laser-based AM method [3]. During the last decade, a number of investigations have been undertaken to elucidate the microstructure and mechanical properties of AlSi10Mg, which show the negative effect of L-PBF on ductility and fatigue resistance [4]. Therefore, some researchers have attempted to overcome this problem using post-treatments, including both heat and deformation treatments [5–8].

Improving the ductility of AlSi10Mg by post-treatment has often been found in conjunction with a decrease in strength [1]. One of the most promising methods for post-treatment of AlSi10Mg is friction stir processing (FSP) [9]. During FSP, which was developed based on the friction stir welding (FSW) concept, the heat and severe plastic deformation induced by a rotational tool cause microstructural changes, such as grain refinement and the fragmentation of secondary phases [10,11]. Thus, FSP can be used to modify the microstructure and mechanical properties of metals and alloys [10]. In the case of L-PBF AlSi10Mg, FSP causes less reduction in strength when ductility is improved [1]. For instance, Macías et al. [9] reported that FSP can increase the ductility and fatigue life time by 448% and 200%, respectively, with only a reduction in yield strength by 34%.

In ref. [9,12–21], most microstructural studies of friction-stir-processed (FSPed) AlSi10Mg produced by L-PBF have focused on defect elimination, the evolution of the Si structure, and precipitation phenomena. However, fundamental information regarding the grain structure formation during FSP and the corresponding grain boundaries and crystallographic texture components is lacking, which is necessary to be understood to control the final properties of the processed materials [10]. On the other hand, the parameters of FSW and FSP, such as tool geometry, pin to shoulder ratio, shoulder surface features, tool eccentricity, tool rotational speed, tool traverse speed, number of passes, etc., influence the microstructure and mechanical performance [11–14]. For example, Hou et al. [11] investigated the role of pin eccentricity on microstructural evolution and mechanical properties of FSWed Al-Mg-Si alloy sheets. They found that pin eccentricity had a positive effect on promoting the material flow and grain refinement in the stir zone. The texture intensity was increased as the pin eccentricity was introduced, and the dominant texture component converted from $C\{001\}\langle[110]\rangle$ (without eccentricity) to $B/\overline{B}\{112\}\langle[110]\rangle$ when eccentric pin was used. Senthilkumar et al. [14] studied the effect of the number of passes in FSP of AA6082 alloy. They revealed that increasing the number of passes with the specific rotational speed, the tunnel void at stir zone could be minimized due to the recrystallization mechanism. In addition, it increased the grain size with more dissolution and re-precipitation along with intense fragmentation of second-phase particles. Therefore, in this study, the microstructural evolution in FSPed zones in single-pass and multi-pass AlSi10Mg was studied in detail. The grain structure formation and changes in the eutectic Si structure and texture were characterized using scanning electron microscopy and electron backscattered diffraction (EBSD) to reveal the governing microstructural mechanisms during FSP. Consequently, a strategy has been proposed for grain structure formation and the modification of crystallographic textures in L-PBF AlSi10Mg to enhance the performance of the final products.

2. Materials and Methods

AlSi10Mg plates with dimensions of 80 mm length × 40 mm width × 2 mm thickness were produced by the L-PBF method. Gas-atomized AlSi10Mg powders with an average diameter smaller than 65 μm were used in this study. A Noura M100P printer with the following process conditions was employed to fabricate the samples: powder consumption of 80 g, layer thickness of 30 μm, energy density of 60 J/mm^3, hatch space of 0.19 mm, scan rotation of 67°, number of slices of 260, and overall build time of 2 h. Subsequently, three different friction stir post-processing conditions were employed on the AlSi10Mg L-PBF plates with single, double, and multi-pass FSP passes. To achieve this, the FSP was performed perpendicular to the building direction of the plate at a traverse speed of 85 mm/min, a rotational speed of 950 rpm, a plunge depth of 0.1 mm, and a tilt angle of 2° using a tool made of H13 tool steel, employing a conventional milling machine (Vertical WMW Heckert FSS 315 V/2). It is worth noting that the plunge depth was applied once the tool head touched the surface of the plates. The FSP tool is composed of a pin (with a diameter of 3 mm and a length of 1.7 mm) and a shoulder with a diameter of 12 mm. Single-, double-, and multi-pass FSP were conducted with a 100% overlap to produce different samples denoted here as 1-pass, 2-pass, and 3-pass samples, respectively.

Metallographic samples were cut from the processed samples perpendicular to the FSP direction and prepared using a standard metallographic procedure. The optical macrostructure of the joint was etched with a solution of 5 cc HNO_3, 3 cc HCl, 2 cc HF, and 190 cc distilled water. The EBSD analysis was performed using a scanning electron microscope (SEM) with a step size of 0.2 μm, and all the corresponding data were processed using TSL-OIM software with a threshold of 15° between low-angle grain boundaries (LABs) and high-angle grain boundaries (HABs). Brandon's criterion was used for coincidence site lattice (CSL) boundary classification in TSL-OIM software. Moreover, SEM was employed to observe the different structures of the Si-rich phase in the L-PBF as-built and FSP-treated states.

3. Results

3.1. Grain Structure Formation

The inverse pole figure (IPF) and image quality (IQ) maps of the as-built L-PBF AlSi10Mg, which are treated here as the base material (BM) of the FSP, are presented in Figure 1. The BM is composed of two types of grains: large elongated grains (black arrows in Figure 1a) and fine equiaxed grains (white arrows in Figure 1a) within the melt pools, as indicated by the red arrows in Figure 1b. The average grain size of BM was 16.1 μm. The elongated grains are primarily formed parallel to the BD, which is against the heat transfer direction during the solidification of melt pools [1]. The formation of fine equiaxed grains occurs when MP is consumed by elongated grains, and the temperature gradient decreases to a certain value for the nucleation of equiaxed grains [22]. From Figure 1, the BM contains 78 and 22% HABs and LABs, respectively. It is well documented that FSP destroys the initial microstructure of BMs owing to various restoration mechanisms [10,15]. However, in the case of L-PBF AlSi10Mg, the grain structure formation, restoration mechanisms, grain boundaries, and texture evolution during FSP have not yet been disclosed, which is discussed in detail in the following sections.

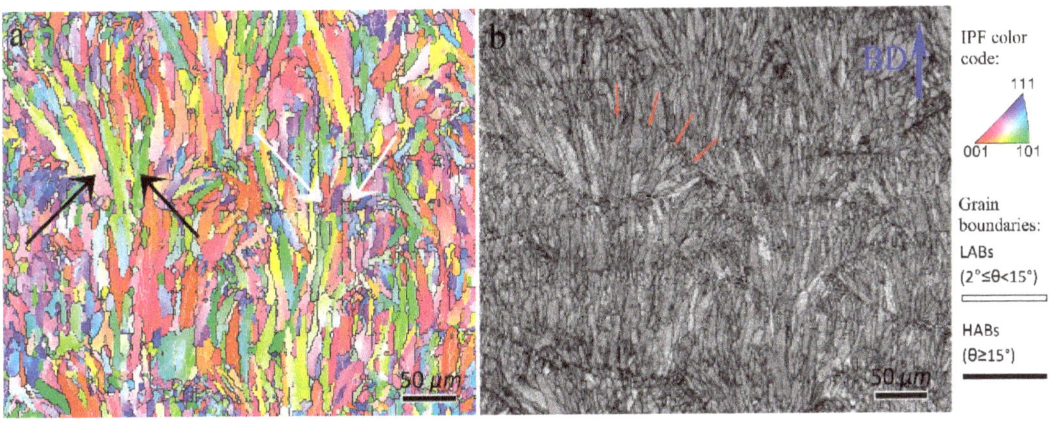

Figure 1. (**a**) Inverse pole figure and (**b**) image quality maps of as-built L-PBF AlSi10Mg. High and low angle grain boundaries are drawn in black and white colors on all inverse pole figure maps through paper, respectively. Black and white in (**a**) arrows show large-elongated and fine-equiaxed grains. Red arrows in (**b**) refer to the region of the melt pool boundary.

3.1.1. Single Pass FSP

The cross-sectional macrostructure of the 1-pass sample is shown in Figure 2a, indicating the presence of distinct zones including the BM, transition zone, and stir zone (SZ). The BM was not affected by heat or deformation during FSP, exhibiting a typical microstructure of L-PBF AlSi10Mg (Figure 1) containing overlapping melt pool boundaries. However, the transition zone between the BM and SZ (Figure 2b), which is usually called the thermomechanically affected zone (TMAZ), is affected during FSP. Upon reaching the center of the SZ, the temperature, strain, and strain rate induced by the rotational tool increased to their maximum values. Hence, microstructural evolution can be completed in the SZ, whereas it occurs partially in the transition zone owing to inadequate heat and deformation. As shown in Figure 2b, the SZ was divided into two regions, SZ-I and SZ-II, which have different dark and bright contrasts. The four blue rectangles in Figure 2b are the different areas of the EBSD scans from TMAZ (rectangle 1) to SZ-I (rectangle 4). It is notable that the characterization of the TMAZ can be used to reveal the microstructural evolution during grain structure formation because it is the transition zone including incomplete mechanisms between the BM and SZ [23,24].

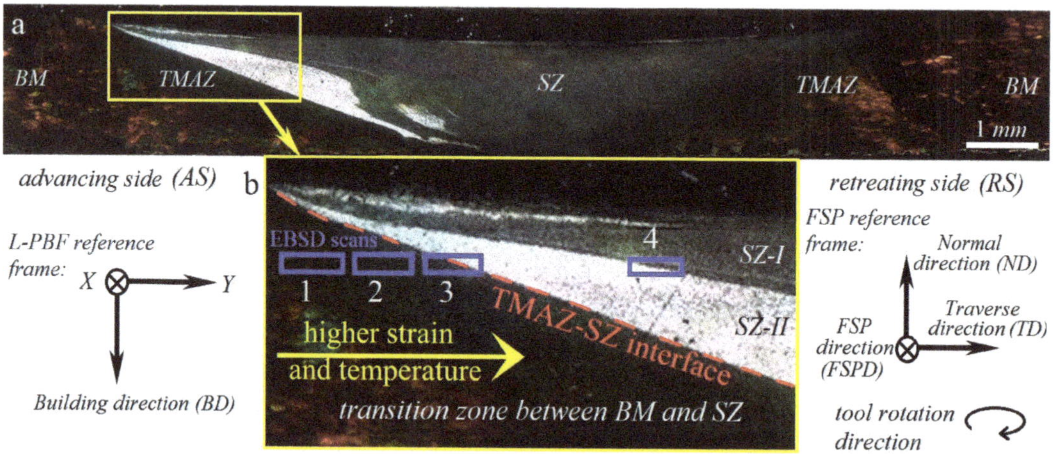

Figure 2. (**a**) Cross-sectional macrostructure of the 1-pass friction stir processed sample. (**b**) The higher magnification of the transition zone between BM and SZ on the advancing side and corresponding details of the L-PBF reference frame, FSP reference frame, and tool rotation direction. The blue rectangles indicate the different areas of EBSD scans.

The inverse pole figure map of rectangle 1 in Figure 2b is shown in Figure 3. The higher magnified views of the three different zones in conjunction with the grain boundaries (HABs: black color and LABs: white color) are illustrated in Figure 3b–d. There was a gradient of strain, temperature, and strain rate from the BM to the center of the SZ during the FSP [15], which caused the formation of different microstructural zones, as shown in Figure 3. At low strain and temperature (Figure 3b), the grains were elongated owing to the shear deformation induced by FSP, and the melt pool boundaries of the BM disappeared. The presence of subgrains surrounded by LABs are confirmed by black arrows in Figure 3b. Evidently, the grain structure formation can be started with a mechanism accelerated by subgrains. In the zones with higher temperature and strain during FSW (Figure 3c,d), the LABs were gradually transformed to HABs, as indicated by the blue arrows, which confirms the occurrence of continuous dynamic recrystallization (CDRX). Upon increasing the strain value, more dislocations were absorbed into the LABs, resulting in a gradual increase in their misorientations. When the misorientation of the LABs is larger than 15° (the threshold for identifying HABs), they transform into HABs; hence, new DRX grains are formed. In Figure 3d, in addition to CDRX, geometric dynamic recrystallization (GDRX) was detected in the TMAZ, as indicated by red arrows. The higher temperature and strain values in this zone compared to those in (Figure 3b,c) caused the formation of more elongated/fibrous grains. When the width of the elongated grains was approximately equal to the grain size of the SZ, the serrations of the old grain boundaries touched each other, and GDRX grains could be formed [25]. Notably, GDRX is usually categorized as a type of CDRX. Thus, it can be concluded that at the initial stage of hot deformation in the 1-pass sample during FSP, the grain structure formation begins with DRV, followed by CDRX through LABs→HABs transformation and GDRX. The other point in Figure 3 is the heterogeneous structure within the TMAZ, which indicates non-uniform plastic deformation during FSP causing different amounts of dynamic recrystallization in different zones (Figure 3c,d).

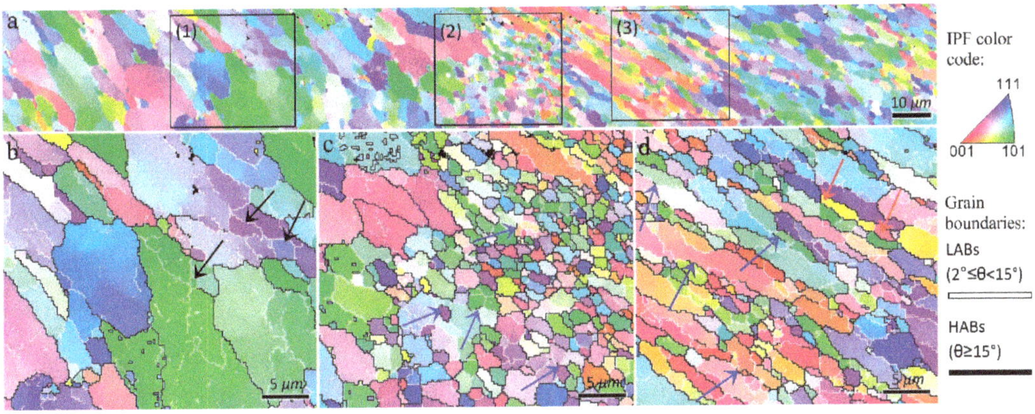

Figure 3. (**a**) Inverse pole figure maps of the rectangle 1 in Figure 2b. (**b–d**) Higher magnified views of the zones 1–3 indicated in (**a**). Black arrows in (**b**) refer to the LABs. Blue arrows in (**c**,**d**) show the continuous dynamic recrystallization mechanism occurring by transformation of LABs to HABs. The red arrows in (**d**) mention the fibrous grains in which the geometrically dynamic recrystallization (GDRX) arises.

The inverse pole figure map of rectangle 2 in Figure 2b is shown in Figure 4. The enlarged views of zones 1 and 2 in Figure 4a in conjunction with the HABs and LABs are illustrated in Figure 4b,c. From Figure 4, at higher levels of strain and temperature during FSP, more DRX grains were generated within the TMAZ because the DRV and CDRX mechanisms can occur in higher quantities compared to the areas with lower strain and temperature (Figure 3). The inverse pole figure map of the rectangle 3 in Figure 2b are shown in Figure 5. A further increase in strain and temperature in the TMAZ, that is, by approaching the TMAZ and SZ-II interface in Figure 2b, the DRX was completed, as shown in Figure 5b. As shown in Figure 5c, SZ-II contained larger grains than the TMAZ, which is attributed to the higher temperature in this area during FSP, which encouraged grain growth. The origin of the different regions in the SZ in the cross-sectional macrostructure (Figure 2), that is, SZ-I and SZ-II, is shown in the inverse pole figure maps in Figure 6, which illustrates the interface between these zones. SZ-II (Figure 6a) was composed of larger grains than SZ-I (Figure 6b), resulting in a bimodal grain structure in the SZ. This effect can be explained by the correlation between the thermomechanical behavior of the material using the Zener–Holloman parameter (Z) [26]:

$$Z = \frac{d\varepsilon}{dt} exp\left(\frac{Q_{def}}{RT}\right) \quad (1)$$

where $\frac{d\varepsilon}{dt}$ is the strain rate, Q_{def} is the activation energy required for the deformation, R is the gas constant, and T is the deformation temperature. The Z value indicates the grain size of metals and alloys that undergo plastic deformation during thermomechanical processing. The relationship between Z value and average grain size (d_{Rx}) can be expressed as follows [27,28]:

$$Z d_{Rx}{}^{m} = A \quad (2)$$

where A, d_{Rx}, and m are the constant, DRX grain size, and grain-size exponent, respectively. Equations (1) and (2) reveal that the strain rate and temperature are critical in determining the final grain size of FSPed metals and alloys, which have opposite effect on final grain size (d_{Rx}). From Figure 6 and considering Equations (1) and (2), the strain rate ($\frac{d\varepsilon}{dt}$) and temperature (T) are not the same in SZ-I and SZ-II. The finer grain size in SZ-I (Figure 6c) is related to lower temperature and higher strain rate compared to those in SZ-II. In contrast, the larger grain size in SZ-II (Figure 6b) is attributed to the higher temperature and lower the strain rate in this area. From literature [10,15], the complex material flow during FSW/P causes the formation of distinct zones with different local deformation and temperature values, and hence different microstructures such as onion rings. Similarly, as mentioned before, the existence of two distinct zones (SZ-I and SZ-II) with different grain size and sharpness can be explained by qualitative analysis using Equations (1) and (2). Therefore, a combination of temperature and strain rate can be applied to achieve the grain size revealed by the EBSD measurements in SZ-I and SZ-II. For this aim, combined simulation/experimental research is needed to measure the exact values of strain, strain rate and temperature in different zones during FSW/P, which can be a topic for future study.

3.1.2. Double and Multiple Pass FSP

The cross-sectional macrostructure of 2-pass FSPed samples is shown in Figure 7, which indicates a similar behavior to the 1-pass FSPed sample (Figure 2). The fractions of SZ-II and SZ-I did not change considerably after the second pass of the FSP. The EBSD scan areas are shown by blue rectangles in Figure 7. The inverse pole figure maps corresponding to the TMAZ are shown in Figure 8. As shown in Figure 8b,c, the second pass of FSP caused a fully equiaxed grain structure in the TMAZ. During the second pass of the FSP, the non-DRX parts of the TMAZ after the single-pass FSP (Figures 3 and 4) underwent CDRX, and hence the DRX was completed in this area. When approaching the SZ-II region in Figure 7, coarser grains appeared, as shown in the inverse pole figure maps in Figure 9. Therefore, according to Figures 7–9, the second pass of the FSP completes the partial restoration mechanisms during the single-pass FSP. The interfacial area between SZ-II and SZ-I is similar to that of the 1-pass samples, and for the sake of brevity, the corresponding inverse pole figures are not presented.

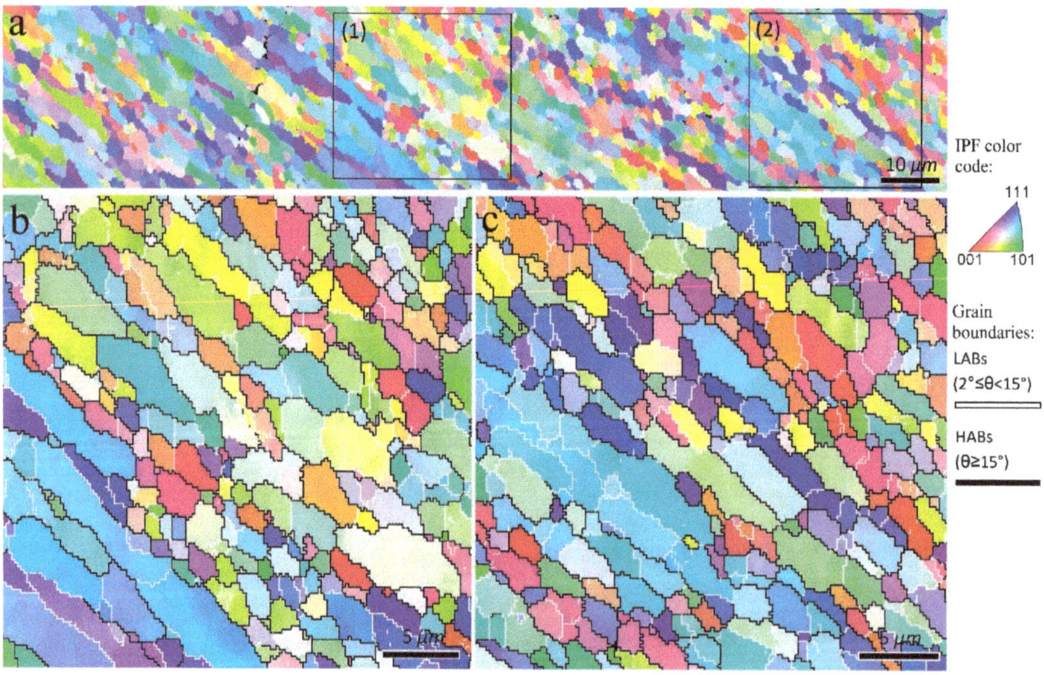

Figure 4. (**a**) Inverse pole figure maps of the TMAZ indicated by rectangle 2 in Figure 2b. Higher magnification of zones 1 and 2 indicated in (**a**) are illustrated in (**b**,**c**), respectively.

Figure 5. (**a**) Inverse pole figure maps of the interfacial zone between TMAZ and SZ-II regions in 1-pass sample indicated by rectangle 3 in Figure 2b. (**b**,**c**) Higher magnification of zones 1 and 2 indicated in (**a**), respectively.

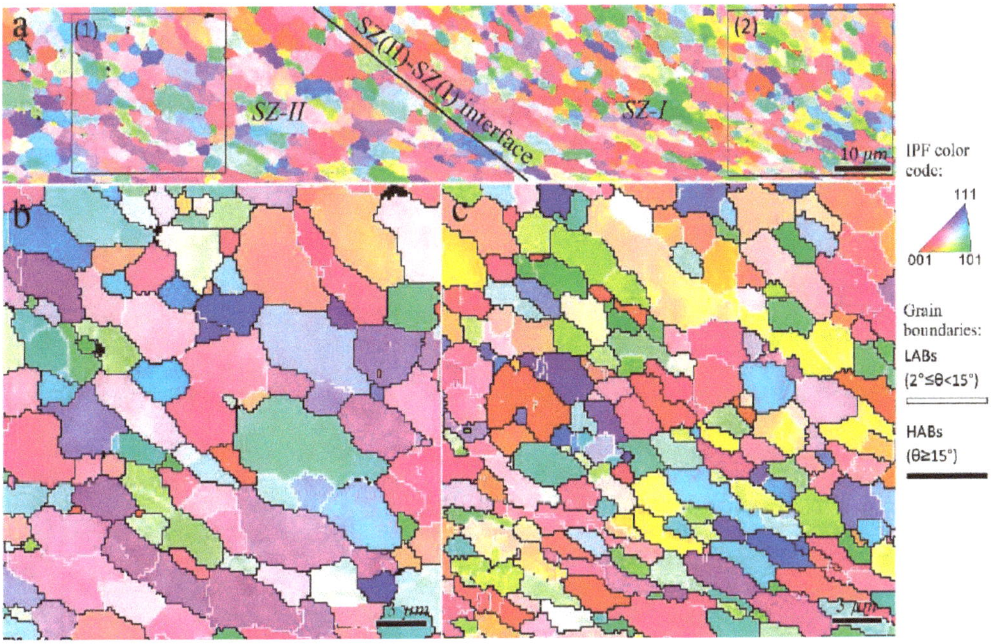

Figure 6. (**a**) Inverse pole figure maps of the interfacial zone between SZ-II and SZ-I regions in 1-pass sample indicated by rectangle 4 in Figure 2b. Higher magnification of zones 1 and 2 indicated in (**a**) are illustrated in (**b**,**c**), respectively.

Figure 7. Cross-sectional macrostructure of 2-pass friction stir processed sample. The blue rectangles indicate areas 1 and 2 of EBSD scans. The details of the FSP reference frame are the same as Figure 2. The interface between TMAZ and SZ is depicted by a red-dashed line.

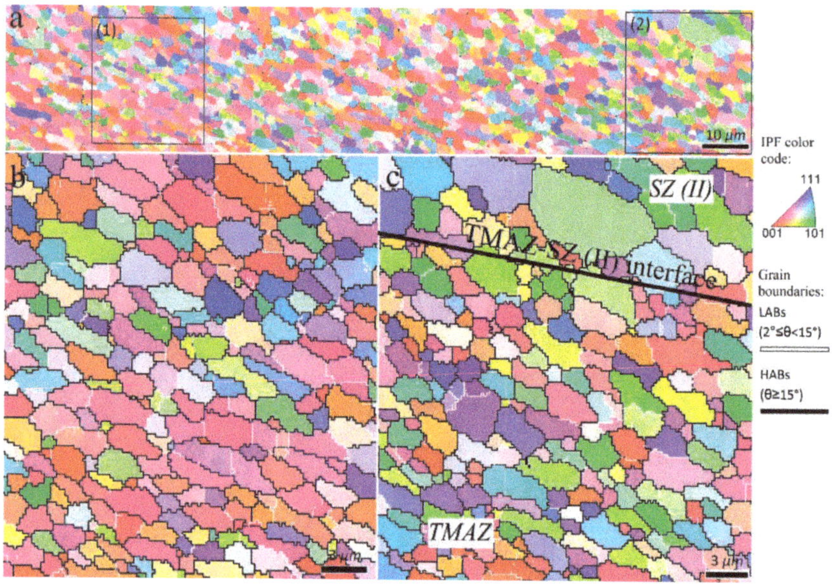

Figure 8. Inverse pole figure maps of the area indicated by rectangle 1 in Figure 7. Higher magnified views of zones 1 and 2 indicated in (**a**) are illustrated in (**b**) TMAZ and (**c**) interfacial area between TMAZ and SZ-II, respectively.

The cross-sectional macrostructure of the 3-pass sample is presented in Figure 10, indicating the presence of different microstructural zones similar to those of the 1- and 2-pass samples (Figures 2 and 7, respectively). However, unlike the 2-pass sample (Figure 7), the third pass of FSP resulted in a considerable increase in the volume fraction of the SZ-II region.

The inverse pole figure maps of the interfacial area between the TMAZ and SZ-II (indicated by rectangle 1 in Figure 10) are presented in Figure 11. From the perspective of grain structure, the TMAZ and SZ-II in the 3-pass sample had similar characteristics to the 2-pass sample. The existence of considerable LABs (Figure 11) confirmed that the grains underwent deformation after their formation by DRX. The inverse pole figure map of the interfacial area between SZ-II and SZ-I (area indicated by rectangle 2 in Figure 10) is shown in Figure 12a. The higher magnification inverse pole figure maps of SZI and SZ-II are shown in Figure 12b,c, respectively. From Figure 12a–c, SZ-I contained finer grains compared to those of SZ-II, similar to those of the 1- and 2-pass samples. Similar to the grain structure, SZ-I and SZ-II exhibit strong texture formation with a bimodal nature. The texture components formed in different samples are discussed in detail in the following sections.

3.2. Evolution of Si Structure

The SEM images of BM, SZ-I, and SZ-II for different samples perpendicular to the BD are illustrated in Figure 13 to reveal the evolution of the Si structure during the FSP. As shown in Figure 13(a1), the melt pools were composed of three distinct zones: coarse melt pool (Figure 13(a2)), heat-affected zone (Figure 13(a3)), and fine melt pool (Figure 13(a3)), which is in agreement with those reported in the literature [1]. As shown in Figure 13b,c, FSP caused fragmentation of eutectic Si cells into Si particles. In addition, by increasing the number of FSP passes, a slight increase in Si particle size was detected. Moreover, in all cases, SZ-II (Figure 13(b1–d1)) contained larger Si particles than SZ-I (Figure 13(b2–d2)) did. During FSP, which induces heat into the material, the Si particles coalesce or join to the surface of the larger Si particles, resulting in the continuous growth of Si particles. The areas indicated by the blue arrows in Figure 13(c1–c3) confirm the growth mechanism of the Si particles. Moreover, the presence of Si particles at the grain boundaries (indicated by the yellow arrow in Figure 3b) showed the Zener pinning effect [25], which has a negative effect on the mobility of grain boundaries, and hence, the grain growth mechanism. The presence of large Si particles in SZ-II, with diameters larger than 1 μm, can be a suitable condition for the occurrence of particle stimulated nucleation [25]. For example, in the case of the 3-pass sample, fine grains were observed around the non-indexed zones in the inverse pole figure maps of SZ-II, as shown by the circles in Figure 11c, which can be an indication of the particle-stimulated nucleation mechanism.

Figure 9. Inverse pole figure maps of the area indicated by rectangle 2 in Figure 7. Higher magnification of zones 1 and 2 indicated in (**a**) are illustrated in (**b**) interfacial area between TMAZ and SZ-II and (**c**) SZ-II, respectively.

Figure 10. Cross-sectional macrostructure of 3-pass friction stir processed sample. The blue rectangles indicate the areas of EBSD scans. The details of the FSP reference frame are the same as Figure 2. Interface between TMAZ and SZ is depicted by a red-dashed line.

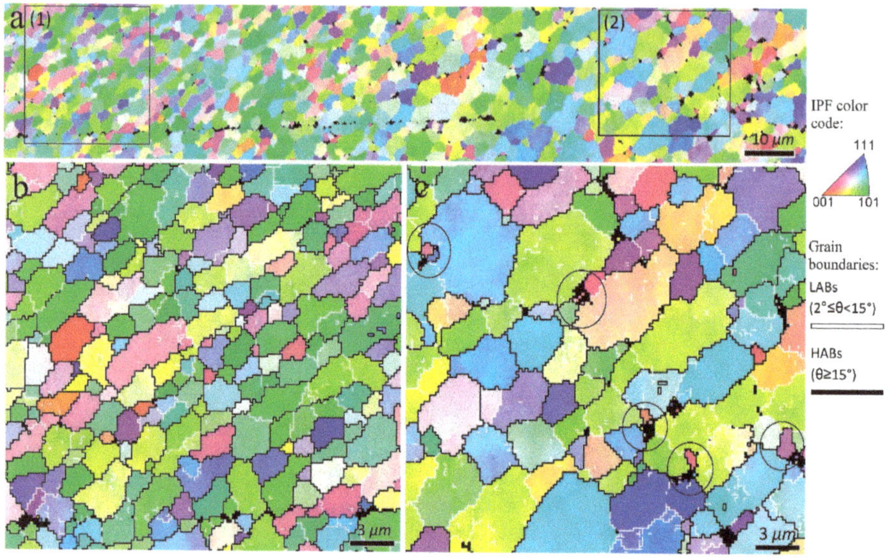

Figure 11. Inverse pole figure maps of the interfacial area between TMAZ and SZ-II in the 3-pass sample indicated by rectangle 1 in Figure 10. Higher magnification of zones 1 and 2 indicated in (**a**) are illustrated in (**b**,**c**), respectively. The circles in (**c**) refer to the non-index regions containing fine DRX grains, i.e., particle stimulated nucleation mechanism.

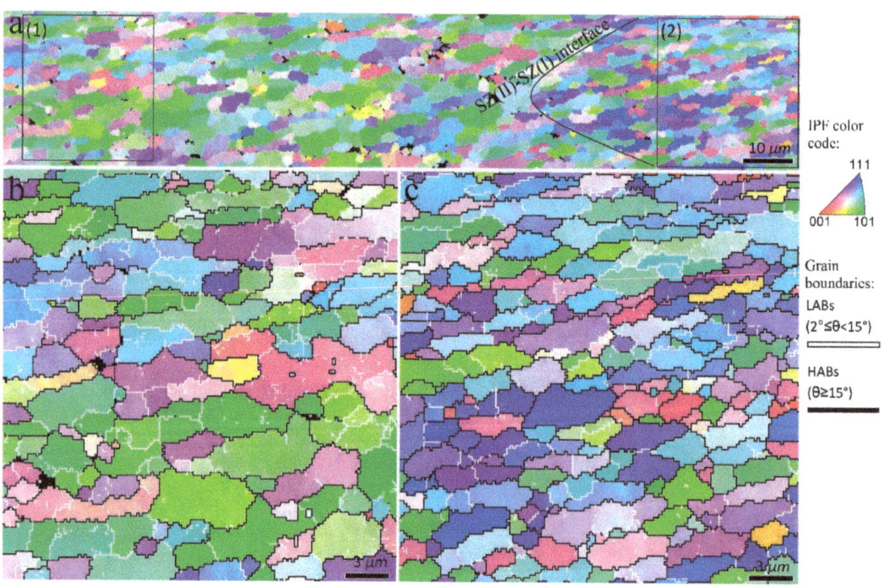

Figure 12. Inverse pole figure maps of the interfacial area between SZ-I and SZ-II in the 3-pass sample indicated by rectangle 2 in Figure 10). Higher magnification of zones 1 and 2 indicated in (**a**) are illustrated in (**b**,**c**), respectively.

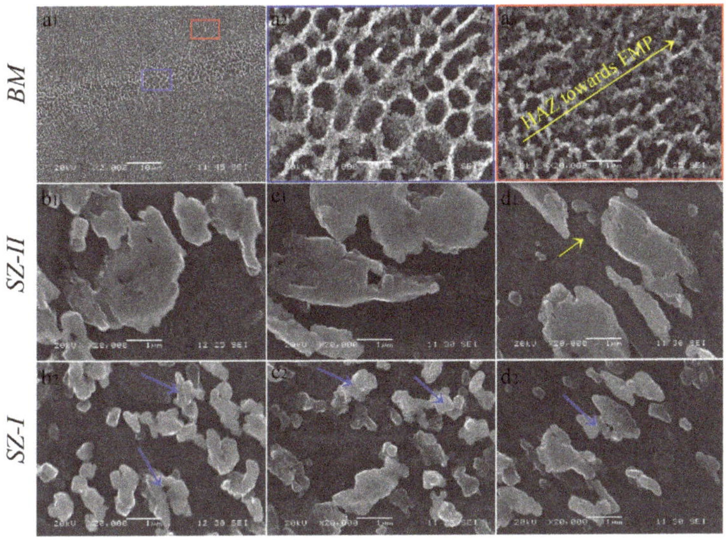

Figure 13. (**a**) SEM micrographs of as-built L-PBF AlSi10Mg at different magnification. The higher magnification of zones indicated by both blue and red rectangles in (**a1**) are shown in (**a2**,**a3**), respectively. (**b1–d1**) SEM images of SZ-I and (**b2–d2**) SZ-II in 1–3 pass FSPed samples. The yellow arrow in SZ-II of the 3-pass sample (**d1**) shows the grain boundary pinned by Si particles. Blue arrows in images related to SZ-I refer to the coalescence of Si particles during the growth.

4. Discussion

The grain size, grain boundary characterization, and coincidence site lattice boundary distributions in both the SZ-I and SZ-II areas in all the samples are illustrated in Figure 14a–c. From the grain size distribution (Figure 14(a1,a2)), FSP results in the formation of a fine grain structure with an average grain size (d_{av}) between 2.5 and 3.7 µm. Grain refinement is due to the occurrence of CDRX and GDRX.

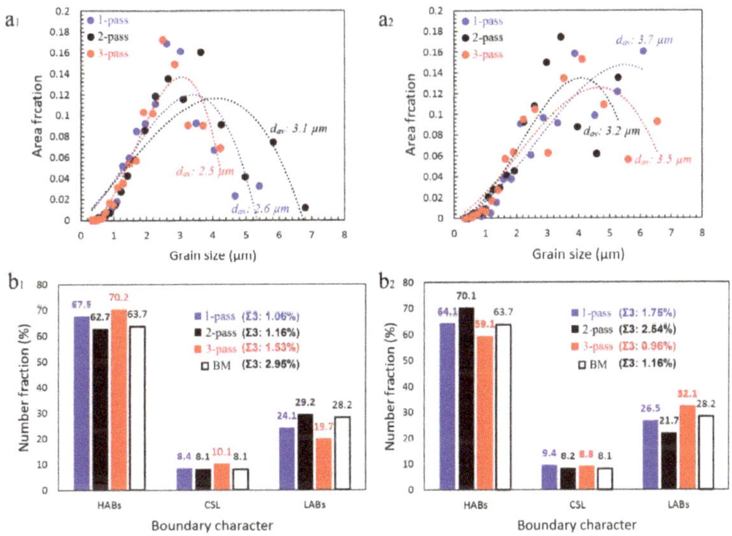

Figure 14. (**a**) Grain size and (**b**) grain boundary characterization distribution: (**a1**,**b1**) SZ-I and (**a2**,**b2**) SZ-II.

Moreover, in the case of SZ-I, the second pass of FSP caused an increase in d_{av} from 2.6 μm to 3.1 μm (Figure 14(a1)) and an enhancement in LABs from 22% to 32% (Figure 14(b1)), which indicated the occurrence of grain growth and formation of LABs through DRV. However, the third pass of FSP consumed the LABs and transformed them to HABs (Figure 14(b1)), which reduced the d_{av} from 3.1 μm to 2.5 μm (Figure 14(a1)). In the case of SZ-II, there was an opposite way compared to SZ-I, in which the second pass of FSP caused a reduction in d_{av} from 3.7 μm to 3.2 μm (Figure 14(a2)) by reducing the amount of LABs from 26% to 20% (Figure 14(b2)). Furthermore, the third pass resulted in an increase in the amount of LABs (34%) and d_{av} of 3.5 μm (Figure 14(a2,b2)). From the perspective of d_{av}, single- pass and multi-pass FSP resulted in a bimodal grain structure in SZ and finer grains (d_{av} = 2.6 μm and 2.5 μm) in SZ-I and coarser grains (d_{av} = 3.7 μm and 3.5 μm) in SZ-II, respectively. However, in the case of the 2-pass sample, there was no considerable difference between the d_{av} values of SZ-I and SZ-II. However, from the viewpoint of the Si phase, in all samples, there was a bimodal distribution of finer (SZ-I) and coarser Si particles (SZ-II) in the SZs (Figure 13). Therefore, the different contrasts of SZ-I and SZ-II in the cross-sectional macrostructures (Figures 2, 7 and 10) were due to the different grain and Si particle sizes of these zones. From Figure 14b, FSP has not a considerable effect on the amount of coincidence site lattice boundaries compared to that of the BM. The largest change in the number of low energy coincidence site lattice boundaries, i.e., $\Sigma \leq 27$, belongs to the SZ-I in 3-pass sample containing 10.1% coincidence site lattice boundaries, which indicates only a 2% enhancement. Moreover, the $\Sigma 3$ boundaries have not a considerable fraction in coincidence site lattice boundaries ranged between 0.96% and 2.54%.

The inverse pole figures parallel to the x, y, and BD directions, in conjunction with the (001), (011), and (111) pole figures (PFs) of the BM, are illustrated in Figure 15a,b, respectively. As shown in Figure 15a,b, a sharp texture of BD//[001] is generated because the preferred solidification direction in face-centered cubic (FCC) metals, that is, the [001] crystallographic direction, aligns along the heat transfer direction, which is almost parallel to BD [22].

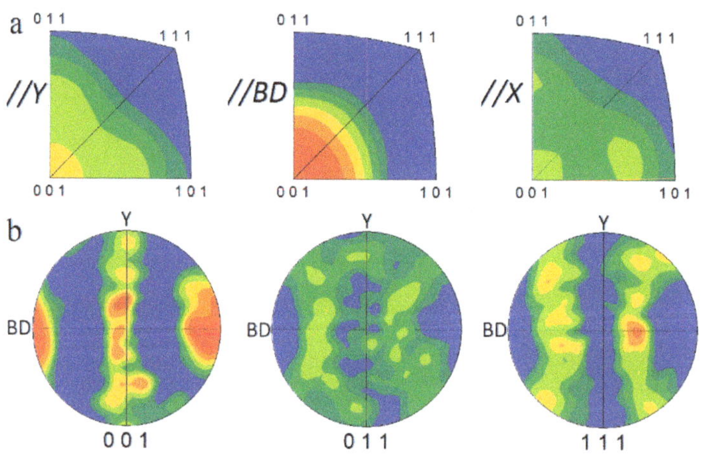

Figure 15. (**a**) Inverse pole figures parallel to x, y, and BD directions in conjunction with (**b**) (001), (011), and (111) pole figures (PFs) of as-built L-PBF AlSi10Mg.

Similar to the grain structure, FSP destroys the crystallographic texture of metals and alloys owing to severe plastic deformation induced by the rotational tool [29]. The deformation of metals and alloys during FSP is shear nature [29]. Therefore, to analyze the texture components of SZs, they are usually compared to those of simple shear deformation, as summarized in Table 1 for FCC materials [10]. It should be noted that the as-acquired

EBSD data should be rotated to adjust it to the standard reference frame of a simple shear illustration [10].

Table 1. Ideal texture components with corresponding Euler angles and miller indices in simple shear deformation of metals with a face centered cubic crystallographic structure [10,30].

Symbol	Euler Angles (°)			Miller Indices (hkl) <uvw>	(111) Pole Figure
	φ_1	Φ	φ_2		
A_1^*	35.26/215.26 125.26	45 90	0/90 45	$(111)[\overline{1}\overline{1}2]$	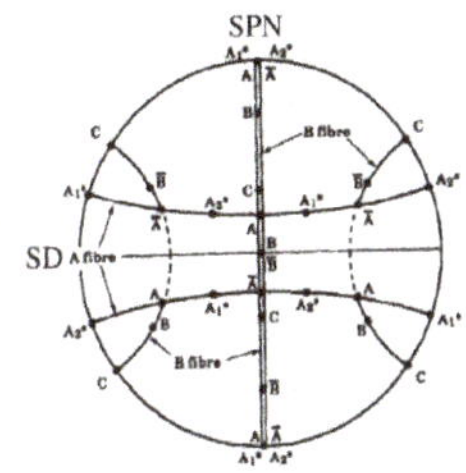
A_2^*	144.74 54.74/234.74	45 90	0/90 45	$(111)[11\overline{2}]$	
A	0	35.26	45	$(1\overline{1}1)[110]$	
$\overline{A}$	180	35.26	45	$(\overline{1}1\overline{1})[\overline{1}\overline{1}0]$	
B	0/120/240	54.74	45	$(1\overline{1}2)[110]$	
$\overline{B}$	60/180	54.74	45	$(\overline{1}1\overline{2})[\overline{1}\overline{1}0]$	
C	90/270 0/180	45 90	0/90 45	$\{001\}\langle 110\rangle$	

The rotated (001), (011), and (111) pole figures of the SZ-I and SZ-II areas in the 1–3 pass FSPed samples are illustrated in Figure 16. In the case of SZ-I, the first pass of FSP formed the shear texture component of $A_1^*(111)[\overline{1}\overline{1}2]$ (Figure 16(a1)), whereas the second and third passes resulted in the formation of $A_2^*(111)[11\overline{2}]$ component in addition to the $A_1^*(111)[\overline{1}\overline{1}2]$ component (Figure 16(b1,c1)). The formation of the $A_1^*(111)[\overline{1}\overline{1}2]$ component might be due to more shear deformation induced by the FSP during the second and third passes, which could activate more crystallographic slip systems. In the case of SZ-II, $A_1^*(111)[\overline{1}\overline{1}2]$ and $B(1\overline{1}2)[110]$ were the main shear texture components in the 1-pass sample (Figure 16(a2)). The second pass of FSP (Figure 16(b2)) generated trimodal texture components in SZ-II composed of $A_1^*(111)[\overline{1}\overline{1}2]$, $B(1\overline{1}2)[110]$, and $\overline{B}(\overline{1}1\overline{2})[\overline{1}\overline{1}0]$, indicating a reduction in the fraction of A_1^* in the total crystallographic texture by the onset of $\overline{B}$ component. From Figure 16(c2), interestingly, the multipass FSP caused the complete elimination of A_1^* by the generation of $B(1\overline{1}2)[110]$ and $\overline{B}(\overline{1}1\overline{2})[\overline{1}\overline{1}0]$ components with a considerable texture intensity of 12.9. It is worth noting that the formation of sharp shear texture components in SZs (Figure 16) confirms the suggested mechanisms, as CDRX and GDRX inherited the texture components induced by shear deformation during FSP [10]. The inverse pole figure maps of SZ-I and SZ-II after the applied rotations for texture analysis are shown in Figure 17a,b, respectively. In SZ-II, the dense (111) planes were approximately parallel to the surface of the sample (Figure 17b), whereas in SZ-I, the dense planes were not aligned with the surface (Figure 17a).

The multi-pass FSP increased the fraction of SZ-II containing fine grains, coarse Si particles, a large number of LABs, and $B/\overline{B}$ shear texture components. These findings indicate that by controlling the number of passes during the FSP of an L-PBF AlSi10Mg alloy, the grain structure and crystallographic texture can be modified in conjunction with the Si structure. This strategy can be a platform for future investigations to enhance the performance of L-PBF AlSi10Mg in various sectors, according to the corresponding conditions of industrial applications. For instance, it is well documented that aluminum alloys containing more low-energy grain boundaries (LABs) and coincidence site lattice boundaries ($\Sigma \leq 27$, especially $\Sigma 3$) exhibit higher corrosion resistance [31]. Moreover, crystallographic planes with lower surface energies, such as the (111) plane, have higher corrosion resistance [32–34]. Efforts to disclose the effect of FSP on the electrochemical response, as an important final property, of L-PBF AlSi10Mg are very limited. Rafieazad et al. [21] employed a single-pass FSP on an L-PBF AlSi10Mg alloy and studied the corrosion behavior of the as-built and processed samples. They reported that a single-pass FSP improved the corrosion resistance by eliminating the porosity, grain refinement, and uniform distribution of Si particles in the SZ. The grain structure and texture characteristics obtained in this study can be used to further improve corrosion resistance, which requires further investigation in the future. This strategy can also be applied to modify the physical and mechanical responses of the L-PBF Al-Si10Mg alloy according to the requirements of industrial applications.

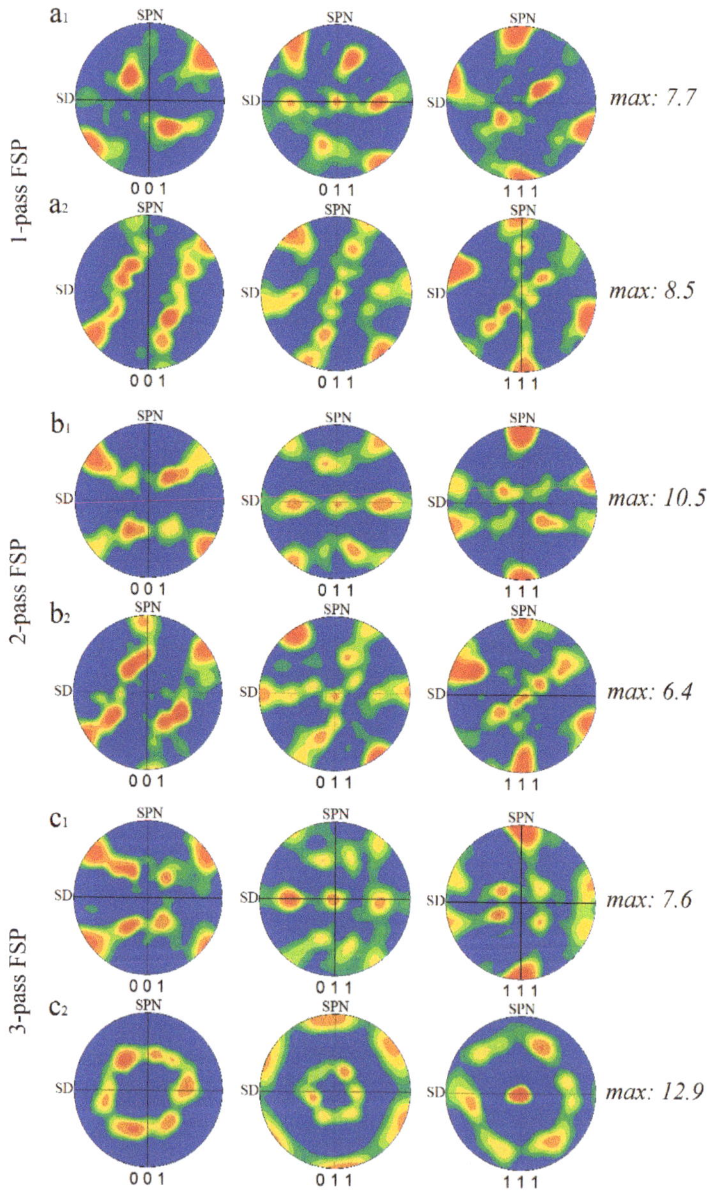

Figure 16. Rotated (001), (011), and (111) pole figures of SZ-I and SZ-II areas in (**a**) 1-pass, (**b**) 2-pass, and (**c**) 3-pass FSPed samples, respectively: (**a1,b1,c1**) SZ-I and (**a2,b2,c2**) SZ-II. The texture intensity values are written at the right-hand side of PFs.

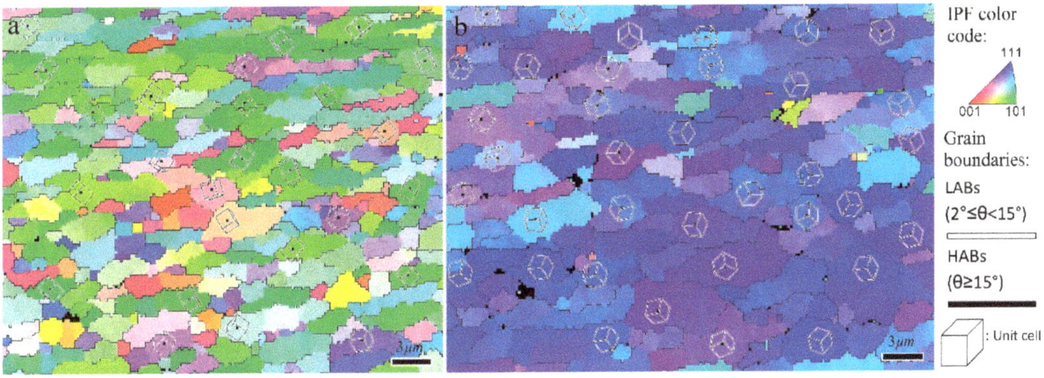

Figure 17. Inverse pole figure maps of SZ-I (**a**) and SZ-II (**b**) after applied rotations for texture analysis. The drawn cubes refer to the orientation of unit cells in individual grains.

5. Conclusions

Single-, double-, and multi-pass FSPs were used to modify the microstructure and texture of the L-PBF AlSi10Mg alloy. The following conclusions can be drawn:

(1) During the first pass of FSP, dynamic recovery, continuous dynamic recrystallization and geometric dynamic recrystallization caused the formation of a fine grain structure in the SZ with an average grain size of 2.6–3.7 μm. The bimodal grain structure results from the formation of two distinct zones: SZ-I with a fine grain size and SZ-II with a coarse grain size in the SZ.

(2) The second pass of FSP completed the recrystallization in the TMAZ. However, in the case of SZ, it causes the rotation of grains, which generates more shear texture components in SZ-I and SZ-II compared with single-pass FSP.

(3) The third pass of FSP increased the volume fraction of SZ-II, which possessed unique characteristics of higher values of LABs. During the third pass, more rotation of DRX grains in SZ occurred, which resulted in the generation of $B(1\bar{1}2)[110]$ and $\bar{B}(\bar{1}1\bar{2})[\bar{1}\bar{1}0]$ components with a considerable texture intensity of 12.9 in SZ-II. The produced texture components in SZ-II aligned the crystallographic planes with lower surface energy, such as {111} planes parallel to the surface of samples.

(4) In addition, in all cases, the Si morphology changed from cell to particle structure in SZs. Similar to grain size distribution, the SZ-II contained larger Si particles compared to the SZ-I.

(5) The outcome of this study can be used for the modification of the microstructure (grain size, grain boundaries, and texture) of L-PBF AlSi10Mg alloy in a manner that, by controlling the number of passes during FSP, the desired microstructural features can be obtained for potential applications.

Author Contributions: A.H.: methodology, investigation, formal analysis, writing—original draft; M.J.: conceptualization, formal analysis, writing—review and editing; M.M.: methodology, investigation; P.M.-n.: methodology, investigation; R.M.: methodology, investigation, formal analysis; H.J.: methodology, investigation, formal analysis; X.-G.C.: conceptualization, validation, review and editing, project administration. All authors have read and agreed to the published version of the manuscript.

Funding: This research was partially funded by Natural Sciences and Engineering Research Council of Canada (NSERC) under Grant No. CRDPJ 514651-17.

Data Availability Statement: The raw/processed data required to reproduce these findings cannot be shared at this time as the data also forms part of an ongoing study.

Conflicts of Interest: The authors declare no conflict of interest.

References

1. Zhao, L.; Song, L.; Macías, J.G.S.; Zhu, Y.; Huang, M.; Simar, A.; Li, Z. Review on the correlation between microstructure and mechanical performance for laser powder bed fusion AlSi10Mg. *Addit. Manuf.* **2022**, *56*, 102914. [CrossRef]
2. Hadadzadeh, A.; Baxter, C.; Amirkhiz, B.S.; Mohammadi, M. Strengthening mechanisms in direct metal laser sintered AlSi10Mg: Comparison between virgin and recycled powders. *Addit. Manuf.* **2018**, *23*, 108–120. [CrossRef]
3. Zhang, J.; Song, B.; Wei, Q.; Bourell, D.; Shi, Y. A review of selective laser melting of aluminum alloys: Processing, microstructure, property and developing trends. *J. Mater. Sci. Technol.* **2019**, *35*, 270–284. [CrossRef]
4. Macías, J.G.S.; Douillard, T.; Zhao, L.; Maire, E.; Pyka, G.; Simar, A. Influence on microstructure, strength and ductility of build platform temperature during laser powder bed fusion of AlSi10Mg. *Acta Mater.* **2020**, *201*, 231–243. [CrossRef]
5. Maamoun, A.H.; Elbestawi, M.; Dosbaeva, G.K.; Veldhuis, S.C. Thermal post-processing of AlSi10Mg parts produced by Selective Laser Melting using recycled powder. *Addit. Manuf.* **2018**, *21*, 234–247. [CrossRef]
6. Aboulkhair, N.T.; Maskery, I.; Tuck, C.; Ashcroft, I.; Everitt, N.M. The microstructure and mechanical properties of selectively laser melted AlSi10Mg: The effect of a conventional T6-like heat treatment. *Mater. Sci. Eng. A* **2016**, *667*, 139–146. [CrossRef]
7. Larrosa, N.; Wang, W.; Read, N.; Loretto, M.; Evans, C.; Carr, J.; Tradowsky, U.; Attallah, M.; Withers, P. Linking microstructure and processing defects to mechanical properties of selectively laser melted AlSi10Mg alloy. *Theor. Appl. Fract. Mech.* **2018**, *98*, 123–133. [CrossRef]
8. Hirata, T.; Kimura, T.; Nakamoto, T. Effects of hot isostatic pressing and internal porosity on the performance of selective laser melted AlSi10Mg alloys. *Mater. Sci. Eng. A* **2020**, *772*, 138713. [CrossRef]
9. Macías, J.G.S.; Elangeswaran, C.; Zhao, L.; Van Hooreweder, B.; Adrien, J.; Maire, E.; Buffière, J.-Y.; Ludwig, W.; Jacques, P.J.; Simar, A. Ductilisation and fatigue life enhancement of selective laser melted AlSi10Mg by friction stir processing. *Scr. Mater.* **2019**, *170*, 124–128. [CrossRef]
10. Heidarzadeh, A.; Mironov, S.; Kaibyshev, R.; Çam, G.; Simar, A.; Gerlich, A.; Khodabakhshi, F.; Mostafaei, A.; Field, D.P.; Robson, J.D.; et al. Friction stir welding/processing of metals and alloys: A comprehensive review on microstructural evolution. *Prog. Mater. Sci.* **2021**, *117*, 100752. [CrossRef]
11. Hou, W.; Ding, Y.; Huang, G.; Huda, N.; Shah, L.H.; Piao, Z.; Shen, Y.; Shen, Z.; Gerlich, A. The role of pin eccentricity in friction stir welding of Al-Mg-Si alloy sheets: Microstructural evolution and mechanical properties. *Int. J. Adv. Manuf. Technol.* **2022**, *121*, 7661–7675. [CrossRef]
12. Hou, W.; Shah, L.H.; Huang, G.; Shen, Y.; Gerlich, A. The role of tool offset on the microstructure and mechanical properties of Al/Cu friction stir welded joints. *J. Alloys Compd.* **2020**, *825*, 154045. [CrossRef]
13. Anand, R.; Sridhar, V.G. Studies on process parameters and tool geometry selecting aspects of friction stir welding—A review. *Mater. Today Proc.* **2020**, *27*, 576–583. [CrossRef]
14. Senthilkumar, R.; Prakash, M.; Arun, N.; Jeyakumar, A.A. The effect of the number of passes in friction stir processing of aluminum alloy (AA6082) and its failure analysis. *Appl. Surf. Sci.* **2019**, *491*, 420–431. [CrossRef]
15. Mishra, R.S.; Ma, Z.Y. Friction stir welding and processing. *Mater. Sci. Eng. R Rep.* **2005**, *50*, 1–78. [CrossRef]
16. Du, Z.L.; Tan, M.J.; Guo, J.F.; Wei, J.; Chua, C.K. Dispersion of CNTs in selective laser melting printed AlSi10Mg composites via friction stir processing. *Mater. Sci. Forum* **2017**, *879*, 1915–1920. [CrossRef]
17. Yang, T.; Wang, K.; Wang, W.; Peng, P.; Huang, L.; Qiao, K.; Jin, Y. Effect of friction stir processing on microstructure and mechanical properties of AlSi10Mg aluminum alloy produced by selective laser melting. *JOM* **2019**, *71*, 1737–1747. [CrossRef]
18. Maamoun, A.H.; Veldhuis, S.C.; Elbestawi, M. Friction stir processing of AlSi10Mg parts produced by selective laser melting. *J. Mater. Process. Technol.* **2019**, *263*, 308–320. [CrossRef]
19. Zhao, L.; Macías, J.G.S.; Ding, L.; Idrissi, H.; Simar, A. Damage mechanisms in selective laser melted AlSi10Mg under as built and different post-treatment conditions. *Mater. Sci. Eng. A* **2019**, *764*, 138210. [CrossRef]
20. Zhao, L.; Macías, J.G.S.; Dolimont, A.; Simar, A.; Rivière-Lorphèvre, E. Comparison of residual stresses obtained by the crack compliance method for parts produced by different metal additive manufacturing techniques and after friction stir processing. *Addit. Manuf.* **2020**, *36*, 101499. [CrossRef]
21. Rafieazad, M.; Mohammadi, M.; Gerlich, A.; Nasiri, A. Enhancing the corrosion properties of additively manufactured AlSi10Mg using friction stir processing. *Corros. Sci.* **2021**, *178*, 109073. [CrossRef]
22. Aboulkhair, N.T.; Simonelli, M.; Parry, L.; Ashcroft, I.; Tuck, C.; Hague, R. 3D printing of Aluminium alloys: Additive Manufacturing of Aluminium alloys using selective laser melting. *Prog. Mater. Sci.* **2019**, *106*, 100578. [CrossRef]
23. Suhuddin, U.; Mironov, S.; Sato, Y.; Kokawa, H.; Lee, C.-W. Grain structure evolution during friction-stir welding of AZ31 magnesium alloy. *Acta Mater.* **2009**, *57*, 5406–5418. [CrossRef]
24. Mironov, S.; Sato, Y.; Kokawa, H. Microstructural evolution during friction stir-processing of pure iron. *Acta Mater.* **2008**, *56*, 2602–2614. [CrossRef]
25. Humphreys, F.J.; Hatherly, M. *Recrystallization and Related Annealing Phenomena*; Elsevier: Amsterdam, The Netherlands, 2012.
26. Razmpoosh, M.H.; Zarei-Hanzaki, A.; Imandoust, A. Effect of the Zener–Hollomon parameter on the microstructure evolution of dual phase TWIP steel subjected to friction stir processing. *Mater. Sci. Eng. A* **2015**, *638*, 15–19. [CrossRef]
27. Xu, S.; Kamado, S.; Honma, T. Recrystallization mechanism and the relationship between grain size and zener–hollomon parameter of mg–al–zn–ca Alloys During Hot Compression. *Scr. Mater.* **2010**, *63*, 293–296. [CrossRef]

28. Commin, L.; Dumont, M.; Masse, J.E.; Barrallier, L. Friction stir welding of AZ31 magnesium alloy rolled sheets: Influence of processing parameters. *Acta Mater.* **2009**, *57*, 326–334. [CrossRef]
29. Meng, X.; Huang, Y.; Cao, J.; Shen, J.; dos Santos, J.F. Recent progress on control strategies for inherent issues in friction stir welding. *Prog. Mater. Sci.* **2021**, *115*, 100706. [CrossRef]
30. Heidarzadeh, A.; Laleh, H.M.; Gerami, H.; Hosseinpour, P.; Shabestari, M.J.; Bahari, R. The origin of different microstructural and strengthening mechanisms of copper and brass in their dissimilar friction stir welded joint. *Mater. Sci. Eng. A* **2018**, *735*, 336–342. [CrossRef]
31. Kim, S.; Erb, U.; Aust, K.; Palumbo, G. Grain boundary character distribution and intergranular corrosion behavior in high purity aluminum. *Scr. Mater.* **2001**, *44*, 835–839. [CrossRef]
32. Song, G.-L.; Mishra, R.; Xu, Z. Crystallographic orientation and electrochemical activity of AZ31 Mg alloy. *Electrochem. Commun.* **2010**, *12*, 1009–1012. [CrossRef]
33. Wolf, D. Correlation between energy, surface tension and structure of free surfaces in fcc metals. *Surf. Sci.* **1990**, *226*, 389–406. [CrossRef]
34. Rodríguez, A.M.; Bozzolo, G.; Ferrante, J. Multilayer relaxation and surface energies of fcc and bcc metals using equivalent crystal theory. *Surf. Sci.* **1993**, *289*, 100–126. [CrossRef]

materials

MDPI

Article

Metallographic Mechanism of Embrittlement of 15 μm Ultrafine Quaternary Silver Alloy Bonding Wire in Chloride Ions Environment

Jun-Ren Zhao, Fei-Yi Hung * and Che-Wei Hsu

Department of Materials Science and Engineering, National Cheng Kung University, Tainan 701, Taiwan
* Correspondence: fyhung@mail.ncku.edu.tw

Abstract: Chloride ions contained in the sealing compound currently used in the electronic packaging industry not only interact with intermetallic compounds but also have a serious impact on silver alloy wires. A 15 μm ultrafine quaternary silver-palladium-gold-platinum alloy wire was used in this study. The wire and its bonding were immersed in a 60 °C saturated sodium chloride solution (chlorination experiment), and the strength and elongation before and after chlorination were measured. Finally, the fracture surface and cross-section characteristics were observed using a scanning electron microscope and focused ion microscope. The results revealed that chloride ions invade the wire along the grain boundary, and chlorides have been generated inside the cracks to weaken the strength and elongation of the wire. In addition, chloride ions invade the interface of the wire bonding to erode the aluminum substrate after immersing it for enough long time, causing galvanic corrosion, which in turn causes the bonding joint to separate from the aluminum substrate.

Keywords: silver wire; wire bonding; chloride ions; ion migration; galvanic corrosion

Citation: Zhao, J.-R.; Hung, F.-Y.; Hsu, C.-W. Metallographic Mechanism of Embrittlement of 15 μm Ultrafine Quaternary Silver Alloy Bonding Wire in Chloride Ions Environment. *Materials* **2023**, *16*, 1066. https://doi.org/10.3390/ma16031066

Academic Editors: Pan Gong, Maojun Li, Guangchao Han and Xin Wang

Received: 21 December 2022
Revised: 17 January 2023
Accepted: 18 January 2023
Published: 25 January 2023

1. Introduction

Wire bonding is the most widely used technology in the electronic packaging industry [1–3]. Previously, gold wire was the most commonly used material in wire bonding, but with the high price of gold, many alternative materials have been proposed [4–6]. Currently, the alternative materials with the most potential are copper-based and silver-based wires [7–10]. The most attention has been paid to palladium-coated copper wire [11,12]. Pure copper wires easily oxidize and corrode, causing reliability concerns. This shortcoming has been overcome by modifying the surface of the palladium coating. Silver-based wires are mainly used in the form of silver alloy [13,14] and gold-coated silver wires [15,16]. According to research [11,17,18], adding palladium to silver-based wires results in a palladium-rich single-phase layer at the bonding interface, which can inhibit the formation of intermetallic compounds (IMCs). In addition, adding gold and palladium simultaneously helps improve the electromigration effect and increases the life cycle of silver-based wires.

This study used the newly developed quaternary silver-palladium-gold-platinum alloy (coded APAP) wire. The solid solution of the platinum element can further improve the efficiency of wire drawing [19,20], resulting in a wire diameter as low as 15 μm. This is the finest wire diameter in the microelectronic packaging industry at present [21]. A fine wire diameter has the advantages of a higher pin count and lower feeding costs. This is a new-generation material with great potential.

Currently, the sealant used in the microelectronic packaging industry contains chloride ions [22,23]; thus, the wire is prone to reliability problems in relation to electrical and mechanical properties and disconnection. Studies [24,25] have also noted that after the copper-based wire and aluminum substrate have been welded, IMCs are susceptible to corrosion caused by chloride ions and decomposition, leading to bonding peeling, which greatly reduces reliability. Few studies have investigated the effect of chloride ions on alloy wires [1,23], and investigations into the influence of chloride ions on a silver-based wire are lacking.

The chlorination of silver-based wires has rarely been studied, and the development of related test methods and mechanisms remains insufficient. Therefore, a newly developed quaternary Ag-2.5 Pd-1.5 Au-0.15 Pt (wt.%) alloy wire was used in this study, as well as an independently developed chlorination test method to analyze and clarify the chlorination mechanism. The results can be used as a reference for the microelectronic packaging industry.

2. Experimental Procedures

A 15 μm diameter APAP wire from the wire drawing process was the research material in this study (general silver wire diameter of 20 μm). After the wire was soaked in a saturated sodium chloride solution (concentration of chloride ions: 16.5 wt.%) at 60 °C for 4 h, it was washed with deionized water and dried. The wire surfaces were observed under a scanning electron microscope (SEM; HITACHI SU-5000, HITACHI, Tokyo, Japan). A micro-tensile tester was used to investigate the changes in the tensile mechanical properties of the silver alloy wire during immersion in the saturated sodium chloride solution at different times. The initial strain rate and gauge length were 5×10^{-3} s^{-1} and 50 mm, respectively. The tensile data were the average of five tests.

The tensile fracture and cross-section surface were observed, and the profile was analyzed using a dual-beam focus ion beam (FIB; FEI Nova 200, FEI, Hillsboro, OR, USA), and energy dispersive spectroscopy (EDS) was employed to analyze the inside of the cracks to clarify the chlorine embrittlement fracture mechanism of the wire. Finally, a chlorination experiment was conducted on the first and second bonds of the APAP wire on the aluminum substrate (the downforce of the first and second bonds was 20 and 60 gf, respectively). The chlorination holding time was 10 min and 30 min, and the temperature was 60 °C. Subsequently, the chlorination condition of the fracture and the bonding positions were observed through the micro-tensile test, FIB, and EDS element analysis.

3. Results and Discussion

3.1. Fracture Effect of Chloride Ions: APAP Wire

Figure 1 presents the change in the tensile strength (UTS) and elongation of the APAP wires immersed in the saturated sodium chloride solution for 10 s, 10 min, 1 h, 2 h, 4 h, and 8 h. According to the results, no significant difference in the strength and elongation of the wire within 1 h of immersion was detected, but as the immersion time increased to more than 2 h, a significant decrease in strength and elongation was observed, indicating that the corrosion caused by chloride ions has a considerable influence on the reliability of the wire. After being immersed for 8 h, the wire broke when tensile, indicating that the APAP wire cannot be used in a chlorinated environment for long periods.

The tensile fracture surface of the APAP wire before and after chlorination was observed using an SEM (Figure 2). According to the literature [1], the fracture morphology of general metal wires is conical (the fracture mode is dominated by ductility). The APAP wire exhibited ductile failure characteristics before chlorination and after chlorination for 2 h. After chlorination for 4 h, the failure characteristics changed from ductile to brittle. After immersion in the saturated sodium chloride solution for 4 h, numerous fine cracks were distributed on the surface of the wire (Figure 3b,c), which was different from the wire before immersion in the saturated sodium chloride solution (Figure 3a,b). When the wire was subjected to tensile stress, these small cracks gradually connected to form large cracks, which eventually led to brittle fractures.

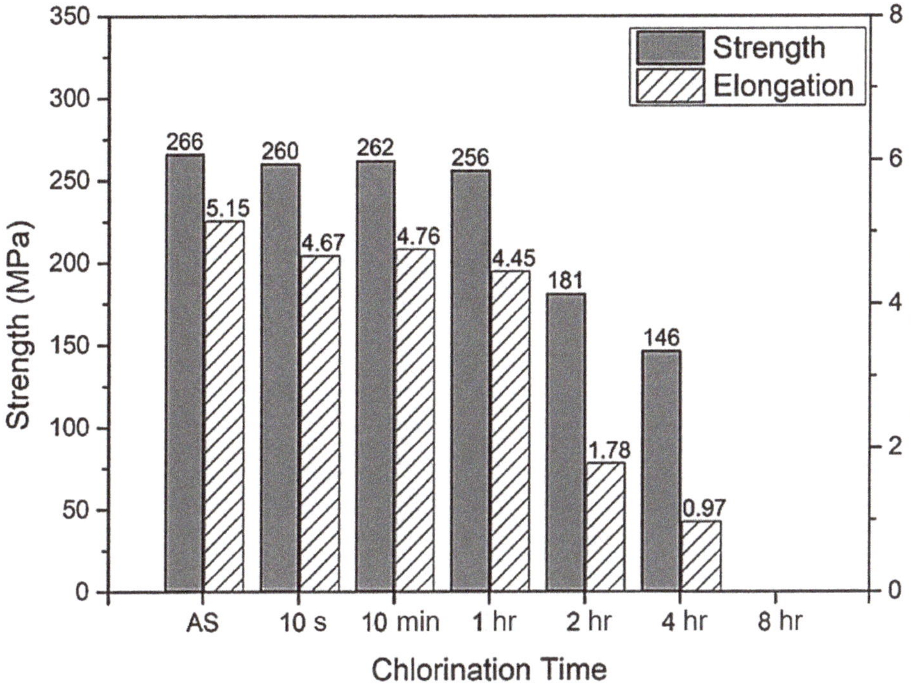

Figure 1. Changes in tensile strength and elongation of the APAP wires after immersion in saturated sodium chloride solution. (The tensile data were the average of five tests).

To clarify the chlorination mechanism of the APAP wire, FIB scanning and profile analysis were performed on the fracture surface after chlorination for 4 h (Figure 4). Figure 4a presents the surface features 100 μM from the fractured surface. In addition to a large number of cracks on the surface of the wire, most cracks are located at the boundary of different color blocks. This indicates that the location of the cracks is at the grain boundary. Figure 4b presents the FIB profile analysis, and the microstructure was similar to our previous coated silver wire study [26]. The results demonstrate that the cracks penetrate deep into the wire along the grain boundary, and the longest crack is approximately 5 μm. Figure 5 presents the EDS element analysis results inside the cracks, revealing that chlorides have been generated inside the cracks. This demonstrates that chloride ions penetrate the wire through the grain boundary to form chlorides, causing the grain boundary to weaken.

During the tensile test, the core not affected by chloride ions remained connected after the surface was broken and continued to deform. The cracks gradually widened from the inside to the outside, gradually penetrating the wire because of stress concentration. Finally, a brittle intergranular fracture occurred. The brittle fracture mechanism of the silver alloy wire is illustrated in Figure 6. The chlorine embrittlement phenomenon that occurs in silver alloy wires is caused by foreign chlorine ions that gradually erode the inside of the wire along the grain boundary or defects. Subsequently, a brittle fracture occurs when the wire is subjected to strain. The weakening of the grain boundary causes decreases in the tensile strength and elongation.

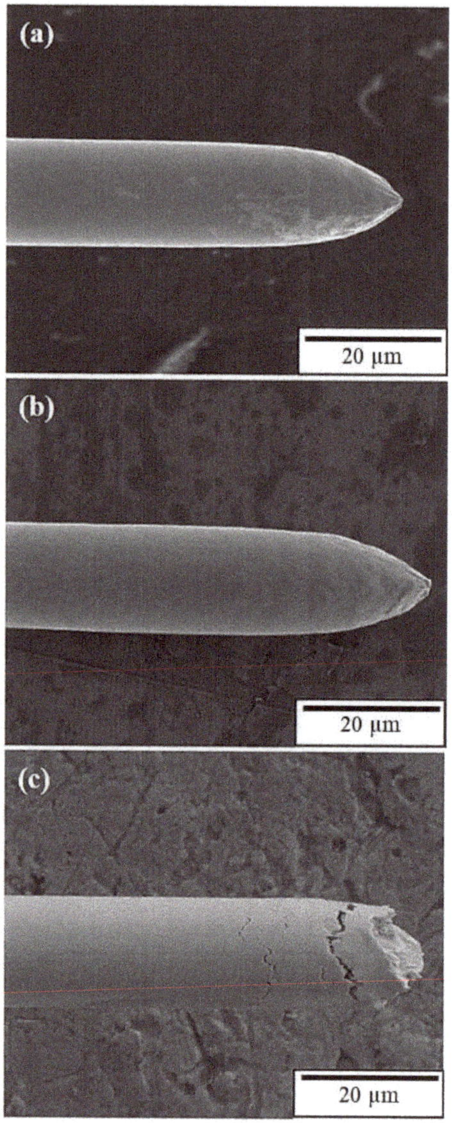

Figure 2. Tensile fracture surface of the APAP wire at different chlorination times: (**a**) before chlorination, (**b**) 2 h, and (**c**) 4 h.

As the wire diameter gradually shrank, subtler phenomena, such as electromigration and chloride ion diffusion, became more evident. Another study [27] revealed that the silver-based wire exhibits a weakened grain boundary and intergranular fracture caused by electromigration during the electrification process. This is because the wires used today are mostly narrower than 20 μm in diameter; thus, even small changes cause major changes to the wire. For example, the longest crack in this study was 5 μm, and the unaffected wire diameter at the center was only 5 μm, too. This highlights the influence of the chlorine embrittlement phenomenon.

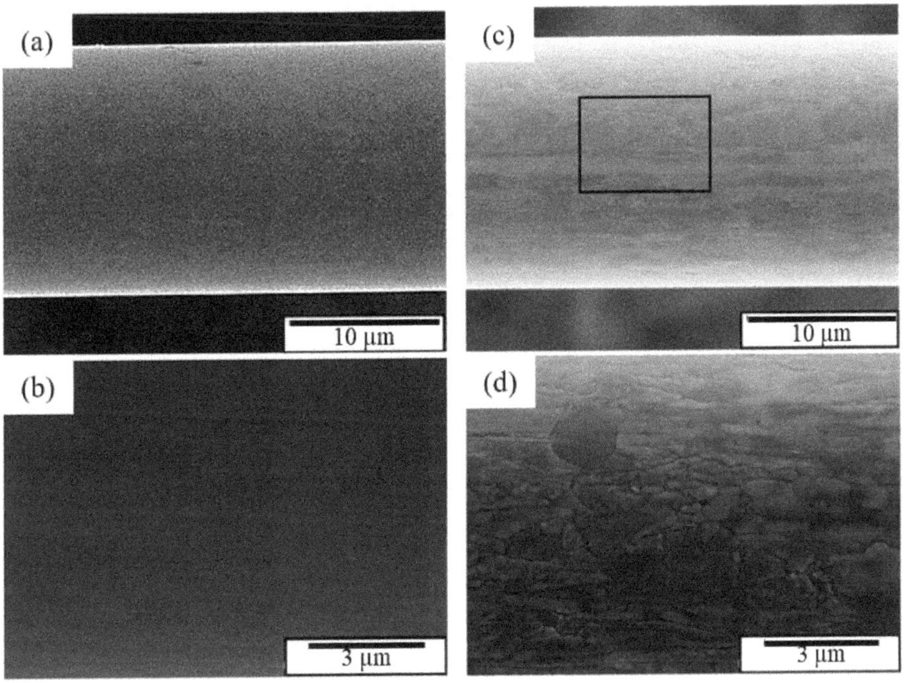

Figure 3. Surface morphology of the APAP wire: (**a**) macroscopic image before chlorination, (**b**) local magnified image before chlorination, (**c**) macroscopic image after chlorination for 4 h, and (**d**) local magnified image after chlorination for 4 h.

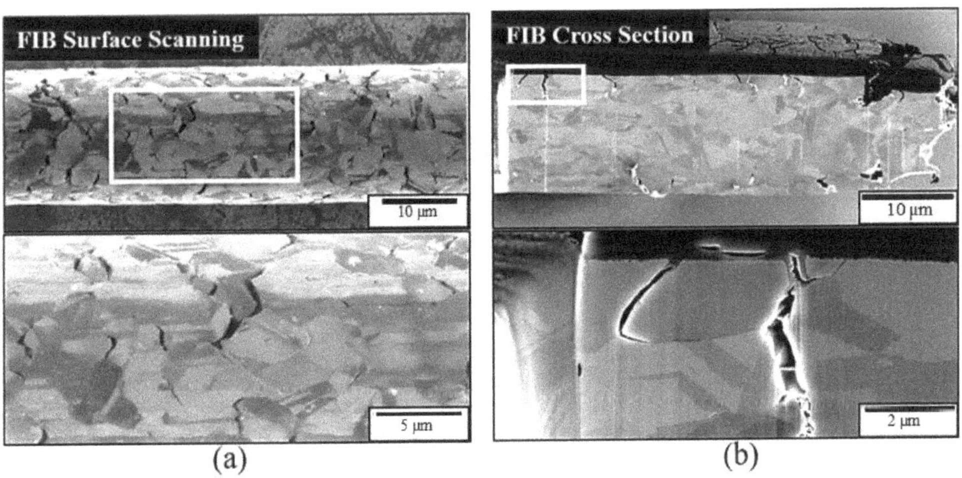

Figure 4. Tensile fracture cross-section of the APAP wire after 4 h of chlorination: (**a**) focus ion beam (FIB) surface scanning, and (**b**) cross-section image.

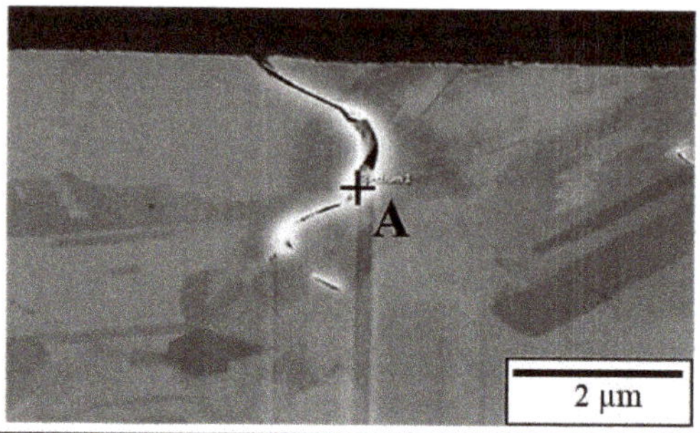

Element (at.%)	Ag	Cl	Na
Point A	77.50	15.90	6.60

Figure 5. Energy dispersive spectroscopy (EDS) element analysis results of the crack.

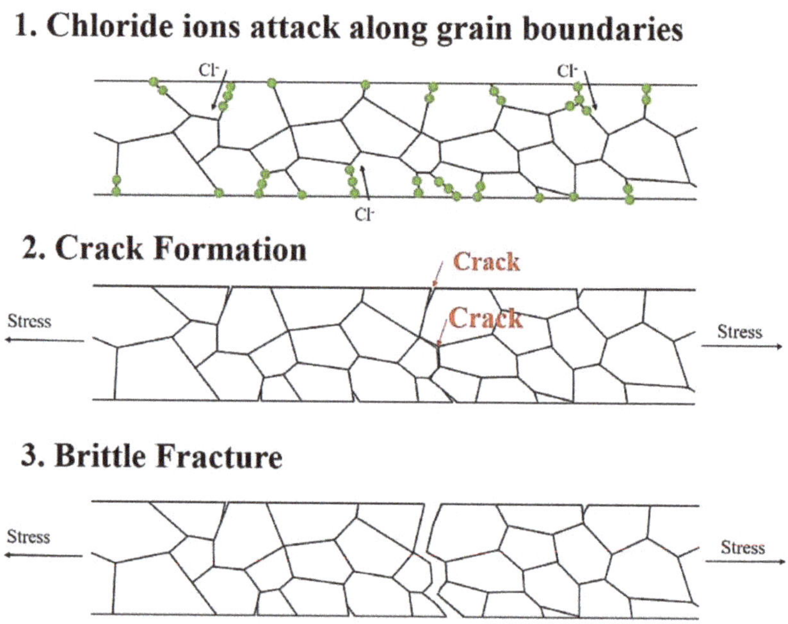

Figure 6. Intergranular brittleness fracture mechanism.

3.2. *Fracture Effect of Chloride Ions: APAP Wire Bondingm*

Figure 7 presents the tensile fracture position of the first bond of the APAP wire immersed in a saturated sodium chloride solution at 60 °C for 10 and 30 min. After 10 min of chlorination, the first bond fracture position occurred in the heat-affected zone (Figure 7a), indicating that it retained some strength under these conditions. However, as the chlorination time increased to 30 min, the first bond joint separated from the aluminum substrate, and some traces remained on the substrate (Figure 7b). This suggests that after a long period of chlorination, chloride ions corrode the joint surface of the first bond, possibly causing the bond joint to separate and reducing reliability. Figure 8 presents the EDS analysis results of the fracture surface after 30 min of chlorination of the first bond joint of the APAP wire. The aluminum film on the substrate has been eroded by chloride ions and has peeled off. Points A and B in the figure denote the residual aluminum film, and point C is the exposed silicon substrate after the aluminum film has peeled off.

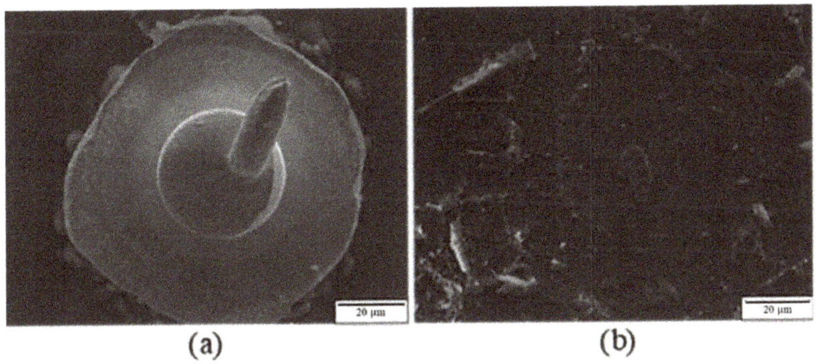

Figure 7. Fracture position of the first bond of the APAP wire after chlorination: (**a**) 10 min and (**b**) 30 min.

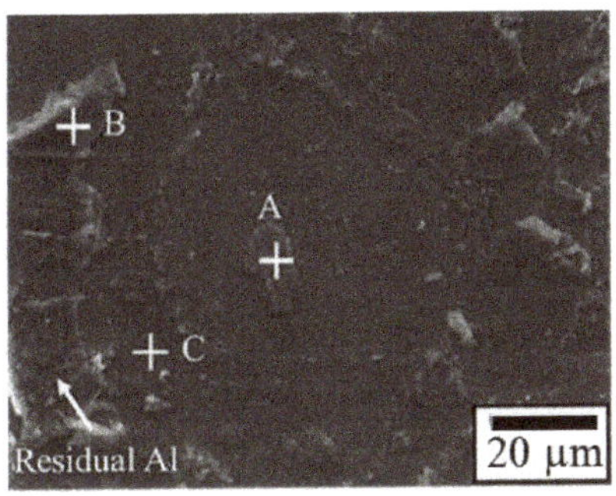

Element(At.%)	A	B	C
Na	4.88	10.01	1.93
Al	39.97	61.72	1.93
Si	54.85	0.61	95.7
Cl	0.30	27.65	0.44

Figure 8. EDS analysis of the fracture surface of the first bond of the APAP wire after 30 min of chlorination. (A, B: residual aluminum film; C: silicon substrate).

The FIB cross-section and EDS element analysis results of the first bond of the APAP wire after chlorination for 30 min are presented in Figure 9. A clear bond boundary can be identified at the center of the bond. At the periphery of the bond (point C), the silver ball and aluminum film have peeled off, reducing the bond strength. The EDS element analysis reveals a very small chlorine signal at the center of the bond (point A) and a larger chlorine signal at the periphery of the bond (point B). This finding indicated that chlorine ions corroded the aluminum film from the outside to the inside of the bond. As the chlorination time increased, the aluminum film at the bond junction was gradually eroded and hollowed out. Finally, the first bond joint was peeled off and failed (Figure 10).

Figure 11 presents the surface morphology of the second bond of the APAP wire after 30 min of chlorination, with no bond joint peeling. Figure 12 presents the FIB cross-section analysis results, which reveal that the second bond is well joined; only the aluminum film at the end of the fishtail joint is eroded and hollowed out at approximately 1.5 μm The downforce of the second bond is usually greater than that of the first bond, maintaining a tighter second bond. Therefore, the joint surface of the second bond has greater reliability. A schematic of the chlorination process is presented in Figure 13.

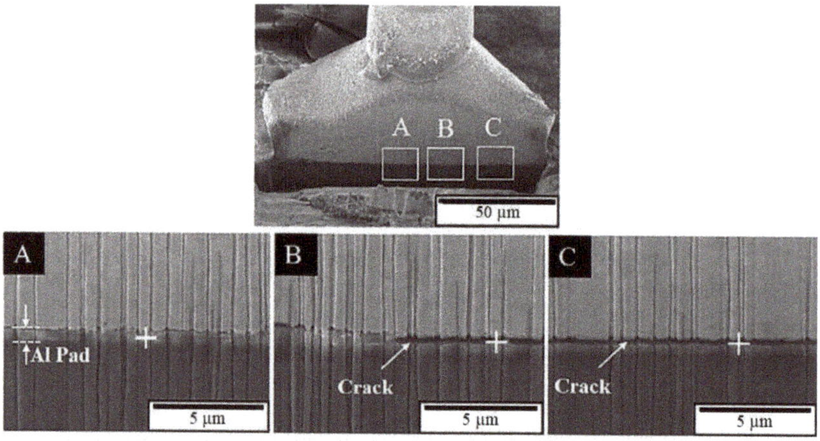

Element (At.%)	A	B	C
Cl	1.17	4.96	7.29
Ag	7.81	7.38	6.84
Al	91.02	87.66	85.87

Figure 9. FIB cross-section and EDS element analysis results of the first bond of the APAP wire after 30 min of chlorination.

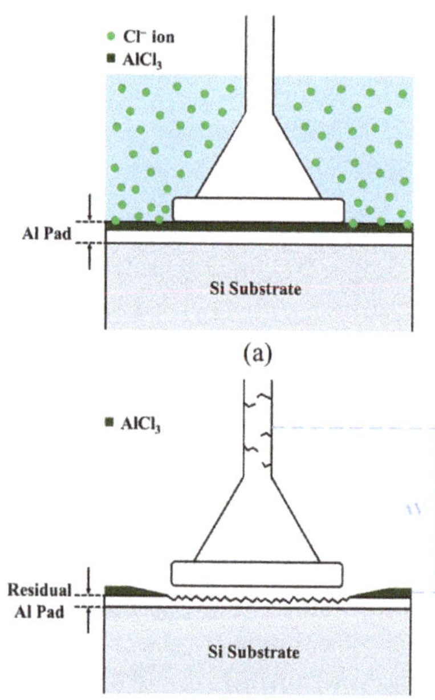

Figure 10. Schematic of the joint of the first bond after chlorination: (**a**) chloride ions corrode the aluminum film; (**b**) the joint of the first bond peels off.

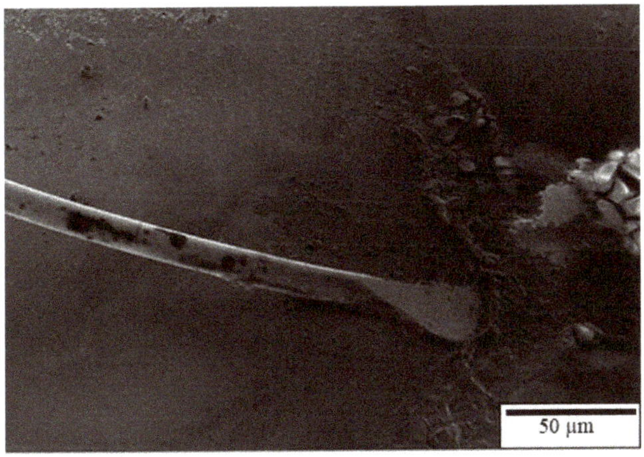

Figure 11. Surface morphology of the second bond of the APAP wire after 30 min of chlorination.

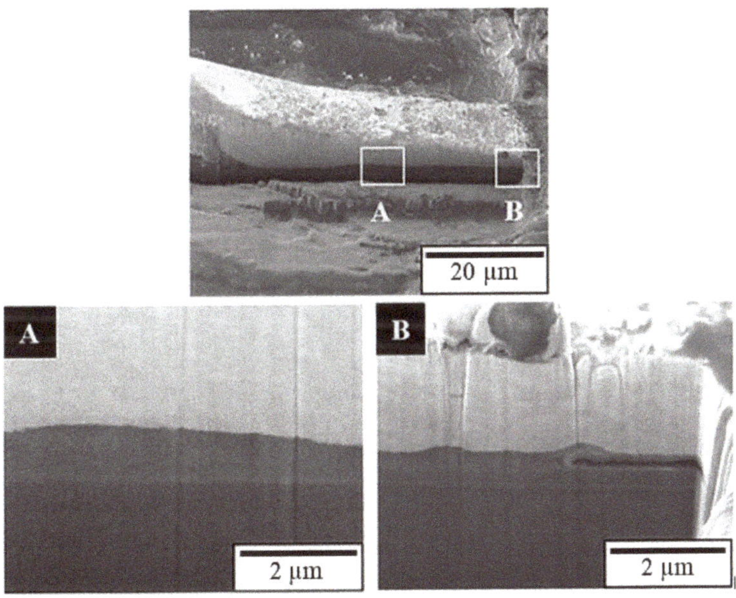

Figure 12. FIB cross-section of the second bond of the APAP wire after 30 min of chlorination. (**A**) Smooth interface; (**B**) aluminum film at the edge is eroded and hollowed out.

This study has three strengths (Figure 14). First, a quaternary alloy APAP wire was used to replace the common coated wire to reduce the galvanic corrosion caused by the multilayered metal when the wire is bonded to the substrate [28]. Furthermore, the diameter of the quaternary alloy APAP in this study was 15 μm, which is thinner than the silver wire with a diameter of 18–20 μm currently available in the market, reducing the use of precious metals and increasing the packaging density. Finally, we investigated the wire's corrosion fracture in a halogen environment to establish application reference data in depth. The results indicate that the sealant used in the microelectronics packaging industry should contain few chloride ions in the future.

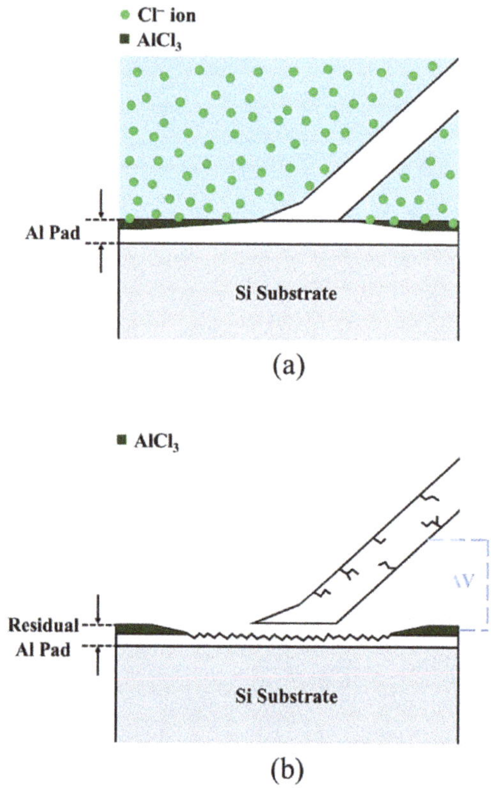

Figure 13. Schematic of the joint of the second bond after chlorination: (**a**) chloride ions corrode the aluminum film; (**b**) partial cracks produced at the joint of the second bond.

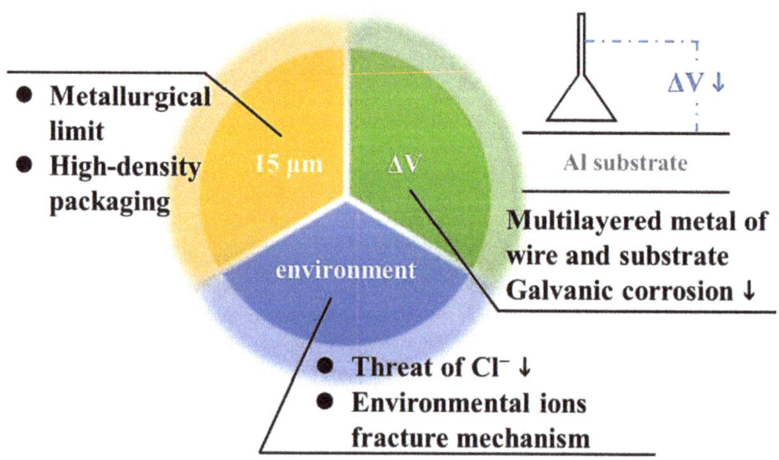

Figure 14. Summary of the study results.

4. Conclusions

1. When the APAP wire is immersed in a saturated sodium chloride solution for a long time, chloride ions diffuse into the wire through the grain boundary causing the intergranular fracture to greatly reduce the mechanical properties of the wire;
2. Chloride ions erode the aluminum substrate, causing the aluminum film to become eroded and hollowed out on the joint surfaces of the first and second bonds;
3. With the lower downforce of the first bond, the joint surface is not tight and separates from the Al substrate after 30 min of chlorination. The second bond still combines with the Al substrate at this time.

Author Contributions: Conceptualization, F.-Y.H.; methodology, J.-R.Z. and C.-W.H.; validation, J.-R.Z., and C.-W.H.; investigation, J.-R.Z. and C.-W.H.; data curation, J.-R.Z.; writing—original draft preparation, J.-R.Z.; writing—review and editing, F.-Y.H.; supervision, F.-Y.H.; project administration, F.-Y.H. All authors have read and agreed to the published version of the manuscript.

Funding: This research received no external funding.

Institutional Review Board Statement: Not applicable.

Informed Consent Statement: Not applicable.

Data Availability Statement: The data presented in this study are available on request from the corresponding author. The data are not publicly available due to privacy.

Acknowledgments: The authors are grateful to the Instrument Center of National Cheng Kung University and the Ministry of Science and Technology of Taiwan for their financial support. They also thank Niche-Tech Group Limited and ASE Group Taiwan for providing the wires and tests used in this study, respectively.

Conflicts of Interest: The authors declare no conflict of interest.

References

1. Wu, Y.H.; Hung, F.Y.; Lui, T.S.; Chen, K.J. Study of wire bonding reliability of Ag-Pd-Au alloy wire with flash-gold after chlorination and sulfidation. *Microelectron. Reliab.* **2019**, *99*, 186–196. [CrossRef]
2. Chen, K.J.; Hung, F.Y.; Chang, C.Y. A Study of the Sulfidation Behavior on Palladium-Coated Copper Wire with a Flash-Gold Layer (PCA) after Wire Bonding. *Electronics* **2019**, *8*, 792. [CrossRef]
3. Yang, G.; Zhou, Z.; Zhang, H.; Zhang, Y.; Peng, Z.; Gong, P.; Wang, X.; Cui, C. Improved Anti-Vulcanization and Bonding Performance of a Silver Alloy Bonding Wire by a Cathodic Passivation Treatment with Palladium. *Materials* **2022**, *15*, 2355. [CrossRef] [PubMed]
4. Manoharan, S.; Patel, C.; McCluskey, P. Advancements in silver wire bonding. In *International Electronic Packaging Technical Conference and Exhibition, 2017 August, 58097, V001T05A004)*; American Society of Mechanical Engineers: New York, NY, USA, 2017.
5. Mazloum-Nejadari, A.; Khatibi, G.; Czerny, B.; Lederer, M.; Nicolics, J.; Weiss, L. Reliability of Cu wire bonds in microelectronic packages. *Microelectron. Reliab.* **2017**, *74*, 147–154. [CrossRef]
6. Alim, M.A.; Abdullah, M.Z.; Aziz, M.A.; Kamarudin, R. Die attachment, wire bonding, and encapsulation process in LED packaging: A review. *Sens. Actuators A Phys.* **2021**, *329*, 112817. [CrossRef]
7. Mancaleoni, A.; Sitta, A.; Colombo, A.; Villa, R.; Mirone, G.; Renna, M.; Calabretta, M. Copper wire bonding process characterization and simulation. In Proceedings of the Conference on Integrated Power Electronics Systems (CIPS), Berlin, Germany, 24–26 March 2020; pp. 1–4.
8. Liu, C.P.; Chang, S.J.; Liu, Y.F.; Su, J. Corrosion-induced degradation and its mechanism study of Cu–Al interface for Cu-wire bonding under HAST conditions. *J. Alloys Compd.* **2020**, *825*, 154046. [CrossRef]
9. Lall, P.; Deshpande, S.; Nguyen, L. Reliability of copper, gold, silver, and pcc wirebonds subjected to harsh environment. In Proceedings of the Electronic Components and Technology Conference (ECTC), Orlando, FL, USA, 30 May–2 June 2023; pp. 724–734.
10. Zhou, W.Y.; Pei, H.Y.; Luo, J.Q.; Wu, Y.J.; Chen, J.L.; Kang, F.F.; Kong, J.W.; Ji, X.Q.; Yang, G.X.; Yang, A.H. Influence of Pd content on microstructure and performance of Au-coated Ag alloy wire. *J. Mater. Sci. Mater. Electron.* **2020**, *31*, 6241–6247. [CrossRef]
11. Ly, N.; Xu, D.E.; Song, W.H.; Mayer, M. More uniform Pd distribution in free-air balls of Pd-coated Cu bonding wire using movable flame-off electrode. *Microelectron. Reliab.* **2015**, *55*, 201–206. [CrossRef]

12. Tajedini, M.; Osmanson, A.T.; Kim, Y.R.; Madanipour, H.; Kim, C.U.; Glasscock, B.; Khan, M. Electromigration effect on the Pd coated Cu wirebond. In Proceedings of the Electronic Components and Technology Conference (ECTC), Orlando, FL, USA, 30 May–2 June 2023; pp. 661–666.
13. Tseng, Y.W.; Hung, F.Y.; Lui, T.S.; Chen, M.Y.; Hsueh, H.W. Effect of annealing on the microstructure and bonding interface properties of Ag-2Pd alloy wire. *Microelectron. Reliab.* **2015**, *55*, 1256–1261. [CrossRef]
14. Chen, C.H.; Lee, P.I.; Chuang, T.H. Microstructure evolution and failure mechanism of electromigration in Ag-alloy bonding wire. *J. Alloys Compd.* **2022**, *913*, 165266. [CrossRef]
15. Tseng, Y.W.; Hung, F.Y.; Lui, T.S. Microstructure, mechanical and high-temperature electrical properties of cyanide-free Au-coated Ag wire (ACA). *Mater. Trans.* **2015**, *56*, 441–444. [CrossRef]
16. Zhou, W.; Chen, J.; Pei, H.; Kang, F.; Wu, Y.; Kong, J.; Yang, G.; Yi, J. Microstructure and bonding performance of an Au-coated Ag alloy wire. *Microelectron. Eng.* **2019**, *217*, 111104. [CrossRef]
17. Liqun, G.; Qiang, C.; Juanjuan, L.; Zhengrong, C.; Jianwei, Z.; Maohua, D.; Chung, M. Comparison of Ag Wire and Cu Wire in Memory Package. *ECS Trans.* **2013**, *52*, 747–751. [CrossRef]
18. Kai, L.J.; Hung, L.Y.; Wu, L.W.; Chiang, M.Y.; Jiang, D.S.; Huang, C.M.; Wang, Y.P. Silver alloy wire bonding. In Proceedings of the Electronic Components and Technology Conference (ECTC), Orlando, FL, USA, 30 May–2 June 2023; pp. 1163–1168.
19. Szymanski, R.; Charcosset, H. Platinum-Zirconium alloy catalysts supported on carbon or zirconia. *Platin. Met. Rev.* **1986**, *30*, 23–27.
20. Mori, K. Manufacture of Platinum Fibre and Fabric. *Platin. Met. Rev.* **2004**, *48*, 56–57.
21. Huang, I.T.; Hung, F.Y.; Lui, T.S.; Chen, L.H.; Hsueh, H.W. A study on the tensile fracture mechanism of 15 μm copper wire after EFO process. *Microelectron. Reliab.* **2011**, *51*, 25–29. [CrossRef]
22. Tsai, H.H.; Lee, J.D.; Tsai, C.H.; Wang, H.C.; Chang, C.C.; Chuang, T.H. An innovative annealing-twinned Ag-Au-Pd bonding wire for IC and LED packaging. In Proceedings of the International Microsystems, Packaging, Assembly and Circuits Technology Conference (IMPACT), Taipei, Taiwan , 23–26 October, 2012; pp. 243–246.
23. Wu, B.D.; Hung, F.Y. Effect of electrification and chlorination on the microstructure and electrical properties of fine Al wires. *Microelectron. Reliab.* **2021**, *124*, 114234. [CrossRef]
24. Yamaji, Y.; Hori, M.; Ikenosako, H.; Oshima, Y.; Suda, T.; Umeki, A.; Kandori, M.; Oida, M.; Goto, H.; Katsumata, A.; et al. IMC study on Cu wirebond failures under high humidity conditions. In Proceedings of the Electronics Packaging Technology Conference (EPTC), Singapore, 7–9 December 2011; pp. 480–485.
25. Uno, T. Bond reliability under humid environment for coated copper wire and bare copper wire. *Microelectron. Reliab.* **2011**, *51*, 148–156. [CrossRef]
26. Tseng, Y.W.; Hung, F.Y.; Lui, T.S. Microstructure, tensile and electrical properties of gold-coated silver bonding wire. *Microelectron. Reliab.* **2015**, *55*, 608–612. [CrossRef]
27. Hsueh, H.W.; Hung, F.Y.; Lui, T.S. A study on electromigration-inducing intergranular fracture of fine silver alloy wires. *Appl. Phys. Lett.* **2017**, *110*, 031902. [CrossRef]
28. Lee, C.S.; Tran, T.; Boyne, D.; Higgins, L.; Mawer, A. Copper versus palladium coated copper wire process and reliability differences. In Proceedings of the Electronic Components and Technology Conference (ECTC), Orlando, FL, USA, 30 May–2 June 2023; pp. 1539–1548.

materials

MDPI

Article

Improving the Thermal Stability of the Fine-Grained Structure in the Cu-15Ni-8Sn Alloy during Solution Treatment by the Additions of Si and Ti

Chao Zhao [1], Daoxi Li [2,3], Xiaotao Liu [1], Minghan Sun [1,*], Zhi Wang [2], Zongqiang Luo [2] and Weiwen Zhang [2,*]

1 School of Materials Science and Engineering, Huazhong University of Science and Technology, Wuhan 430074, China
2 School of Mechanical and Automotive Engineering, South China University of Technology, Guangzhou 510640, China
3 Department of Mechanical and Energy Engineering, Shaoyang University, Shaoyang 422004, China
* Correspondence: sunmh@hust.edu.cn (M.S.); mewzhang@scut.edu.cn (W.Z.)

Abstract: Grain refinement has been found to be an effective method for simultaneously enhancing strength and toughness. To avoid the sharp coarsening of grains in Cu-Ni-Sn alloys during solution treatment and thereby overcoming the tradeoff between strength and ductility, this work attempted to modify the composition and improve the thermal stability of the fine-grained structure in Cu-Ni-Sn alloys. The grain growth behavior during a solution treatment of the Cu-15Ni-8Sn alloys with/without Si and Ti additions was systematically investigated. The result reveals that compared to the grain size of 146 μm in the based alloy (without trace additions) after solution processing at 1073 K for 2 h, the fine-grained structure with a size below 20 μm is maintained owing to the benefit from Si and Ti addition. It was observed that the addition of Si and Ti offer the inhibition effect on the dissolution of the γ phase and $Ni_{16}Si_7Ti_6$ particles after solution treatment. The grain boundary movement is severely hindered by these two aspects: the pinning effect from these particles, and the drag effect induced by additional solute atoms. Based on the analysis of grain growth kinetics, the activation energy of grain growth is increased from 156 kJ/mol to 353 kJ/mol with the addition of Si and Ti.

Keywords: Cu-15Ni-8Sn alloys; grain size; solution treatment; microalloying; microstructural evolution

Citation: Zhao, C.; Li, D.; Liu, X.; Sun, M.; Wang, Z.; Luo, Z.; Zhang, W. Improving the Thermal Stability of the Fine-Grained Structure in the Cu-15Ni-8Sn Alloy during Solution Treatment by the Additions of Si and Ti. *Materials* **2023**, *16*, 1252. https://doi.org/10.3390/ma16031252

Academic Editor: Sergey V. Zherebtsov

Received: 15 October 2022
Revised: 19 January 2023
Accepted: 21 January 2023
Published: 1 February 2023

1. Introduction

Cu-15Ni-8Sn alloys possess excellent mechanical properties and excellent stress relaxation resistance, as well as excellent wear resistance and corrosion resistance, being considered as promising beryllium–copper substitute materials [1,2]. These high-performance copper alloys have great potential applications in the electronic and electrical industries, and simultaneously show wide prospects for major equipment involving aircraft landing gear, heavy loaded vehicles, and marine engineering equipment [3–5]. However, the trade-off between strength and ductility of the alloys is still the critical problem for meeting the increasing performance demands of key components in aerospace, mechanical systems, and the oil and gas industries. For example, after the commonly used pre-cooling deformation–aging treatment of the alloy, its strength can reach up to 1400 MPa, but its elongation rate is only ~1% [6]. One essential approach is to improve the plasticity of the alloy while maintaining its strength.

Grain refinement is commonly deemed an effective method to improve plasticity in polycrystalline metallic materials. Our previous research obtained the fine-grain structure in the Cu-15Ni-8Sn alloy via the hot extrusion process, effectively improving the elongation of the as-extruded alloy to more than 30% [7–9]. On the other hand, the Spinodal decomposition and ordering strengthening could not completely occur in the as-extruded alloy due to the absence of the subsequent solution and aging treatments, so it failed to reap the full performance advantages of the Cu-15Ni-8Sn alloy [10]. High temperature solution treatments are able to promote the formation of supersaturated solid solutions, providing a chemical driving force for Spinodal decomposition and ordering transformation [11,12], but they often lead to uncontrollable grain coarsening, encroaching on the superior ductility induced by the fine-grain structure via the hot extrusion process. A large number of studies have shown that the addition of trace elements can result in the development of the drag effect on solid solution atoms and the pinning effects on the insoluble second phase, which shows an important role of the suppression of grain boundary migration during solution treatment [13–16]. At present, a thorough investigation of the effect of trace elements on the grain growth kinetics and the thermal stability of grains is lacking and is critical to seek a way to realize the synergy of superior strength and high ductility of the Cu-15Ni-8Sn alloy.

In our previous research, the Cu-15Ni-8Sn-0.3Si-0.1Ti alloy was found to be of the optimal composition, as it showed a fine-grained structure even after solution treatment and thereby gained excellent comprehensive mechanical properties [7,8,17]. Accordingly, in the present work, a comprehensive comparison of the grain growth behavior between the Cu-15Ni-8Sn alloy with and without Si and Ti addition has been studied in order to uncover the effect of trace elements (for instance Si and Ti) in the thermal stability of the fine-grained structure during solution treatment. The microstructural features of the based and modified alloys during solution treatment, involving grain morphology and the size and distribution of secondary phase, were investigated in detail. The influences of trace additions on the thermal stability of grains were discussed from the perspective of the drag effect on solute atoms and the pinning effect on secondary particles. The information gained on grain growth behavior would help to maintain the grain refinement and further lay a foundation for the high ductility of the Cu-15Ni-8Sn alloy.

2. Materials and Methods

The Cu-15Ni-8Sn-based alloy and Cu-15Ni-8Sn-0.3Si-0.1Ti modified alloy were prepared by an intermediate frequency induction furnace. The actual chemical compositions of the as-cast materials determined by optical emission spectroscopy (OES) are listed in Table 1. The cast ingots were homogenization-treated at 1113 K for 8 h and subsequently hot-extruded to a rod with a diameter of 12 mm at 1223 K. The extruded rods were cut and annealed at 1073 K, 1093 K, and 1113K for 0.5 to 2 h, followed by quenching in water.

Table 1. Chemical compositions of the as-cast alloys (wt.%).

Alloy Designation	Ni	Sn	Si	Ti	Cu
Cu-15Ni-8Sn (based alloy)	15.03	8.12	-	-	bal.
Cu-15Ni-8Sn-0.3Si-0.1Ti (modified alloy)	15.23	7.92	0.29	0.09	bal.

The microstructure of the annealed samples was characterized by optical microscopy (OM), scanning electronic microscopy (SEM) equipped with energy dispersive X-ray spectroscopy (EDS), and transmission electron microscopy (TEM). The mean grain size was obtained from at least 10 random fields of each specimen by the linear intercept method based on the observation with a LEICA/DMI 5000 M optical microscope. The specimens for OM and SEM observations were prepared by polishing and then etching in a solution of 5 g $FeCl_3$ + 10 mL HCl + 100 mL H_2O. The TEM observation was carried on a FEI TECNAI G2 S-TWIN F20 transmission electron microscope. TEM samples were prepared via the twin jet electro-polishing method in 30% nitric acid and 70% methanol held at 248 K. The hardness measurement was performed on a Vickers microhardness tester using a load of 100 gf and a dwelling time of 10 s. The average values of at least 5 measurements were used to establish the hardness data. The thermal stability of the fine-grained structure in the alloys was assessed through the above microstructural observations.

3. Results

Figure 1 shows the optical images of as-extruded Cu-15Ni-8Sn-0.3Si-0.1Ti and Cu-15Ni-8Sn alloys before the solution treatment. The average grain size of the alloy with additions of Si and Ti (12.4 μm) is slightly smaller than that of the based alloy (15.4 μm) owing to the formation of fine second phase particles [17].

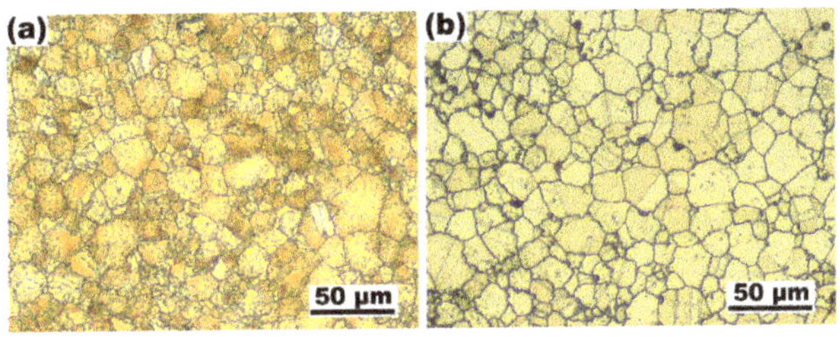

Figure 1. Representative OM images of as-extruded Cu-15Ni-8Sn-0.3Si-0.1Ti (**a**) and Cu-15Ni-8Sn (**b**) alloys.

Figure 2 demonstrates a series of representative OM micrographs of Cu-15Ni-8Sn and Cu-15Ni-8Sn-0.3Si-0.1Ti alloys annealed at 1073 K to 1113 K for different time periods. The relevant average grain sizes are shown in Figure 3. It is evident that the grain size is increased with the increment increase in solution temperature and time, and a more rapid grain growth is induced at a higher temperature. Compared with the excessive grain growth in the Cu-15Ni-8Sn alloy, the velocity of grain growth in the modified alloy is much slower. Even after it was in the solution at 1113 K for 2 h, the increment in grain size in the modified alloy is only about 6 μm (from 12.4 μm to 18.6 μm), while it is measured as 130.7 μm in the based alloy (from 15.4 μm to 146.1 μm).

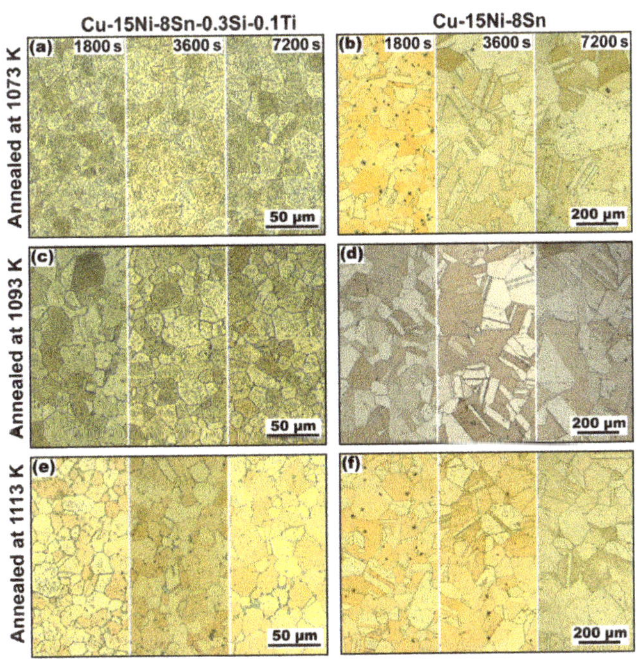

Figure 2. Representative optical images of the microstructure of the studied materials after annealing in different conditions: (**a**,**c**,**e**) Cu-15Ni-8Sn-0.3Si-0.1Ti; (**b**,**d**,**f**) Cu-15Ni-8Sn.

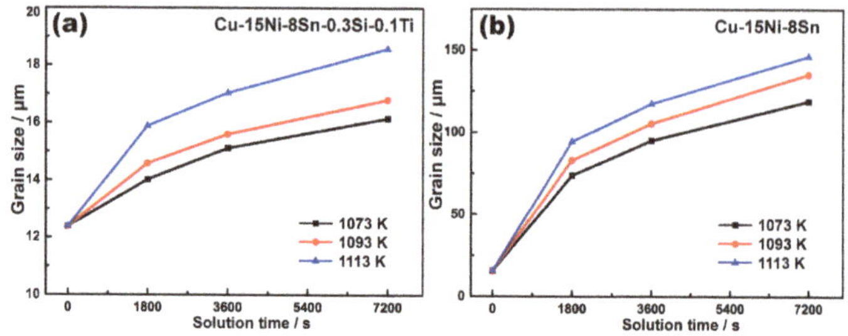

Figure 3. Average grain sizes of the Cu-15Ni-8Sn-0.3Si-0.1Ti (**a**) and Cu-15Ni-8Sn (**b**) alloys after different solution processes.

It can be found from Figure 2a,c,e that the amount of second phase particles in the modified alloy is decreased with the prolongation of the solution time, and the dissolution rate of the second phase becomes faster with a higher solution temperature. After being in the solution at 1113 K for 2 h, only a few large-sized particles survive at the grain boundaries (Figure 2e). This positive relationship between the dissolution degree of the second phase particles and the grain growth rate in the modified alloy indicates that the existence of the second phase may be responsible for restraining the grain growth in the alloy.

For a quantitative comparison of the tendency of grain growth during annealing for 2 h, the grain growth factor (G_f), defined as $(d_f - d_0)/d_0$ [18] (where d_f and d_0 are the final and the initial grain sizes, respectively) is calculated and listed in Table 2. The value of G_f of the based alloy is about 20 times bigger than that of the modified alloy, indicating a distinct suppression of the grain growth in the modified alloy due to the addition of Si and Ti.

Table 2. Grain growth factor (G_f), grain grow exponent (n), and activation energy (Q) for the studied alloys with a solution process of 2 h.

Alloy	T (K)	G_f	n	Q (kJ/mol)
Cu-15Ni-8Sn	1073	6.70	3	156
	1093	7.76		
	1113	8.46		
Cu-15Ni-8Sn-0.3Si-0.1Ti	1073	0.30	10	353
	1093	0.36		
	1113	0.50		

In order to study the changes in the morphology and distribution of the second phase in the alloy with additions of Si and Ti under different solution conditions, further microstructural observation was conducted for the alloy in a 1093 K solution at different times. Figure 4 shows the microstructure of the based and modified alloys annealed at 1093 K for 1 h. According to the results of the selected area diffraction pattern (SADP), the precipitate phase is confirmed as a $Ni_{16}Si_7Ti_6$ intermetallic compound with a size of ~200 nm [7] (Figure 4a). The rod-like γ phase with a DO3 structure was also confirmed by SADP analysis, which is consistent with J.C. Zhao's results [11]. Meanwhile, no precipitates were observed in the based alloy after solution treatment, implying the improvement in thermal stability by the addition of Si and Ti.

Figure 5 shows the SEM images and energy dispersive surface scanning results of the 0.3Si-0.1Ti alloy after solution processing at 1093 K for 5 min. There are a large number of second phases formed with the grains and at grain boundaries. According to the results of the EDS analysis shown in Table 3, these particles are defined as $Ni_{16}Si_7Ti_6$ and γ phases. Figure 5b shows the rod-like $Ni_{16}Si_7Ti_6$ particles (~2.8 μm) distributed at the grain boundaries. It can be seen from the high magnification SEM image (Figure 5c) that small needle-like particles are distributed along the grain boundaries (as marked by the green dotted line), whose average major axis, minor axis, and aspect ratio are about 617 nm, 225 nm, and 2.8, respectively.

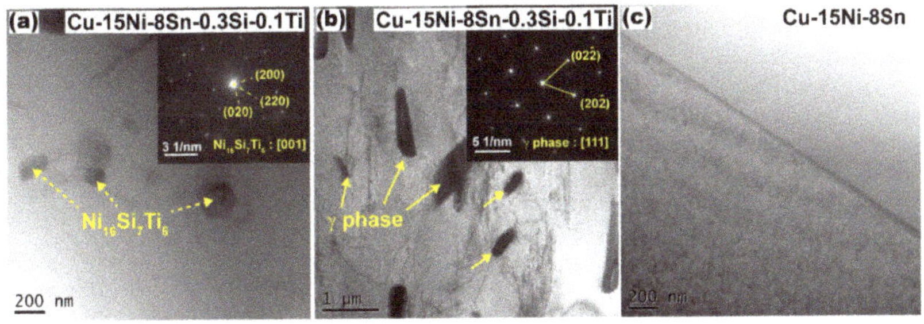

Figure 4. TEM images revealing the microstructures of Cu-15Ni-8Sn-0.3Si-0.1Ti (**a**) and Cu-15Ni-8Sn alloys (**b**,**c**) after annealing at 1093 K for 1 h. The inserts reveal the selected area diffraction patterns (SADP) of related particles.

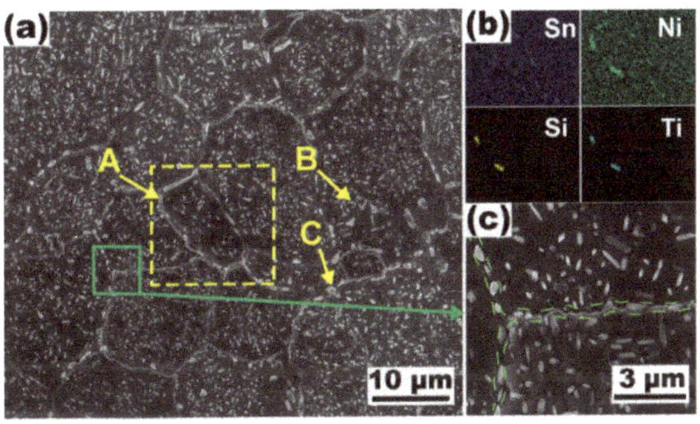

Figure 5. SEM micrographs of the Cu-15Ni-8Sn-0.3Si-0.1Ti alloy after annealing at 1093 K for 5 min. (**a**) Grain morphology in the alloy; (**b**) the EDS-mapping result corresponding to the yellow rectangle in (**a**); (**c**) enlarged view of the green domain in (**a**), small needle-like particles distributed along the grain boundaries are marked by the green dotted line.

Table 3. EDS results of corresponding positions in Figure 5 (wt.%).

Positions	Phases	Cu	Ni	Sn	Si	Ti
A	$Ni_{16}Si_7Ti_6$	33.29	41.02	6.78	8.44	10.47
B	γ	74.7	17.07	9.62	0.35	-
C	γ	37.74	30.63	29.81	1.82	-

Figure 6 indicates the microstructural features of the modified alloy after solution processing at 1093 K for 1 h. A large amount of the $Ni_{16}Si_7Ti_6$ and γ phases can still be observed in the alloy. Figure 6b shows micron particles are distributed around the grains, which implies the inhibiting effect of grain growth induced by these insoluble particles. The amount of fine needle-like second phase which survived after solution processing for 5 min is significantly reduced, while there are still densely distributed micron particles with an average size of 1.2 μm (Figure 6c).

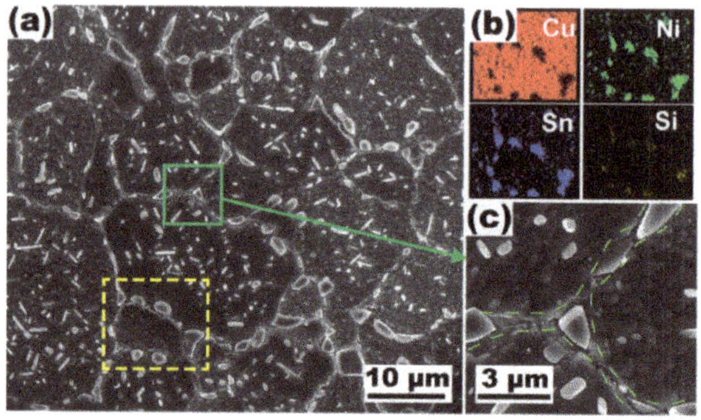

Figure 6. SEM micrographs of the Cu-15Ni-8Sn-0.3Si-0.1Ti alloy after annealing at 1093 K for 1 h. (**a**) Grain morphology in the alloy; (**b**) the EDS-mapping result corresponding to the yellow rectangle in (**a**); (**c**) higher magnification image of the green domain in (**a**), intergranular micron particles are highlighted by the green dotted line.

Figure 7 shows the SEM image of the modified alloy after solution processing at 1093 K for 2 h. The fine-grained structure is still retained in the alloy despite solid solution processing for 2 h (Figure 7a). The amount of intragranular second phase is markedly reduced, while relatively more particles are distributed at grain boundaries (Figure 7b). According to the results of the EDS analysis shown in Table 4, these intergranular particles are of the $Ni_{16}Si_7Ti_6$ and γ phases, indicating that these second phases have good thermal stability, and show indissolubility in the solution process at 1093 K.

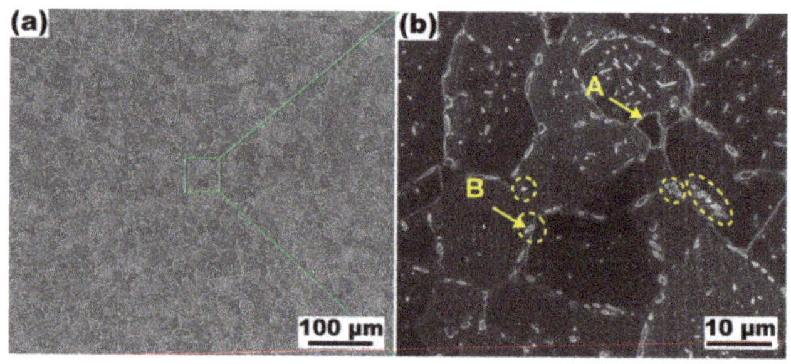

Figure 7. SEM micrographs of the Cu-15Ni-8Sn-0.3Si-0.1Ti alloy after annealing at 1093 K for 2 h. Intergranular micron particles are highlighted by the yellow dotted line.

Table 4. EDS results of corresponding positions in Figure 7 (wt.%).

Positions	Phases	Cu	Ni	Sn	Si	Ti
A	γ	36.34	32.2	29.54	1.92	-
B	$Ni_{16}Si_7Ti_6$	49.87	31.31	4.38	7.27	7.16

According to the TTT diagram summarized by J.C. Zhao and M.R. Notis [11], the Cu-15Ni-8Sn alloy was located in the so-called α single phase area with a temperature above 1073 K. As a result, the γ phase would be dissolved completely in this temperature window, which is consistent with the microstructure of the based alloy after solution processing at 1093 K for 1 h (Figure 4a). On the contrary, the existence of the γ phase in the modified alloy after solution indicates that the γ phase shows a higher thermal stability due to the addition of Si and Ti.

4. Discussion

4.1. Kinetics of Grain Growth of the Alloys with and without Additions

The kinetics of grain growth can be deduced by analyzing the grain size as a function of time, which is depicted as the well-known parabolic kinetic model [19]:

$$d^n - d_0^n = kt \quad (1)$$

where d is the instantaneous grain size, d_0 is the initial grain size that can be considered as the grain size of as-extruded alloys, t is the solution holding time, n is a grain growth exponent, and k is a kinetic constant, which depends primarily on the temperature and grain boundary energy. As shown in Figure 8a,b, the value $1/n$ can be calculated as the slope of linear fitting plots of *lnd-lnt*. The results of the based and modified alloys are 10 and 3, respectively. The relationship between the grain size and solution time can be obtained by introducing the n value into Equation (1), which is shown in Figure 8c,d. The constant k has an Arrhenius form as [12,13]:

$$k = k_0 \exp(-Q/RT) \quad (2)$$

in which k_0 is the pre-exponential constant and Q is the grain growth activation energy. R is the gas constant and T is the absolute temperature. The k values of these two alloys at different solution temperatures can be obtained from Figure 8c,d. By plotting semi-logarithmic plots of ln(k) versus $1/T$, as demonstrated in Figure 8e, the value of the grain growth activation energy (Q) in the based and modified alloys can be measured from the slope of the plotted curves, which is 156 kJ/mol and 353 kJ/mol, respectively. More than twice the activation energy is required in the modified alloy to trigger grain growth, indicating that the higher thermal stability of the fine-grained structure is induced by the addition of Si and Ti.

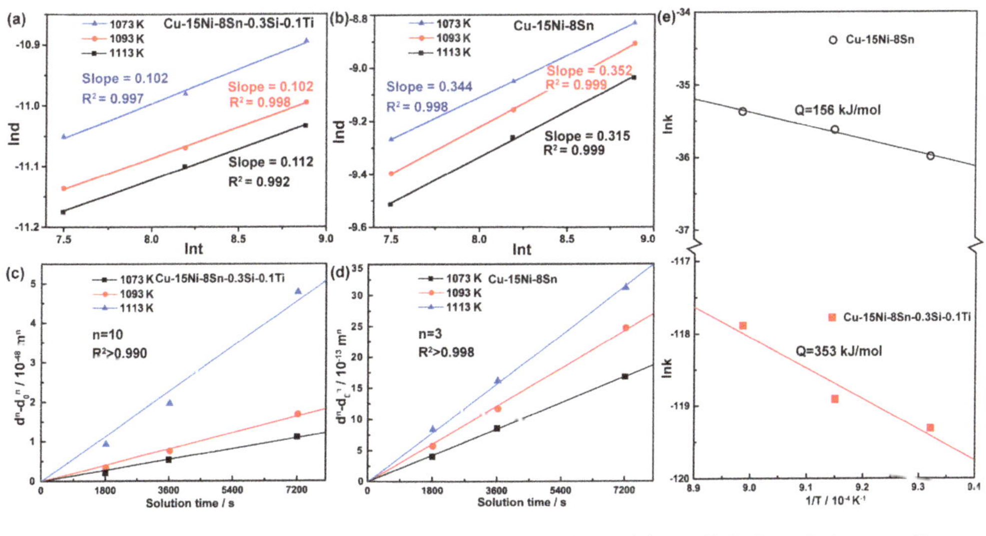

Figure 8. The grain growth kinetics analysis of the studied alloys during annealing treatment. (**a**,**b**) Linear fitted plots of *lnd-lnt*; (**c**,**d**) curve fit plots of Equation (1) used to estimate grain growth exponents; (**e**) grain growth exponents as a function of the reciprocal of absolute temperature used to calculate the grain growth activation energy.

Burke and Turnbull found that if the driving force of grain growth in pure metals under ideal conditions is mainly from the grain boundary curvature, the grain growth exponent (n) in Equation (1) is equal to 2 [20]. Fan and Chen found that when the grain growth is determined by the long-range atomic diffusion in the two-phase solid solution, the grain growth exponent (n) in Equation (1) is equal to 3 [21]. In the study of multi-component alloys, such as magnesium-based alloys, nickel-based alloys, iron-based alloys, and high-entropy alloys, the n value is approximately 3 [12,13], which is consistent with the n value of Cu-15Ni-8Sn. However, the higher value of the grain growth exponent (n) indicates that the grain growth dynamics in the alloy are also affected by other factors, which are often related to the solid solution atoms, the second phase, texture, etc. In this study, the role of trace elements in the modified alloy in increasing the value of n mainly involves the drag effect of solid solution atoms and the pinning effects of the second phase.

4.2. Effects of the Additions of Trace Elements on the Grain Growth

In order to explain the effects of the addition of Si and Ti, the grain boundary migration rate (V_{gb}) was introduced which is related to the diffusion ability of the solute and the interaction between the grain boundary and solute atoms. It is assumed that the interaction energy between the grain boundary and one solute atom is $E(x)$, and the diffusion coefficient of the solute atom perpendicular to the grain boundary is $D(x)$. Both $E(x)$ and $D(x)$ are the functions of the distance x from the center of the grain boundary. When V_{gb} is low, the atomic drag (α) can be expressed as [12]:

$$\alpha = 4N_v kT \int \frac{\sinh^2(E(x)/2kT)}{D(x)} dx \quad (3)$$

where N_v is the number of solute atoms per unit volume, k is the Boltzmann constant, and T is the absolute temperature. According to Equation (3), the solute drag effect (α) shows an inversely proportional relationship to the diffusivity of solute atoms, as $\alpha \propto 1/D(x)$. Thus, V_{gb} can be expressed as:

$$V_{gb} = \frac{P}{\lambda + \alpha C_0} \quad (4)$$

where P is the driving force of grain growth, C_0 is the concentration of solute in the alloy, and λ is a constant. Generally, C_0 is considered as a constant in the same alloy. As a result, V_{gb} is inversely proportional to α, equivalently there is a proportional relationship between V_{gb} and $D(x)$, indicating that the lower the diffusion ability of the solute atoms is, the stronger the drag effect is. The diffusion coefficients of each solute atom in pure copper are used for analysis (as shown in Table 5) [22]. Compared with the diffusion ability of Sn, the diffusion coefficients of Si and Ti in pure copper are much lower, enhancing the drag effect of solute atoms on grain boundaries. Based on microstructural observation, the concentration of Si (0.23 at%) dissolved in the modified alloy is about 3 times higher than that of Ti (0.06 at%), indicating that the addition of Si should play a dominant role in the drag effect.

Table 5. The diffusion coefficients of different solutes in pure copper [22].

Solutes	D (10^{-4} m^2/s)	T-Range (K)
Sn	0.84	1011–1321
Si	0.21	998–1173
Ti	0.693	973–1283

The pinning effect of the second phase on grain boundary migration depends on the radius (r), shape, spacing, and volume fraction (F_v) of the particles. The effect of the second phase on the grain boundary per unit area (P_z) is usually expressed by the following equation [23]:

$$P_Z = \frac{3F_v\gamma}{2r} \quad (5)$$

here γ means the interface energy of the grain boundary per unit area. According to the microstructural observation, there is no precipitation of the second phase particles in the based alloy after solution treatment, while a large number of second phase particles were found in the modified alloy (Figure 4). If we take the modified alloy with solution processing at 1093 K for 1 h as an example, the related parameters of the secondary particles are shown in Table 6. It can be estimated that the pinning effect of the second relative unit area grain boundary in the modified alloy is 0.0165 MPa based on Equation (5), which is a result of the existence of the γ phase and $Ni_{16}Si_7Ti_6$ particles. This pinning effect also plays a significant role in improving the thermal stability of the fine-grained structure in the Cu-15Ni-8Sn alloy during the solution treatment.

Table 6. Parameters of secondary particles in the modified alloy after the solution processing at 1093 K for 1 h.

F_v	r (μm)	γ (J/m^2) [24]	P_z (MPa)
0.0075	0.444	0.65	0.0165

5. Conclusions

In this work, the microstructural features of the Cu-15Ni-8Sn alloy with and without adding Si and Ti were investigated. After the solution treatment at 1073~1113 K for 0.5~2 h, the initial (as-extruded) fine-grain structure with the grain size below 20 μm is able to be maintained in the Cu-15Ni-8Sn-0.3Si-0.1Ti alloy while grains are excessively coarsened in the Cu-15Ni-8Sn alloy. Based on the analysis of grain growth kinetics, the activation energy of the grain growth is increased from 156 kJ/mol to 353 kJ/mol, suggesting a significant improvement in the thermal stability of the fine-grained structure ascribed to additions Si and Ti. Owing to the solid solution Si in the γ phase, a large amount of γ phase would be retained after solution treatment. At the same time, $Ni_{16}Si_7Ti_6$ particles are formed by introducing Si and Ti. Consequently, the grain boundary movement is severely hindered by two aspects: the pinning effect from the γ phase and $Ni_{16}Si_7Ti_6$ particles, and the drag effect induced by solute Si and Ti atoms.

Author Contributions: C.Z. and W.Z. designed the experiments; C.Z., D.L., X.L. and M.S. conducted the data analysis. Z.L. and Z.W. provided the experimental resources and conducted experimental supervision. C.Z. organized the database and wrote the manuscript. M.S. and W.Z. did the writing review and editing. All authors have read and agreed to the published version of the manuscript.

Funding: This research was funded by the National Natural Science Foundation of China (No. 52205367), the China Postdoctoral Science Foundation (No. 2021M691107).

Institutional Review Board Statement: Not applicable.

Informed Consent Statement: Not applicable.

Data Availability Statement: The raw/processed data required to reproduce these findings cannot be shared at this time as the data also forms part of an ongoing study.

Conflicts of Interest: The authors declare no conflict of interest.

References

1. Cribb, W.R. Copper spinodal alloys for aerospace. *Adv. Mater. Process.* **2006**, *6*, 44.
2. Li, D.X.; Wang, Z.; Zhao, C.; Luo, Z.Q.; Yang, C.; Zhang, W.W. Comprehensive characterisation of tribo-layer in a Cu-15Ni-8Sn alloy during dry sliding wear. *Mater. Sci. Technol.* **2022**, *38*, 57–68. [CrossRef]
3. Zhang, K.; Zhang, L.M.; Shu, C.L.; Fan, W.X.; Li, S.; Yuan, X.; Zhao, J.; Wang, Y.; Wang, P. Effect of different Mo addition on microstructure and mechanical properties of Cu-15Ni-8Sn alloy. *Materials* **2022**, *15*, 5521. [CrossRef] [PubMed]
4. Zhao, C.; Wang, Z.; Li, D.X.; Kollo, L.; Luo, Z.Q.; Zhang, W.W.; Prashanth, G.K. Selective laser melting of Cu-Ni-Sn: A comprehensive study on the microstructure, mechanical properties, and deformation behavior. *Int. J. Plast.* **2021**, *138*, 102926. [CrossRef]

5. Johnsen, R.; Lange, T.; Stenerud, G.; Olsen, J.S. Environmentally assisted degradation of spinodal copper alloy C72900. *Corros. Sci.* **2018**, *142*, 45–55. [CrossRef]
6. Plewes, J.T. High-strength Cu-Ni-Sn alloys by thermomechanical processing. *Metall. Mater. Trans. A* **1975**, *6A*, 537–544. [CrossRef]
7. Zhao, C.; Wang, Z.; Li, D.X.; Pan, D.Q.; Lou, B.M.; Luo, Z.Q.; Zhang, W.W. Optimization of strength and ductility in an as-extruded Cu-15Ni-8Sn alloy by the additions of Si and Ti. *J. Alloys Comp.* **2020**, *823*, 153759. [CrossRef]
8. Zhao, C.; Wang, Z.; Pan, D.Q.; Li, D.X.; Luo, Z.Q.; Zhang, D.T.; Yang, C.; Zhang, W.W. Effect of Si and Ti on dynamic recrystallization of high-performance Cu-15Ni-8Sn alloy during hot deformation. *Trans. Nonferrous Met. Soc. China* **2019**, *29*, 2556–2565. [CrossRef]
9. Niu, D.X.; Zhao, C.; Li, D.X.; Wang, Z.; Luo, Z.Q.; Zhang, W.W. Constitutive modeling of the flow stress behavior for the hot deformation of Cu-15Ni-8Sn alloys. *Front. Mater.* **2020**, *7*, 577867. [CrossRef]
10. Cribb, W.R.; Ratka, J.O. Copper spinodal alloys. *Adv. Mater. Process.* **2002**, *12*, 1–4.
11. Zhao, J.C.; Notis, M.R. Spinodal decomposition, ordering transformation, and discontinuous precipitation in a Cu-15Ni-8Sn alloy. *Acta Mater.* **1998**, *42*, 4203–4218. [CrossRef]
12. Zhao, C.; Zhang, W.W.; Wang, Z.; Li, D.X.; Luo, Z.Q.; Yang, C.; Zhang, D.T. Improving the mechanical properties of Cu-15Ni-8Sn alloys by addition of Titanium. *Materials* **2017**, *10*, 1038. [CrossRef]
13. Hoseini-Athar, M.M.; Mahmudi, R.; Babu, R.P.; Hedströmb, P. Effect of Zn content on the microstructural stability and grain growth kinetics of fine-grained extruded Mg-Gd-Zn alloys. *J. Alloys Comp.* **2020**, *831*, 154766. [CrossRef]
14. Klimova, M.V.; Shaysultanov, D.G.; Zherebtsov, S.V.; Stepanov, N.D. Effect of second phase particles on mechanical properties and grain growth in a CoCrFeMnNi high entropy alloy. *Mat. Sci. Eng. A* **2019**, *748*, 228–235. [CrossRef]
15. Qian, X.Y.; Zeng, Y.; Jiang, B.; Yang, Q.R.; Wan, Y.J.; Quan, G.F.; Pan, F.S. Grain refinement mechanism and improved mechanical properties in Mg-Sn alloy with trace Y addition. *J. Alloys Comp.* **2020**, *820*, 153122. [CrossRef]
16. Xing, Q.Y.; Wu, X.H.; Zang, J.X.; Meng, L.G.; Zhang, X.G. Effect of Er on microstructure and corrosion behavior of Al-Zn-Mg-Cu-Sc-Zr aluminum alloys. *Materials* **2022**, *15*, 1040. [CrossRef]
17. Li, D.X.; Wang, Z.; Zhao, C.; Luo, Z.Q.; Zhang, W.W. The role of the transfer layer on the sliding wear behaviour of a Cu-15Ni-8Sn alloy under different loads. *Tribol. Lett.* **2022**, *70*, 9. [CrossRef]
18. Vaidya, M.; Anupam, A.; Bharadwaj, J.V.; Srivastava, C.; Murty, B.S. Grain growth kinetics in CoCrFeNi and CoCrFeMnNi high entropy alloys processed by spark plasma sintering. *J. Alloys Comp.* **2019**, *791*, 1114–1121. [CrossRef]
19. Liu, W.H.; Wu, Y.; He, J.Y.; Nieh, T.G.; Lu, Z.P. Grain growth and the Hall-Petch relationship in a high-entropy FeCrNiCoMn alloy. *Scr. Mater.* **2013**, *68*, 526–529. [CrossRef]
20. Burke, J.E.; Turnbull, D. Recrystallization and grain growth. *Prog. Met. Phys.* **1952**, *3*, 220–292. [CrossRef]
21. Fan, D.; Chen, L.Q. Diffusion-controlled grain growth in two-phase solids. *Acta Metall.* **1997**, *45*, 3297–3310. [CrossRef]
22. Gerhard, N.; Cornelis, T. Self-diffusion and impurity diffusion in Group I metals. *Pergamon Mater. Ser.* **2008**, *14*, 37–97.
23. Guo, F.; Pei, R.S.; Jiang, L.Y.; Zhang, D.F.; Korte-Kerzel, S.; Al-Samman, T. The role of recrystallization and grain growth in optimizing the sheet texture of magnesium alloys with calcium addition during annealing. *J. Magnes. Alloy.* **2020**, *8*, 252–268. [CrossRef]
24. Miki, M.; Ogino, Y. Influence of Solution-Treatment Conditions on the Cellular Precipitation in Si-doped Cu-10Ni-8Sn alloy. *Mater. Trans.* **1991**, *32*, 1135–1140. [CrossRef]

materials

MDPI

Article

Multipass Friction Stir Processing of Laser-Powder Bed Fusion AlSi10Mg: Microstructure and Mechanical Properties

Akbar Heidarzadeh [1,*], Mahsa Khorshidi [1], Roghayeh Mohammadzadeh [1], Rasoul Khajeh [2], Mohammadreza Mofarrehi [3], Mousa Javidani [3] and X.-Grant Chen [3,*]

1 Department of Materials Engineering, Azarbaijan Shahid Madani University, Tabriz P.O. Box 53714-161, Iran
2 School of Metallurgy and Materials Engineering, Iran University of Science and Technology, Tehran P.O. Box 16846-13114, Iran
3 Department of Applied Science, University of Quebec at Chicoutimi, Saguenay, QC G7H 2B1, Canada
* Correspondence: xgrant_chen@uqac.ca (X.-G.C.); ac.heydarzadeh@azaruniv.ac.ir (A.H.)

Abstract: The effect of multipass friction stir processing (FSP) on the microstructure and mechanical properties of an AlSi10Mg alloy produced by laser-powder bed fusion was investigated. FSP was performed at a rotational speed of 950 rpm and traverse speed of 85 mm/min. The results indicated that FSP destroyed the coarse grain structure in the as-built AlSi10Mg by generating fine and equiaxed grain structures with shear texture components of $A_1^*(111)[\bar{1}\bar{1}2]$ and $A_2^*(111)[11\bar{2}]$, in addition to causing fragmentation and refinement of the Si networks. FSP reduced the tensile strength slightly but significantly improved ductility. One-pass FSP exhibited superior mechanical properties compared with the two- and three-pass scenarios. The higher strength of the one-pass sample was attributed to the strengthening mechanisms induced by the Si particles, which were grown by repeated FSP. The higher ductility of the one-pass sample was explained using the kernel and grain average misorientations. Furthermore, the post-FSP microstructural evolution and fracture behavior of the samples were discussed.

Keywords: AlSi10Mg alloy; laser-powder bed fusion; friction stir processing; microstructure; mechanical properties

Citation: Heidarzadeh, A.; Khorshidi, M.; Mohammadzadeh, R.; Khajeh, R.; Mofarrehi, M.; Javidani, M.; Chen, X.-G. Multipass Friction Stir Processing of Laser-Powder Bed Fusion AlSi10Mg: Microstructure and Mechanical Properties. *Materials* **2023**, *16*, 1559. https://doi.org/10.3390/ma16041559

Academic Editors: Pan Gong, Xin Wang, Maojun Li and Guangchao Han

Received: 20 January 2023
Revised: 6 February 2023
Accepted: 7 February 2023
Published: 13 February 2023

1. Introduction

Laser-powder bed fusion (L-PBF), also termed selective laser melting, is an additive manufacturing method and a rapid three-dimensional (3D) production technique that can be applied to a wide range of metals and alloys [1]. Recently, several studies have focused on L-PBF-related topics, looking at how they overcome the restrictions of traditional manufacturing methods, such as machining and casting [2]. From a metallurgical point of view, L-PBF possesses characteristics similar to those of laser welding; however, it includes higher cooling rates. During L-PBF, in accordance with the 3D geometry, a laser (single or multiple beams) selectively scans and melts a layer of powder that was previously laid on a stage. Another layer of powder is subsequently deposited on the stage, and the process is repeated for the new layer. By repeating this procedure, a metal part is produced until the entire 3D geometry is scanned [3].

The AlSi10Mg alloy, a common L-PBF alloy owing to its high printability, has recently attracted research attention in the field of industrial engineering [4]. For example, the characteristics of L-PBF AlSi10Mg, such as its high strength-to-density ratio, make it an appropriate candidate for aerospace applications. However, AlSi10Mg produced by L-PBF has the disadvantage of low ductility owing to the presence of porosity and inhomogeneity in its microstructure [4,5]. Therefore, several studies have explored new methods for improving the ductility of L-PBF AlSi10Mg [4,5].

Among various post-heat and post-deformation treatments, friction stir processing (FSP) has been introduced as a promising method of increasing the ductility of L-PBF

AlSi10Mg [4]. FSP, which is based on the friction stir welding (FSW) concept, is a solid-state deformation method that can be applied to modify the microstructure and mechanical properties of metals and alloys [6]. During FSP, the deformation and heat induced by a non-consumable-rotating tool cause various microstructural evolutions, such as dynamic recrystallization (DRX), which destroys the initial microstructure of the material [5,7]. The final microstructure and mechanical properties of friction stir processed (FSPed) metals and alloys, such as L-PBF AlSi10Mg, depend on process parameters such as tool rotational speed, tool traverse speed, tool geometry, and tool axial force [8].

Yang et al. [9] investigated the effect of tool rotational speed (750 and 1180 rpm, at a constant tool traverse speed of 37.5 mm/min) on the microstructure and mechanical properties of L-PBF AlSi10Mg. They determined that an increase in tool rotational speed results in higher density by eliminating gas and solidification porosities inside the base material (BM). In addition, they confirmed that FSP at an optimal tool rotational speed can improve tensile elongation by 298%. Maamoun et al. [10] employed FSP for the surface processing of an L-PBF AlSi10Mg disk using a tool rotational speed of 1750 rpm and a tool traverse speed of 160 mm/min. They concluded that FSP is a promising method for localized modification of the microstructure and mechanical properties of L-PBF AlSi10Mg alloys. Zhao et al. [11,12] and Macías et al. [13] studied the effect of FSP on the damage mechanisms in L-PBF AlSi10Mg, using a rotational speed of 1000 rpm and a traverse speed of 500 mm/min. They determined that the FSP increases ductility by globularization of Si in the microstructure, which delays the formation of voids on Si particles and hinders void growth in the matrix. Rafieazad et al. [14] employed FSP to enhance the corrosion resistance of L-PBF AlSi10Mg. They demonstrated that FSP modifies the corrosion properties by forming a uniform microstructure, including a uniform distribution of Si particles, fine and equiaxed grains, and a small fraction of low-angle grain boundaries (LAGBs).

The number of passes plays a key role in determining the final properties of FSPed metals and alloys because of their direct effect on microstructural evolution, which has been investigated by several studies on different types of material, including aluminum-based cast alloys [15]. For example, Nakata et al. [16] studied the effect of multipass FSP on the microstructural and mechanical properties of a die-cast aluminum alloy. They reported that multipass FSP causes grain refinement, elimination of casting defects, and uniform distribution of Si particles in the matrix, thereby improving the tensile strength compared to that of the BM. Moharrami et al. [17] employed multipass FSP on Mg_2Si-rich aluminum alloys. They observed that, by increasing the number of passes, more Mg_2Si fragmentation and grain refinement can be achieved. Hence, the hardness and strength of the processed alloy improved, whereas the wear rate and friction coefficient deteriorated. Sing et al. [18] investigated the influence of multipass FSP on the wear properties and machinability of an Al–Si hypoeutectic A356 alloy. They reported that highest wear resistance and machinability can be achieved by employing two-pass FSP.

The results from these studies indicate that multipass FSP has never been implemented on L-PBF AlSi10Mg, which can affect the microstructure and corresponding mechanical properties. Therefore, the aim of this study was to explore the effect of multipass FSP (a total of three passes) on the characteristics (mainly microstructure evolution and mechanical properties) of L-PBF AlSi10Mg alloys.

2. Materials and Methods

The AlSi10Mg (10 wt.% Si, 0.4 wt.% Mg, and 89.6 wt.% Al) plate with dimensions of $120 \times 50 \times 2.5$ mm was produced by L-PBF using a Noura M100P machine (Noura, Isfahan, Iran) within the following parameters: number of slices, 260; powder consumption, 80 g; build time, 2 h; layer thickness, 30 μm; energy density, 60 J/mm^3; hatch space, 0.19 mm; and scan rotation, 67°. Gas-atomized AlSi10Mg powder with an average diameter of <65 μm was used. After L-PBF, multipass FSP was used for post-processing of the AlSi10Mg plate using three passes, denoted as samples S1, S2, and S3. To achieve this, FSP was performed perpendicular to the building direction (BD) of the plate at a rotational speed

(*W*) of 950 rpm, traverse speed (*V*) of 85 mm/min, plunge depth of 0.1 mm, and tilt angle of 2° using an H13 steel tool. The FSP tool was composed of a pin and shoulder with dimensions, as schematically shown in Figure 1a. A schematic of the multipass FSP used in this study is shown in Figure 1b. In order to save processing time, the length of the plate was divided into three parts. The first pass was conducted along the entire length of the plate, which took 1.77 min. The second pass was performed on two-thirds of the plate's length for 0.8 min. The third pass was conducted on the final third of the plate for 0.33 min. Considering 10 s (0.16 min) of shoulder-holing time at the start of each pass, the processing time of three passes totaled 3.38 min.

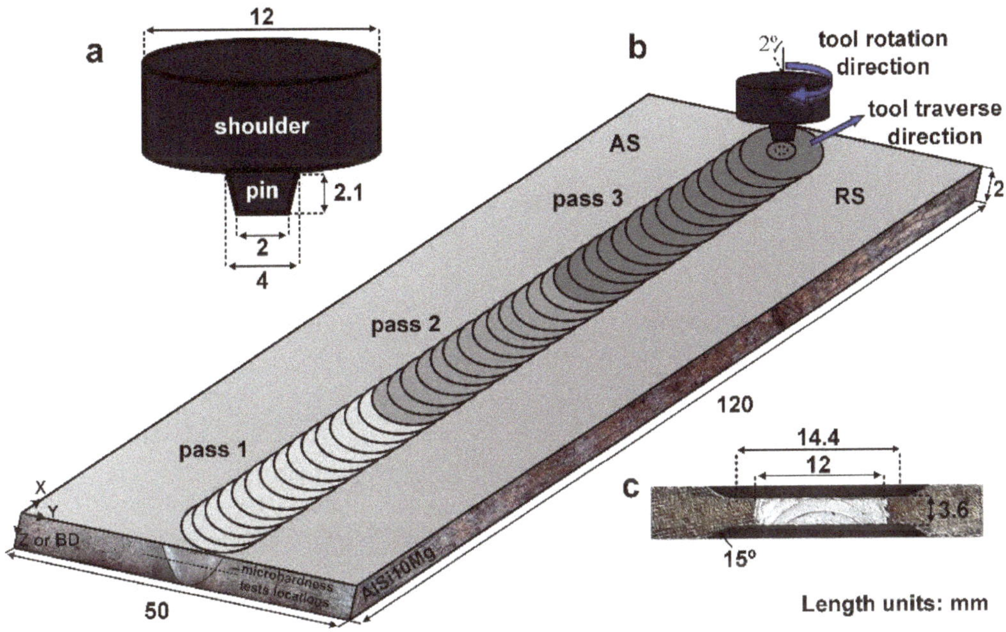

Figure 1. Schematics of (**a**) FSP tool, (**b**) FSP process, and (**c**) tensile test samples.

The macrostructure and microstructure of the joint were characterized using optical microscopy, scanning electron microscopy (SEM) equipped with an electron backscatter diffraction (EBSD) unit, and X-ray diffraction (XRD). Metallographic samples were cut from the joint perpendicular to the FSW direction, polished, and etched with a solution of HNO_3 (5 mL), HCl (3 mL), HF (2 mL), and distilled water (190 mL). EBSD analysis was performed in the SEM with a step size of 0.2 μm, and all corresponding data (that is, inverse pole figure (IPF), grain boundary (GB), grain average misorientation (GAM), and Taylor factor maps) were processed using TSL-OIM software. To identify the grain and sub-grain structures, the boundaries and sub-boundaries were defined based on the misorientation angle as low-angle boundaries (LAGBs: <15°) and high-angle boundaries (HAGBs: >15°). Tensile tests were performed to determine the mechanical properties of the joints and BM. The tensile samples were wire cut according to the JIS no. 7 standard (Figure 1c) and tested at a strain rate of 1 mm/min. Moreover, the fractured surfaces of the tensile samples were characterized using SEM.

3. Results

3.1. Microstructural Zones

The cross-sectional macrostructures of the different FSPed samples are shown in Figure 2, indicating the presence of distinct zones including the BM, transition zone, and stir zone (SZ). It is obvious from the macrostructure images (Figure 3a) that the BM is not affected by heat and deformation during FSP, exhibiting a typical microstructure of L-PBF AlSi10Mg containing overlapping melt pool boundaries (MPBs). However, the transition zone between the BM and SZ (Figure 3b), which is composed of the heat-affected (HAZ) and thermomechanically affected zones (TMAZ), is influenced during FSP. By reaching the center of the SZ, the temperature, strain, and strain rate induced by the rotational tool increased to their maximum values. Hence, the microstructural evolution can be completed in the SZ, whereas it only partially occurs in the transition zone owing to inadequate heat and deformation.

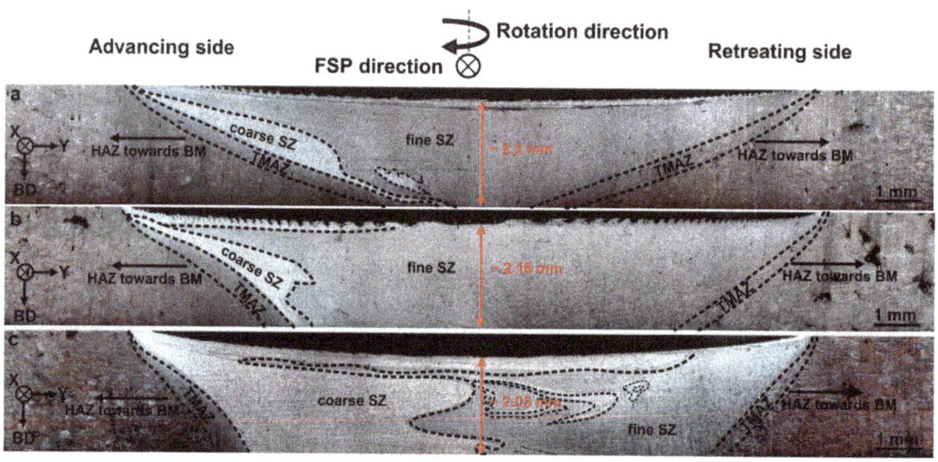

Figure 2. Cross-sectional macrostructure of (**a**) S1, (**b**) S2, and (**c**) S3.

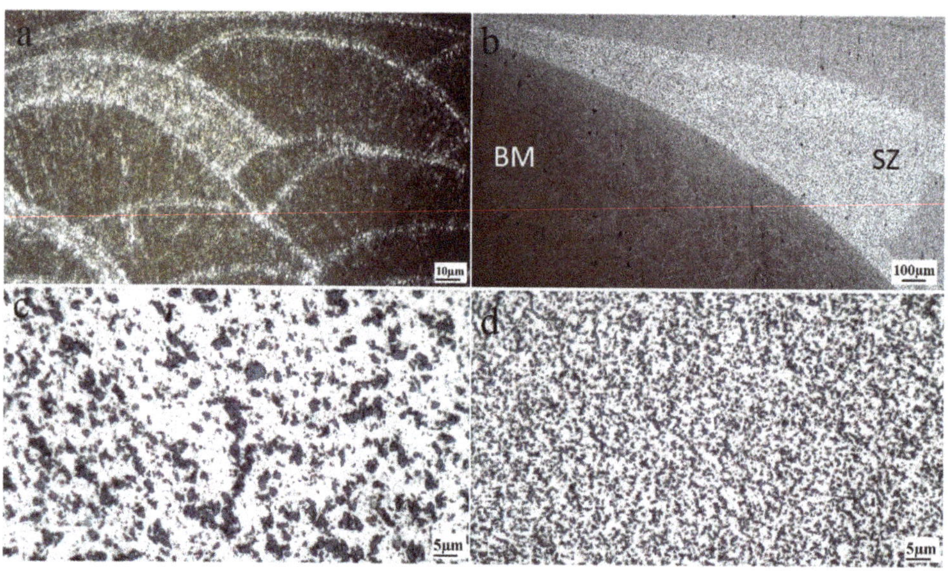

Figure 3. Optical microscopy images showing typical microstructural zones S1: (**a**) BM, (**b**) transition zone between SZ and BM including TMAZ and HAZ, (**c**) coarse SZ, and (**d**) fine SZ. The fine and coarse SZ refer to the zones with fine and coarse Si particles.

From Figure 2, based on the size of the Si particles, the SZ can be divided into two regions: coarse and fine SZs. The coarse SZ is found to be created in the advancing side (between TMAZ and fine SZ) and under the shoulder, where the temperature is higher than other regions [7]. Moreover, the size of the coarse SZ area is increased by multipass FSP (Figure 2). This can be related to the repeated heat input derived from multipass FSP. The microstructures of the coarse and fine SZs (for example in S1) are shown in Figure 3c,d, indicating higher growth of Si particles in the coarse SZ than the fine SZ. As shown in Figure 2, the width and depth of the FSP zone was increased by the multipass FSP. Furthermore, the large L-PBF porosities were eliminated after FSP (Figure 2). The peak temperature in SZ during FSP directly affects the microstructure of the corresponding zones, which depends on the rotational and traverse speeds based on the following equation [6,7]:

$$T = KT_m(\frac{W^2}{10^4 \times V})^{\alpha} \tag{1}$$

where T_m stands for the melting point of the AlSi10Mg (~600 °C), and K and α are the constants in the ranges of 0.04–0.06 and 0.65–0.75, respectively. By inserting the corresponding values to Equation (1), the temperature in SZ was estimated to be ~315 °C, an appropriate temperature for the occurrence of restoration and dynamic softening mechanisms in aluminum alloys [6]. Therefore, the material can flow easily during FSP to fill the porosities.

The SEM micrographs of the melt pool interior (Figure 3) and fine SZs are shown in Figure 4, which confirm the fragmentation of fibrous Si networks (indicated by yellow arrows in Figure 4a) into individual particles (Figure 4b–d) by FSP. In addition, one-pass FSP (S1) caused the finest Si particles among the three FSPed samples, whereas there was no considerable difference between S2 and S3. Importantly, from Figure 4, the presence of fine Si precipitate in the BM is observed (white arrows in Figure 4a in the Al matrix surrounded by Si networks), which can also be found in the SZs (Figure 4b–d).

The XRD patterns (Figure 5) indicate the existence of Mg_2Si precipitates in the BM and SZs in addition to the Si phase, which is well-described in literature [4]. Moreover, the change in (111) Al peak intensity suggests the texture formation during FSP, which is discussed in detail based on EBSD pole figures and inverse pole figures.

3.2. EBSD Maps

As shown in Figure 6a, the BM is composed of two types of grain, including large elongated grains (black arrows) and fine equiaxed grains (white arrows) within a single melt pool, as indicated by the black line area. The average grain size of the BM was 15.8 μm. The elongated grains are mainly formed parallel to the BD, going against the heat transfer direction during the solidification of the melt pools [4]. The formation of fine equiaxed grains occurs when the melt pool is consumed by elongated grains and the temperature gradient decreases to a certain value for the nucleation of equiaxed grains [19]. From Figure 6b, the BM contained 81% and 19% HAGBs and LAGBs, respectively. The grain average misorientation (GAM) map (Figure 6c) for misorientation between 0° and 2° can be employed to qualitatively estimate the dislocation density, with higher GAM values indicating a higher dislocation density in the material [20]. As shown in Figure 6c, in the BM, the elongated grains have larger GAM values than the fine equiaxed grains, and the average GAM value was calculated as 0.53°. The Taylor factor in polycrystalline metals and alloys represents the effect of crystallographic texture on yield strength [21]. A higher Taylor factor indicates higher strength of the metals and alloys. In Figure 6d, the Taylor factor map of the BM is shown, in which the average value of the Taylor factor was calculated to be 2.84.

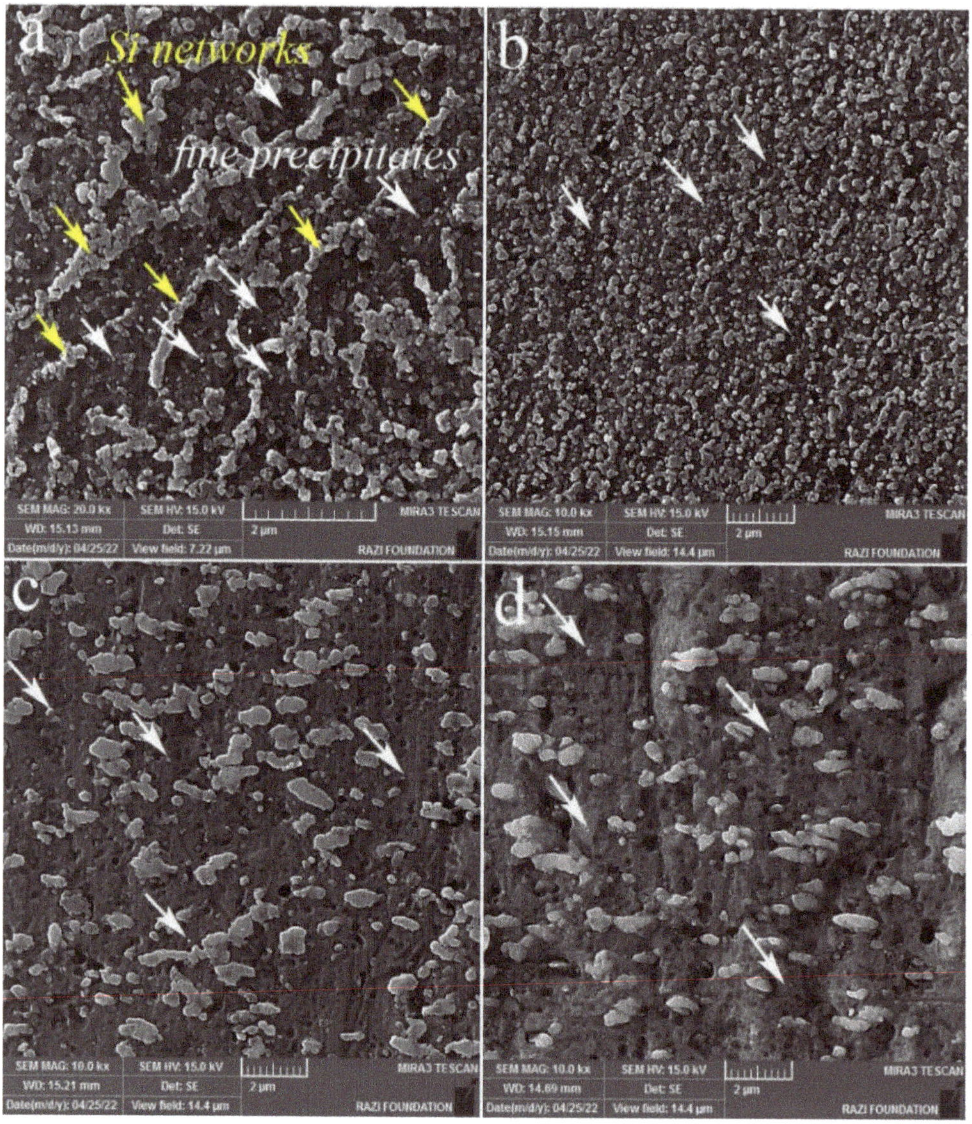

Figure 4. Secondary electron SEM images: (**a**) Si networks in BM, and (**b–d**) Si particles at the center of the SZs (fine SZs) for S1, S2, and S3, respectively. The yellow arrows in (**a**) indicate the fibrous Si networks. The white arrows denote the fine precipitates.

The EBSD maps of S1, S2, and S3 SZs are shown in Figures 7 and 8. From Figure 7, FSP leads to grain refinement in the SZs, and the average grain sizes of S1, S2, and S3 were calculated as 3.0, 3.1, and 2.5 μm, respectively. The formation of fine-grained structures in the SZs of FSPed metals and alloys is related to the occurrence of different DRX mechanisms [6]. According to the GB map of the samples in Figure 7, a large number of grain boundaries (80%, 71%, and 83% for S1, S2, and S3, respectively) have the characteristics of HAGBs.

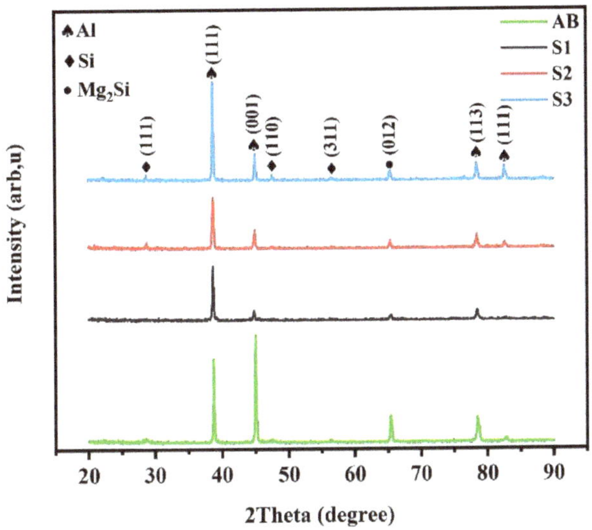

Figure 5. XRD patterns for BM (as-built material: AB) and SZ of different FSPed samples.

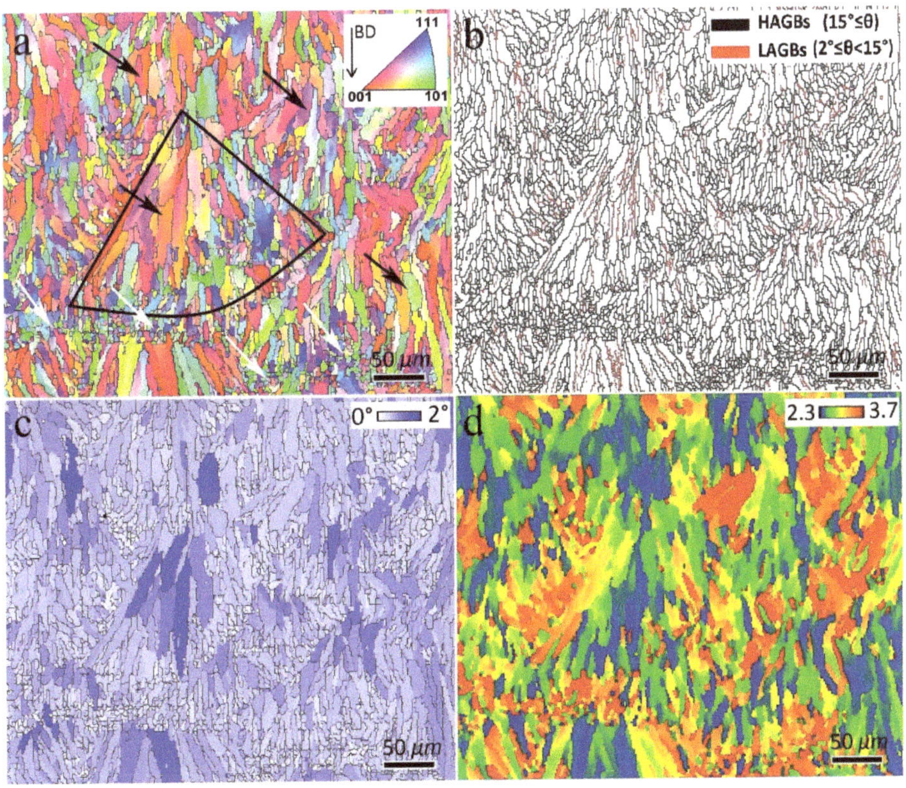

Figure 6. (**a**) Inverse pole figure (IPF), (**b**) Grain boundary (GB), (**c**) Grain average misorientation (GAM), and (**d**) Taylor factor maps of BM. In (**a**,**c**), HAGBs are superimposed as black boundaries. In (**d**), the {111}<110> family of slip systems was considered for calculation of Taylor factor values.

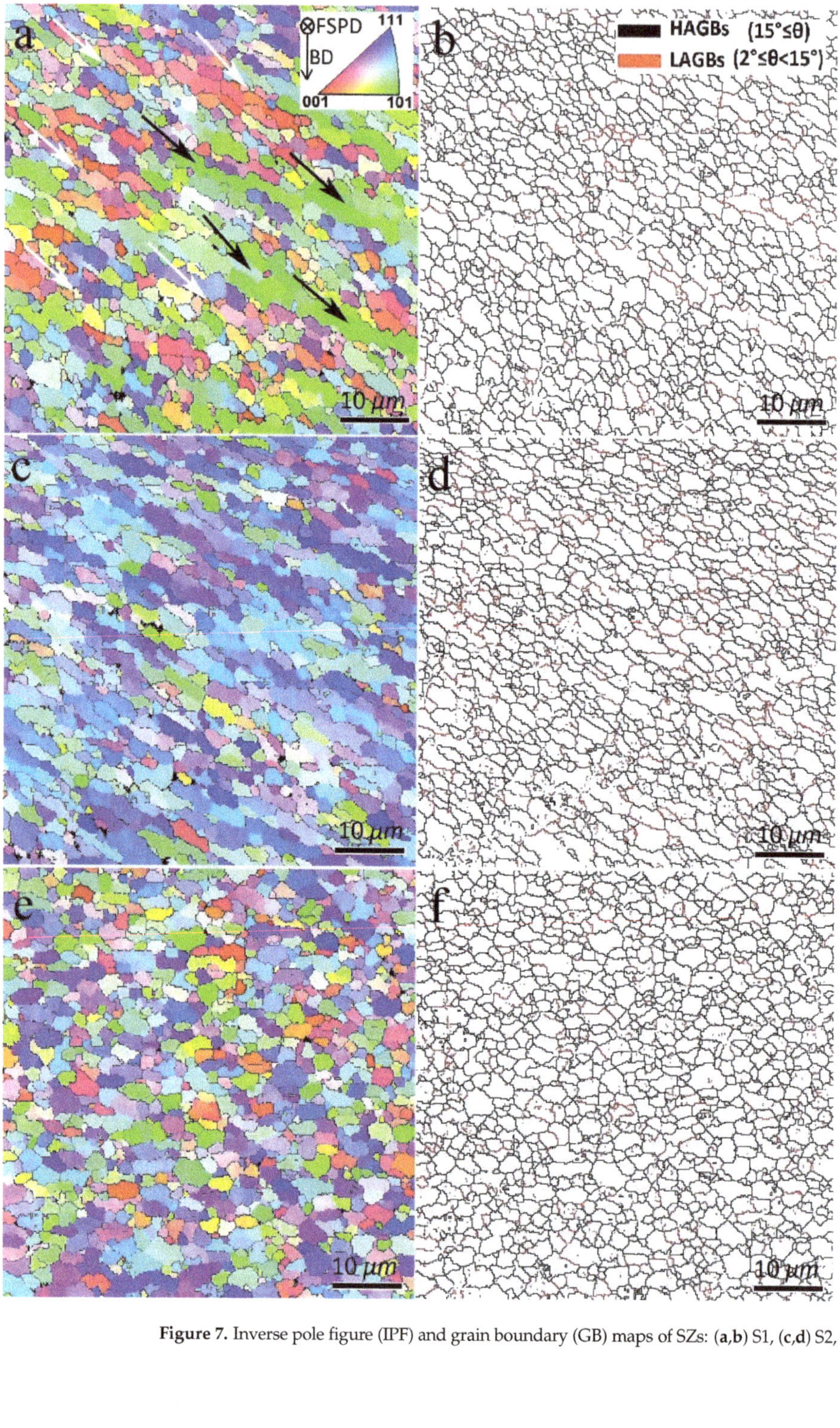

Figure 7. Inverse pole figure (IPF) and grain boundary (GB) maps of SZs: (**a**,**b**) S1, (**c**,**d**) S2, and (**e**,**f**) S3.

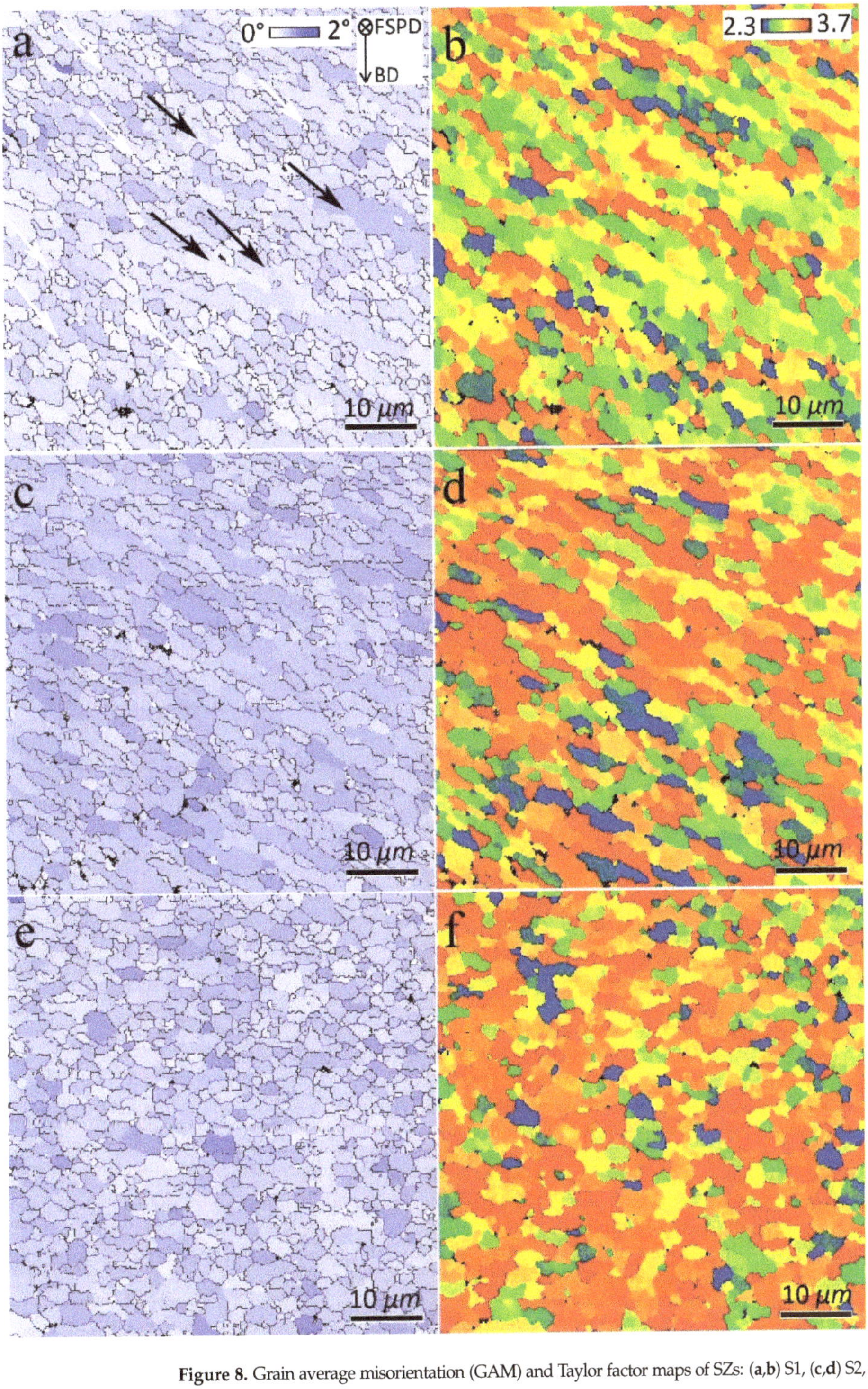

Figure 8. Grain average misorientation (GAM) and Taylor factor maps of SZs: (**a**,**b**) S1, (**c**,**d**) S2, and (**e**,**f**) S3.

The GAM maps of the FSPed samples in Figure 8 and their calculations indicated 0.45°, 0.59°, and 0.54° values for S1, S2, and S3, respectively. Therefore, the first FSP pass resulted in reduced dislocation density. However, more passes of FSP resulted in larger GAM values than those of the BM (Figure 6). Furthermore, from Figures 7 and 8, the formation of a bimodal grain structure in S1 is evident, including fine grains (white arrows in Figures 7 and 8) and elongated large grains (black arrows in Figures 7 and 8), which decreased in S2 and disappeared in S3. According to the GB maps in Figure 7 and the GAM maps in Figure 8, the larger grains in the bimodal structures of S1 and S2 contained substructures such as LAGBs and higher dislocation densities compared to those of sample S3. Analysis of the Taylor factor maps and values (Figure 8) indicated that the bimodal structure was enhanced compared to that of the BM. The Taylor factor values for S1, S2, and S3 are calculated as 2.98, 3.12, and 3.18, respectively.

3.3. Texture

The (001) pole figure (PF) and inverse pole figure (IPF) along the [001] direction of L-PBF AlSi10Mg are shown in Figure 9. It presents the strong texture formation in which the [001] crystallographic direction is aligned with the BD, indicated by red arrows at the center and black arrows at the periphery of PF. The formation of such a texture component is related to the solidification nature of metals and alloys with face-centered cubic (FCC) crystallographic structures. During the solidification of these alloys, the preferred growth directions (<100> directions) are aligned parallel and opposite to the heat transfer direction (that is, //BD) in the melt pools, hence a <100>//BD can be formed (Figure 9b). Texture weakening may occur due to the thermal gradient that is not parallel to the BD, and, depending upon the L-PBF parameters, may have a radial nature [19]; hence, the elongated grains (Figure 6) can be tilted outwards toward the center of the melt pools.

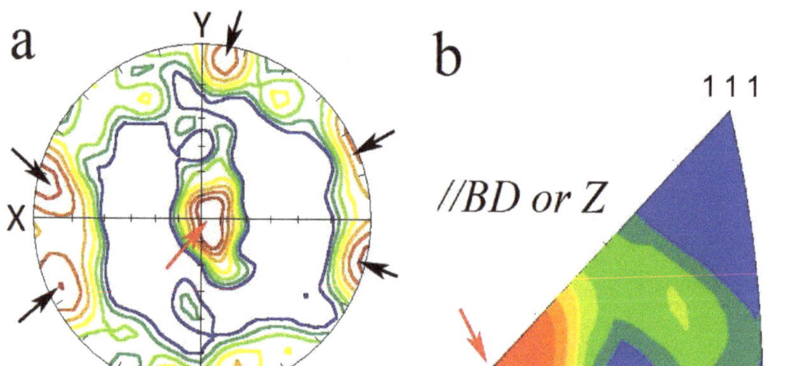

Figure 9. (001) pole figure (**a**) and (001) inverse pole figure (**b**) of BM. X and Y in (**a**) belong to the BM coordination system as indicated in Figure 1b. BD denotes the building direction of the BM during L-PBF. Inverse pole figure in (**b**) was generated parallel to BD.

It is known that FSP causes the formation of shear texture components in the SZ [6]. Therefore, for the analysis of texture components in SZs, they are typically compared with simple shear texture components for FCC metals and alloys, as listed in Table 1. However, owing to the complex deformation in the SZ during FSP, the texture components exhibit a mismatch with those in simple shear deformation. To overcome this problem, the as-acquired PFs should be rotated about the X, Y, and Z axes, which aligns them with a standard frame of reference composed of shear and shear plane normal directions (SD and

SPN) [6]. In Figure 10, the rotated (111) PFs for S1, S2, and S3 and the ideal simple shear texture components of metals with FCC crystallographic structures are illustrated.

Table 1. Ideal texture components with corresponding Euler angles and Miller indices in simple shear deformation of metals with a face-centered cubic (FCC) crystallographic structure [6].

Symbol	Euler Angles (°)			Miller Indices (hkl)<uvw>
	φ_1	Φ	φ_2	
A_1^*	35.26/215.26 125.26	45 90	0/90 45	$(111)[\bar{1}\bar{1}2]$
A_2^*	144.74 54.74/234.74	45 90	0/90 45	$(111)[11\bar{2}]$
A	0	35.26	45	$(1\bar{1}1)[110]$
$\overline{A}$	180	35.26	45	$(\bar{1}1\bar{1})[\bar{1}\bar{1}0]$
B	0/120/240	54.74	45	$(1\bar{1}2)[110]$
$\overline{B}$	60/180	54.74	45	$(\bar{1}1\bar{2})[\bar{1}\bar{1}0]$
C	90/270 0/180	45 90	0/90 45	$\{001\}\langle 110\rangle$

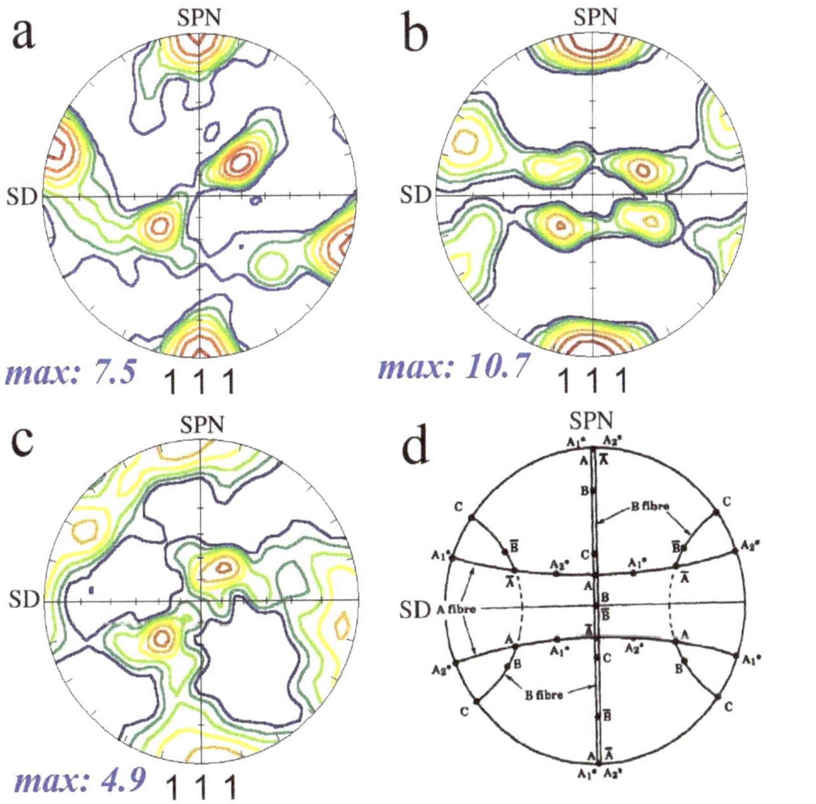

Figure 10. (111) pole figures of different SZs: (**a**) S1, (**b**) S2, and (**c**) S3. SD and SPN denote the shear and shear plane normal directions. (**d**) Ideal simple shear texture components of metals with face-centered crystallographic (FCC) structures [22].

As shown in Figure 10a and Table 1, the first pass of FSP resulted in the formation of the $A_1^*((111)[\overline{1}\overline{1}2])$ shear texture component in the SZ, with an intensity 7.5 times larger than the random texture. The second pass of FSP (Figure 10b) resulted in more intense shear texture components (intensity = 10.7) for both the $A_1^*(111)[\overline{1}\overline{1}2]$ and $A_2^*(111)[11\overline{2}]$. The third pass of FSP (Figure 10c) led to the consumption of A_2^* and a reduction in the texture intensity to 4.9. This texture analysis reveals that, at the first and second passes of the FSP, the grains inherit the shear texture induced by the rotational tool. By repeating the FSP, the texture intensity increases and other types of shear texture component can be formed. However, the third pass of FSP has a negative effect on texture intensity, which implies that DRX gradually becomes the dominant microstructural mechanism by further increasing the number of FSP passes.

3.4. Mechanical Properties

The engineering stress-strain curves of the BM and FSPed samples are illustrated in Figure 11. The results indicated that the as-built BM exhibited high tensile strength (~354 MPa) and low fracture elongation (~8.4%), which is consistent with literature [4,5]. FSP caused a moderate reduction in the ultimate tensile strength (UTS) of the L-PBF AlSi10Mg alloy; however, it resulted in significantly higher fracture elongation in all three passes. Moreover, the first pass of FSP resulted in higher UTS and elongation compared to the second and third passes. It is evident that, by increasing the number of passes, both the UTS and elongation decreased.

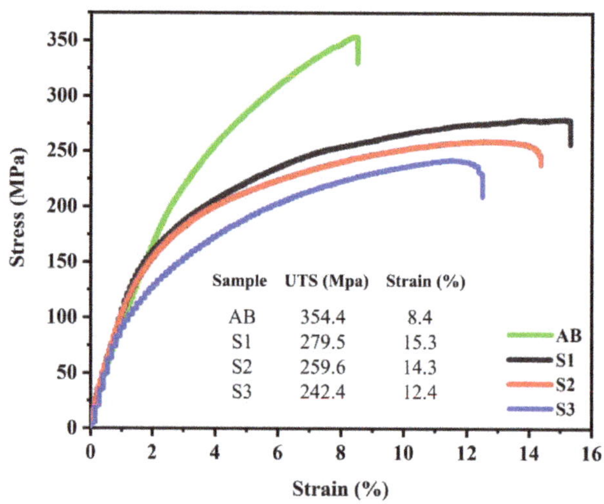

Figure 11. Engineering stress–strain plots of BM and different FSPed samples.

The SEM images of the FSPed fractured surfaces are illustrated in Figure 12, which confirm the ductile fracture of the samples owing to the presence of dimples. In addition, the fracture surface of S1 (Figure 12a,b) contained finer dimples than those of the other samples (Figure 12c–f), which confirms its higher elongation. Moreover, in all cases, the fractured surface was composed of different-sized dimples, owing to the different grain (Figure 7) and Si particle (Figure 4) sizes in the SZs. Figure 13 revealed the presence of large (white arrows in Figure 13a) and fine (yellow arrows in Figure 13a) Si particles on the fractured surfaces of all samples and interior walls of the dimples (yellow arrows in Figure 13b). Thus, it can be concluded that the damage sites during tensile loading were composed mainly of the AlSi10Mg matrix and, to some extent, Si particles. The correlation between microstructural features and tensile properties is discussed in the next section.

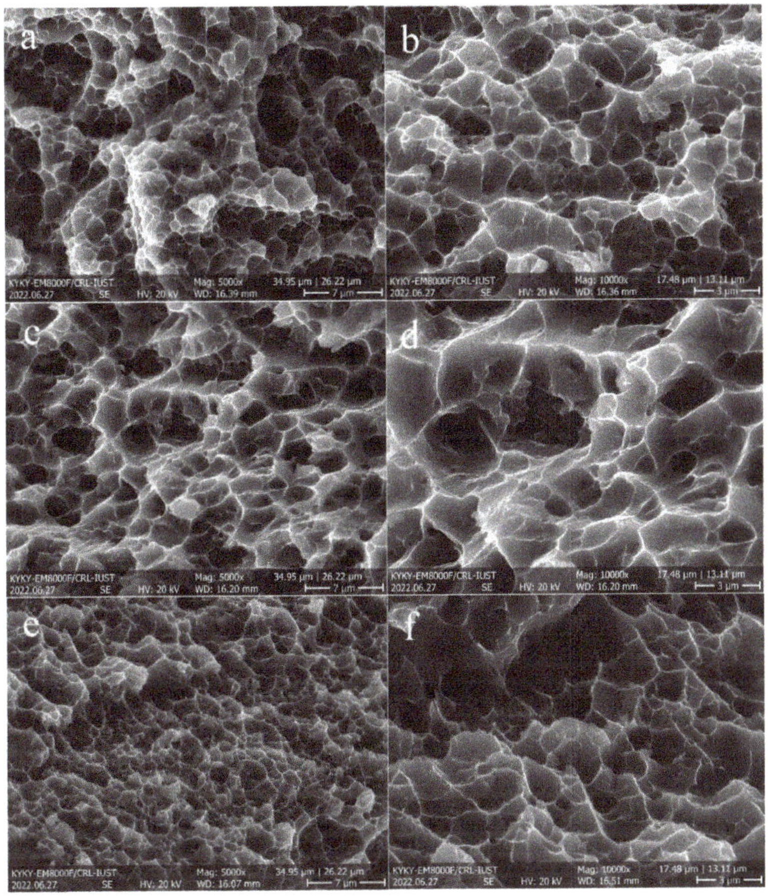

Figure 12. Secondary electron SEM images of fractured surfaces at different magnifications: (**a**,**b**) S1, (**c**,**d**) S2, and (**e**,**f**) S3.

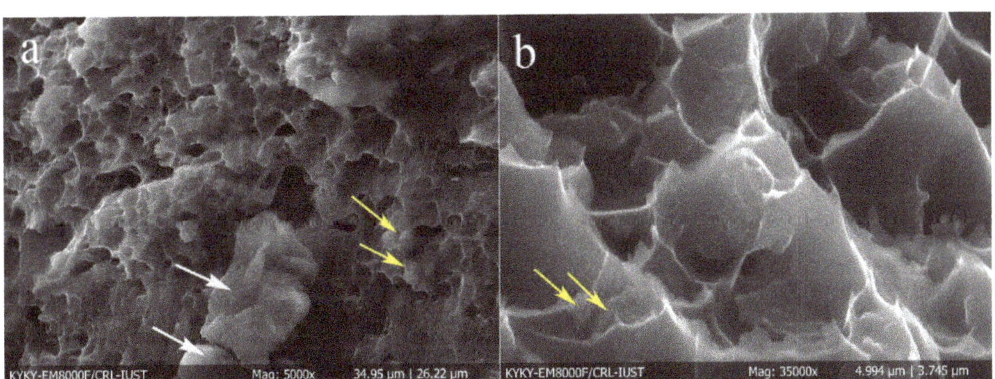

Figure 13. Secondary electron SEM images of fractured surfaces: (**a**) S2, showing the presence of large Si particles, and (**b**) S3, revealing the existence of fine Si particles on the dimple walls.

4. Discussion

The FSP reduced the strength but enhanced the ductility of L-PBF AlSi10Mg, while one pass of FSP exhibited higher strength and ductility compared to higher numbers (two and three) of FSP passes. The origin of this behavior is discussed in this section. The strength of a polycrystalline metal can be attributed to various mechanisms, such as grain boundary, solid solution, secondary phase or precipitation, dislocation density, and texture-strengthening mechanisms [21]. For brevity, the EBSD data are summarized in Table 2, distinguishing the strengthening mechanisms.

Table 2. EBSD data of the grain and texture characteristics in the BM and SZs. The data comes to the fine SZs as shown in cross-sectional macrostructures (Figure 2).

Sample	D_{av} (μm)	HAGBs (%)	LAGBs (%)	GAM Value (°)	Taylor Factor	Texture Components	Texture Intensity
BM	15.8	81	19	0.53	2.84	BD//[001]	3.0
S1 (1 pass)	3.0	80	20	0.45	2.98	A_1^*	7.6
S2 (2 pass)	3.1	71	29	0.59	3.12	A_1^* and A_2^*	10.7
S3 (3 pass)	2.5	83	17	0.54	3.18	A_1^*	4.9

According to Table 2, the as-built BM of the L-PBF AlSi10Mg alloy has a larger grain size (15.8 μm) and lower Taylor factor (2.84) compared to the FSPed samples; however, they exhibit higher strength values (Figure 11). The main reason for the higher strength of the BM is the effect of the eutectic Si networks (Figure 4a), which act as barriers to dislocation movement. The strengthening effect of these Si networks is similar to that of the grain boundaries [23,24]; that is, narrower Si networks are more effective in strengthening the L-BPF AlSi10Mg.

It is evident from Table 2 that, despite the lower GAM and Taylor factor values of S1, it possesses a higher strength than S2 and S3 (Figure 11). The grain size of S1 is similar to that of S2, and it is larger than that of S3, suggesting that the effect of grain size cannot be the dominant factor on the strength of the FSPed samples. In addition, the same trend was observed in the case of HAGB values. Thus, it can be concluded that the main strengthening mechanism in the FSPed sample is not attributed to those Si particles activated in the Al matrix but is related to their nature. As shown in Figure 4b, the first pass of FSP resulted in the formation of finer Si particles compared to the second and third passes. The finer Si particles are more effective in hindering dislocation movement [25], thus increasing the strength of S1. In the SZ of the S2 and S3 FSPed samples, larger Si particles were formed (Figure 4c,d), indicating Si growth and coarsening. The activation energy for the growth and coarsening of Si particles is the high number of solute Si atoms trapped in the Al matrix because of the high cooling rate during the non-equilibrium solidification of L-PBF AlSi10Mg [15]. During the FSP, the temperature induced by the rotational tool caused the precipitation of supersaturated Si particles. The newly precipitated Si in conjunction with fine Si particles (primarily present in the Al matrix) join together (coalescence mechanism) or join the surface of the larger Si particles, resulting in the continuous growth of Si particles. The fine Si particles (red arrows) on the surface of a large particle and the coalescence mechanism (blue arrows) on the fractured surface of the FSPed samples confirm the suggested growth mechanism (Figure 14).

This growth mechanism has also been suggested by other studies for the heat treatment of L-PBF AlSi10Mg [2]. More FSP passes provide more time at high temperatures for the diffusion of Si solute atoms, which helps explain the growth mechanism. It is worth noting that the growth of Si particles also has an indirect negative effect on the solid solute strengthening mechanism, as the amount of solute Si atoms is reduced in the Al matrix. For example, energy-dispersive X-ray spectroscopy (EDS) point analysis showed that the amount of solute Si in the matrix was reduced from 2.7 wt.% (for as-built material) to 1.9 wt.% (for the S1 sample). Notably, the lower strength of S3 compared to the other samples is related to the cross-sectional macrostructures (Figure 2), in which the coarse SZ of S3 is larger than that of the other samples. All the tensile samples were broken from the center of the SZs, while for S3 the large coarse SZ adversely affected tensile strength and ductility. This coarse SZ effect was negligible in the cases of S1 and S2 because of the existence of fine SZs in the broken areas of their tensile samples. It is worth noting that the results of hardness measurements confirmed the measured tensile strengths of the samples. The hardness profiles of the samples are illustrated in Figure 15, showing that S1 exhibited the highest SZ hardness values among the samples. In addition, in all cases, FSP induced a sharp reduction in hardness from BM to the SZs.

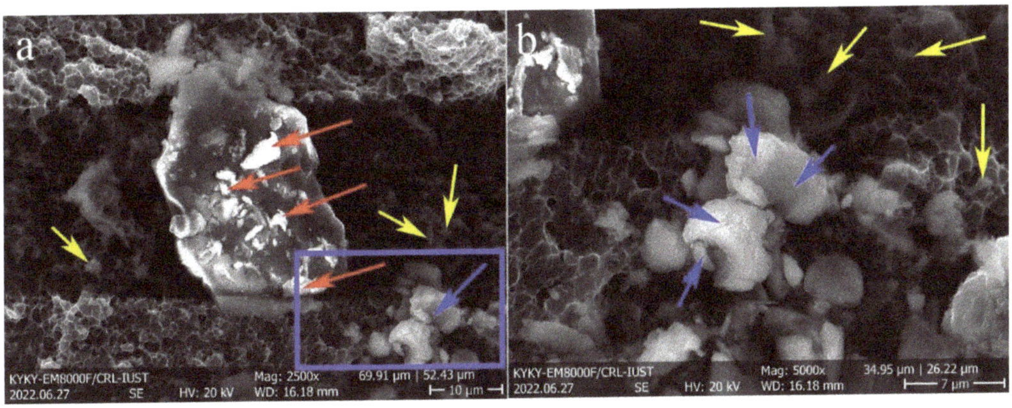

Figure 14. Secondary electron SEM images of the (**a**) fractured surface of FSPed samples, and (**b**) magnified view of the blue rectangle area in (**a**). The red, blue, and yellow arrows refer to the sticking of fine Si particles to the surface of a large one, coalescence of particles, and fine Si particles preset in dimples, respectively.

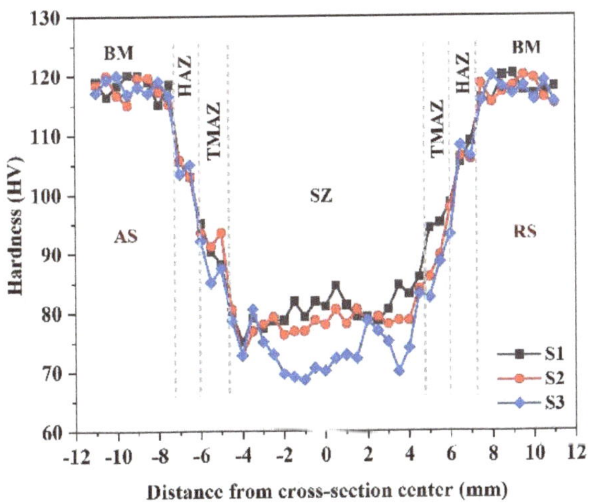

Figure 15. Hardness profiles of BM and different FSPed samples.

The higher ductility of FSPed samples compared to that of the BM (Figure 11) is related to the finer grain sizes of the SZs (Table 2) and elimination of large L-PBF porosities by FSP, as shown in the cross-sectional macrostructures (Figure 2). For more detail, the higher magnified OM images of porosities in BM and their absence after one-pass FSP are shown in Figure 16.

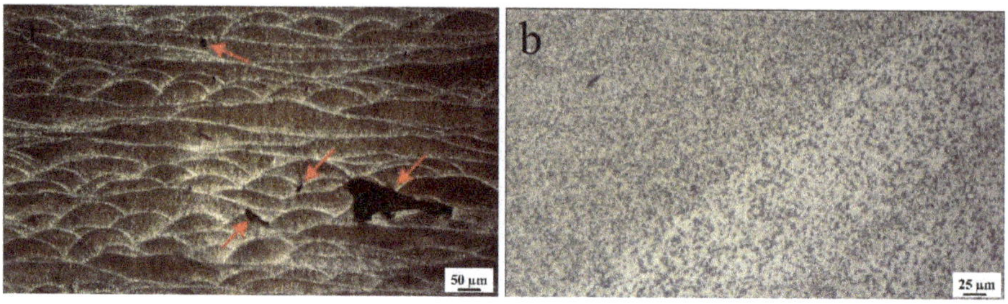

Figure 16. Optical microscope (OM) images of (**a**) large porosities in BM and (**b**) SZ of one-pass FSPed samples showing the absence of large porosities. The red arrows refer to the large porosities in BM (**a**).

In the case of the FSPed samples, two main reasons can be explored for the higher ductility of S1, as listed in Table 2. Raising the number of FSP passes increased the GAM value from 0.45° (S1) to 0.59° (S2). The higher GAM values indicate a qualitatively higher dislocation density in S2 and S3. Thus, dislocation movement can be hindered owing to their interaction during tensile loading more rapidly than in S1, thus ductility decreases. Notably, the GAM value refers to the effect of the Al matrix on ductility. The kernel average misorientation (KAM) map can be employed to determine the effect of Si particles on ductility. In KAM maps, the misorientation among a grain at the center of the kernel and all points at the border of the kernel is measured. The KAM value of the center point is the average of these misorientations, which indicates the amount of stress concentration [26,27]. The more uniform distribution and lower KAM values indicate less-concentrated stresses in the samples, which causes delayed necking and higher ductility [15]. The KAM maps of the SZs in different samples are illustrated in Figure 17, which shows the existence of the initial concentrated stress in all samples. However, for S1, the KAM values were lower and more uniform than those of the other samples. During tensile loading, the different deformations of the hard Si particles and soft Al matrix cause inhomogeneity in the plastic strain. To accommodate this inhomogeneity, geometrically necessary dislocations (GNDs) are formed, hence a strain gradient or stress concentration will develop in the interfacial zones. The primary high and non-uniform KAM values in S2 and S3 can promote the formation of a strain gradient during tensile loading, hence lower ductility. Finally, from the perspective of the Al matrix, owing to the lower amount of GAM in S1, it has a higher strain-hardening capacity compared to other samples, therefore causing higher ductility.

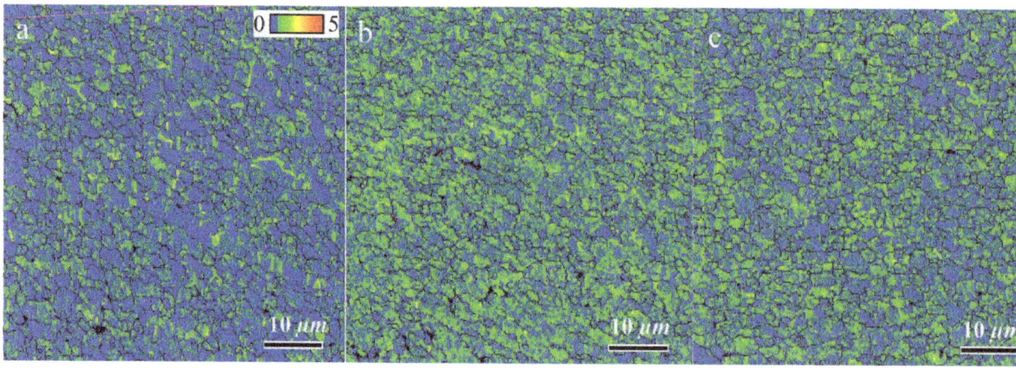

Figure 17. Kernel average misorientation (KAM) maps of SZs in different samples: (**a**) S1, (**b**) S2, and (**c**) S3.

5. Conclusions

Multipass FSP was used to modify the microstructure and mechanical properties of the L-PBF AlSi10Mg plates. The following conclusions were made:

(1) FSP causes the formation of different microstructural zones, such as the BM, HAZ, TMAZ, and SZ, in various metals and alloys, as reported in literature. However, in the case of L-PBF AlSi10Mg, the SZs comprised two distinct fine and coarse regions in accordance with the size of the Si particles. The amount of coarse SZ in the one- and two-pass FSPed samples was negligible. However, in the three-pass FSPed sample, a significant part of the SZ belonged to the coarse SZ.

(2) The elongated grains and BD//[001] texture in the as-built AlSi10Mg disappeared after FSP, and fine equiaxed grains with shear texture components were formed. In addition, the fiber-like Si networks were replaced with Si particles because of their fragmentation during the FSP.

(3) The main reason for the higher strength of the as-built AlSi10Mg compared to those of the FSPed samples is the effect of eutectic Si networks, which act as barriers to dislocation movement, such as the effect of grain boundaries on the strength of metals and alloys based on the Hall–Petch relationship.

(4) The main strengthening mechanisms in the FSPed samples were not related to the Al matrix; however, they were attributed to the Si particles. By repeating the FSP (two and three passes), the Si particles grow by consuming the fine Si particles and Si solute atoms in the Al matrix, resulting in larger and fewer particles. The finer Si particles and higher solute Si atoms in the one-pass FSPed sample led to higher tensile strength compared with the other samples.

(5) Finer grain sizes of SZs and elimination of large L-PBF porosities by FSP are the main reasons for the higher ductility of the FSPed samples compared to that of the as-built AlSi10Mg. The higher ductility of the one-pass FSPed sample can be attributed to the lower and uniform distribution of KAM values and lower GAM values. Notably, in the case of three-pass FSP, the existence of a severely coarse SZ in the macrostructure is the additional origin of lower ductility.

(6) The outcome of this study can be useful for the material science community, particularly for studies investigating the post-treatment of L-PBF AlSi10Mg.

Author Contributions: A.H.: methodology, investigation, formal analysis, writing—original draft; M.K.: methodology, investigation; formal analysis; R.M.: methodology, investigation, formal analysis; R.K.: methodology, investigation, formal analysis; M.M.: methodology, investigation; M.J.: conceptualization, formal analysis, writing—review and editing; X.-G.C.: conceptualization, validation, writing—review and editing, project administration. All authors have read and agreed to the published version of the manuscript.

Funding: This research was partially funded by the Natural Sciences and Engineering Research Council of Canada (NSERC) under Grant No. CRDPJ 514651-17.

Institutional Review Board Statement: Not applicable.

Informed Consent Statement: Not applicable.

Data Availability Statement: The raw/processed data required to reproduce these findings cannot be shared at this time as the data also forms part of an ongoing study.

Conflicts of Interest: The authors declare no conflict of interest.

References

1. Ngo, T.D.; Kashani, A.; Imbalzano, G.; Nguyen, K.T.Q.; Hui, D. Additive manufacturing (3D printing): A review of materials, methods, applications and challenges. *Compos. Part B Eng.* **2018**, *143*, 172–196. [CrossRef]
2. Zhang, J.; Song, B.; Wei, Q.; Bourell, D.; Shi, Y. A review of selective laser melting of aluminum alloys: Processing, microstructure, property and developing trends. *J. Mater. Sci. Technol.* **2019**, *35*, 270–284. [CrossRef]
3. Dowling, L.; Kennedy, J.; O'Shaughnessy, S.; Trimble, D. A review of critical repeatability and reproducibility issues in powder bed fusion. *Mater. Des.* **2020**, *186*, 108346. [CrossRef]

4. Zhao, L.; Song, L.; Macías, J.G.S.; Zhu, Y.; Huang, M.; Simar, A.; Li, Z. Review on the correlation between microstructure and mechanical performance for laser powder bed fusion AlSi10Mg. *Addit. Manuf.* **2022**, *56*, 102914. [CrossRef]
5. Blakey-Milner, B.; Gradl, P.; Snedden, G.; Brooks, M.; Pitot, J.; Lopez, E.; Leary, M.; Berto, F.; du Plessis, A. Metal additive manufacturing in aerospace: A review. *Mater. Des.* **2021**, *209*, 110008. [CrossRef]
6. Heidarzadeh, A.; Mironov, S.; Kaibyshev, R.; Çam, G.; Simar, A.; Gerlich, A.; Khodabakhshi, F.; Mostafaei, A.; Field, D.P.; Robson, J.D.; et al. Friction stir welding/processing of metals and alloys: A comprehensive review on microstructural evolution. *Prog. Mater. Sci.* **2021**, *117*, 100752. [CrossRef]
7. Mishra, R.S.; Ma, Z.Y. Friction stir welding and processing. *Mater. Sci. Eng. R Rep.* **2005**, *50*, 1–78. [CrossRef]
8. Meng, X.; Huang, Y.; Cao, J.; Shen, J.; dos Santos, J.F. Recent progress on control strategies for inherent issues in friction stir welding. *Prog. Mater. Sci.* **2021**, *115*, 100706. [CrossRef]
9. Yang, T.; Wang, K.; Wang, W.; Peng, P.; Huang, L.; Qiao, K.; Jin, Y. Effect of friction stir processing on microstructure and mechanical properties of AlSi10Mg aluminum alloy produced by selective laser melting. *JOM* **2019**, *71*, 1737–1747. [CrossRef]
10. Maamoun, A.H.; Veldhuis, S.C.; Elbestawi, M. Friction stir processing of AlSi10Mg parts produced by selective laser melting. *J. Mater. Process. Technol.* **2019**, *263*, 308–320. [CrossRef]
11. Zhao, L.; Macías, J.G.S.; Ding, L.; Idrissi, H.; Simar, A. Damage mechanisms in selective laser melted AlSi10Mg under as built and different post-treatment conditions. *Mater. Sci. Eng. A* **2019**, *764*, 138210. [CrossRef]
12. Zhao, L.; Macías, J.G.S.; Dolimont, A.; Simar, A.; Rivière-Lorphèvre, E. Comparison of residual stresses obtained by the crack compliance method for parts produced by different metal additive manufacturing techniques and after friction stir processing. *Addit. Manuf.* **2020**, *36*, 101499. [CrossRef]
13. Macías, J.G.S.; Elangeswaran, C.; Zhao, L.; van Hooreweder, B.; Adrien, J.; Maire, E.; Buffière, J.-Y.; Ludwig, W.; Jacques, P.J.; Simar, A. Ductilisation and fatigue life enhancement of selective laser melted AlSi10Mg by friction stir processing. *Scr. Mater.* **2019**, *170*, 124–128. [CrossRef]
14. Rafieazad, M.; Mohammadi, M.; Gerlich, A.; Nasiri, A. Enhancing the corrosion properties of additively manufactured AlSi10Mg using friction stir processing. *Corros. Sci.* **2021**, *178*, 109073. [CrossRef]
15. Nouri, Z.; Taghiabadi, R. Tribological properties improvement of conventionally-cast Al-8.5 Fe-1.3 V-1.7 Si alloy by multi-pass friction stir processing. *Trans. Nonferrous Met. Soc. China* **2021**, *31*, 1262–1275. [CrossRef]
16. Nakata, K.; Kim, Y.G.; Fujii, H.; Tsumura, T.; Komazaki, T. Improvement of mechanical properties of aluminum die casting alloy by multi-pass friction stir processing. *Mater. Sci. Eng. A* **2006**, *437*, 274–280. [CrossRef]
17. Moharrami, A.; Razaghian, A.; Paidar, M.; Šlapáková, M.; Ojo, O.O.; Taghiabadi, R. Enhancing the mechanical and tribological properties of Mg2Si-rich aluminum alloys by multi-pass friction stir processing. *Mater. Chem. Phys.* **2020**, *250*, 123066. [CrossRef]
18. Singh, S.K.; Immanuel, R.J.; Babu, S.; Panigrahi, S.K.; Janaki Ram, G.D. Influence of multi-pass friction stir processing on wear behaviour and machinability of an Al-Si hypoeutectic A356 alloy. *J. Mater. Process. Technol.* **2016**, *236*, 252–262. [CrossRef]
19. Aboulkhair, N.T.; Simonelli, M.; Parry, L.; Ashcroft, I.; Tuck, C.; Hague, R. 3D printing of Aluminium alloys: Additive Manufacturing of Aluminium alloys using selective laser melting. *Prog. Mater. Sci.* **2019**, *106*, 100578. [CrossRef]
20. Tian, Y.Z.; Gao, S.; Zhao, L.J.; Lu, S.; Pippan, R.; Zhang, Z.F.; Tsuji, N. Remarkable transitions of yield behavior and Lüders deformation in pure Cu by changing grain sizes. *Scr. Mater.* **2018**, *142*, 88–91. [CrossRef]
21. Starink, M.J.; Wang, S.C. A model for the yield strength of overaged Al–Zn–Mg–Cu alloys. *Acta Mater.* **2003**, *51*, 5131–5150. [CrossRef]
22. Heidarzadeh, A.; Laleh, H.M.; Gerami, H.; Hosseinpour, P.; Shabestari, M.J.; Bahari, R. The origin of different microstructural and strengthening mechanisms of copper and brass in their dissimilar friction stir welded joint. *Mater. Sci. Eng. A* **2018**, *735*, 336–342. [CrossRef]
23. Macías, J.G.S.; Douillard, T.; Zhao, L.; Maire, E.; Pyka, G.; Simar, A. Influence on microstructure, strength and ductility of build platform temperature during laser powder bed fusion of AlSi10Mg. *Acta Mater.* **2020**, *201*, 231–243. [CrossRef]
24. Hadadzadeh, A.; Baxter, C.; Amirkhiz, B.S.; Mohammadi, M. Strengthening mechanisms in direct metal laser sintered AlSi10Mg: Comparison between virgin and recycled powders. *Addit. Manuf.* **2018**, *23*, 108–120. [CrossRef]
25. Fiocchi, J.; Tuissi, A.; Biffi, C.A. Heat treatment of aluminium alloys produced by laser powder bed fusion: A review. *Mater. Des.* **2021**, *204*, 109651. [CrossRef]
26. Rui, S.-S.; Niu, L.-S.; Shi, H.-J.; Wei, S.; Tasan, C.C. Diffraction-based misorientation mapping: A continuum mechanics description. *J. Mech. Phys. Solids* **2019**, *133*, 103709. [CrossRef]
27. Lehto, P. Adaptive domain misorientation approach for the EBSD measurement of deformation induced dislocation sub-structures. *Ultramicroscopy* **2021**, *222*, 113203. [CrossRef]

materials

MDPI

Article

Formation of TiB_2–$MgAl_2O_4$ Composites by SHS Metallurgy

Chun-Liang Yeh * and Fu-You Zheng

Department of Aerospace and Systems Engineering, Feng Chia University, Taichung 40724, Taiwan
* Correspondence: clyeh@fcu.edu.tw; Tel.: +886-4-2451-7250 (ext. 3963)

Abstract: TiB_2–$MgAl_2O_4$ composites were fabricated by combustion synthesis involving metallothermic reduction reactions. Thermite reagents contained Al and Mg as dual reductants and TiO_2 or B_2O_3 as the oxidant. The reactant mixtures also comprised elemental Ti and boron, as well as a small amount of Al_2O_3 or MgO to serve as the combustion moderator. Four reaction systems were conducted and all of them were exothermic enough to proceed in the mode of self-propagating high-temperature synthesis (SHS). The reaction based on B_2O_3/Al/Mg thermite and diluted with MgO was the most exothermic, while that containing TiO_2/Al/Mg thermite and Al_2O_3 as the diluent was the least. Depending on different thermites and diluents, the combustion front temperatures in a range from 1320 to 1720 °C, and combustion wave velocity from 3.9 to 5.7 mm/s were measured. The XRD spectra confirmed in situ formation of TiB_2 and $MgAl_2O_4$. It is believed that $MgAl_2O_4$ was synthesized through a combination reaction between Al_2O_3 and MgO, both of which can be totally or partially produced from the metallothermic reduction of B_2O_3 or TiO_2. The microstructure of the TiB_2–$MgAl_2O_4$ composite exhibited fine TiB_2 crystals surrounded by large densified $MgAl_2O_4$ grains. This study demonstrated an energy-saving and efficient route for fabricating $MgAl_2O_4$-containing composites.

Keywords: $MgAl_2O_4$; TiB_2; metallothermic reduction; SHS powder metallurgy; combustion wave propagation

Citation: Yeh, C.-L.; Zheng, F.-Y. Formation of TiB_2–$MgAl_2O_4$ Composites by SHS Metallurgy. *Materials* **2023**, *16*, 1615. https://doi.org/10.3390/ma16041615

Academic Editor: Dina V. Dudina

Received: 30 January 2023
Revised: 13 February 2023
Accepted: 13 February 2023
Published: 15 February 2023

1. Introduction

TiB_2 has been one of the most studied ultra-high temperature ceramics (UHTCs) due to its unique properties, including a high melting point (3225 °C), high hardness (33 MPa), high Young's modulus (530 MPa), excellent wear and oxidation resistance, thermal shock resistance, chemical inertness, and good electric conductivity [1–3]. Combination of these properties makes TiB_2 an ideal candidate for use in ballistic armors, crucibles, metal evaporation boats, cutting tools, wear resistance parts, and cathodes for alumina smelting [4–6]. Many ceramic phases, such as Al_2O_3, SiC, B_4C, and $MgAl_2O_4$, have been considered as the reinforcement to improve fracture toughness, oxidation resistance, heat resistance, and mechanical strength of the TiB_2-based composites [7–10]. Moreover, a recent study showed that TiB_2–Al_2O_3–$MgAl_2O_4$ composite possesses temperature insensitive and enhanced microwave absorption properties [11]. Magnesium aluminate spinel, $MgAl_2O_4$, as an additive has been rarely studied, possibly because the fabrication of $MgAl_2O_4$ via either the direct solid-state reaction of oxides or wet chemical methods requires multiple steps that are complicated and time-consuming [12–14]. However, $MgAl_2O_4$ is an attractive component due to its high melting point, chemical inertness, high hardness, corrosion resistance, high mechanical strength, and low cost [12].

Among various fabrication routes for preparing multiphasic ceramics, metallothermic reduction reactions (i.e., thermite reactions) combined with combustion synthesis have been recognized as a promising technique for in situ formation of $MgAl_2O_4$-containing composites [15]. Combustion synthesis in the mode of self-propagating high-temperature synthesis (SHS), which is based on strongly exothermic reactions, has merits of low energy consumption, short reaction time, simple equipment and operation, high-purity products, and in situ formation of composite components [16–18]. Moreover, aluminothermic and magnesiothermic reduction reactions have Al_2O_3 and MgO as their respective by-products, both of which are precursors for the formation of $MgAl_2O_4$. Consequently, Omran et al. [19] applied the reduction-based SHS technique to produce $MgAl_2O_4$–W–W_2B composites through magnesiothermic reduction of B_2O_3 and WO_3 in the presence of Al_2O_3. Zaki et al. [20] synthesized $MoSi_2$– and Mo_5Si_3–$MgAl_2O_4$ composites by SHS with a reducing stage from raw materials consisting of MoO_3, SiO_2, and Al as aluminothermic reagents and MgO as a precursor. Similarly, MgO was added into the reactive mixture of TiO_2, B_2O_3, and Al to fabricate TiB_2–$MgAl_2O_4$ composites by thermitic combustion synthesis [21]. Generally, most of the previous studies on the formation of $MgAl_2O_4$-containing composites had prior addition of one of two precursors (Al_2O_3 or MgO) in the green samples.

This study represents the first attempt to prepare TiB_2–$MgAl_2O_4$ composites from the SHS powder metallurgy simultaneously involving aluminothermic and magnesiothermic reduction of TiO_2 or B_2O_3. Only a small amount of Al_2O_3 or MgO was included in the reactive mixture to serve as the combustion moderator and part of the precursors for the formation of $MgAl_2O_4$. Four SHS reaction systems formulated with different metallothermic reagents and combustion diluents were investigated. In this work, combustion exothermicity and kinetics of the combustion wave of the SHS process, as well as compositions and microstructures of the final products were explored.

2. Materials and Methods

The starting materials adopted by this study included TiO_2 (Acros Organics, Geel, Belgium, 99.5%), B_2O_3 (Acros Organics, 99%), Al_2O_3 (Alfa Aesar, Haverhill, MA, USA, 99%), MgO (Acros Organics, 99.5%), Al (Showa Chemical Co., Tokyo, Japan, <45 μm, 99.9%), Mg (Alfa Aesar, <45 μm, 99.8%), Ti (Alfa Aesar, <45 μm, 99.8%), and amorphous boron (Noah Technologies, San Antonio, TX, USA, <1 μm, 93.5%). Four SHS reactions were formulated for the synthesis of $3TiB_2$–$MgAl_2O_4$ composites. Two metallothermic reagents (i.e., thermites) were considered; one is composed of TiO_2, Al, and Mg, as shown in Equations (1) and (2), and the other comprises B_2O_3, Al, and Mg, as in Equations (3) and (4). Due to strong exothermicity of combustion, Al_2O_3 with an amount of 0.3 mol. was included in Equations (1) and (3) as the combustion moderator (or combustion diluent) in order to attain stable propagation of the combustion wave. The pre-added Al_2O_3 also acted as part of the precursor for the synthesis of $MgAl_2O_3$. Likewise, an equal amount of MgO was adopted by Equations (2) and (4) and MgO played the same role as Al_2O_3 in Equations (1) and (3).

$$(1.55TiO_2 + 1.4Al + Mg) + 0.3Al_2O_3 + 1.45Ti + 6B \rightarrow 3TiB_2 + MgAl_2O_4 \quad (1)$$

$$(1.85TiO_2 + 2Al + 0.7Mg) + 0.3MgO + 1.15Ti + 6B \rightarrow 3TiB_2 + MgAl_2O_4 \quad (2)$$

$$(1.033B_2O_3 + 1.4Al + Mg) + 0.3Al_2O_3 + 3Ti + 3.934B \rightarrow 3TiB_2 + MgAl_2O_4 \quad (3)$$

$$(1.233B_2O_3 + 2Al + 0.7Mg) + 0.3MgO + 3Ti + 3.534B \rightarrow 3TiB_2 + MgAl_2O_4 \quad (4)$$

Combustion exothermicity of the above four reactions, Equations (1)–(4), was evaluated by calculating their adiabatic combustion temperatures (T_{ad}) from the following energy balance equation [17,22] with thermochemical data taken from [23].

$$\Delta H_r + \int_{298}^{T_{ad}} \sum n_j c_p(P_j)dT + \sum_{298-T_{ad}} n_j L(P_j) = 0$$

where ΔH_r is the reaction enthalpy at 298 K, c_p and L are the heat capacity and latent heat, n_j is the stoichiometric coefficient, and P_j refers to the product component.

The SHS experiments were conducted in a windowed combustion chamber filled with Ar at 0.25 MPa. Reactant powders were well mixed in a tubular ball mill and then uniaxially compressed into cylindrical test specimens with a diameter of 7 mm, a height of

12 mm, and a relative density of 55%. The sample compact was ignited by an electrically heated tungsten coil. An R-type bare-wire thermocouple (Pt/Pt-13%Rh) with a bead size of 125 μm was used to measure the combustion temperature. The propagation velocity of combustion wave (V_f) was determined by the time derivative of the flame-front trajectory constructed from the recorded series of combustion images. Phase compositions of the products were identified by an X-ray diffractometer (XRD, Bruker D2 Phaser, Karlsruhe, Germany). Microstructures and constituent elements of the products were examined by the scanning electron microscopy (SEM, Hitachi, Tokyo, Japan, S3000H) and energy dispersive spectroscopy (EDS). Details of the experimental methods were reported elsewhere [24].

3. Results and Discussion

3.1. Combustion Exothermicity of Reduction-Based SHS Reactions

Figure 1 presents the calculated ΔH_r and T_{ad} of reactions Equations (1)–(4) and shows that Equation (4) has the highest values while Equation (1) has the lowest ones. Both ΔH_r and T_{ad} increase from Equations (1)–(4). Specifically, the values of T_{ad} are 2530 K, 2595 K, 2783 K, 2897 K for Equation (1)–(4), respectively. A comparison between Equations (1) and (2) revealed that the combustion moderator Al_2O_3 appeared to impose a stronger dilution effect on combustion than MgO, which led to a lower T_{ad} for Equation (1) than Equation (2). Similar results were observed in Equations (3) and (4). These findings could also be explained by the fact that metallothemic reduction of TiO_2 or B_2O_3 by Al is more exothermic than that by Mg [15,25].

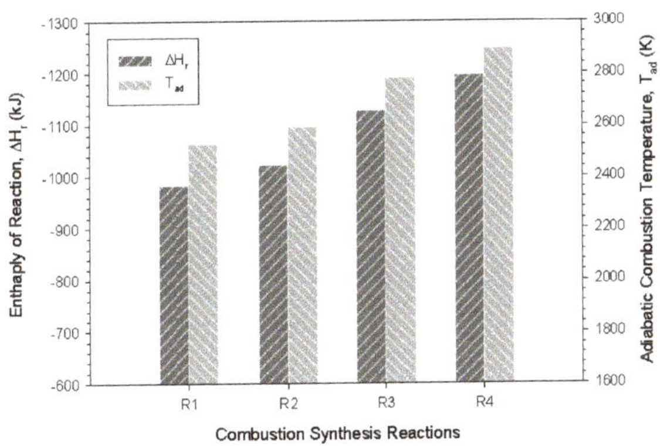

Figure 1. Enthalpies of reaction (ΔH_r) and adiabatic combustion temperatures (T_{ad}) of Equation (1)–(4) for the synthesis of $3TiB_2$–$MgAl_2O_4$ composites.

On the other hand, because the B_2O_3-based thermite is more energetic than the one using TiO_2 [25], Equation (3) has a higher T_{ad} than Equation (1). Similarly, Equation (4) has a higher T_{ad} than Equation (2). According to the calculated T_{ad}, it is realized that the thermite oxidants (i.e., B_2O_3 versus TiO_2) have a more pronounced influence on combustion exothermicity than the diluent oxides (i.e., Al_2O_3 versus MgO).

3.2. Combustion Temperature and Self-Propagating Velocity

Two series of the SHS processes recorded from reactions Equations (1) and (3) are illustrated in Figure 2a,b, respectively. It is apparent that upon ignition, the reaction was initiated and characterized by a self-sustaining combustion wave. More intense combustion accompanied with a faster combustion wave was observed in Figure 2b, when compared with that in Figure 2a. Combustion luminosity and flame spreading speed reflected the degree of reaction exothermicity. As mentioned above, B_2O_3/Al/Mg-based Equation (3) is more energetic than TiO_2/Al/Mg-based Equation (1). Similar combustion behavior was also noticed in Equations (2) and (4).

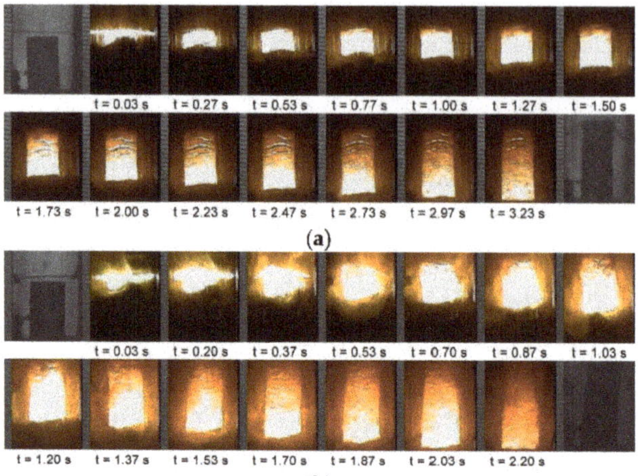

Figure 2. Time sequences of recorded SHS images illustrating self-sustaining combustion wave of (**a**) Equation (1) and (**b**) Equation (3).

Figure 3 depicts typical combustion temperature profiles measured from four different reactions. All profiles exhibit a steep temperature rise followed by a rapid descent, which is characteristic of the SHS reaction that features a fast combustion wave and a thin reaction zone. The peak value is considered as the combustion front temperature (T_c). When compared with pinnacles in the contours of Equations (1) and (2), sharper peaks were detected in the profiles of Equations (3) and (4). This implied a faster combustion wave in Equations (3) and (4). As shown in Figure 3, the values of T_c from Equation (1)–(4) in ascending order are 1348 °C, 1445 °C, 1660 °C, and 1736 °C. It should be noted that the measured combustion front temperatures are in agreement with the calculated reaction exothermicity.

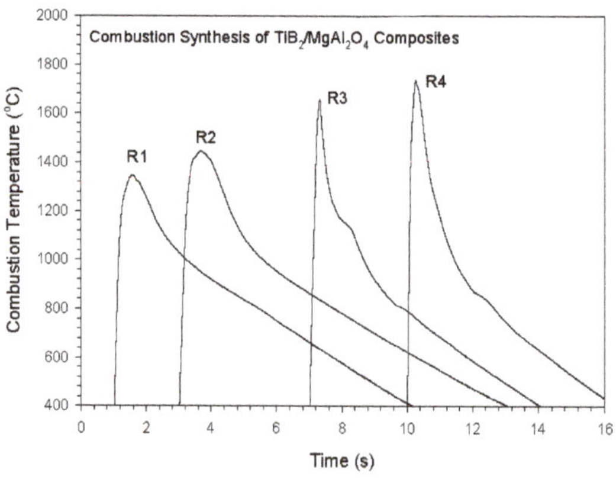

Figure 3. Typical combustion temperature profiles measured from Equation (1)–(4) for the synthesis of $3TiB_2$–$MgAl_2O_4$ composites.

It is useful to note in Figure 3 that the curves of Equations (3) and (4) have a shape peak with a pronounced shoulder. The shape peak was a result of the fast combustion wave. The pronounced shoulder could be caused by the occurrence of volumetric synthesis reactions after the passage of the rapid combustion wave.

Figure 4 plots the measured combustion wave propagation velocities (V_f) and temperatures (T_c) of four reactions. The rising trend of V_f from Equations (1)–(4) is consistent with that of T_c. This can be understood by the fact that the propagation of combustion wave is essentially governed by layer-by-layer heat transfer from the reaction zone to unreacted region, and therefore, is subject to the combustion front temperature. As presented in Figure 4, the average combustion velocities are 3.9, 4.7, 5.1, and 5.7 mm/s for Equations (1)–(4), respectively. It is worth noting that the measured combustion temperature not only justified the reaction exothermic analysis, but confirmed the temperature dependence of combustion wave velocity.

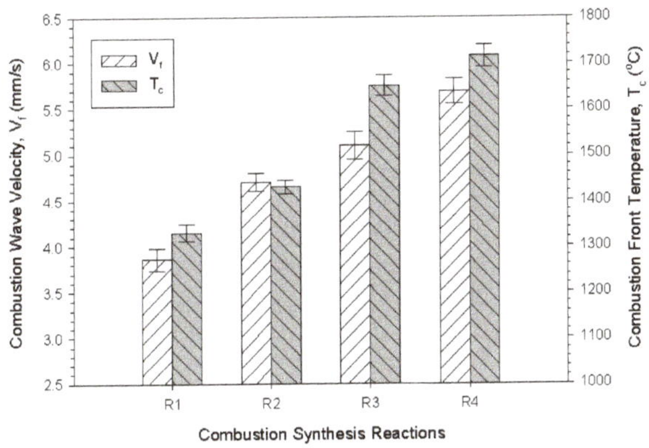

Figure 4. Combustion wave velocity (V_f) and combustion front temperature (T_c) measured from Equation (1)–(4) for the synthesis of $3TiB_2$–$MgAl_2O_4$ composites.

3.3. Composition and Microstructure Analyses of Synthesized Products

The XRD spectra of the final products synthesized from Equations (1) and (2) are shown in Figure 5a,b, respectively. Both indicated the formation of TiB_2 and $MgAl_2O_4$ along with two minor phases, magnesium titanate ($MgTiO_3$) and MgO. It is believed that $MgAl_2O_4$ was synthesized through a combination reaction between Al_2O_3 and MgO. Equation (1) and (2) were formulated with the same thermite reagents of TiO_2, Al, and Mg, but diluted by different metal oxides. That is, Al_2O_3 was partly pre-added and partly thermite-produced, while MgO was completely generated from the reduction of TiO_2 by Mg in Equation (1). In contrast, the required Al_2O_3 in Equation (2) was entirely produced from the reduction of TiO_2 by Al, but MgO was supplied in part from prior addition and in part from the reduction of TiO_2 by Mg. For both Equations (1) and (2), TiB_2 was synthesized from the reaction of elemental boron with reduced and metallic Ti.

Traces of MgO suggested an incomplete reaction due probably to the relatively low reaction temperatures of Equations (1) and (2). The presence of $MgTiO_3$ in the final products of Equations (1) and (2) could be attributed to the reaction of MgO with the thermite oxidant TiO_2 [26,27]. The formation of $MgTiO_3$ in the SHS-produced TiB_2–$MgAl_2O_4$ composites was also observed by Radishevskaya et al. [10] using Ti, boron, and $MgAl_2O_4$ as their starting materials and a partial decomposition of $MgAl_2O_4$ during combustion synthesis was considered as a possible route resulting in the formation of $MgTiO_3$.

The presence of MgO along with no detection of Al_2O_3 in the final products of Equations (1) and (2) suggested that the as-synthesized $MgAl_2O_4$ is an Al_2O_3-rich spinel. The formation of $MgTiO_3$ could also result in the production of Al_2O_3-rich spinel. According to Naghizadeh et al. [28], magnesium titanate compounds ($MgTiO_3$ and Mg_2TiO_4) were identified in the phase evolution of $MgAl_2O_4$ produced from TiO_2-containing samples and stoichiometric $MgAl_2O_4$ spinel shifted toward the Al_2O_3-rich type. Due to the formation of $MgTiO_3$, the amount of TiB_2 formed in the composite should be less than the stoichiometric amount.

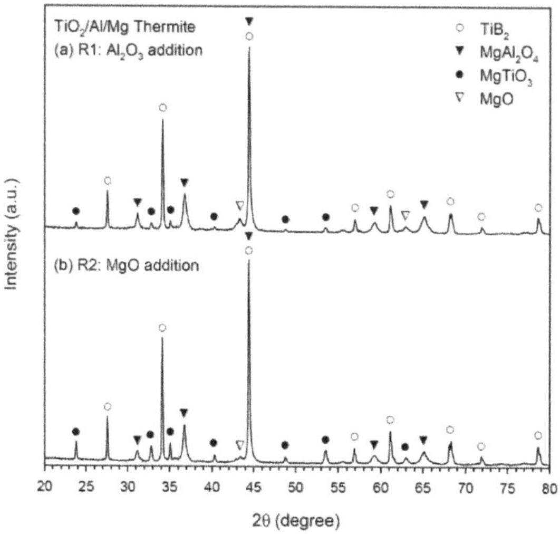

Figure 5. XRD patterns of $TiB_2/MgAl_2O_4$ composites obtained from SHS reactions of (**a**) Equation (1) and (**b**) Equation (2).

Figure 6a,b exhibits the XRD patterns of the synthesized composites from Equations (3) and (4), respectively. In addition to TiB_2 and $MgAl_2O_4$, a small amount of $MgTiO_3$ was identified. The formation of $MgAl_2O_4$ from a combination reaction between Al_2O_3 and MgO was proved. Both Al_2O_3 and MgO can be totally or partially produced from the reduction of B_2O_3 by Al and Mg. For Equations (3) and (4) containing B_2O_3/Al/Mg-based thermite, TiB_2 was produced from the reaction of metallic Ti with reduced and elemental boron. Moreover, the formation of $MgTiO_3$ in Equations (3) and (4) might involve some interaction of Ti with B_2O_3 to form TiO_2 which further reacted with MgO. Unlike that in Equations (1) and (2), MgO was no longer detected in the final products of Equations (3) and (4).

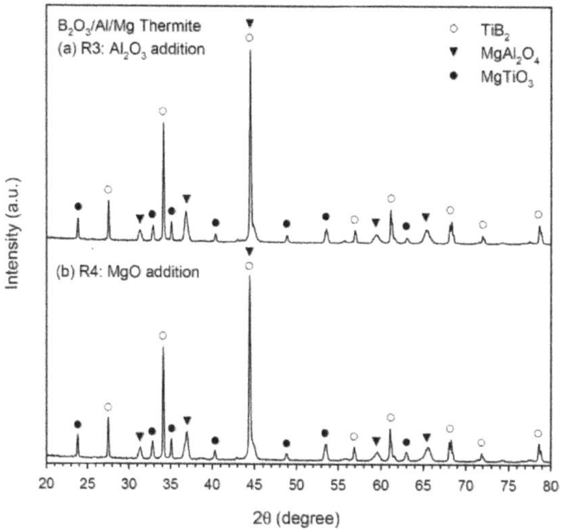

Figure 6. XRD patterns of $TiB_2/MgAl_2O_4$ composites obtained from SHS reactions of (**a**) Equation (3) and (**b**) Equation (4).

$MgTiO_3$ ceramic has been proved to be an excellent dielectric material, owing to its high dielectric constant, low dielectric loss, high value of quality factors, and good temperature stability [29,30]. It is believed that a trace amount of $MgTiO_3$ as a minor phase existed in the as-synthesized TiB_2-$MgAl_2O_4$ products has no effect on the refractory properties of the composites. However, removal of $MgTiO_3$ from the TiB_2-$MgAl_2O_4$ composite would be difficult, since it could combine with $MgAl_2O_4$ in a solid solution form [28].

The SEM image shown in Figure 7 illustrates the microstructure of fracture surface of the product synthesized from Equation (1) which contains a TiO_2/Al/Mg-based thermite. The morphology displays several large and solidified $MgAl_2O_4$ aggregates of 5–15 μm surrounded by fine-grain TiB_2 crystals with a particle size of about 1–2 μm. Moreover, EDS analysis of two characteristic regions in the product surface indicates that the atomic ratios of Ti:B = 35.2:64.8 and Mg:Al:O = 13.1:28.1:58.8 match well with the stoichiometries of TiB_2 and $MgAl_2O_4$, respectively.

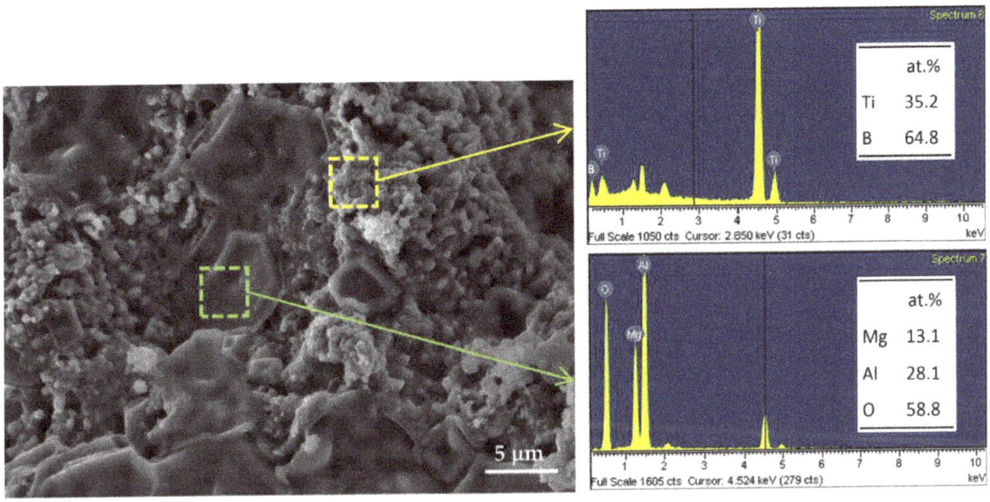

Figure 7. SEM image and EDS spectra of TiB_2/$MgAl_2O_4$ composite obtained from the SHS reaction of Equation (1).

For the final product of B_2O_3/Al/Mg-based Equation (4), the microstructure and elemental ratios of the components are presented in Figure 8. As can be seen, $MgAl_2O_4$ was formed as large densified aggregates of around 20 μm and TiB_2 crystals were in a short-rod form with a length of 2–4 μm or in a shape of fine grains of 1–2 μm. Based on the EDS analysis, the atomic ratio of the selected area in an aggregate is Mg:Al:O = 15.1:30.6:54.3 that is reasonably close to $MgAl_2O_4$. Short-rod crystals have a composition of Ti:B = 34.3:65.7, which certainly is TiB_2.

In summary, the addition of $MgAl_2O_4$ into TiB_2 enhanced the refractory properties, such as the high-temperature oxidation and corrosion resistance and thermal shock resistance [10,31,32]. Like many sintering aids, $MgAl_2O_4$ as an additive could improve densification of TiB_2 ceramics and reduce sintering temperatures [33,34]. The abnormal grain growth could be efficiently prevented during the sintering process. As a result, it is more likely to obtain a uniform grain distribution.

Moreover, the TiB_2-$MgAl_2O_4$ composite is a promising high-temperature microwave absorption material with a reflection loss less than −5 dB at 8.2–18.0 GHz in the temperature range of 25 °C to 1100 °C [11]. The composite also exhibits an extremely high tolerance against intense irradiation in harsh environments [35,36]. Therefore, the potential uses of the TiB_2-$MgAl_2O_4$ composite might include heat-resistant coatings, nozzles and nose cones of supersonic jets, microwave absorption components, diagnostic or detector windows in fusion devices, target materials in the nuclear applications, etc., [11,35,36].

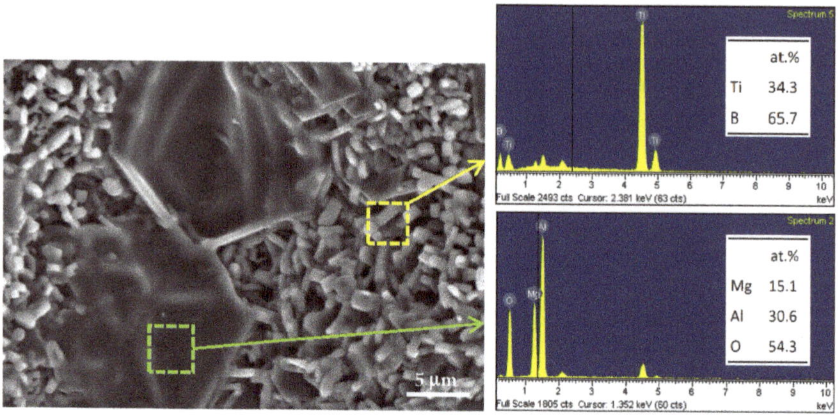

Figure 8. SEM image and EDS spectra of $TiB_2/MgAl_2O_4$ composite obtained from the SHS reaction of Equation (4).

4. Conclusions

In situ formation of $3TiB_2–MgAl_2O_4$ composites was conducted by combustion synthesis combined with metallothermic reduction reactions involving Al and Mg as dual reductants. Thermite reagents with different oxidants were considered; one utilized TiO_2 and the other B_2O_3. The reactant mixtures also contained elemental Ti and boron. This study in total completed four SHS reactions, within which a small amount of Al_2O_3 or MgO was included in the reactive mixture to serve as the combustion moderator and part of the precursors for the formation of $MgAl_2O_4$. The overall synthesis reaction was exothermic enough to proceed in the SHS mode. An energy-saving and efficient fabrication route for the formation $MgAl_2O_4$-containing composites was demonstrated.

The analysis of combustion exothermicity indicated that the SHS reaction containing $B_2O_3/Al/Mg$-based thermites was more energetic than that adopting TiO_2 as the oxidant. Prior addition of Al_2O_3 had a greater cooling effect on combustion than that of MgO. Depending on different thermites and diluents, the measured combustion front temperatures ranged from 1320 to 1720 °C, and combustion wave velocity from 3.9 to 5.7 mm/s. The temperature dependence of combustion wave velocity was justified. The XRD analysis confirmed in situ formation of TiB_2 and $MgAl_2O_4$. A small amount of $MgTiO_3$ was found as the impurity. It is believed that $MgAl_2O_4$ was synthesized through a combination reaction between Al_2O_3 and MgO, both of which can be totally or partially produced from the metallothermic reduction of B_2O_3 or TiO_2. The microstructure of the synthesized composite exhibited that $MgAl_2O_4$ was surrounded by closely packed TiB_2 grains. $MgAl_2O_4$ was formed as densified aggregates with a size of 5–20 µm. TiB_2 crystals were produced in a shape of short rods of 2–4 µm and fine grains of 1–2 µm.

Author Contributions: Conceptualization, C.-L.Y.; methodology, C.-L.Y. and F.-Y.Z.; validation, C.-L.Y. and F.-Y.Z.; formal analysis, C.-L.Y. and F.-Y.Z.; investigation, C.-L.Y. and F.-Y.Z.; resources, C.-L.Y.; data curation, C.-L.Y. and F.-Y.Z.; writing—original draft preparation, C.-L.Y. and F.-Y.Z.; writing—review and editing, C.-L.Y.; supervision, C.-L.Y.; project administration, C.-L.Y.; funding acquisition, C.-L.Y. All authors have read and agreed to the published version of the manuscript.

Funding: This research work was funded by the National Science and Technology Council of Taiwan under the grant of NSTC 110-2221-E-035-042-MY2.

Institutional Review Board Statement: Not applicable.

Informed Consent Statement: Not applicable.

Data Availability Statement: Data presented in this study are available in the article.

Conflicts of Interest: The authors declare no conflict of interest.

References

1. Wilkins, J.M.L. *Boron and Refractory Borides*; Matkovich, V.I., Ed.; Springer: New York, NY, USA, 1977; p. 633.
2. Balcı, Ö.; Burkhardt, U.; Schmidt, M.; Hennicke, J.; Yağcı, M.B.; Somer, M. Densification, microstructure and properties of TiB_2 ceramics fabricated by spark plasma sintering. *Mater. Charact.* **2018**, *145*, 435–443. [CrossRef]
3. Ramesh, B.; Showman, E.; Abraar, S.A.M.; Saxena, K.K.; Tharwan, M.Y.; Alsaadi, N.; Al Sofyani, S.; Elsheikh, A.H. Microstructure, mechanical characteristics, and wear performance of spark plasma sintered TiB_2–Si_3N_4 as affected by B_4N doping. *Materials* **2022**, *15*, 7096. [CrossRef] [PubMed]
4. Munro, R.G. Material properties of titanium diboride. *J. Res. Natl. Inst. Stand. Technol.* **2000**, *105*, 709–720. [CrossRef] [PubMed]
5. Vaferi, K.; Nekahi, S.; Vajdi, M.; Moghanlou, F.S.; Shokouhimehr, M.; Motallebzadeh, A.; Sha, J.; Asl, M.S. Heat transfer, thermal stress and failure analyses in a TiB_2 gas turbine stator blade. *Ceram. Int.* **2019**, *45*, 19331–19339. [CrossRef]
6. Sulima, I.; Hyjek, P.; Podsiadło, M.; Boczkal, S. Effect of zirconium diboride and titanium diboride on the structure and properties of 316L steel-based composites. *Materials* **2023**, *16*, 439. [CrossRef]
7. Popov, A.Y.; Sivak, A.A.; Borodianska, H.Y.; Shabalin, I.L. High toughness TiB_2–Al_2O_3 composite ceramics produced by reactive hot pressing with fusible components. *Adv. Appl. Ceram.* **2015**, *114*, 178–182. [CrossRef]
8. Kozień, D.; Czekaj, I.; Gancarz, P.; Ziąbka, M.; Wieczorek, W.; Pasiut, K.; Zientara, D.; Pędzich, Z. Ceramic matrix composites obtained by the reactive sintering of boron carbide with intermetallic compounds from the Ti-Si system. *Materials* **2022**, *15*, 8657. [CrossRef]
9. Li, Z.; Chen, Q.; Jiao, B.; Wang, Q.; Zhang, H.; Jia, Q.; Zhang, S.; Liu, J. Direct laser powder-bed fusion additive manufacturing of complex-shaped TiB_2-B_4C composite with ultra-fine eutectic microstructure and outstanding mechanical performances. *J. Eur. Ceram. Soc.* **2023**, *43*, 1230–1236. [CrossRef]
10. Radishevskaya, N.; Lepakova, O.; Karakchieva, N.; Nazarova, A.; Afanasiev, N.; Godymchuk, A.; Gusev, A. Self-propagating high temperature synthesis of TiB_2–$MgAl_2O_4$ composites. *Metals* **2017**, *7*, 295. [CrossRef]
11. Yang, J.; Liu, X.; Gong, W.; Wang, T.; Wang, X.; Gong, R. Temperature-insensitive and enhanced microwave absorption of $TiB_2/Al_2O_3/MgAl_2O_4$ composites: Design, fabrication, and characterization. *J. Alloys Compd.* **2022**, *894*, 162144. [CrossRef]
12. Ganesh, I. A review on magnesium aluminate ($MgAl_2O_4$) spinel: Synthesis, processing and applications. *Int. Mater. Rev.* **2013**, *58*, 63–112. [CrossRef]
13. Corlett, C.A.; Frontzek, M.D.; Obradovic, N.; Watts, J.L.; Fahrenholtz, W.G. Mechanical activation and cation site disorder in $MgAl_2O_4$. *Materials* **2022**, *15*, 6422. [CrossRef]
14. Yuan, L.; Tian, C.; Yan, X.; He, X.; Liu, Z.; Wen, T.; Jin, E.; Yu, J. Preparation of porous $MgAl_2O_4$ ceramics by a novel pectin gel-casting process. *J. Aust. Ceram. Soc.* **2021**, *57*, 1049–1055. [CrossRef]
15. Yeh, C.L.; Chen, M.C.; Shieh, T.H. Formation of silicide/spinel ceramic composites via Al- and Mg-based thermitic combustion synthesis. *J. Aust. Ceram. Soc.* **2022**, *58*, 1275–1282. [CrossRef]
16. Levashov, E.A.; Mukasyan, A.S.; Rogachev, A.S.; Shtansky, D.V. Self-propagating high-temperature synthesis of advanced materials and coatings. *Int. Mater. Rev.* **2017**, *62*, 203–239. [CrossRef]
17. Xu, J.; Ma, P.; Zou, B.; Yang, X. Reaction behavior and formation mechanism of ZrB_2 and ZrC from the Ni-Zr-B_4C system during self-propagating high-temperature synthesis. *Materials* **2023**, *16*, 354. [CrossRef]
18. Zakaryan, M.K.; Zurnachyan, A.R.; Amirkhanyan, N.H.; Kirakosyan, H.V.; Antonov, M.; Rodriguez, M.A.; Aydinyan, S.V. Novel pathway for the combustion synthesis and consolidation of boron carbide. *Materials* **2022**, *15*, 5042. [CrossRef]
19. Omran, J.G.; Afarani, M.S.; Sharifitabar, M. Fast synthesis of $MgAl_2O_4$–W and $MgAl_2O_4$–W–W_2B composite powders by self-propagating high-temperature synthesis reactions. *Ceram. Int.* **2018**, *44*, 6508–6513. [CrossRef]
20. Zaki, Z.I.; Mostafa, N.Y.; Rashad, M.M. High pressure synthesis of magnesium aluminate composites with $MoSi_2$ and Mo_5Si_3 in a self-sustaining manner. *Ceram. Int.* **2012**, *38*, 5231–5237. [CrossRef]
21. Zaki, Z.I.; Ahmed, Y.M.Z.; Abdel-Gawad, S.R. In-situ synthesis of porous magnesia spinel/TiB_2 composite by combustion technique. *J. Ceram. Soc. Jpn.* **2009**, *117*, 719–723. [CrossRef]
22. Liang, Y.H.; Wang, H.Y.; Yang, Y.F.; Zhao, R.Y.; Jiang, Q.C. Effect of Cu content on the reaction behaviors of self-propagating high-temperature synthesis in Cu-Ti-B_4C system. *J. Alloys Compd.* **2008**, *462*, 113–118. [CrossRef]
23. Binnewies, M.; Milke, E. *Thermochemical Data of Elements and Compounds*; Wiley-VCH Verlag GmbH: Weinheim, Germany, 2002.
24. Yeh, C.L.; Lin, J.Z. Combustion synthesis of Cr-Al and Cr-Si intermetallics with Al_2O_3 additions from Cr_2O_3-Al and Cr_2O_3-Al-Si reaction systems. *Intermetallics* **2013**, *33*, 126–133. [CrossRef]
25. Wang, L.L.; Munir, Z.A.; Maximov, Y.M. Thermite reactions: Their utilization in the synthesis and processing of materials. *J. Mater. Sci.* **1993**, *28*, 3693–3708. [CrossRef]
26. Singh, J.; Bahel, S. Synthesis of single phase $MgTiO_3$ and influence of Sn^{4+} substitution on its structural, dielectric and electrical properties. *J. Alloys Compd.* **2020**, *816*, 152679. [CrossRef]

27. Cheng, L.; Liu, P.; Qu, S.X.; Cheng, L.; Zhang, H. Microwave dielectric properties of Mg_2TiO_4 ceramics synthesized via high energy ball milling method. *J. Alloys Compd.* **2015**, *623*, 238–242. [CrossRef]
28. Naghizadeh, R.; Rezaie, H.R.; Golestani-Fard, F. Effect of TiO_2 on phase evolution and microstructure of $MgAl_2O_4$ spinel in different atmospheres. *Ceram. Int.* **2011**, *37*, 349–354. [CrossRef]
29. Jo, H.J.; Kim, J.S.; Kim, E.S. Microwave dielectric properties of $MgTiO_3$-based ceramics. *Ceram. Int.* **2015**, *41*, S530–S536. [CrossRef]
30. Thatikonda, S.K.; Goswami, D.; Dobbidi, P. Effects of CeO_2 nanoparticles and annealing temperature on the microwave dielectric properties of $MgTiO_3$ ceramics. *Ceram. Int.* **2014**, *40*, 1125–1131. [CrossRef]
31. Li, Y.; Nan, L.; Ruan, G.; Li, X. Reaction path in the aluminothermic reduction nitridation reaction to synthesize $MgAl_2O_4$/TiN composite. *Ceram. Int.* **2005**, *31*, 825–829. [CrossRef]
32. Liu, X.; Li, K.; Bao, K.; Chen, J.; Zhang, H.; Zhang, S. In-situ synthesis of magnesium aluminate spinel–Zirconium diboride composite powder in magnesium chloride melt. *Ceram. Int.* **2022**, *48*, 11869–11871. [CrossRef]
33. Ma, L.; Yu, J.; Guo, X.; Xie, B.; Gong, H.; Zhang, Y.; Zhaie, Y.; Wu, X. Preparation and sintering of ultrafine TiB_2 powders. *Ceram. Int.* **2018**, *44*, 4491–4495. [CrossRef]
34. Sarkar, R. Additives for magnesium aluminate spinel: A review. *Interceram Refract. Manual* **2011**, *1*, 28–32.
35. Lushchik, A.; Feldbach, E.; Kotomin, E.A.; Kudryavtseva, I.; Kuzovkov, V.N.; Popov, A.I.; Seeman, V.; Shablonin, E. Distinctive features of diffusion-controlled radiation defect recombination in stoichiometric magnesium aluminate spinel single crystals and transparent polycrystalline ceramics. *Sci. Rep.* **2020**, *10*, 7810. [CrossRef] [PubMed]
36. Ananchenko, D.V.; Nikiforov, S.V.; Kuzovkov, V.N.; Popov, A.I.; Ramazanova, G.R.; Batalov, R.I.; Bayazitovc, R.M.; Novikov, H.A. Radiation-induced defects in sapphire single crystals irradiated by a pulsed ion beam. *Nucl. Instrum. Methods Phys. Res. B* **2020**, *466*, 1–7. [CrossRef]

materials

Article

A Comparative Study on Microstructural Characterization of Thick High Strength Low Alloy Steel Weld by Arc Welding and Laser Welding

Yunxia Chen [1,*], Xiao Xu [1], Yanjing Liu [2] and Haichao Cui [2]

1 School of Mechanical Engineering, Shanghai Dianji University, Shanghai 201306, China
2 Shanghai Key Laboratory of Materials Laser Processing and Modification, Shanghai Jiao Tong University, Shanghai 200240, China
* Correspondence: chenyx@sdju.edu.cn; Tel.: +86-21-38223822

Abstract: Welding and the behavior of the weldments are important, since welding of high strength low alloy (HSLA) steels is a conventional method for manufacturing industrial parts. This work conducts a comparative investigation of microstructural characteristics and mechanical properties for joints of 16-mm-thick HSLA Q890 steel produced by multi-layer multi-pass shielded metal arc welding (SMAW) with filler wire and single-layer autogenous laser beam welding (LBW). The mechanical properties of the welded joints were assessed in terms of tensile and impact using butt joints. The results show that tensile failure occurred in the base metal during the tensile tests for most of the trials. The ultimate tensile strength and percent elongation of the LBW welded joint (973.5 MPa and 10%) are higher than those of the SMAW joint (951 MPa and 2.9%) due to the filler filling process of the SMAW process. The Charpy impact energy of the weld metal (16.4 J and 15.1 J) is lower than that of the heat-affected zone (18.5 J and 19.5 J) in the LBW joint and the SMAW joint.

Keywords: HSLA steel; microstructural characterization; laser welding; thick plate

Citation: Chen, Y.; Xu, X.; Liu, Y.; Cui, H. A Comparative Study on Microstructural Characterization of Thick High Strength Low Alloy Steel Weld by Arc Welding and Laser Welding. *Materials* **2023**, *16*, 2212. https://doi.org/10.3390/ma16062212

Academic Editor: Antonino Squillace

Received: 12 February 2023
Revised: 26 February 2023
Accepted: 8 March 2023
Published: 9 March 2023

1. Introduction

Welded structures of thick plate are broadly applied to pipelines to transmit gas and oil in large ships, offshore oil drilling platforms, pressure vessels, shipbuilding [1,2], bridges and storage tanks, railway components, nuclear power plants, pressure, and naval vessels [3,4]. HSLA steels have gained widespread use because they exhibit higher strength and better weldability than conventional high alloy steels [5].

The weldability of HSLA steel is reasonably good due to its reduced carbon equivalent [6]. The equipment used for these processes is quite large, and for medium and heavy plate thicknesses, workpieces can only be jointed piece by piece using fusion welding processes, such as submerged arc welding (SAW) [7,8], gas tungsten arc welding (GTAW), gas metal arc welding (GMAW) [9,10], and SMAW [11]. Arc welding has productivity limits due to its low penetration depths. Because of this, multi-pass welding is required for these thick plate structures, which are cut with large-sized single V grooves and require multi-layer multi-pass arc welding, potentially presenting some disadvantages, such as gas porosity, slag inclusion, large levels of distortion and residual stress, poor welding quality defects, and low productivity [12,13]. In addition, other welding processing methods of HSLA steels joints include electron beam welding (EBW), LBW, and friction stir welding (FSW). However, the size of parts for EBW is still a challenge because the weld environment needs to be set in a vacuum chamber, the FSW of thick plates is constrained due to the size of the rotating tool, and the required rigidity of the welding machine [14–16]. In contrast, LBW is particularly attractive for innovative and cost-effective applications because of its high precision, low heat input, narrow weld pool, small welding deformation, and the narrow width of the heat-affected zone (HAZ) [17,18]. Nevertheless, the application of

LBW to HSLA steels has not been widely adopted for practical applications due to the short cooling times after welding [19–21].

Published comparative studies on SMAW and LBW of HSLA steel plates with higher thicknesses are very scant. The behavior of weldments has been related to microstructural features [22,23]. Therefore, studying the effect of different welding techniques and their parameters on thick Q890 HSLA steel is essential. The present study was carried out to compare the microstructural characteristics and mechanical properties of 16 mm thick Q890 HSLA steel weldment fabricated by SMAW and LBW techniques.

2. Experimental Details

The 16-mm-thick HSLA Q890 steel used in this investigation was provided by Baoshan Iron and Steel Co., Ltd., (Shanghai, China). The main chemical compositions (in wt.%) of the steel and X90 filling wire are given in Table 1. The equivalent carbon content (CE) for this steel grade amounted to about 0.55 according to SS-EN1011-2 [24], indicating a relatively high hardenability. Its parent microstructure was of a predominantly high temperature tempering structure of martensite (M). In addition, 200 mm × 200 mm welded blanks were fabricated by butt welding. LBW equipment consisted of an IPG Photonics 10 kW fiber laser system (IPG-10000, Burbach, Germany) mounted on a Fronius Transpuls Synergic 5000 wire feeding machine (Fronius, Pettenbach, Austria), and a system of shielding gas 99.99% pure Argon was used as the protective gas, which had a flow rate of 15 L/min. The laser beam was perpendicular to the steel sheet being welded. The weld geometries and dimensions are presented in Figure 1a,b. Prior to welding, mechanical polishing of the BM was performed and the oil on the BM surface was cleaned using acetone.

Table 1. Composition and content of chemical elements in HSLA Q890 and X90 filling wire (wt.%).

	C	Si	Mn	Cr	Mo	Nb	V	Ni	W	B	P	S
Q890	0.18	0.47	1.10	0.78	0.05	0.05	0.07	0.16	0.53	0.01	0.01	0.01
X90	0.10	0.80	1.80	0.35	0.60	-	-	2.30	-	-	-	-

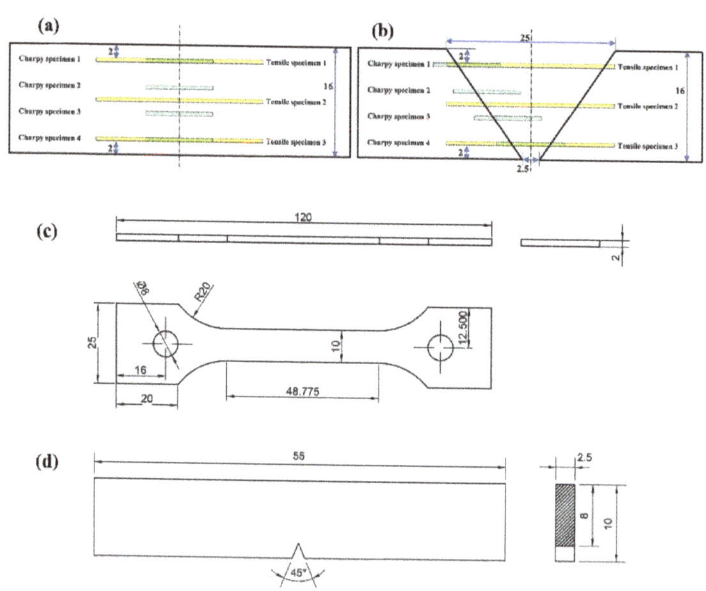

Figure 1. Dimensions of grooves shape (**a**) LBW joint (**b**) SMAW joint (**c**) Tensile specimen (**d**) Charpy specimen (Unit: mm).

The LBW experiments were carried out with a laser power of 8 kW, a welding velocity of 0.3 m/min, and a negative defocusing length of 4 mm.

The welding process of SMAW consisted of six layers and twenty-one passes. The basic parameters of SMAW were set as: a welding current of 250 A; a welding velocity of 0.11 m/min; a welding voltage of 28 V. To prevent cracks, the preheating temperature before welding was 120 °C, the inter-pass temperature was strictly controlled below 200 °C, and post-weld heat treatments were adopted. After welding, the weldment was cooled in air for 20 min, heated to 360 °C in 60 min, and kept constant for 30 min in a heat treatment furnace before being cooled in air.

Nine replications of each level of two joints were made to achieve measurement accuracy. Out of those, three were selected for tensile tests, four were employed in impact tests, and two were used for microhardness and microstructure observations. Following GB/T 228-2002 [25], tensile test specimens were sectioned from the welded blanks, as presented in Figure 1a,b. Tensile tests were performed on a Zwick/Roell Z100 (Zwick Roell Group, Ulm, Germany) universal testing facility with a constant strain rate of 1 mm/min at room temperature. The tensile test recorded for each joint was the average of three trials. The schematic of tensile test specimens is presented in Figure 1c.

The Charpy V-notch impact tests were performed at −40 °C (according to GB/T 229-2007 [26]). Four different test sampling locations in each joint were designed in view of the effect of the sampling position on the low temperature impact property, as shown in Figure 1a,b. The absorbed energy value recorded for each joint was the average of four trials. The schematic of the unstandardized 2.5 × 10 × 55 mm specimen with a V-notch is presented in Figure 1d.

The metallographic specimens were prepared by cutting along the cross section of the weld followed by a conventional procedure, such as mounting, grinding, polishing, or etching. Thereafter, the etched samples were observed using OM (Leica DM4000 M LED, Leica Camera, Wetzlar, Germany). Microstructures of the unetched cross sections were characterized by using SEM (JEOL JSM-7800F Prime, JEOL Ltd., Tokyo, Japan) with EDX.

Vickers microhardness tests (Zwick/Roell ZHμ, Zwick Roell Group) were utilized at the mid-thickness of the weld cross section to evaluate the microhardness of the weld metal (WM) and HAZ. According to ASTM:E384, an indentation load of 500 g with a dwell time of 15 s was employed for each test.

3. Results and Discussion

3.1. LBW Weld Microstructure

The HAZ can be described by four subzones at a fine-grained level: the coarse-grained heat affected zone (CGHAZ); the fine-grained heat affected zone (FGHAZ); the two-phase zone (TPZ); and the over-tempering zone (OTZ) in detail. The definition of each zone is as follows: the fusion line is used as the boundary between the weld metal (WM) and HAZ; the boundary between the CGHAZ and the FGHAZ is about 10 μm in width with obvious grain size growth; the boundary between the FGHAZ and the TPZ is where the tempered sorbite of the base metal (BM) disappears; the boundary between the TPZ and the OTZ is based on the disappearance of white striped microstructures and the appearance of obvious granular carbides. Each zone width was calculated using the average width method that was equal to the zone area divided by its thickness using Image-Pro Plus software: IPP 7.0.

Figure 2a shows the optical morphology of the LBW joint. It can be seen that the weld morphology is good without any obvious defects. Although the carbon equivalent (0.55%) of the Q890 HSLA steel is a little higher, the LBW performance of the welded joint had been greatly improved by the preheating treatment before welding. It can also be seen that the maximum and minimum bead width of LBW are about 9.5 mm and 3.6 mm, respectively. The maximum and minimum bead width of the HAZ are 4 mm and 2.2 mm, respectively, and its average bead width is 2.9 mm. The weld penetration is about 13 mm. The LBW weld can be divided by ① WM, ② CGHAZ, ③ FGHAZ, ④ TPZ, ⑤ OTZ, and ⑥ BM.

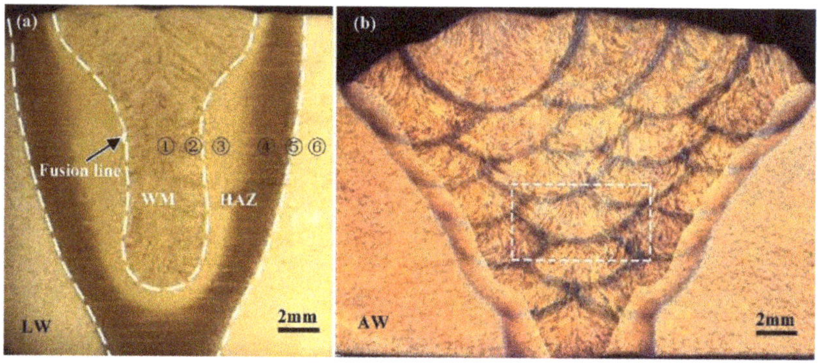

Figure 2. Optical morphology of joints (**a**) the LBW (**b**) the SMAW.

① Weld Metal

In the LBW joint, the weld seam was formed by the solidification of partial base metal after high temperature melting. It was mainly composed of a certain amount of equiaxed crystal phases surrounded by coarse columnar crystal phases. The columnar crystal phases grew along the perpendicular direction to the side wall of the molten pool from the base metal to the weld seam, as shown in Figure 3. The width of columnar crystal phases is about 10 μm, and their maximum length can be greater than 200 μm. The mechanical properties of the joint significantly decreased because of the existence of the columnar crystal phases. The WM was mainly composed of ferrite (F) and bainite (B). The ferrite was mainly proeutectoid ferrite, which is an important component of columnar crystal phases, while the equiaxed crystal phases were almost entirely composed of polygonal ferrite phases. In addition, there were a few acicular ferrite phases and a certain amount of martensite phases. Acicular ferrite phases formed with some inclusions as the core and most martensite phases were distributed in the fusion line, at the root or top of the weld due to faster cooling rate.

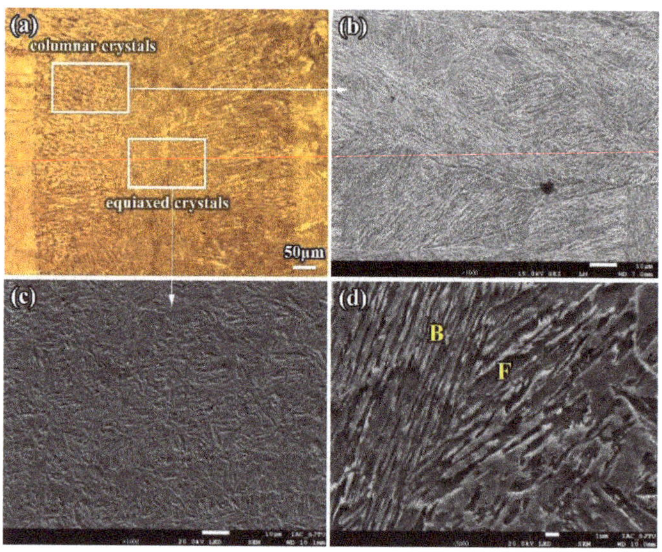

Figure 3. Microstructure of WM of the LBW joint (**a**) Macro-photo in weld center area (**b**) Columnar crystal (**c**) Equiaxed crystal (**d**) Ferrite (F) and bainite (B).

② Coarse-Grained Heat-Affected Zone

The CGHAZ is adjacent to the WM, and its higher peak temperature of the thermal cycle ranges from 1100 °C to the liquidus temperature. The austenite phases can stay in this temperature range for a long time and dramatically grow, and then form larger overheating phases during cooling, which can worsen the mechanical performance of the whole joint. The average grain size of the CGHAZ can reach about 50 μm; even 100 μm for the grains near to the fusion line.

The microstructure features of the CGHAZ are closely related to the heat input. When the heat input increases, acicular ferrite begins to gradually form in the CGHAZ. Acicular ferrite with low hardness and good toughness can significantly improve the impact behavior of the CGHAZ. If the heat input keeps increasing and the cooling rate decreases, polygonal ferrite phases will increase and acicular ferrite will decrease. Meanwhile, there is a small amount of pearlite that can possibly appear in this area.

Due to the lower heat input and faster cooling rate, there are more and more strip-shape bainite, ferrite, and low-carbon martensite phases formed in the CGHAZ. Most of the strip-shape phases initially form from the original austenite grain boundary and grow toward the crystal at a certain angle; each strip maintains a certain phase relationship and a fine structure inside the strip. With lower heat input (9.6 kJ/cm) in the LBW, the internal phases in the CGHAZ are low-carbon martensite and a few bainite and ferrite, as shown in Figure 4.

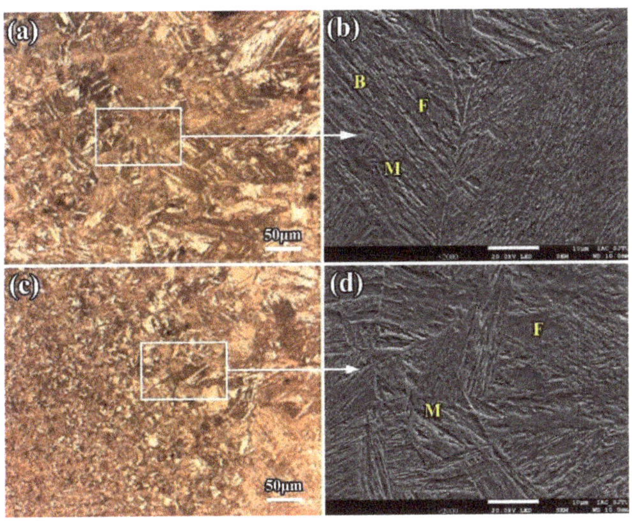

Figure 4. Microstructure of the CGHAZ of the LBW joint (**a**) CGHAZ close to fusion line (**b**) Martensite and bainite (**c**) CGHAZ far from fusion line (**d**) Martensite and ferrite. F: ferrite; M: martensite; B: bainite.

③ Fine-Grained Heat Affected Zone

During the LBW process, for the FGHAZ, the peak temperature of the welding thermal cycle is between 1100 °C and A_{C3} temperature, and its mechanical properties performance is the best out of all the weld joints. The reason for this is that the microstructure in the FGHAZ undergoes a recrystallization and new grain refinement during heating and cooling; more fine grains can be obtained, which is equivalent to the base metal normalized. In general, most grain sizes of the fine grains are less than 10 μm. Observed under a high magnification microscope, the microstructure of the FGHAZ can be described by some finer martensite, bainite, and few fine carbides precipitated between ferrite blocks, as shown in Figure 5.

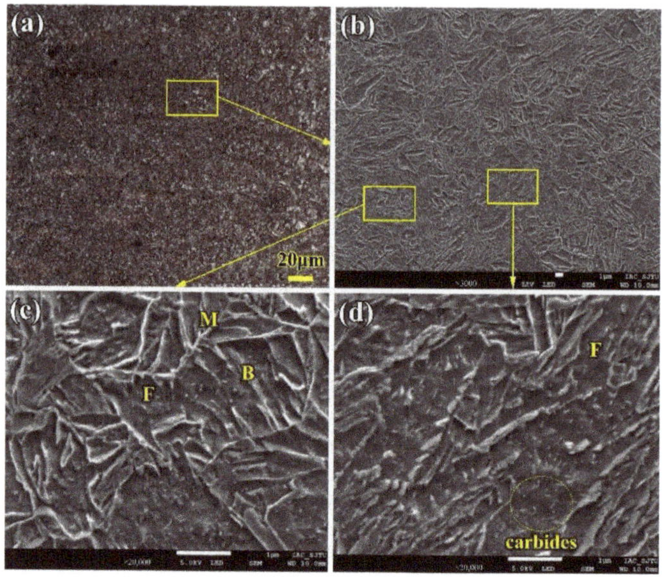

Figure 5. Microstructure of the FGHAZ of the LBW joint (**a**) Macro-photo of the FGHAZ (**b**) Micro-photo of the FGHAZ (**c**) Ferrite and bainite (**d**) Carbides in the FGHAZ. F: ferrite; M: martensite; B: bainite.

④ Two-Phase Zone

The range of the TPZ in the LBW joints is wider (especially in the center of the joint), and its peak temperature of the welding thermal cycle is between A_{C3} and A_{C1}. During this temperature range, only partial base metal can obtain sufficient heat to finish phase transformation and form complex phases based on sorbite and the partial transformation structure.

It can be seen from Figure 6a,b that there are some obvious striped microstructures in the TPZ. The reason for these striped microstructures is due to interlacing ferrite and other phases along the deformation direction when heating. The appearance of striped microstructures makes the steel anisotropic, and has a negative effect on the mechanical properties of the steel. In the base metal, there are few striped microstructures can be less obviously seen. However, striped microstructures can be easily found after welding because the secondary cementite of the austenite with partial phase transformation in the TPZ precipitates along the original strip direction and forms duplex phases of ferrite and cementite. These striped microstructures can be seen more clearly due to their own fine structures, making most of the incomplete phase transformation turn black after corrosion.

The striped microstructures in the TPZ are shown in Figure 6c,d, under the scanning electron microscope. The black part is similar to the tempered sorbite duplex phase, which is mainly composed of striped ferrite, carbides, and few polygonal ferrite. However, there are more granular carbides among the ferrite phases, and black striped microstructures increase with the precipitation of carbides, which results in a finer microstructure and a stronger black color. The white part of the striped microstructures is mainly irregular tangled ferrite bands and a few precipitated carbides. Compared with the black part, this part looks white and its hardness is lower due to obvious ferrite characteristics; the reason is that when carbon-poor austenite reaches the austenitizing temperature, there are more proeutectoid ferrite precipitated, and then ferrite and cementite are precipitated, which results in the formation of some complex ferrite morphologies.

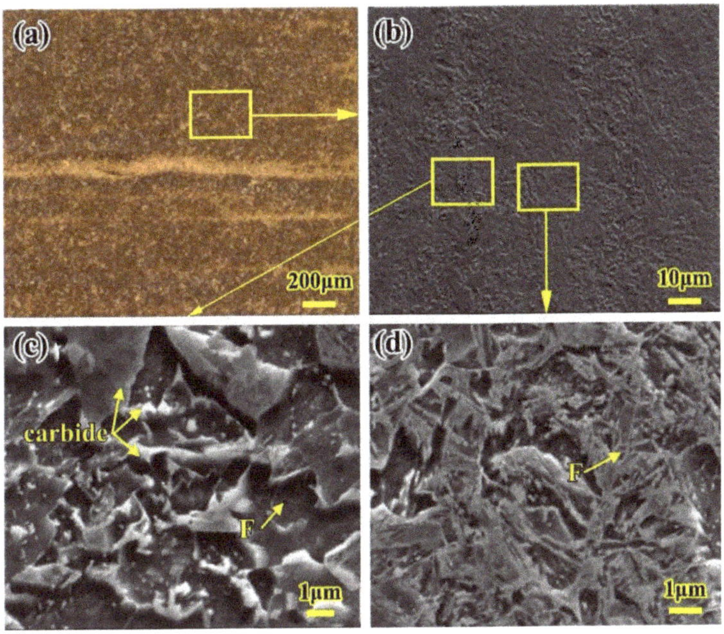

Figure 6. Microstructure of the TPZ of the LBW joint (**a**) Segregation band in the TPZ (**b**) Microstructure of the segregation band in SEM (**c**) Microstructure in black part (**d**) Microstructure in white part. F: ferrite.

⑤ Over-Tempering Zone

Due to the OTZ being far away from the WM, its lower peak temperature of the welding thermal cycle is not more than A_{C1} and its cooling speed for welding is less. Therefore, the base metal in the over tempered zone does not undergo austenitizing transformation in the welding thermal cycle; it seems to be double tempered at higher temperature than that of the quenching and tempering. The higher temper temperature is beneficial for the carbide particles to aggregate and grow up in this area, which results in more formations of polygonal ferrite phases, which weaken the microhardness. Therefore, the OTZ has a significant responsibility for the softening of the welded joint.

From the discussion of the OTZ in the welded joint, it can be observed that there are more granular carbides dispersed on the ferrite matrix; these granular carbides probably originate from the sorbite phase carbides in the base metal, where they aggregate and grow up and are distributed on the ferrite grain boundary after double tempering, as shown in Figure 7. The area with more coarse granular carbides is black, while the polygonal ferrite phase is white. In the OTZ, there are more carbides precipitated, grown, and coarsened, which causes the hardness of the OTZ to decrease. Therefore, it is the most softened zone in the whole welded joint.

In order to figure out the phase constituents of the joint softening zone, EBSD was used and the results are shown in Figure 8. The EBSD results show that the Fe_3C fraction of the phase constituents is 0.77% and much higher, as shown in Table 2. The reason is that there is no post-weld heat treatment after welding, which results in some residual stress in the welded joint; the lower resolution rate is 75.23%. The face-centered cubic (FCC) austenite fraction of the phase constituents is only 0.01%, which means that the amount of retained austenite in the laser-welded joint is very small, and the body-centered cubic (BCC) ferrite fraction of the phase constituents is 74.44%. According to the previous analysis, this zone consists of some polygonal ferrite and few sorbite phases. These polygonal ferrite phases are responsible for the hardness decreasing.

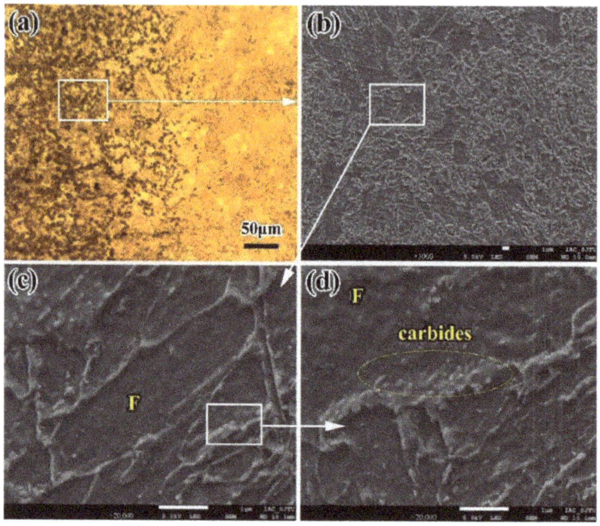

Figure 7. Microstructure of the OTZ of the LBW joint (**a**) Microstructure in the OTZ (**b**) SEM in the OTZ (**c**) Polygonal ferrite (**d**) Carbides on ferrite grain boundaries. F: ferrite.

Figure 8. Results of EBSD in the OTZ of the LBW joint (**a**) Location of EBSD (**b**) Phase distribution.

Table 2. Phase content in the OTZ of the LBW joint.

Phase Content	FCC	BCC	Fe_3C	Unidentified Resolution
Phase percentage	0.01	74.44	0.77	24.77
Phase count	3	16749	174	5574

3.2. SMAW Weld Microstructure

Figure 2b shows the microstructure of the whole arc welded joint. It can be seen that 21 passes of SMAW were need to fill the V-groove for 16-mm thick weldment. The heat input was higher than 16.4 kJ/cm and the maximum bead width was approximately 30 mm. The joint morphology was good and the maximum bead width was about 30 mm; the average width of HAZ between the BM and the WM was nearly 2.5 mm; while the width of HAZ in the inner WM was only 0.5 mm. This can be explained by the fact that the HAZ is produced by reheating the previous part of the weld with the following weld in the area where the welds overlap with each other. The formation of HAZ leads to the change in microstructure, which in turn has an impact on the performance of the joint. Finally, the typical solidification structure, such as the columnar crystal and the equiaxed crystal, appears in each layer and in each pass; most of them are the column crystal.

The microstructure in the SMAW joint is almost similar to that in the LBW joint overall. However, if a higher heat input (16.4 kJ/cm) and slower cooling rate were used, the microstructure in the SMAW joint would show some differences that the LBW joint would not have. The following discussion focuses on the microstructure differences of the WM and HAZ.

In the multi-layer multi-pass SMAW, there are many HAZs that are heated many times in the WM, which results in obvious secondary HAZs existing among each layer and each pass of the WM, as shown in Figure 9a. The reason for this is similar to the formation of the black microstructures in the TPZ of the LBW HAZ. Due to the higher heat input and slower cooling rate, carbides in the secondary HAZ have sufficient time to precipitate and grow up among each layer of the WM, forming complex phases of ferrite and cementite with finer microstructures and a darker color. Figure 9c is a local magnification photo of Figure 9a. The typical secondary HAZ and WM can be easily seen in the 2-mm weld pass. The WM is mainly composed of equiaxed crystal and columnar crystal phases, and the equiaxed crystal phases are composed of some acicular and polygonal ferrite and bainite. The columnar crystal is mainly made up proeutectoid ferrite and bainite ferrite, as shown in Figure 9c. The size of the equiaxed crystal phase is around several microns, and the width of the columnar crystal phase is 150 μm; the width of the black secondary HAZ is almost 250 μm. Typical bainite ferrite clusters can be seen in the equiaxed crystal phase in Figure 10b; the length of ferrite is 70 μm and the width is 3 μm.

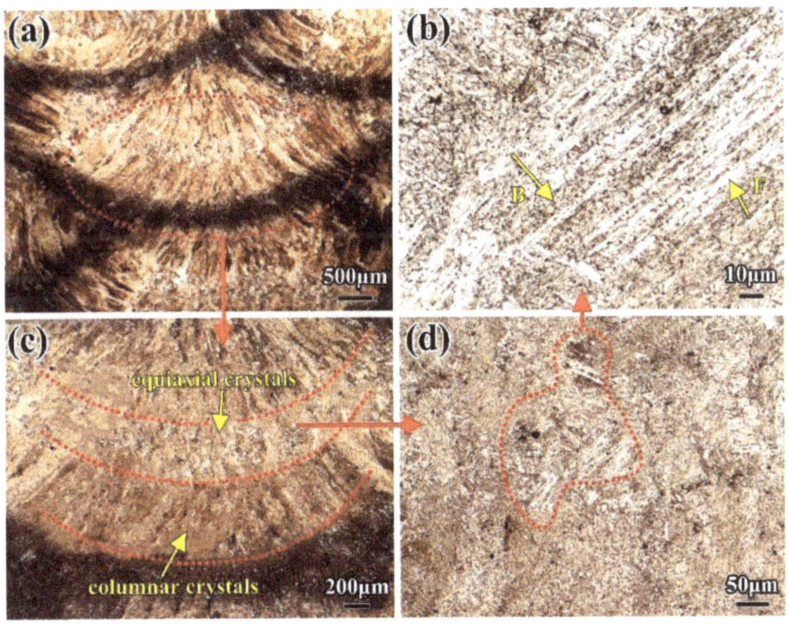

Figure 9. Microstructure of SMAW joint (**a**) Macro-photo in weld metal (**b**) Bainite ferrite in weld metal (**c**) Equiaxial and columnar crystals (**d**) Equiaxial crystals. F: ferrite; B: Bainite.

The HAZ of the SMAW is also composed of a CGHAZ, a FGHAZ, a TPZ, and an OTZ. The grain size of the CGHAZ is nearly 100 μm and consists of coarse martensite and bainite. Because of the higher heat input and slower cooling rate, most phases of the CGHAZ are bainite with a few polygonal ferrite. In Figure 10, the grain size in the FGHAZ is nearly 10 μm. The TPZ consists of some duplex microstructures and ferrite phases. The more the HAZ is away from the WM, the lower the decrease in the peak temperature and cooling rate of the thermal cycle, and the less carbide precipitate. The precipitated carbide is discontinuously distributed and becomes a tempered sorbite phase when it reaches the TPZ. In view of the post-weld heat treatment, it is not easy to find black and white striped phases of the TPZ in SMAW.

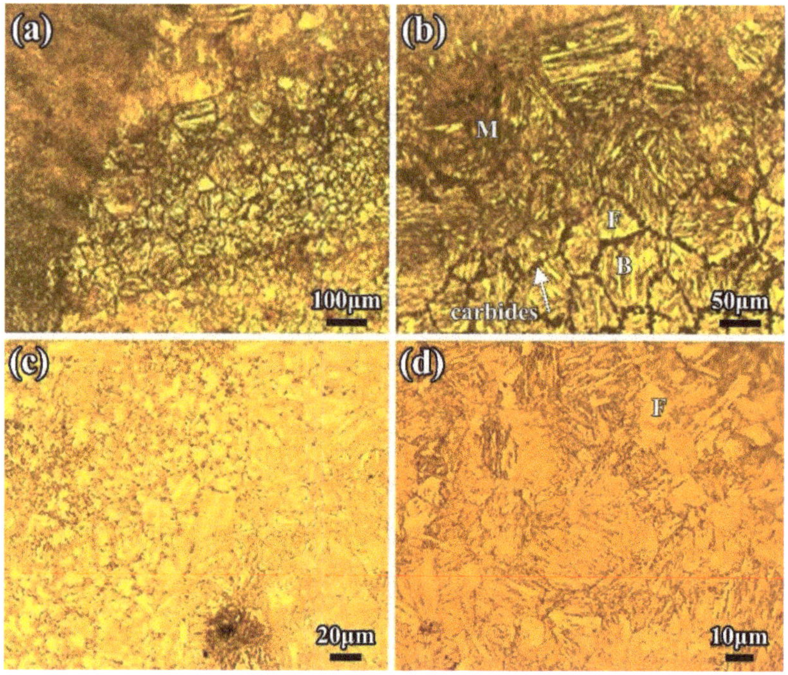

Figure 10. Microstructure in HAZ of SMAW joint (**a**) Macro-photo in HAZ (**b**) Chain-distribution of carbides in HAZ (**c**) OTZ (**d**) OTZ close to BM. F: ferrite; M: martensite; B: bainite.

The total heat input of the SMAW with filler wire was greater than that of the LBW; the difference in microstructure resulted in different performances of their mechanical properties. The comparison results are shown in Table 3.

Table 3. Microstructure comparison between two processes.

Welded Joint	OTZ	TPZ	FGHAZ	CGHAZ	WM
LBW (9.6 kJ/cm)	multipase + polygon F	multipase + band F	Fine M + B + few F	Coarse M + B	F + B + few M
SMAW (16.4 kJ/cm)	unobvious	unobvious band	Majority B + F, few M	Majority B, few F	F + B, Complex

Meanwhile, the differences in microstructure can be described through the content of phases and the width of each zone in particular. In order to find the difference between each welding, the abovementioned average statistical method was used and the statistical results are shown in Figure 11.

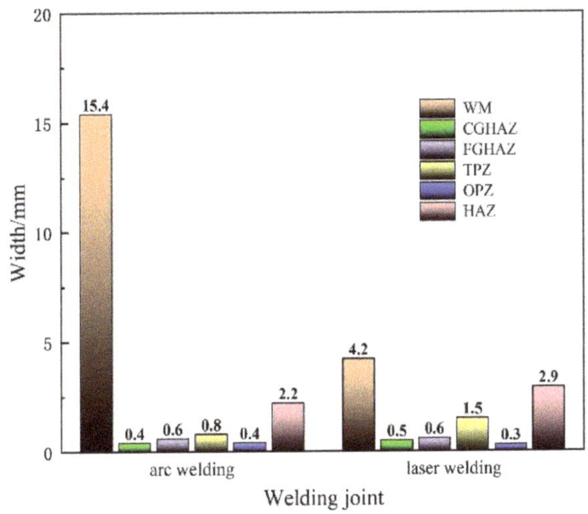

Figure 11. Dimensions of each zone in different welding joints.

The results show that significant differences between two joints are as follows:

The joint weld by multi-layer multi-pass SMAW results in more complex microstructures and double-HAZ, which are composed of some black duplex phases, like those of the two-phase zone, and there are much more proeutectoid ferrite phases in the WM. The average bead widths of the LBW joint and the SMAW joint are 4.2 mm and 15.4 mm, respectively. The average bead width of the LBW is significantly smaller than that of the SMAW. The performance of mechanical properties in the WM is always poorest in the whole joint. The decrease in bead width in LBW is helpful in order to improve the performance of the whole welded joint.

There is minor difference in the width of the CGHAZ and the FGHAZ for the two joints. The difference is that there is more bainite phases in the CGHAZ and the FGHAZ in the SMAW joint, and several chained carbide particles along the grain boundary; the most probably reason for this is the larger heat input used.

The TPZ is not obvious and its width is 0.8 mm in the SMAW joint, while the width in the LBW joint can reach 1.5 mm, or even more than 2.5 mm, and the performance of the TPZ is slightly lower than that of the base metal. Therefore, the performance of the LBW joint could worsen due to its wider TPZ.

The width of the OTZ in the two joints are almost same and equal to 0.4 mm. The microstructure has no obvious change and it has improved to a degree because of the overheating treatment in the SMAW. However, it is not good for the performance of the OTZ to improve the LBW without any post-weld heat treatment.

3.3. Mechanical Properties

- Microhardness test results

In Figure 12a, the microhardness variations across the middle of the SMAW joint are presented. The hardness of the BM is about 360 $HV_{0.3}$, which is similar to that of the LBW, indicating that post-weld heat treatment had little effect on the hardness of the BM. Because of multi-layer and multi-pass welding, the transverse hardness fluctuation of the arc-welded joint is obvious. It can be seen that the whole hardness spot experienced four weld bead ups and downs, and each weld bead had some degree of hardening and softening, but due to the effect of post-weld heat treatment, the hardening and softening effect was not obvious. The highest hardness is 430 $HV_{0.3}$ in the HAZ between each pass, which is about 30 $HV_{0.3}$ lower than that of the LBW joint, which is the result of the post-weld heat treatment.

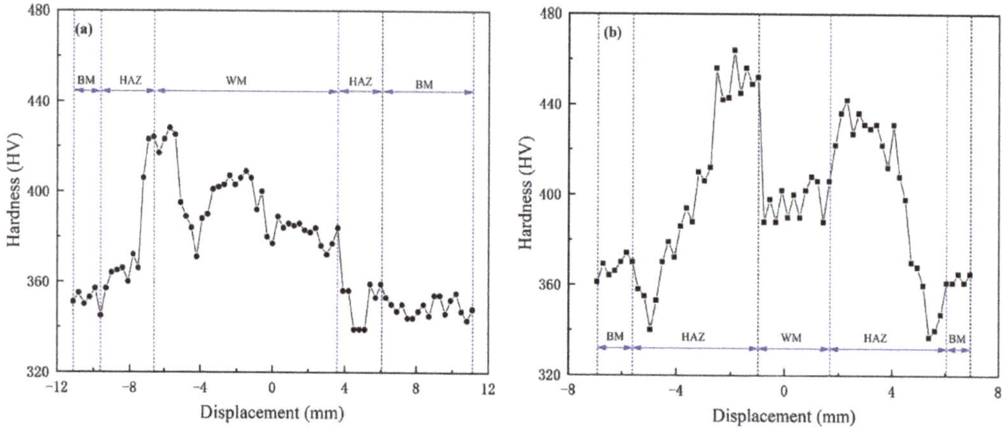

Figure 12. Vickers microhardness distributions (**a**) Hardness area of the SMAW joint (**b**) Hardness area of the LBW joint.

Figure 12b depicts the Vickers microhardness variations across the middle of the LBW joint. The hardness of the BM is about 360 $HV_{0.22}$, and the WM are formed by the remelting of the BM. Because the WM had not undergone any post-weld heat treatment, its hardness is higher than that of the BM, which is about 400 $HV_{0.22}$. There is a certain degree of softening in the OTZ; the lowest hardness is only 334 $HV_{0.22}$, and the width of the whole softening zone is about 1.2 mm.

- Tensile test results

The stress–strain curves for the LBW and SMAW joints are shown in Figure 13, and the average values of the tensile tests for the LBW and SMAW joints are summarized in Table 4. The average yield strength $R_{p0.2}$, the average ultimate tensile strength R_m, and the average elongation A of LBW joints are 898 MPa, 973.5 MPa, and 10%, respectively. Compared to those of the LBW joints, the decline in the average yield strength (896 MPa) and the average ultimate tensile strength (951 MPa) in the SMAW joints is not glaring, except that the average elongation (2.9%) dramatically decreases. It could be that increasingly more weld microstructures are involved in the tensile behavior from the root of the weld to the top of the SMAW joint, resulting in a smaller elongation of the tensile coupon. In Table 4, it can be found that the fracture of most of specimens occurred in the BM for the LBW and SMAW joints; only one fracture occurred in the WM of LBW joint due to weld defects.

Table 4. Results of tensile test.

Welding Joint	Notch Position	Impact Energy (J, −40 °C)			
		①	②	③	④
BM	BM	32.21	unbroken	unbroken	30.59
LBW	WM	15.40	3.16	17.66	16.23
	HAZ	14.92	24.10	19.82	20.87
SMAW	WM	13.46	16.96	15.84	14.37
	HAZ	20.40	19.52	18.86	31.96

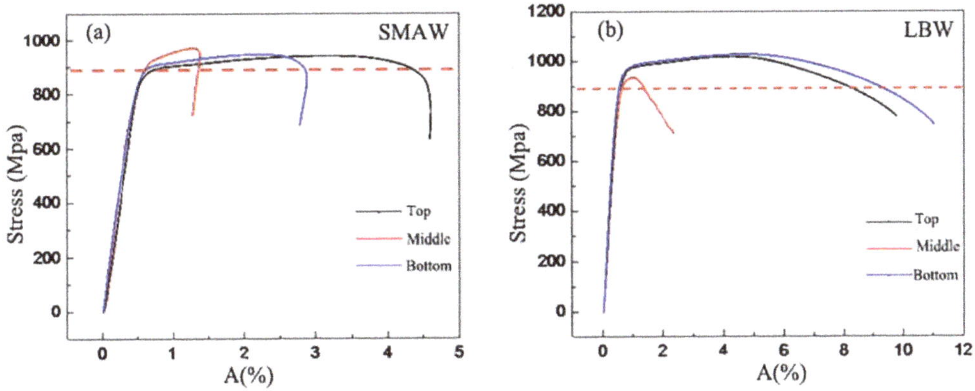

Figure 13. Stress–strain curves of the joints (**a**) the SMAW (**b**) the LBW.

- Impact test results

Table 5 depicts results of the impact tests at −40 °C, showing that the Charpy impact energy of the WM (16.4 J) is less than the HAZ (18.5 J) samples employed in the LBW joints. Additionally, the Charpy impact energy of the WM (15.1 J) is significantly less than the HAZ (19.5 J) for all SMAW samples. The Charpy impact energy of the WM for the SMAW samples is lower than that of the LBW samples. The presented post-weld heat treatment in the SMAW process could influence part of the HAZ and improve HAZ toughness. Because of this, the Charpy impact energy of the HAZ for the SMAW samples is a little higher than that of the LBW samples.

Table 5. Results of the Charpy impact test.

Sample Position		*E* (GPa)	$R_{p0.2}$ (MPa)	R_m (MPa)	*A* (%)	Fracture Position
LBW	top	175	901	969	9.3	BM
	middle	189	896	935	2.0	WM
	bottom	198	895	978	10.7	BM
SMAW	top	193	900	955	1.4	BM
	middle	170	894	950	2.8	BM
	bottom	175	895	948	4.5	BM

Additionally, the prescribed table values came from the unstandardized impact specimens (2.5 × 10 × 55 mm) compared to ISO standard samples (10 × 10 × 55 mm); an equivalent Charpy impact energy value is equal to the presented value multiplies 2 or 4 [27]. Therefore, even though the minimum value of the WM for the SMAW sample is equivalent to 26.92 J, it meets the requirement for the Charpy impact energy value (27 J).

Figure 14 shows the impact fracture photo of the WM of the LBW joint. From Figure 14a, it can be seen that there is almost no fiber area, the fracture surface is relatively flat, and pits occasionally exist, showing brittle fracture characteristics. Figure 14b is an enlarged view of the fracture fiber area. It can be seen that there are relatively few dimples and a pore with a size of 10 μm. Figure 14c is the morphology of the fiber region at higher magnification, and dimples vary in size between 0.5–3 μm. Figure 14d shows a high-magnification dimple photo in the radiative region, where the shear characteristics are more significant compared with Figure 14c. On the whole, the fracture impact energy of the WM is low, with 11.50 J, and the mode of failure is brittle.

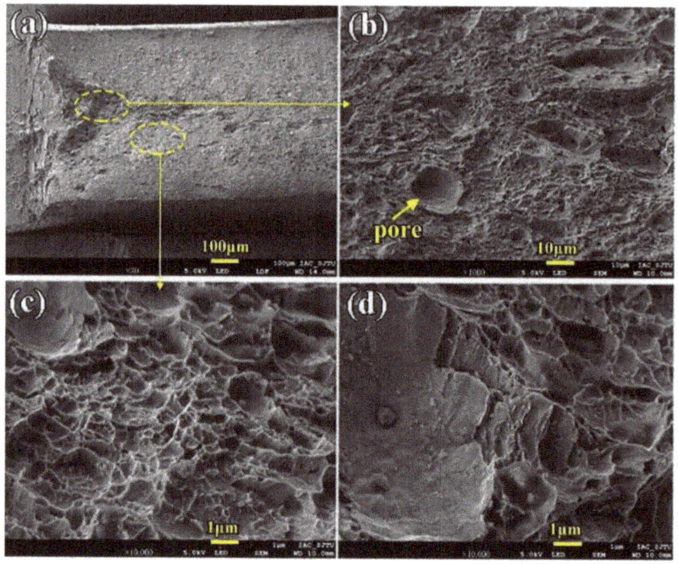

Figure 14. The WM impact fracture surface morphology (**a**) Macro-morphology (**b**) Fiber area morphology (**c**) Dimples details in fiber area (**d**) Dimples details in radiation area.

4. Conclusions

In the present work, the mechanical properties and microstructural features of SMAW and LBW welded joints for 16-mm-thick HSLA Q890 steel were discussed. The microhardness of the welded joint results showed that the LBW was slightly higher than the SMAW, meaning that the post-weld heat treatment associated with the SMAW process can improve microhardness. Furthermore, the tensile results showed that tensile failure occurred in the base metal; the ultimate tensile strength and percent elongation of the LBW welded joint (973.5 MPa and 10%) were higher than those of the SMAW joint (951 MPa and 2.9%) due to the filler filling of the SMAW process. Finally, the impact results showed that the Charpy impact energy of the weld metal (16.4 J and 15.1 J) was lower than that of the heat-affected zone (18.5 J and 19.5 J) in the LBW joint and the SMAW joint.

Author Contributions: Conceptualization, H.C.; validation, Y.L.; writing—original draft preparation, Y.L. and Y.C.; writing—review and editing, Y.C. and X.X. All authors have read and agreed to the published version of the manuscript.

Funding: This research was funded by National Natural Science Foundation of China grant number 51809161 and 52005315.

Institutional Review Board Statement: This research did not require ethical approval.

Informed Consent Statement: This research did not involve humans.

Data Availability Statement: We are willing to share our research data in MDPI journals.

Acknowledgments: The authors gratefully acknowledge the financial support by the National Natural Science Foundation of China (No. 51809161, No. 52005315), Shanghai local colleges and universities construction project supported by the Shanghai Committee of Science and Technology (No. 23010501100) to carry out the present research work. The authors would like to particularly acknowledge to Lu Fenggui from Shanghai Jiaotong University for fruitful discussions and suggestions.

Conflicts of Interest: The authors declare that they have no conflict of competing financial interests in this paper.

References

1. Alipooramirabad, H.; Ghomashchi, R.; Paradowska, A.; Reid, M. Residual stress-microstructure-mechanical property inter relationships in multi pass HSLA steel welds. *J. Mater. Process. Technol.* **2016**, *231*, 456–467. [CrossRef]
2. Hu, X.; Feng, G.; Wang, Y.; Zhang, C.; Deng, D. Influence of lumping passes on calculation accuracy and efficiency of welding residual stress of thick-plate butt joint in boiling water reactor. *Eng. Struct.* **2020**, *222*, 111136. [CrossRef]
3. Fisher, C.R.; Nygren, K.E.; Beaudoin, A.J. Validation of materials-informed digital twin: Mapping residual strains in HSLA steel weldment using high energy X-rays. *J. Manuf. Process.* **2022**, *74*, 75–87. [CrossRef]
4. Hariprasath, P.; Sivaraj, P.; Balasubramanian, V.; Pilli, S.; Sridhar, K. Effect of the welding technique on mechanical properties and metallurgical characteristics of the naval grade high strength low alloy steel joints produced by SMAW and GMAW. *CIRP J. Manuf. Sci. Technol.* **2022**, *37*, 584–595. [CrossRef]
5. Kim, Y.; Hwang, W. Effect of weld seam orientation and welding process on fatigue fracture behaviors of HSLA steel weld joints. *Int. J. Fatigue* **2020**, *137*, 105644. [CrossRef]
6. Palanivel, R.; Dinaharan, I.; Laubscher, R. Microstructure and mechanical behavior of Nd:YAG laser beam welded high strength low alloy steel joints. *Optik Int. J. Light Electron Opt.* **2020**, *208*, 164050. [CrossRef]
7. Mohanty, U.K.; Abe, Y.; Fujimoto, T.; Nakatani, M.; Kitagawa, A.; Tanaka, M.; Suga, T.; Sharma, A. Performance Evaluation of Alternating Current Square Waveform Submerged Arc Welding as a Candidate for Fabrication of Thick Welds in 2.25Cr-1Mo Heat-Resistant Steel. *J. Press. Vessel Technol.* **2020**, *142*, 041506. [CrossRef]
8. Zhou, H.; Zhang, Q.; Yi, B.; Wang, J. Hardness prediction based on microstructure evolution and residual stress evaluation during high tensile thick plate butt welding. *Int. J. Nav. Arch. Ocean Eng.* **2020**, *12*, 146–156. [CrossRef]
9. Anant, R.; Ghosh, P.K. Advancement in narrow gap GMA weld joint of thick section of austenitic stainless steel to HSLA steel. *Mater. Today Proc.* **2017**, *4*, 10169–10173. [CrossRef]
10. Alipooramirabad, H.; Paradowska, A.; Ghomashchi, R.; Reid, M. Investigating the effects of welding process on residual stresses, microstructure and mechanical properties in HSLA steel welds. *J. Manuf. Process.* **2017**, *28*, 70–81. [CrossRef]
11. Banik, S.D.; Kumar, S.; Singh, P.K.; Bhattacharya, S.; Mahapatra, M.M. Distortion and residual stresses in thick plate weld joint of austenitic stainless steel: Experiments and analysis. *J. Mater. Process. Technol.* **2021**, *289*, 116944. [CrossRef]
12. Li, R.; Wang, T.; Wang, C.; Yan, F.; Shao, X.; Hu, X.; Li, J. A study of narrow gap laser welding for thick plates using the multi-layer and multi-pass method. *Opt. Laser Technol.* **2014**, *64*, 172–183. [CrossRef]
13. Long, J.; Zhang, L.-J.; Zhuang, M.-X.; Bai, L.-A.; Na, S.-J. Narrow-gap laser welding with beam wobbling and filler wire and microstructural performance of joints of thick TC4 titanium alloy plates. *Opt. Laser Technol.* **2022**, *152*, 108089. [CrossRef]
14. Bunaziv, I.; Akselsen, O.M.; Frostevarg, J.; Kaplan, A.F. Deep penetration fiber laser-arc hybrid welding of thick HSLA steel. *J. Mater. Process. Technol.* **2018**, *256*, 216–228. [CrossRef]
15. Sun, Y.; Fujii, H.; Morisada, Y. Double-sided friction stir welding of 40 mm thick low carbon steel plates using a pcBN rotating tool. *J. Manuf. Process.* **2020**, *50*, 319–328. [CrossRef]
16. Šebestová, H.; Horník, P.; Mrňa, L.; Doležal, P. Microstructure and mechanical properties of hybrid LasTIG welds of HSLA steel. *Procedia CIRP* **2018**, *74*, 743–747. [CrossRef]
17. Coelho, R.S.; Corpas, M.; Moreto, J.A.; Jahn, A.; Standfuß, J.; Kaysser-Pyzalla, A.; Pinto, H. Induction-assisted laser beam welding of a thermomechanically rolled HSLA S500MC steel: A microstructure and residual stress assessment. *Mater. Sci. Eng. A* **2013**, *578*, 125–133. [CrossRef]
18. Moshtaghi, M.; Loder, B.; Safyari, M.; Willidal, T.; Hojo, T.; Mori, G. Hydrogen trapping and desorption affected by ferrite grain boundary types in shielded metal and flux-cored arc weldments with Ni addition. *Int. J. Hydrog. Energy* **2022**, *47*, 20676–20683. [CrossRef]
19. Yang, J.; Oliveira, J.; Li, Y.; Tan, C.; Gao, C.; Zhao, Y.; Yu, Z. Laser techniques for dissimilar joining of aluminum alloys to steels: A critical review. *J. Mater. Process. Technol.* **2022**, *301*, 117443. [CrossRef]
20. Yang, J.; Su, J.; Gao, C.; Zhao, Y.; Liu, H.; Oliveira, J.P.; Tan, C.; Yu, Z. Effect of heat input on interfacial microstructure, tensile and bending properties of dissimilar Al/steel lap joints by laser Welding-brazing. *Opt. Laser Technol.* **2021**, *142*, 107218.
21. Ali, M.; Khosravifard, A.; Hamada, A.; Mattar, T.; Eissa, M.; Kömi, J. Promotion of thermomechanical processing of 2-GPa low-alloyed ultrahigh-strength steel and physically based modelling of the deformation behaviour. *Mater. Sci. Eng. A* **2023**, *867*, 144747. [CrossRef]
22. Khalaj, G.; Nazari, A.; Pouraliakbar, H. Prediction of martensite fraction of microalloyed steel by artificial neural networks. *Neural Netw. World* **2013**, *23*, 117–130. [CrossRef]
23. Khalaj, G.; Yoozbashizadeh, H.; Khodabandeh, A.; Tamizifar, M. Austenite grain growth modelling in weld heat affected zone of Nb/Ti microalloyed linepipe steel. *Mater. Sci. Technol.* **2014**, *30*, 424–433. [CrossRef]
24. *SS-EN 1011-2*; Welding—Recommendations for Welding of Metallic Materials, Part 2: Arc Welding of Ferritic Steels. Beuth Verlag GmbH: Berlin, Germany, 2001.
25. *GB/T228-2002*; Metallic Materials—Tensile Testing at Ambient Temperature. China Standard Press: Beijing, China, 2002. (In Chinese)

26. *GB/T 229-2007*; Metallic Materials—Charpy Pendulum Impact Test Method. China Standard Press: Beijing, China, 2007. (In Chinese)
27. Cao, S.J. Study on equivalent ratio of absorbed energies from Charpy impact tests for specimens of different sizes. *Heat Treat.* **2010**, *25*, 60–64.

materials

MDPI

Article

A Lightweight $AlCrTiV_{0.5}Cu_x$ High-Entropy Alloy with Excellent Corrosion Resistance

Zhen Peng [1,*], Baowei Li [1], Zaibin Luo [1], Xuefei Chen [1], Yao Tang [2], Guannan Yang [3,*] and Pan Gong [4]

1 School of Materials Science and Engineering, Jiangsu University, Zhenjiang 212013, China
2 China Steel Development Research Institute, No.95 Longfusi Street, Dongcheng District, Beijing 100009, China
3 State Key Laboratory of Precision Electronic Manufacturing Technology and Equipment, Guangdong University of Technology, Guangzhou 510006, China
4 State Key Laboratory of Materials Processing and Die & Mould Technology, School of Materials Science and Engineering, Huazhong University of Science and Technology, No. 1037 Luoyu Road, Wuhan 430074, China; pangong@hust.edu.cn
* Correspondence: peng@ujs.edu.cn (Z.P.); ygn@gdut.edu.cn (G.Y.)

Abstract: Lightweight high-entropy alloys (HEAs) are a new class of low-density, high strength-to-weight ratio metallic structural material. Understanding their corrosion behavior is crucial for designing microstructures for their practical applications. This work investigates the electrochemical corrosion behavior of lightweight HEAs $AlCrTiV_{0.5}Cu_x$ (x = 0, 0.2, 0.4, 0.6, 0.8, and 1.0) in a 0.6 M NaCl solution. These HEAs were produced by vacuum arc melting. In contrast to 304L stainless steel, all of the alloys exhibited lower current density levels caused by self-corrosion, with $AlCrTiV_{0.5}$ demonstrating the highest corrosion resistance (0.131 $\mu A/cm^2$). Corrosion resistance decreased along with the content of copper because copper segregation accelerated local corrosion throughout the alloy.

Keywords: corrosion resistance; high entropy alloys; microstructure; lightweight

Citation: Peng, Z.; Li, B.; Luo, Z.; Chen, X.; Tang, Y.; Yang, G.; Gong, P. A Lightweight $AlCrTiV_{0.5}Cu_x$ High-Entropy Alloy with Excellent Corrosion Resistance. *Materials* **2023**, *16*, 2922. https://doi.org/10.3390/ma16072922

Academic Editor: Elena Villa

Received: 28 February 2023
Revised: 28 March 2023
Accepted: 4 April 2023
Published: 6 April 2023

1. Introduction

High-entropy alloys, also known as HEAs, are generally composed of five or more elements in equimolar or near-equimolar ratios. These alloys are stable in a disordered solid solution state owing to a high mixing entropy [1,2]. Over the past twenty years, there has been an explosion of interest in a novel concept of alloy design that allows for the creation of materials with superior mechanical or physical properties. These properties include high yield strengths, ductility, and fracture toughness, in addition to superior high-temperature oxidation resistance and corrosion resistance. As a result of these features, HEAs have found applications in a huge range of different fields [3–13].

However, despite the application potential being high, the high density of many HEAs greatly inhibits the actual deployment of these applications. For example, the density of a CoCrFeNiMn HEA is 8.1 g/cm^3, while the density of refractory high-enthalpy alloys (HEAs) is more than 10 g/cm^3. Both of these HEAs have a mass density that is higher than that of steel, which has a density of 7.9 g per cubic centimeter. For the purpose of increasing the effectiveness of systems that convert energy, it is necessary to reduce the weight. This necessitates the use of building and engineering materials that have a low density [14–20]. Recent studies have shown a significant interest in the development of high-entropy alloys that are also lightweight. While making lightweight HEAs, it is common practice to incorporate significant amounts of light components such as aluminium, titanium, magnesium, and lithium. The incorporation of Al (also for the purpose of alloy strengthening) changes the phase composition and microstructure of HEAs to produce a multiphase microstructure that can include FCC, BCC, and B2 structures [21,22], and the incorporation of Ti can easily lead to the production of secondary phases. Single-phase equiatomic AlCrTiV HEA has a low density of 5.06 g/cm^3 and a hardness of 498 HV. The AlCrTiV quaternary system was selected by Huang et al. [23], who further increased the material's hardness by including lightweight microalloying components. Microalloyed AlCrTiV alloys have a maximum hardness of 710 HV and a density of around 4.5 g per cubic centimeter. In addition to their mechanical advantages, high-performance additives provide excellent resistance to corrosion, which is an important factor in the application of engineering and structural materials.

The microstructure, alloying components, and production procedure of lightweight HEAs are the primary factors that decide the corrosion behavior in general. Tseng and colleagues [24] have developed a lightweight HEA material with the formula $Al_{20}Be_{20}Fe_{10}Si_{15}Ti_{35}$ that has a low density of 3.91 g/cm^3 and a high hardness of 911 HV. At temperatures of 700 and 900 °C, the alloy had an oxidation resistance that was remarkable and was much superior to that of Ti6Al4V. Ji et al. [25] conducted a study in which they manufactured a low-cost $Al_{35}Mg_{30\text{-}x}Zn_{30}Cu_5Si_x$ HEA that had a low density. They found that the resistance of as-cast alloys to corrosion in a solution containing 3.5 wt% NaCl may be improved by increasing the Si/Mg ratio. They also found that corrosion occurred primarily at the grain boundaries and gradually spread into the eutectic and intermetallic phases that included magnesium. This was a new finding for them. According to the findings of Qiu et al. [26], AlTiVCr HEA exhibited remarkable resistance to corrosion in 0.6 M NaCl, much higher than that of pure Al and 304SS.

It is possible to improve the qualities of a material by designing the formation of secondary phases and intermetallic phases inside the substance. The same strategy for strengthening materials, namely alloying, was used in the development of HEAs. The addition of Cu to HEAs is a practical method that may be used to achieve improved properties [15,27]. Yet, when exposed to aqueous environments, the heterogeneous microstructures of HEAs have the potential to cause localized corrosion. It is very necessary to investigate the impact that Cu concentration has on the corrosion resistance of HEAs in order to guarantee the consistency of industrial applications. In this study, lightweight HEAs made of $AlCrTiV_{0.5}Cu_x$ were created by vacuum arc melting. Their electrochemical corrosion behavior in 0.6 M NaCl solution, composition homogeneity, and multi-scale microstructure were investigated.

2. Experimental Materials and Methods

2.1. Material Preparation

The $AlCrTiV_{0.5}Cu_x$ lightweight HEAs were manufactured in a graphite crucible by the process of vacuum arc melting under the protection of an Ar atmosphere. We picked raw minerals that had an extremely high level of purity, such as aluminum (99.5%), chromium (99.95%), titanium (99.95%), vanadium (99.95%), and copper (99.95%). Each sample was remelted at least six times in an ingot to ensure that the chemical composition remained consistent throughout the melting process and to decrease the amount of oxidation that occurred. A specimen measuring 5 mm × 5 mm × 5 mm was extracted from the selected area in the middle of the ingot by using a wire cutter that was coupled to an electric discharge machine. Every sample was given a grit-2000 grinding on SiC sandpaper before being subjected to ultrasonic cleaning in ethanol.

2.2. Microstructural Characterization

In order to determine the phase structure of the samples, a Rigaku SMARTLAB9 X-ray diffractometer was used with Cu-Kα radiation across a range of 10° to 90° and at a scanning rate of 10°/min. The microstructure and corrosion morphologies were investigated using scanning electron microscopy (SEM) and X-ray energy dispersive spectroscopy (EDS), respectively.

2.3. Electrochemical Corrosion Measurements

Electrochemical corrosion tests were carried out at a temperature of 25 °C in 0.6 M NaCl in the presence of air. In order to carry out these measurements, a diamond polishing paste with particles of 1.5 μm in size was applied to specimens that had a cross-sectional area of 0.25 cm^2. Using a CHI760 electrochemical workstation that was supplied with a standard three-electrode setup, polarization curves and electrochemical impedance spectra (EIS) were measured. The corroded samples were used as working electrodes, a AgCl electrode was used as the reference electrode, and a sheet of platinum was used as the counter electrode in this experiment. The open circuit potential (OCP) was monitored for a period of thirty minutes prior to the EIS and polarization studies in order to guarantee a consistent potential for the duration of the tests. The EIS measurements were collected with a potential amplitude of 5 mV and with frequencies ranging from 100 kHz to 10 MHz. Potentiodynamic polarization curves were obtained with a scan rate of 3 mV/s, with a starting potential of −0.9 V_{SHE} and an ending potential of 1.1 V_{SHE}. In order to guarantee the accuracy of the results, more than three measurements were obtained for each different test situation. When the potentiodynamic polarization studies were finished, the samples were wiped clean with ethanol and then left to dry in the air. When the electrochemical tests were completed, the surface morphology of the samples was analyzed using a field emission scanning electron microscope (FEI Nova Nano450) (FESEM).

3. Results and Discussion

3.1. Microstructure Characterization

Cu has a positive binary enthalpy of mixing and relatively high valence electron concentration (VEC) with other constitutional elements. Figure 1 shows that the mixing enthalpy, VE, and mixing entropy ΔS_{mix} of the $AlCrTiV_{0.5}Cu_x$ HEAs increased with the increase in Cu content, while the atomic size difference δ decreased with Cu addition. Hence, the microstructure and phase stability of $AlCrTiV_{0.5}Cu_x$ HEAs may vary with the composition. Cu tends to segregate out of the matrix due to a positive binary enthalpy of mixing of Cu with other constitutional elements. The $AlCrTiV_{0.5}Cu_x$ HEAs have a relatively large atomic size difference. Based on the solid-solution phase formation rules of high entropy alloys [28], there are probably ordered phases such as intermetallic compounds in the matrix. According to the VEC criterion [29], the phase composition of $AlCrTiV_{0.5}$ is mainly BCC phase. The content of FCC phase increases with the addition of Cu, and $AlCrTiV_{0.5}Cu$ is mainly FCC phase. Figure 2 shows XRD patterns of $AlCrTiV_{0.5}Cu_x$ samples. In summary, with the addition of Cu, the phase composition of $AlCrTiV_{0.5}Cu_x$ HEAs may gradually change from BCC phase to FCC phase, and there may be intermetallic compounds. The segregation of Cu between grains increases with the copper content.

Figure 3 shows the SEM images of the $AlCrTiV_{0.5}Cu_x$ lightweight HEAs. $AlCrTiV_{0.5}$ exhibits a typical equiaxed crystal structure with uniform composition and without obvious precipitation phases. With the addition of Cu, the alloys $AlCrTiV_{0.5}Cu_{0.2}$, $AlCrTiV_{0.5}Cu_{0.4}$, $AlCrTiV_{0.5}Cu_{0.6}$, and $AlCrTiV_{0.5}Cu_{0.8}$ show an equiaxed dendritic structure with more obvious grain boundaries. Moreover, the grain size gradually decreased with increasing Cu content, When the Cu content is further increased, $AlCrTiV_{0.5}Cu$ shows a non-equiaxial dendritic structure with a further reduction in grain size. We presented the X-ray diffraction pattern of lightweight $AlCrTiV_{0.5}Cu_x$ HEAs in a previous work [15].

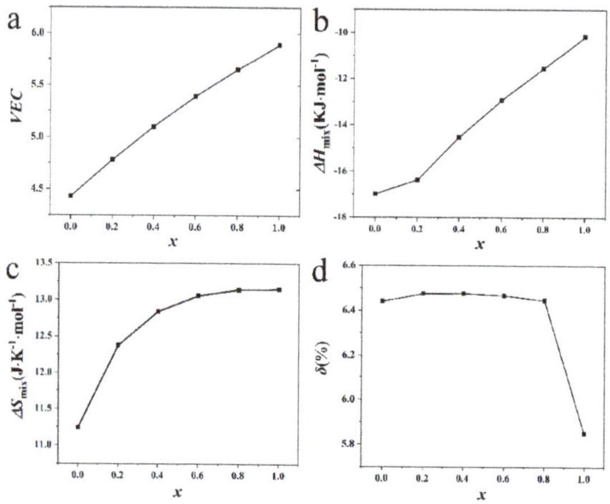

Figure 1. Variation in (**a**)VEC; (**b**) mixing enthalpy ΔH_{mix}; (**c**) mixing entropy ΔS_{mix}; (**d**) atomic size difference δ versus Cu content $x(x = 0, 0.2, 0.4, 0.6, 0.8, 1)$ calculated for the $AlCrTiV_{0.5}Cu_x$ lightweight HEAs.

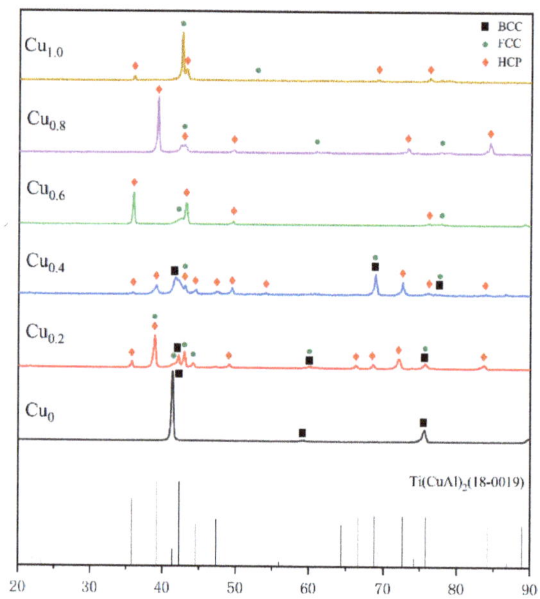

Figure 2. XRD patterns of $AlCrTiV_{0.5}Cu_x$ lightweight HEAs.

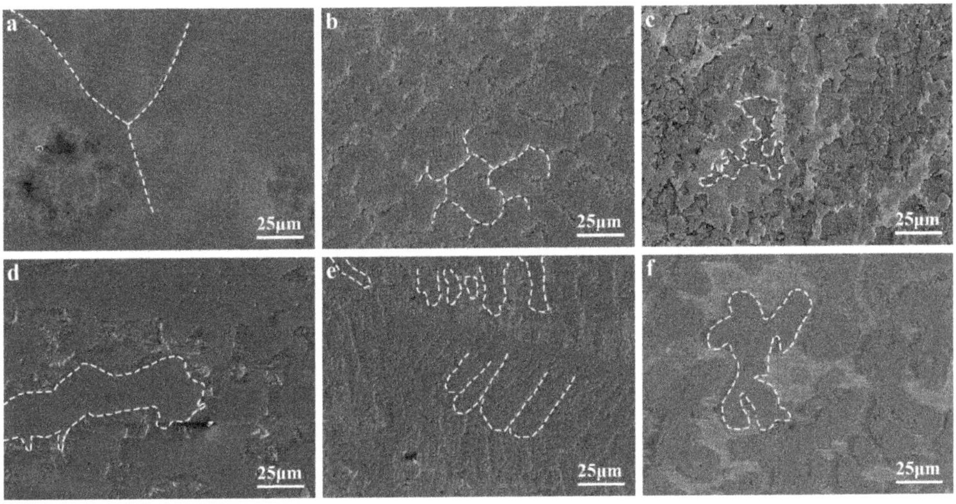

Figure 3. SEM micrographs of the $AlCrTiV_{0.5}Cu_x$ lightweight HEAs. (**a**) $x = 0$; (**b**) $x = 0.2$; (**c**) $x = 0.4$; (**d**) $x = 0.6$; (**e**) $x = 0.8$; (**f**) $x = 1.0$.

Further in-depth research on the elemental composition and dispersion of lightweight $AlCrTiV_{0.5}Cu_x$ HEAs was conducted. The EDS maps of an enlarged portion of the samples are shown in Figure 4 (which shows both the grains and grain borders). The findings indicate that the components are dispersed throughout the samples in an even manner on a microscopic scale. Since Cu is missing, there is no segregation of the elements, and they are all distributed in the same manner. Nevertheless, the results of the XRD investigation reveal that the segregation gets more prominent with increasing Cu concentration. This results in the development of an Al-Ti-Cu-rich HCP phase as well as a V-Cr-rich FCC phase. The inclusion of copper is largely responsible for this kind of alloy having the dendritic structure that is so distinctive in other alloys of this type.

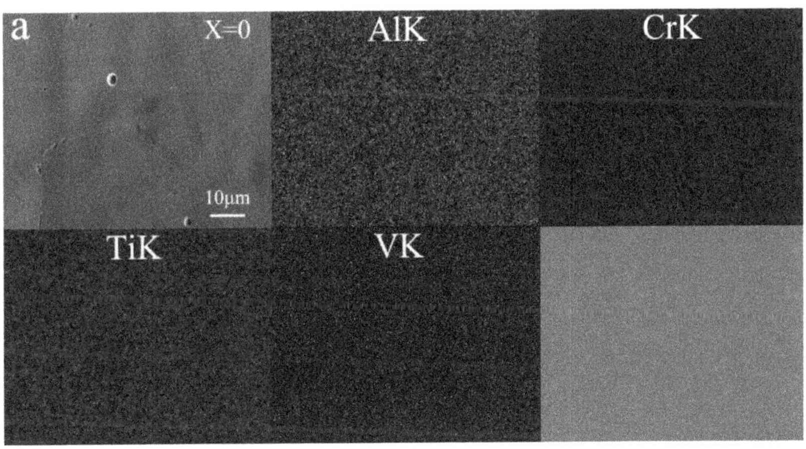

Figure 4. *Cont.*

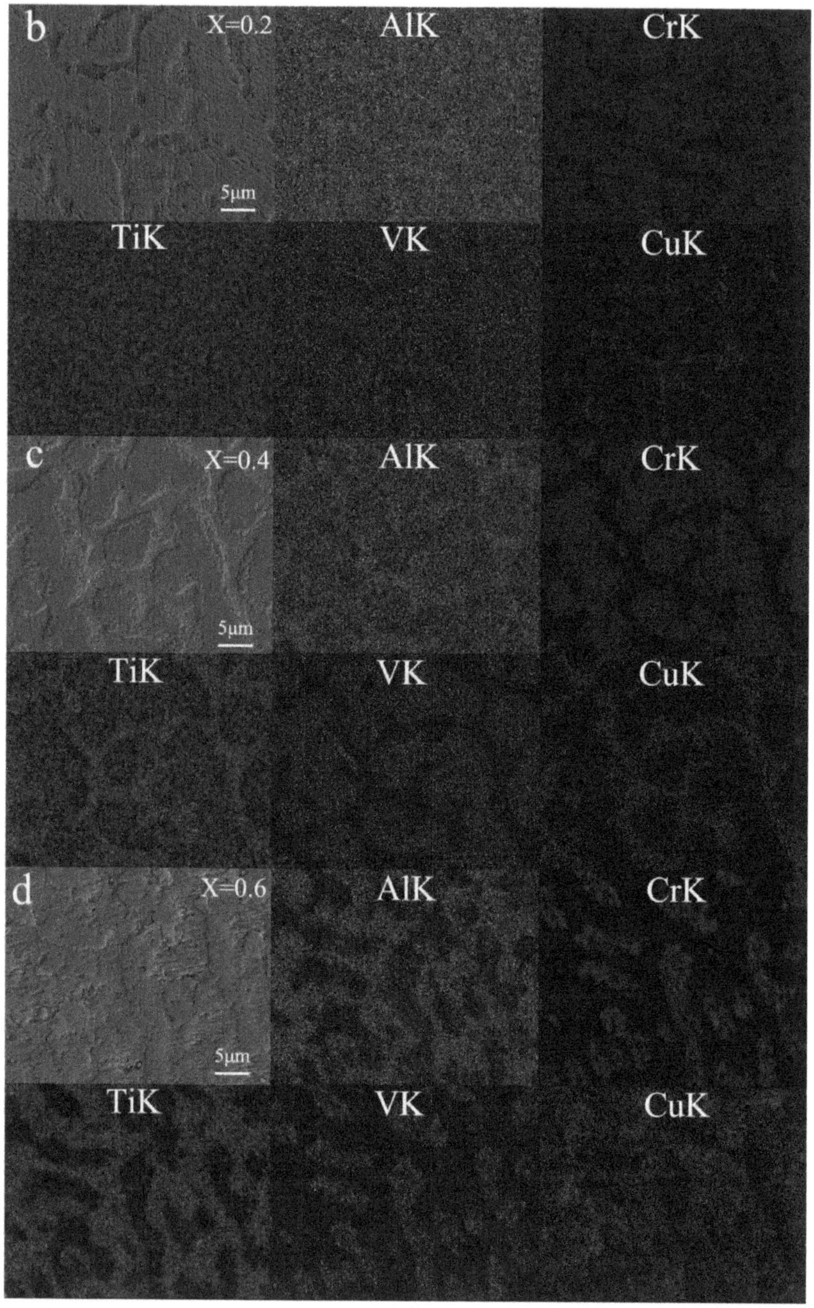

Figure 4. *Cont.*

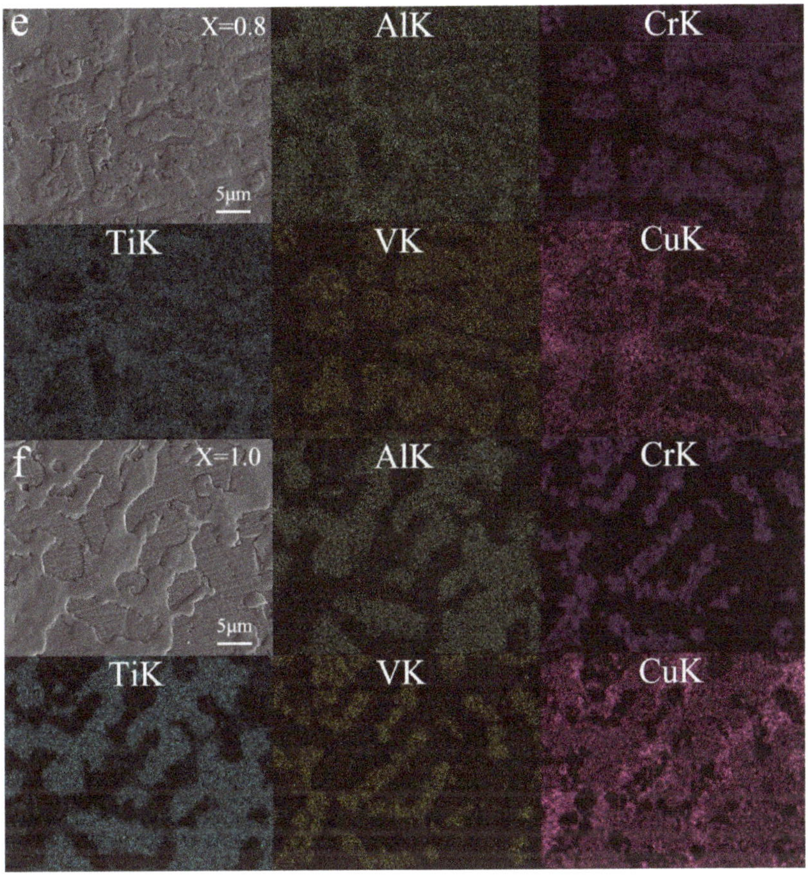

Figure 4. Elemental EDS mapping of $AlCrTiV_{0.5}Cu_x$ lightweight HEAs. (**a**) x = 0; (**b**) x = 0.2; (**c**) x = 0.4; (**d**) x = 0.6; (**e**) x = 0.8; (**f**) x = 1.0.

3.2. Electrochemical Corrosion Behavior in NaCl solution

The potentiodynamic polarization curves of the $AlCrTiV_{0.5}Cu_x$ lightweight HEAs in 0.6 M NaCl solution are shown in Figure 5 and the results of the calculations used to determine the electrochemical corrosion parameters are shown in Table 1. The open-circuit potential of a material is referred to as its corrosion potential, or $E_{corr.}$ It is possible to determine the corrosion rates of materials by utilizing the corrosion current density (I_{corr}). In order to calculate I_{corr} from the Tafel diagram, the linear portion of the polarization curve located close to $E_{corr.}$ is extrapolated. It is clear from both the fitting parameters and the potentiodynamic polarization curves of the alloys that, as the concentration of copper increases, the corrosion current density (I_{corr}) also increases, but the resistance to corrosion decreases. $AlCrTiV_{0.5}$ (0.131 A/cm^2) had the lowest value of I_{corr} and the maximum corrosion resistance out of the six alloys that were tested. This is because the addition of Cu to the alloy will cause segregation at the grain boundaries. Dendrites may be produced by combining a Cu-depleted region with a Cu-rich region. If there is a significant potential difference between the two phases, galvanic coupling can result in corrosion, with the interdendritic area corroding first and increasing the local corrosion of the alloy [30,31]. If there is a significant potential difference between the two phases, galvanic coupling can also result in corrosion. Because of this, the corrosion resistance of the $AlCrTiV_{0.5}Cu_x$ lightweight HEAs in NaCl solution gradually decreases as the content of Cu increases.

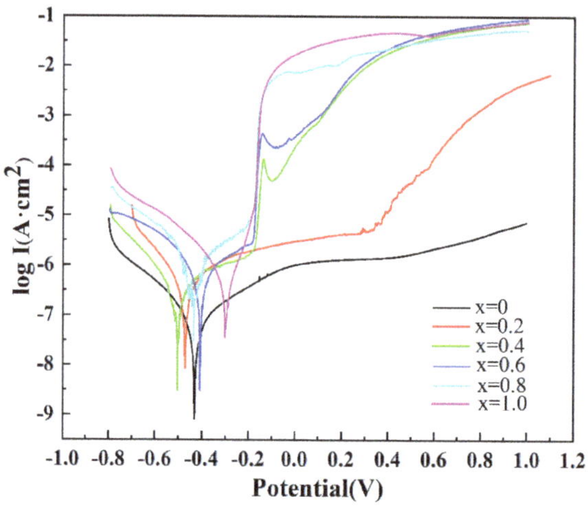

Figure 5. Potentiodynamic polarization curves of AlCrTiV$_{0.5}$Cu$_x$ lightweight HEAs in 0.6 M NaCl solution.

Table 1. Electrochemical corrosion parameters of AlCrTiV$_{0.5}$Cu$_x$ lightweight HEAs obtained from potentiodynamic polarization curve measured in 0.6 M NaCl solution.

Samples	E_{corr}(V)	I_{corr}(μA/cm^2)
$x = 0$	−0.411	0.131
$x = 0.2$	−0.572	0.499
$x = 0.4$	−0.526	0.695
$x = 0.6$	−0.389	1.145
$x = 0.8$	−0.407	2.182
$x = 1.0$	−0.239	2.778
304 L	−0.415	4.7

Figure 6 displays the Nyquist, Bode, and phase-angle charts of the lightweight AlCrTiV$_{0.5}$Cu$_x$ HEAs at a range of temperatures corresponding to varied operating conditions. Z-view was used in order to evaluate the fitted parameters, and Figure 6 presents the equivalent circuit that was produced as a consequence.

A larger semicircle radius indicates that the interface has a higher resistance for the charge transfer and is a more protective passive film [32–35]. Figure 6a depicts the Nyquist plot, and it has the shape of an incomplete semicircle, which indicates that the charge transfer process is in control of the corrosion process. In the high-frequency range, incomplete capacitive arcs are shown by the AlCrTiV$_{0.5}$Cu$_x$ lightweight HEAs. This is evidenced by Figure 6a, which makes it abundantly clear that the incomplete capacitance arcs are caused by charge transfer at inhomogeneous surfaces. As the concentration of copper in the alloy increases, its resistance to corrosion diminishes, and the radius of the capacitive semicircle becomes smaller.

Figure 6b,c show the Bode and phase angle charts of the alloys' electrochemical corrosion, respectively. As can be seen, the impedance modulus and phase-angle change with frequency. The Bode plot shown in Figure 6b demonstrates that the passive films have a pseudocapacitive character since the slopes are less than −1 and the phase angle is smaller than −90°. In Figure 6c, the phase angle is getting close to 90°, and the value of the impedance modulus is linear from 1 Hz all the way up to 103 Hz. The phase angle of the AlCrTiV$_{0.5}$Cu$_x$ alloys decreases from 103 Hz, achieves a tiny value at 104 Hz, and then continues to rise and reaches a high value at 105 Hz. A low value is reached again at 10^4 Hz, and a high value again at 10^5 Hz.

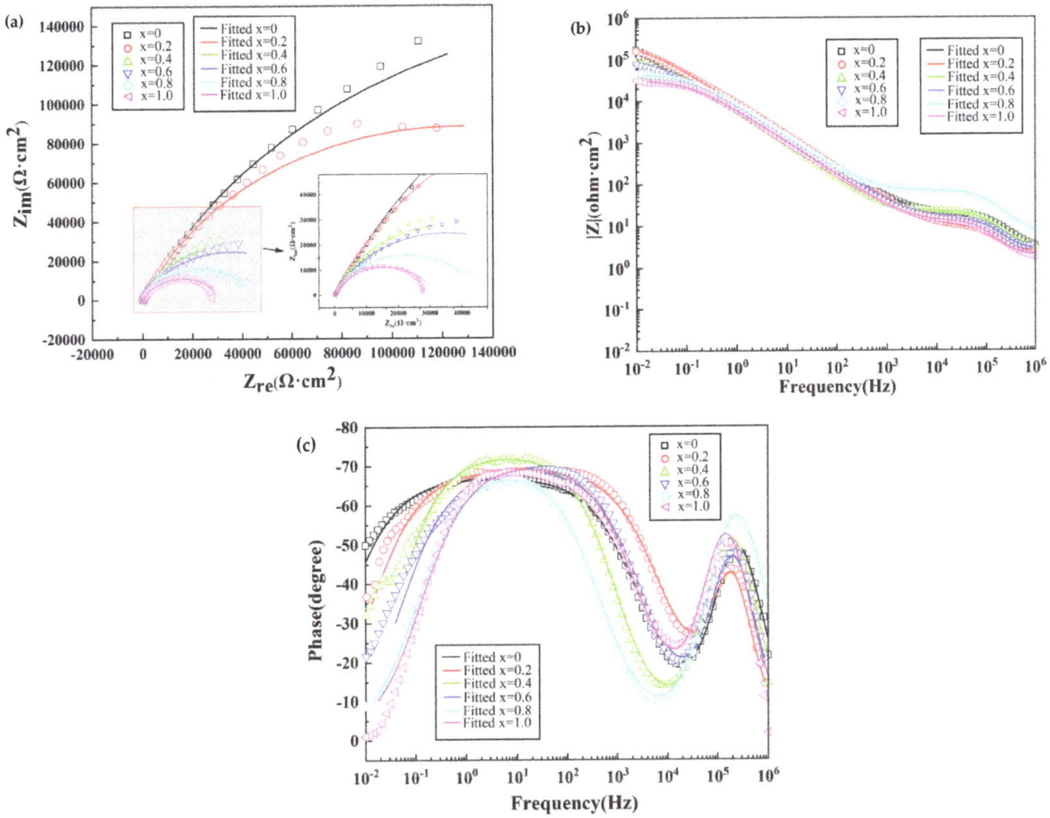

Figure 6. Fitted electrochemical AC impedance of $AlCrTiV_{0.5}Cu_x$ lightweight HEAs in 0.6 M NaCl solution. (**a**) Nyquist plot; (**b**) Bode plot; (**c**) Phase angle plot.

The data on the impedance are used in the production of an equivalent electric circuit (EEC) (shown in Figure 7). R1 represents the resistance of the solution in the EEC model, R2 represents the resistance of the passivated layer, and R3 represents the charge transfer resistance of the limited corrosion zone. In the zone that has been passivated, the capacitance is represented by CPE1, and in the zone that has been partly corroded, it is represented by CPE2. A constant phase-angle element, also known as a CPE, is used in place of a traditional capacitor so that flaws in capacitive components, such as surface inhomogeneities, may be taken into consideration. If the value of the charge transfer resistance, or R3, is lowered, then a greater number of electrons and ions will be able to flow through the passivation layer created by the alloy, which will result in a reduction in the alloy's resistance to corrosion [36–39].

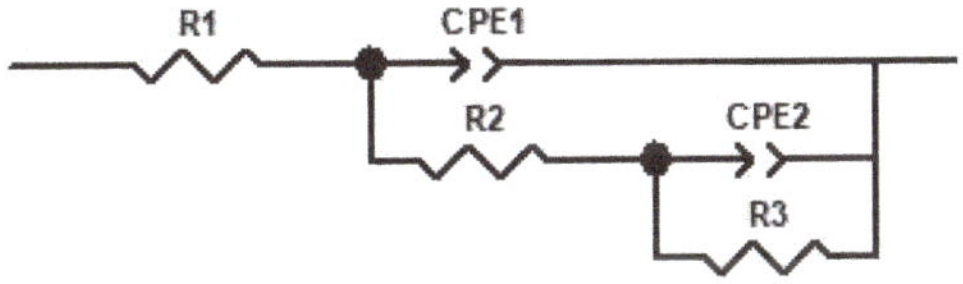

Figure 7. The equivalent electric circuit for impedance data.

Table 2 presents some example parameters for EEC fitting that were generated while operating under OCP conditions. When the value of CPE1-P is close to 0.9, this indicates that the properties of the component are between those of a typical capacitor and those of a Warburg impedance [40–44]. The charge transfer resistance, denoted by R3, will decrease in proportion to the amount of copper that is present. We discover that $AlCrTiV_{0.5}$ has a high charge transfer resistance of 437,920 $\Omega \cdot cm^2$ (and strong corrosion resistance in 0.6 M NaCl solution), while $AlCrTiV_{0.5}Cu$ possesses the lowest charge transfer resistance of 31,276 $\Omega \cdot cm^2$. According to the data presented above, the corrosion process of $AlCrTiV_{0.5}Cu_x$ HEAs is regulated not only by the transfer of charge but also by the regulation of diffusion.

Table 2. Impedance fitting parameters of $AlCrTiV_{0.5}Cu_x$ lightweight HEAs.

Samples	$R_1(\Omega \cdot cm^2)$	CPE_1-T ($\mu F \cdot cm^{-2}$)	CPE_1-P	$R_2(\Omega \cdot cm^2)$	CPE_2-T ($\mu F \cdot cm^{-2}$)	CPE_2-P	$R_3(\Omega \cdot cm^2)$
$x = 0$	3.183	0.0383	0.967	15.09	34.954	0.747	437,920
$x = 0.2$	2.444	0.0359	0.962	6.416	28.217	0.770	255,620
$x = 0.4$	3.106	0.0659	0.994	18.17	38.641	0.817	75,311
$x = 0.6$	2.818	0.0462	0.959	10.8	33.676	0.775	70,653
$x = 0.8$	5.808	0.0344	0.977	59.19	30.418	0.782	45,098
$x = 1.0$	1.934	0.0862	0.871	10.56	37.656	0.790	31,276

3.3. *Corrosion Morphology Analysis*

Figure 8 depicts the surface morphology of the $AlCrTiV_{0.5}Cu_x$ lightweight HEAs after they were subjected to polarization tests in a solution containing 0.6 M NaCl. The corrosion has reached such an advanced stage that the surfaces are severely exfoliated and are in no way uniform. The fact that there are just a few pits on the surface is evidence of limited corrosion; hence, the samples were not fully ruined. The surface roughness was not on the nanoscale scale in the regions of the material that did not include corrosion pits.

Figure 8a depicts the surface morphology of the lightweight $AlCrTiV_{0.5}$ HEA, which demonstrates that the surface is smooth and flat after corrosion and does not exhibit any obvious corrosion pits. The prior results that the $AlCrTiV_{0.5}$ alloy exhibited greater corrosion resistance than the other five HEAs are supported by these new data, which are consistent with those findings. $AlCrTiV_{0.5}$ lightweight HEA has high resistance to corrosion as a result of its uniform composition, single BCC phase structure, and lack of composition segregation at grain boundaries. These characteristics contribute to the material's outstanding uniformity. When $AlCrTiV_{0.5}Cu_x$ is corroded (where x may be any of the values 0.2, 0.4, 0.6, 0.8, or 1.0), substantial aggregation of Cu elements occurs in the dendritic intergranular region (see Figure 8), and enormous corrosion pits also develop. When there is a higher percentage of copper in the alloy, the surface corrosion of the alloy becomes worse.

In general, the $AlCrTiV_{0.5}$ lightweight HEAs show good corrosion resistance in the testing environment. Increased Cu content enhanced Cu segregation, which could induce localized corrosion susceptibility in Cu-added HEAs. The addition of Cu also increases the mixing entropy of the system, thus the effect of the high-entropy effect on the corrosion resistance of the alloys should be considered. The corrosion behavior also seemed to relate to the sluggish diffusion. Dissolution of Cu was dominant in the competitive process of formation of the passive film. Sluggish diffusion can inhibit the migration of Cu, reduce the defects in the passive film, and inhibit the formation of cation vacancies, thereby improving corrosion behavior [35]. In this study, due to the segregation of Cu and Cu-rich areas exposed to the surface, the effect of slow diffusion was not obvious. Only in $AlCrTiV_{0.5}Cu_{0.2}$ HEA, in which the segregation of Cu structure was lower, was the corrosion resistance reduced slowly. When the content of Cu is higher and the degree of segregation is greater, the high-entropy effect has little effect, and the corrosion resistance is significantly reduced.

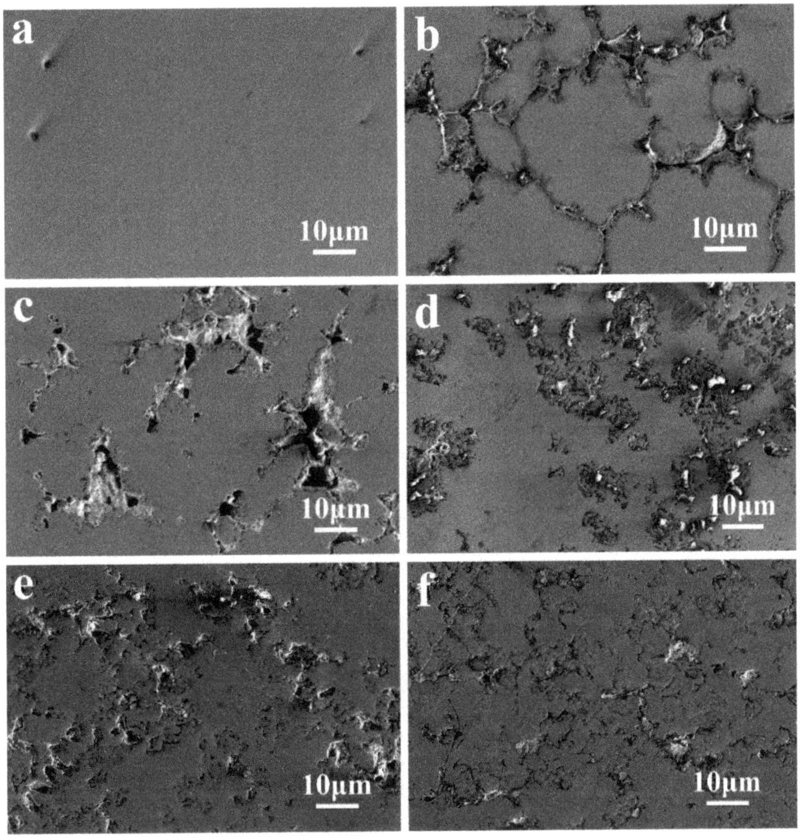

Figure 8. Surface morphology of $AlCrTiV_{0.5}Cu_x$ lightweight HEAs after kinetic potential polarization tests in 0.6 M NaCl solution. (**a**) $x = 0$; (**b**) $x = 0.2$; (**c**) $x = 0.4$; (**d**) $x = 0.6$; (**e**) $x = 0.8$; (**f**) $x = 1.0$.

4. Conclusions

(1) In a solution of 0.6 M NaCl, the electrochemical behavior of $AlCrTiV_{0.5}Cu_x$ lightweight HEAs changes depending on the proportion of Cu. While $AlCrTiV_{0.5}$ has the best corrosion resistance and the lowest self-corrosion current density at 0.131 $\mu A/cm^2$, $AlCrTiV_{0.5}Cu$ has the greatest self-corrosion current density at 2.778 $\mu A/cm^2$ and the poorest corrosion resistance. $AlCrTiV_{0.5}$ has the lowest self-corrosion current density and the best corrosion resistance.

(2) When Cu is added to the HEAs, it is polarised between the dendrites, forming a Cu-rich phase. A higher Cu content leads to more pronounced polarization. The segregation of Cu lead to a large potential difference between the Cu-rich and Cu-poor phases that formed between and within the dendrites, making the area between the dendrites more susceptible to galvanic coupling corrosion.

Author Contributions: Z.P. organized the work, supervised the research progress, and wrote the manuscript. B.L.; Z.L. and Y.T. conducted the sample preparation, performed the microstructural analysis. X.C., G.Y. and P.G. revised the manuscript, and contributed to the discussion. All authors have read and agreed to the published version of the manuscript.

Funding: This research was supported by the National Natural Science Foundation of China (No. 52271147, and No. 12261160364) and the Knowledge Innovation Program of Wuhan-Basic Research.

Institutional Review Board Statement: Not Applicable.

Informed Consent Statement: Not Applicable.

Data Availability Statement: The data presented in this work are available on request from the corresponding authors.

Conflicts of Interest: The authors declare no conflict of interest.

References

1. Yeh, J.W.; Chen, S.K.; Lin, S.J.; Gan, J.Y.; Chin, T.S.; Shun, T.T.; Tsau, C.H.; Chang, S.Y. Nanostructured High-Entropy Alloys with Multiple Principal Elements: Novel Alloy Design Concepts and Outcomes. *Adv. Eng. Mater.* **2004**, *6*, 299–303. [CrossRef]
2. Cantor, B.; Chang, I.T.H.; Knight, P.; Vincent, A.J.B. Microstructural development in equiatomic multicomponent alloys. *Mater. Sci. Eng. A* **2004**, *375–377*, 213–218. [CrossRef]
3. Miracle, D.B.; Senkov, O.N. A critical review of high entropy alloys and related concepts. *Acta Mater.* **2017**, *122*, 448–511. [CrossRef]
4. Otto, F.; Dlouhý, A.; Somsen, C.; Bei, H.; Eggeler, G.; George, E.P. The influences of temperature and microstructure on the tensile properties of a CoCrFeMnNi high-entropy alloy. *Acta Mater.* **2013**, *61*, 5743–5755. [CrossRef]
5. Niu, S.Z.; Kou, H.C.; Wang, J.; Li, J.S. Improved tensile properties of Al0.5CoCrFeNi high-entropy alloy by tailoring microstructures. *Rare Met.* **2021**, *40*, 1–6. [CrossRef]
6. Lei, Z.F.; Liu, X.J.; Wu, Y.; Wang, H.; Jiang, S.; Wang, S.D.; Hui, X.D.; Wu, Y.D.; Gault, B.; Kontis, P.; et al. Enhanced strength and ductility in a high-entropy alloy via ordered oxygen complexes. *Nature* **2018**, *563*, 546–550. [CrossRef] [PubMed]
7. Lu, Z.P.; Wang, H.; Chen, M.W.; Baker, I.; Yeh, J.W.; Liu, C.T.; Nieh, T.G. An assessment on the future development of high-entropy alloys: Summary from a recent workshop. *Intermetallics* **2015**, *66*, 67–76. [CrossRef]
8. Bhalla, S.; Melnekoff, D.T.; Aleman, A.; Leshchenko, V.; Restrepo, P.; Keats, J.; Onel, K.; Sawyer, J.R.; Madduri, D.; Richter, J.; et al. Patient similarity network of newly diagnosed multiple myeloma identifies patient subgroups with distinct genetic features and clinical implications. *Sci. Adv.* **2021**, *7*, eabg9551. [CrossRef] [PubMed]
9. Luan, H.W.; Zhang, X.; Ding, H.Y.; Zhang, F.; Luan, J.H.; Jiao, Z.B.; Yang, Y.C.; Bu, H.T.; Wang, R.; Gu, J.L.; et al. High-entropy induced a glass-to-glass transition in a metallic glass. *Nat. Commun.* **2022**, *13*, 2183. [CrossRef] [PubMed]
10. Peng, Z.; Sun, J.; Luan, H.W.; Chen, N.; Yao, K.F. Effect of Mo on the high temperature oxidation behavior of Al19Fe20-xCo20-xNi41Mo2x high entropy alloys. *Intermetallics* **2023**, *155*, 107845. [CrossRef]
11. Zhang, Y.; Zuo, T.T.; Tang, Z.; Gao, M.C.; Dahmen, K.A.; Liaw, P.K.; Lu, Z.P. Microstructures and properties of high-entropy alloys. *Prog. Mater. Sci.* **2014**, *61*, 1–93. [CrossRef]
12. Chen, X.F.; Wang, Q.; Cheng, Z.Y.; Zhu, M.L.; Zhou, H.; Jiang, P.; Zhou, L.L.; Xue, Q.; Yuan, F.P.; Zhu, J.; et al. Direct observation of chemical short-range order in a medium-entropy alloy. *Nature* **2021**, *592*, 712–716. [CrossRef] [PubMed]
13. George, E.P.; Raabe, D.; Ritchie, R.O. High-entropy alloys. *Nat. Rev. Mater.* **2019**, *4*, 515–534. [CrossRef]
14. Qiu, Y.; Hu, Y.J.; Taylor, A.; Styles, M.J.; Marceau, R.K.W.; Ceguerra, A.V.; Gibson, M.A.; Liu, Z.K.; Fraser, H.L.; Birbilis, N. A lightweight single-phase AlTiVCr compositionally complex alloy. *Acta Mater.* **2017**, *123*, 115–124. [CrossRef]
15. Peng, Z.; Luo, Z.B.; Li, B.W.; Li, J.F.; Luan, H.W.; Gu, J.L.; Wu, Y.; Yao, K.F. Microstructure and mechanical properties of lightweight AlCrTiV0.5Cux high-entropy alloys. *Rare Met.* **2022**, *41*, 2016–2020. [CrossRef]
16. Stepanov, N.D.; Yurchenko, N.Y.; Skibin, D.V.; Tikhonovsky, M.A.; Salishchev, G.A. Structure and mechanical properties of the AlCrxNbTiV (x = 0, 0.5, 1, 1.5) high entropy alloys. *J. Alloy. Compd.* **2015**, *652*, 266–280. [CrossRef]
17. Senkov, O.N.; Senkova, S.V.; Miracle, D.B.; Woodward, C. Mechanical properties of low-density, refractory multi-principal element alloys of the Cr–Nb–Ti–V–Zr system. *Mater. Sci. Eng. A* **2013**, *565*, 51–62. [CrossRef]
18. Youssef, K.M.; Zaddach, A.J.; Niu, C.; Irving, D.L.; Koch, C.C. A Novel Low-Density, High-Hardness, High-entropy Alloy with Close-packed Single-phase Nanocrystalline Structures. *Mater. Res. Lett.* **2015**, *3*, 95–99. [CrossRef]
19. Stepanov, N.D.; Shaysultanov, D.G.; Salishchev, G.A.; Tikhonovsky, M.A. Structure and mechanical properties of a light-weight AlNbTiV high entropy alloy. *Mater. Lett.* **2015**, *142*, 153–155. [CrossRef]
20. Jia, Y.F.; Jia, Y.D.; Wu, S.W.; Ma, X.D.; Wang, G. Novel Ultralight-Weight Complex Concentrated Alloys with High Strength. *Materials* **2019**, *12*, 1136. [CrossRef]
21. Sun, Z.J.; Tan, X.P.; Wang, C.C.; Descoins, M.; Mangelinck, D.; Tor, S.B.; Jägle, E.A.; Zaefferer, S.; Raabe, D. Reducing hot tearing by grain boundary segregation engineering in additive manufacturing: Example of an AlxCoCrFeNi high-entropy alloy. *Acta Mater.* **2021**, *204*, 116505. [CrossRef]
22. Qiu, Y.; Thomas, S.; Fabijanic, D.; Barlow, A.J.; Fraser, H.L.; Birbilis, N. Microstructural evolution, electrochemical and corrosion properties of AlxCoCrFeNiTiy high entropy alloys. *Mater. Des.* **2019**, *170*, 107698. [CrossRef]
23. Huang, X.; Miao, J.; Luo, A.A. Lightweight AlCrTiV high-entropy alloys with dual-phase microstructure via microalloying. *J. Mater. Sci.* **2018**, *54*, 2271–2277. [CrossRef]
24. Tseng, K.; Yang, Y.; Juan, C.; Chin, T.; Tsai, C.; Yeh, J. A light-weight high-entropy alloy Al20Be20Fe10Si15Ti. *Sci. China Technol. Sci.* **2018**, *61*, 184–188. [CrossRef]

25. Ji, C.W.; Ma, A.B.; Jiang, J.H. Mechanical properties and corrosion behavior of novel Al-Mg-Zn-Cu-Si lightweight high entropy alloys. *J. Alloy. Compd.* **2022**, *900*, 163508. [CrossRef]
26. Qiu, Y.; Thomas, S.; Gibson, M.A.; Fraser, H.L.; Pohl, K.; Birblis, N. Microstructure and corrosion properties of the low-density single-phase compositionally complex alloy AlTiVCr. *Corros. Sci.* **2018**, *133*, 386–396. [CrossRef]
27. Yuan, Y.; Feng, K.; Liu, J.J.; Cheng, J.; Qiao, Z.H.; Yang, J.; Li, J.S.; Liu, W.M. Effects of Ti and Cu on the Microstructure Evolution of AlCoCrFeNi High-Entropy Alloy During Heat Treatment. *Acta Metall. Sin. (Engl. Lett.)* **2020**, *33*, 1077–1090.
28. Zhang, Y.; Zhou, Y.J.; Lin, J.P.; Chen, G.L.; Liaw, P.K. Solid-Solution Phase Formation Rules for Multi-component Alloys. *Adv. Eng. Mater.* **2008**, *10*, 534–538. [CrossRef]
29. Guo, S.; Ng, C.; Lu, J.; Liu, C.T. Effect of valence electron concentration on stability of fcc or bcc phase in high entropy alloys. *J. Appl. Phys.* **2011**, *109*, 103505.
30. Shi, Y.Z.; Yang, B.; Xie, X.; Brechtl, J.; Dahmen, K.A.; Liaw, P.K. Corrosion of Al xCoCrFeNi high-entropy alloys: Al-content and potential scan-rate dependent pitting behavior. *Corros. Sci.* **2017**, *119*, 33–45. [CrossRef]
31. Lu, C.W.; Lu, Y.S.; Lai, Z.H.; Yen, H.W.; Lee, Y.L. Comparative corrosion behavior of Fe50Mn30Co10Cr10 dual-phase high-entropy alloy and CoCrFeMnNi high-entropy alloy in 3.5 wt% NaCl solution. *J. Alloy. Compd.* **2020**, *842*, 155824. [CrossRef]
32. Muangtong, P.; Rodchanarowan, A.; Chaysuwan, D.; Chanlek, N.; Goodall, R. The corrosion behaviour of CoCrFeNi-x (x = Cu, Al, Sn) high entropy alloy systems in chloride solution. *Corros. Sci.* **2020**, *172*, 108740. [CrossRef]
33. Xu, Z.L.; Zhang, H.; Du, X.J.; He, Y.Z.; Luo, H.; Song, G.S.; Mao, L.; Zhou, T.W.; Wang, L.L. Corrosion resistance enhancement of CoCrFeMnNi high-entropy alloy fabricated by additive manufacturing. *Corros. Sci.* **2020**, *177*, 108954. [CrossRef]
34. Luo, H.; Li, Z.M.; Mingers, A.M.; Raabe, D. Corrosion behavior of an equiatomic CoCrFeMnNi high-entropy alloy compared with 304 stainless steel in sulfuric acid solution. *Corros. Sci.* **2018**, *134*, 131–139. [CrossRef]
35. Gong, P.; Wang, D.L.; Zhang, C.; Wang, Y.; Shirvan, Z.J.; Yao, K.F.; Wang, X.Y. Corrosion behavior of TiZrHfBeCu(Ni) high-entropy bulk metallic glasses in 3.5 wt. % NaCl. *NPJ Mater. Degrad.* **2022**, *6*, 77. [CrossRef]
36. Pao, L.; Muto, I.; Sugawara, Y. Pitting at inclusions of the equiatomic CoCrFeMnNi alloy and improving corrosion resistance by potentiodynamic polarization in H_2SO. *Corros. Sci.* **2021**, *191*, 109748. [CrossRef]
37. Hsu, K.M.; Chen, S.H.; Lin, C.S. Microstructure and corrosion behavior of FeCrNiCoMnx (x = 1.0, 0.6, 0.3, 0) high entropy alloys in 0.5 M H_2SO. *Corros. Sci.* **2021**, *190*, 109694. [CrossRef]
38. Shi, Y.Z.; Yang, B.; Liaw, P.K. Corrosion-Resistant High-Entropy Alloys: A Review. *Metals* **2017**, *7*, 43. [CrossRef]
39. Chou, Y.L.; Yeh, J.W.; Shih, H.C. The effect of molybdenum on the corrosion behaviour of the high-entropy alloys Co1.5CrFeNi1.5Ti0.5Mox in aqueous environments. *Corros. Sci.* **2010**, *52*, 2571–2581. [CrossRef]
40. Qiu, Y.; Thomas, S.; Gibson, M.A.; Fraser, H.L.; Birbilis, N. Corrosion of high entropy alloys. *Npj Mater. Degrad.* **2017**, *1*, 15. [CrossRef]
41. Hermas, A.A.; Morad, M.S. A comparative study on the corrosion behaviour of 304 austenitic stainless steel in sulfamic and sulfuric acid solutions. *Corros. Sci.* **2008**, *50*, 2710–2717. [CrossRef]
42. Kocijan, A.; Merl, D.K.; Jenko, M. The corrosion behaviour of austenitic and duplex stainless steels in artificial saliva with the addition of fluoride. *Corros. Sci.* **2011**, *53*, 776–783. [CrossRef]
43. Wang, Y.; Jin, J.S.; Zhang, M.; Liu, F.M.; Wang, X.Y.; Gong, P.; Tang, X.F. Influence of plastic deformation on the corrosion behavior of CrCoFeMnNi high entropy alloy. *J. Alloy. Compd.* **2022**, *891*, 161822. [CrossRef]
44. Yao, Y.H.; Jin, Y.H.; Gao, W.; Liang, X.Y.; Chen, J.; Zhu, S.D. Corrosion Behavior of AlFeCrCoNiZrx High-Entropy Alloys in 0.5 M Sulfuric Acid Solution. *Metals* **2021**, *11*, 1471. [CrossRef]

materials

MDPI

Article

On the Aging Kinetics of a Flame-Resistant AZ91D-1.5%Ca Magnesium Alloy Processed with Ultrasonic Vibration

Inês V. Gomes [1,2,*], Fabrizio D'Errico [3], José L. Alves [1,2] and Hélder Puga [1,2]

1 CMEMS—UMinho, Department of Mechanical Engineering, Campus of Azurém, University of Minho, 4800-058 Guimarães, Portugal; puga@dem.uminho.pt (H.P.)
2 LABBELS—Associate Laboratory, 4800-058 Guimarães, Portugal
3 Department of Mechanical Engineering, Politecnico di Milano, Via La Masa 34, 20156 Milan, Italy
* Correspondence: inesvarela@dem.uminho.pt

Abstract: The Mg-Al-Zn-Ca system has demonstrated excellent flame resistance and mechanical properties in the as-cast condition. However, the potential of these alloys to be heat-treated, e.g., by aging, as well as the influence of the initial microstructure on the precipitation kinetics, is yet to be comprehensively explored. Ultrasound treatment was applied during the solidification of an AZ91D-1.5%Ca alloy to promote microstructure refinement. Samples from treated and non-treated ingots were subjected to solution treatment at 415 °C for 480 min, followed by aging at 175 °C for up to 4920 min. The results showed that the ultrasound-treated material could reach the peak-age condition in a shorter period than the non-treated one, suggesting accelerated precipitation kinetics and, thus, enhanced aging response. However, the tensile properties showed a decrease in the peak age compared to the as-cast condition, probably due to the formation of precipitates at the grain boundaries that promote the formation of microcracks and intergranular early fracture. This research shows that tailoring the material's as-cast microstructure may positively affect its aging response, shortening the heat treatment duration, thereby making the process less expensive and more sustainable.

Keywords: magnesium alloys; ultrasound treatment; solution; aging; $Mg_{17}Al_{12}$

Citation: Gomes, I.V.; D'Errico, F.; Alves, J.L.; Puga, H. On the Aging Kinetics of a Flame-Resistant AZ91D-1.5%Ca Magnesium Alloy Processed with Ultrasonic Vibration. *Materials* **2023**, *16*, 3152. https://doi.org/10.3390/ma16083152

Academic Editors: Pan Gong, Maojun Li, Guangchao Han and Xin Wang

Received: 13 March 2023
Revised: 14 April 2023
Accepted: 14 April 2023
Published: 17 April 2023

1. Introduction

The trade-off between magnesium alloys' strength and ductility is still a challenge that has been addressed by several authors in the last few years [1,2]. Significant attention has been paid to wrought magnesium alloys, resulting in substantial advances [3–5], while casting ones have been neglected concerning their processing and optimization [6].

A considerable fraction of intermetallic phases form during the solidification process of the alloys, which can play a role in the material's behavior. The intermetallics' morphology and distribution can thus strongly influence the material's mechanical and corrosion properties, possibly optimizing its performance by tailoring it. As the predominant intermetallic phase in AZ91D magnesium alloy, β-$Mg_{17}Al_{12}$ precipitates may enhance the material's strength, but at the expense of its ductility. The precipitation of this phase from the supersaturated matrix may assume two different and competitive modes: discontinuous and continuous precipitation. Discontinuous precipitates exhibit an elliptical or lamellar shape composed of β-$Mg_{17}Al_{12}$ and aluminum-enriched α-Mg and form mainly at the grain boundaries, growing inward [7]. Continuous precipitates, on the other hand, are characterized by a lozenge-shaped plate phase known as the Widmanstätten phase, which nucleates and grows inside the α-Mg grains with a primary habit plane parallel to the [0001] basal plane of the matrix [8]. The contribution of each precipitate type to the material's mechanical properties is not yet fully understood, and the literature offers conflicting views. Some authors [9,10] claimed that continuous precipitates could endow the material with a superior age-hardening response, given that discontinuous precipitate morphology could lead to early fracture. From a different perspective, discontinuous precipitates were suggested to play the leading role in aging hardening due to the continuous precipitates presenting thin plates parallel to the basal plane, easing the dislocation gliding [7,11].

Solution treatment followed by artificial aging has proven to be able to tailor the morphology of the β-$Mg_{17}Al_{12}$ precipitates by controlling both the temperature and duration of the heat treatment [12–14]. However, it has been noted that magnesium alloys aged at temperatures below 200 °C require long aging periods to attain their peak-aged condition [15], making the process energy-intensive and expensive. On the other hand, ultrasound treatment has been intensively explored for microstructural modification, and it has been demonstrated that it can promote peak-aging conditions in a shorter period in aluminum alloys [16]. Moreover, ultrasonic excitation may grant a refined microstructure with a smaller grain size [17] as well as the formation of vacancies [18], features deeply related to the formation of discontinuous and continuous precipitates, respectively [9]. The potential of ultrasonic vibration to assist in several manufacturing techniques is known, namely in additive and hybrid manufacturing areas. It has been shown that ultrasonic vibration promotes higher powder utilization efficiency, lower surface roughness and microstructural improvement [19], the reason why it can be a path to boost the application of additive and hybrid manufacturing in high-quality standards industries [20].

Although some research has already been published on the aging kinetics of different magnesium alloys, there is still a lack of information regarding the Mg-Al-Zn-Ca system. These alloys have demonstrated excellent results of ignition resistance allied to excellent strength-to-weight ratio, granting an increasing interest in them recently [21,22]. However, adding calcium is known to lead to the formation of thermally stable Al-Ca intermetallics, changing the fraction of the β-$Mg_{17}Al_{12}$ phase formed during solidification and heat treatment [23].

This work details the effect of the as-cast microstructure of an AZ91D-1.5%Ca (wt.%) on its age-hardening response. In this sense, ultrasound treatment was applied to the material during its solidification, refining its microstructure. The heat-treated samples' hardness evolution was investigated for non- and ultrasound-treated (US-treated) conditions, and the precipitates formed were comprehensively characterized. In addition, the tensile mechanical properties of the material in the as-cast, solution-treated and peak-aging states were determined at room temperature.

We hypothesize that ultrasound treatment can accelerate the precipitation kinetics, helping to reduce the heat treatment duration and temperature and thus contributing to a more economical and environmentally sustainable manufacturing chain.

2. Materials and Methods

A SiAlON crucible was used to melt an AZ91D-1.5%Ca alloy (Table 1) in a resistance furnace under an Ar protective atmosphere.

Table 1. Chemical composition of AZ91D-1.5% Ca alloy (wt.%).

Alloy	Mg	Al	Zn	Mn	Ca
AZ91D-1.5%Ca	Bal.	9.7	0.5	0.2	1.5

A previous stage at 450 °C was carried out to prevent melt contamination and eliminate humidity and moisture from the crucible and tools used. The melt was heated to 620 ± 5 °C and held for 20 min to form a CaO protective layer, after which the temperature was increased to 660 ± 5 °C and held for 10 min for homogenization. The melt was then poured into a pre-heated to 350 ± 5 °C metallic mold coupled to an ultrasound device that transmitted acoustic energy to the medium until it reached 525 ± 5 °C. Experiments were also conducted without the use of ultrasonic vibration as a comparison. The samples were subjected to a 480 min heat treatment at 415 °C followed by quenching in water at room temperature to avoid additional phase transition induced by residual heat. Artificial aging was then carried out at 175 °C for 240 to 4920 min.

The microstructural characterization and hardness testing samples were ground with gradually finer SiC papers and polished with a 1 μm polycrystalline diamond solution, followed by oxide polishing with 0.02 μm colloidal silica. A 4% solution of HNO_3 in ethanol was used to etch the samples before optical microscope examination (LEICA DM2500 M). A deep etching technique based on the selective dissolution of the matrix was applied to study the intermetallic shape. The morphology and composition of the phases present in the microstructure were detailed further through scanning electron microscopy with energy dispersive spectroscopy (SEM-EDS Phenom XL2, Thermo Fisher Scientific, Massachusetts, USA). Hardness measurements were conducted in an Officine Galileo Mod using a D200 tester under a load of 50 gf. A minimum of five indentations were averaged for each reported hardness measurement. Casted samples were machined into cylindrical tensile specimens (type D, according to ISO 6892-1) with an 8.00 ± 0.08 mm diameter and a 50.00 ± 0.50 mm proof length. Universal testing equipment (INSTRON 8874) was used to perform tensile testing with a 1 mm/s displacement rate until the specimens' fracture occurred and a load drop was observed. Three samples were tested for non- and US-treated alloy in the as-cast, solutionized, and peak-aging conditions.

3. Results and Discussion

Figure 1a,b present the optical micrographs of the non- and US-treated samples in the as-cast condition.

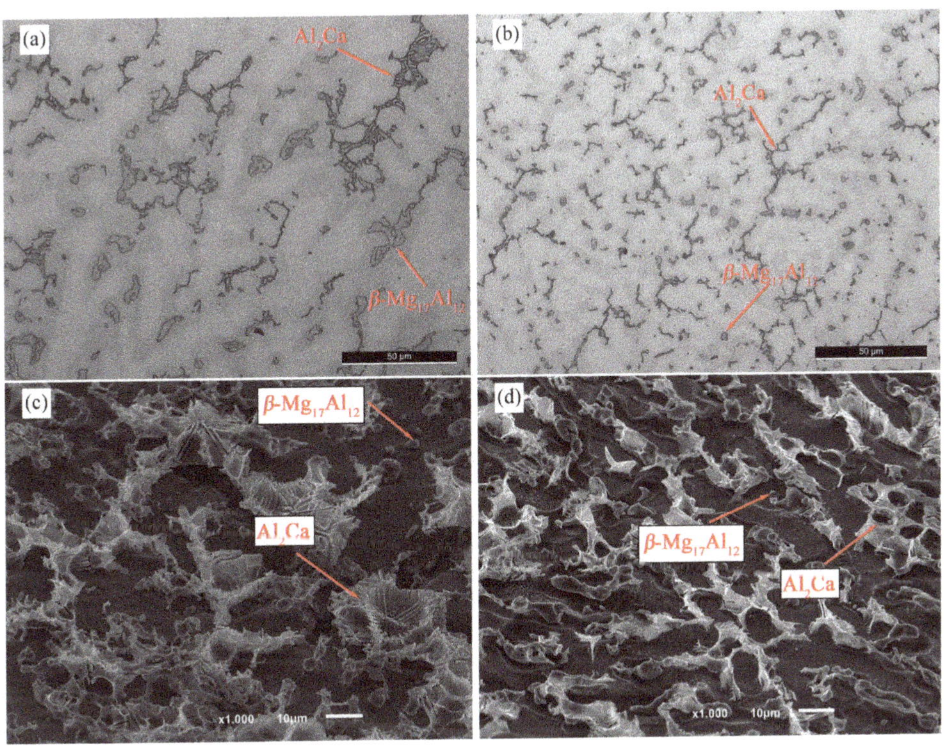

Figure 1. Microstructures of the (**a**,**c**) non- and (**b**,**d**) US-treated samples in the as-cast condition.

Regardless of the processing procedure, the microstructure of the as-cast samples is composed of α-Mg, β-$Mg_{17}Al_{12}$, Al_2Ca, and Al-Mn rich phases, as observed in our previous work [23]. However, significantly different morphologic characteristics could be observed, which may be attributable to the application of ultrasonic vibrations to the solidifying melt. It is suggested that ultrasound treatment has stimulated the refinement of the intermetallic phases, as previously described [24–26]. Indeed, compared to the non-treated sample (Figure 1a), β-$Mg_{17}Al_{12}$ appeared smaller and more rounded in the US-treated sample (Figure 1b), while Al_2Ca was fragmented, exhibiting a script-like disconnected morphology and a more uniform distribution.

Color differences between the dendrite interior and exterior indicate aluminum segregation during the solidification of the alloy. Hence, the Al-rich outer regions showed a lower etching rate compared to the interior areas, which has also been referred to by Esgandari et al. [27].

Deeply etched microstructures (Figure 1c,d) enabled a more detailed examination of the intermetallic morphology, revealing that both non- and US-treated samples exhibited a lamellar Al_2Ca phase with distinct morphological features. While the Al_2Ca intermetallic displayed a rosette-like structure in the non-treated sample, the US-treated sample had a platelet morphology that developed epitaxially from the α-Mg phase.

The microstructure of the solution-treated samples is presented in Figure 2. After solution treatment for 480 min at 415 °C, β-$Mg_{17}Al_{12}$ was nearly fully dissolved in the US-treated sample, whereas several particles could still be observed in the non-treated one. Thus, it is hypothesized that the dissolution process progressed differently depending on the processing conditions. The discrepancies observed in the materials' response to the solution treatment may have been promoted by their different as-cast microstructures. The non-treated sample exhibited β-$Mg_{17}Al_{12}$ coarse bulk particles, which led to its reduced dissolution due to a slower dissolving rate, conversely to the fine intermetallic found in the US-treated sample.

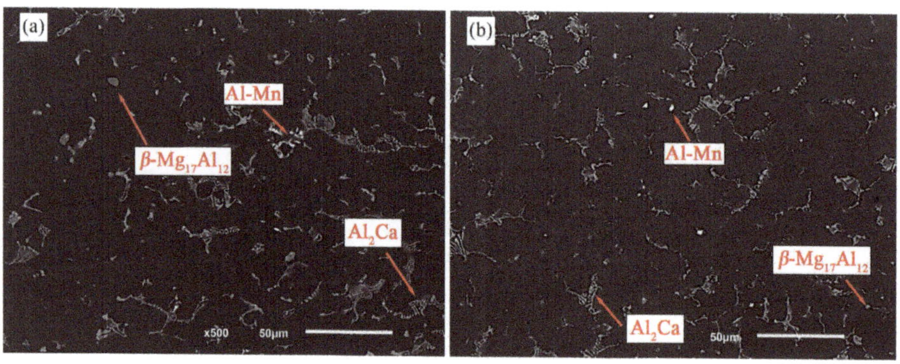

Figure 2. Microstructure of (**a**) non- and (**b**) US-treated samples after solution treatment at 415 °C for 480 min.

Therefore, differences in the size and distribution of intermetallic particles may explain their distinct dissolution kinetics. The non-treated sample presented larger intermetallic particles and a higher aluminum concentration in their vicinity, leading to a slower dissolution rate promoted by the lower gradient. In addition, once the intermetallic volumetric fraction is comparable across the processing conditions, the intermetallics' interfacial area of the non-treated sample is smaller and hinders the aluminum atoms' diffusion to the matrix, slowing down the dissolution process.

On the contrary, the US-treated sample was characterized by smaller and uniformly dispersed β-$Mg_{17}Al_{12}$ particles, shortening the distance over which diffusion occurred and thereby the time required for the dissolution to occur.

The Al_2Ca phase appeared not to suffer significant dissolution, possibly because of its excellent thermal stability [28,29], which limited the aluminum atoms available to migrate to the α-grains.

SEM micrographs of the non- and US-treated samples at the peak-age condition at 175 °C (1440 and 960 min, respectively) are shown in Figure 3, along with the EDS analysis results of the identified precipitates.

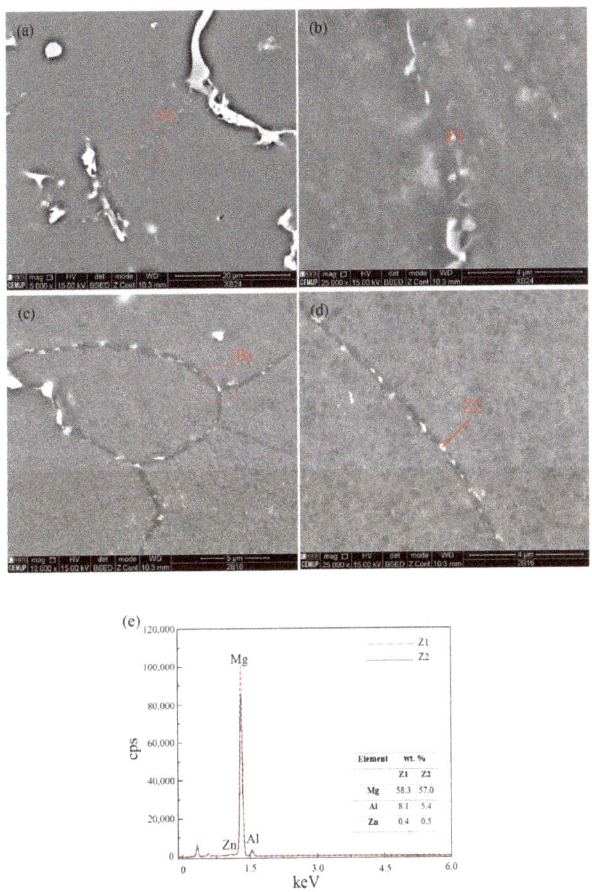

Figure 3. SEM micrographs of the discontinuous β-$Mg_{17}Al_{12}$ precipitates found in (**a**,**b**) non- and (**c**,**d**) US-treated samples, after aging at 175 °C for 1440 and 960 min, respectively; (**b**,**d**) higher magnification images of areas A and B, respectively, (**e**) EDS analysis of the particles identified as Z1 and Z2 in (**b**,**d**).

During the aging treatment, two types of precipitates—continuous and discontinuous—may form and develop competitively as they nucleate and grow at different rates [30,31]. Although their chemical composition is the same, their morphology differs, making their strengthening effect distinct. Both non- and US-treated samples exhibited only discontinuous precipitates, which may be explained by the low aging temperature adopted [27]. Most precipitates were found along the grain boundaries—intergranular precipitates—and presented an elliptical shape. The different processing conditions are suggested to result in the modification of the precipitates, not primarily of their shape, but of their size and number density. Indeed, the non-treated sample was characterized by a lower number of precipitates whose size was larger than that of the US-treated samples. In addition, it is possible that the precipitates grew and coalesced to form massive bulk structures, as shown in Figure 4.

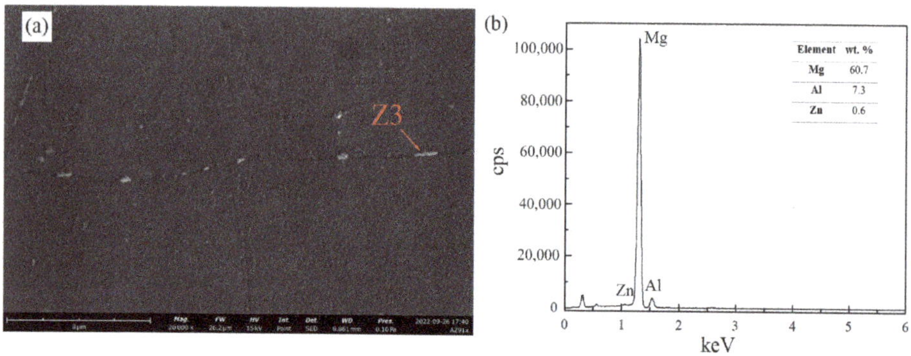

Figure 4. SEM micrograph of β-$Mg_{17}Al_{12}$ bulk precipitates found at grain boundaries of non-treated material after aging at 175 °C for 1440 min; (**b**) EDS analysis of the Z3 particle identified in (**a**).

The US-treated sample showed a higher number of smaller β-$Mg_{17}Al_{12}$ precipitates distributed along the grain boundaries, which may result from the finer microstructure, i.e., the materials' smaller grain size. The grain refinement of the alloy results in increased grain boundaries, providing more nucleation sites for β-$Mg_{17}Al_{12}$ discontinuous precipitates and thereby promoting faster-aging kinetics [9,32].

Figure 5 shows the hardness evolution under as-cast, solution-treated, and artificially aged conditions for non- and US-treated samples.

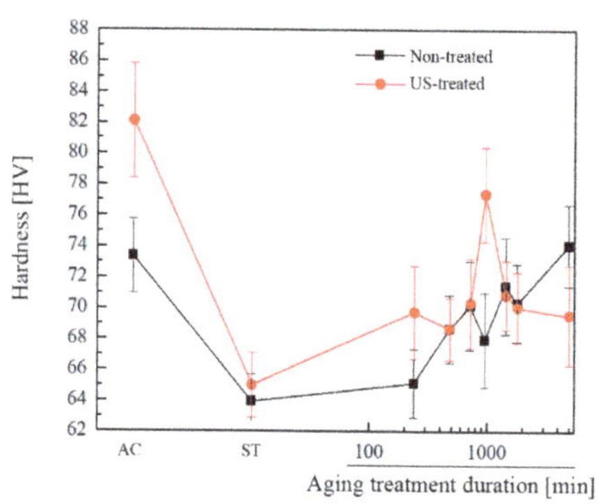

Figure 5. Average Vickers hardness of the non- and US-treated samples during artificial aging.

In the as-cast condition, the US-treated alloy exhibited higher hardness than the non-treated one, possibly due to its finer microstructure. Indeed, the refinement of the α-Mg and intermetallic phases substantially affects the material's mechanical properties, namely its hardness. Increasing grain boundaries led to an increase in the density of small precipitates, thus improving the mechanical properties. In the non-treated samples, however, the coarser morphology of the precipitates resulted in an increased interspacing between them. Under these conditions, there is a reduced capacity to inhibit dislocation movement, so the material's hardness is lower [33]. Similar findings were reported by Lai et al. [15], who studied the effect of aging conditions on the morphology of the β-$Mg_{17}Al_{12}$ precipitates and concluded that numerous smaller precipitates provided higher peak hardness.

Due to the precipitates' dissolution after solution treatment, which promotes material softening [34], the material's hardness dropped dramatically, regardless of the processing conditions. The US-treated alloy hardness remained higher than that of the non-treated one in the solutionized condition due to the grain boundary strengthening effect granted by the grain refinement. At this stage, no contribution of precipitation strengthening was considered since most precipitates were dissolved after solution treatment. On the other hand, a more gradual decline in the hardness in the non-treated material may be attributable to the presence of undissolved β-$Mg_{17}Al_{12}$ intermetallic, whose size has decreased throughout the procedure, favoring its ability to prevent dislocation movement.

The hardness curves' behavior of the non- and US-treated alloys indicates that the processing conditions may profoundly influence the material's aging response. Although the typical hardness increase was noticeable in both non- and US-treated samples, the former presented a longer incubation period, with the hardness value remaining almost constant until aging for 240 min. In contrast, the short incubation period of the US-treated alloy suggests accelerated aging kinetics of β-$Mg_{17}Al_{12}$ precipitates compared to the non-treated sample. Similar results were reported by Wang et al. [32]. The authors investigated whether the manufacturing process—high-pressure die casting or rheo-diecasting—influenced the material's aging response and determined that the latter showed faster dissolution during the solution treatment and accelerated aging kinetics.

After the incubation stage, the non- and US-treated samples' hardness increased, although at a higher rate for the latter. The US-treated alloy reached its peak-aged condition at 960 min, following which its hardness value decreased sharply, whereas the non-treated alloy hardness increased throughout the time range considered. The larger grain size is suggested to significantly reduce the nucleation rate, delaying the material's aging response and resulting in the non-treated sample not achieving the peak-aging condition within the time interval considered [33]. Even so, and for comparison's sake, aging treatment for 1440 min was hereafter referred to as the peak-aging condition of the non-treated sample, corresponding to the highest hardness value observed within the heat treatment period considered. After 4920 min of aging, the non-treated samples exhibited a value comparable to that of the as-cast condition (74 HV vs. 73 HV) and still showed a tendency toward increasing hardness, which indicates that peak-aging has not been reached yet and could be higher. However, the US-treated sample showed the highest hardness in the as-cast condition (82 HV), with peak-aged hardness 5.7% lower (77 HV). According to Zhang et al. [35] and Dumpala et al. [36], the coarsening of grains during the solution treatment may be responsible for this behavior.

Nonetheless, the ultrasound treatment of magnesium alloy has demonstrated the potential to refine the microstructure, which has proven effective in shortening the time needed for reaching the peak-aged state. In addition, the difference between the non- and US-treated samples aged for 240 min is superior to that observed in the solution-treated condition. This behavior supports the hypothesis that ultrasonic processing plays a role in precipitation behavior by anticipating it. In this sense, the increased hardness value obtained for the US-treated sample may have resulted from accelerated precipitation and grain boundary strengthening combined, consistent with Kim et al.'s findings [37]. Even so, no increase in hardness was seen after the aging treatment compared to the as-cast condition for either processing method. These results align with those of Suzuki et al. [30] and Bamberger et al. [31] for Mg-Al-Ca and Mg-Ca-Zn alloys, respectively, demonstrating the poor age-hardening response of these alloys.

Furthermore, due to the formation of the Al_2Ca phase during the alloy solidification, there is a low fraction of β-$Mg_{17}Al_{12}$ phase in the as-cast state and reduced aluminum content on the α-Mg phase, limiting the solute availability after solution heat treatment. In this sense, given the direct relationship between the transformation rate during the aging treatment and the solute atoms amount available in the supersaturated matrix, lower peak hardness values are obtained by calcium-containing magnesium alloys compared with those without this element [27].

Several studies show the ability of ultrasound treatment to enhance the mechanical properties of magnesium alloys, namely AZ91D [17]. However, whether as-cast microstructure affects the tensile behavior of heat-treated materials is still unknown.

The results of the tensile tests performed on the non- and US-treated AZ91D-1.5%Ca (wt.%) in the as-cast, solutionized (T4), and peak-aged conditions are presented in Table 2.

Table 2. Mechanical properties of the non- and US-treated material in the as-cast, solution-treated and peak-aged conditions.

Test	Condition	Yield Strength (MPa)	Tensile Strength (MPa)	Elongation at Break (%)
Non-treated	As-cast	89 ± 5	110 ± 7	1.75 ± 0.34
	T4	137 ± 7	146 ± 8	2.04 ± 0.93
	T6–Peak-age condition (1440 min)	100 ± 4	115 ± 4	1.73 ± 0.71
US-treated	As-cast	125 ± 8	164 ± 6	3.02 ± 0.25
	T4	158 ± 6	204 ± 8	4.31 ± 1.34
	T6–Peak-age condition (960 min)	142 ± 6	169 ± 8	2.69 ± 0.89

The results demonstrate that ultrasound treatment significantly increased mechanical performance under all the tested conditions. In the as-cast state, the tensile strength of the US-treated material was about 50% higher than that of the non-treated one, stressing the grain refinement's role in the material's mechanical resistance. In addition, the yield strength was higher than in non-treated samples, but the difference was less pronounced. The increased elongation at fracture constitutes the most impressive result, enhancing by about 72.5% when ultrasound treatment is applied. The microstructure characteristics may thus be considered a critical factor for the static mechanical properties of the alloy. The conjunction of smaller grain size and refined and more uniform distributed intermetallic phases grant remarkably improved mechanical properties. Conversely, the continuous network of brittle β-$Mg_{17}Al_{12}$ phase detected in the non-treated sample leads to poorer mechanical properties [24]. According to Du et al. [38], such modifications decrease the stress concentration points, improving ductility by preventing early fracture.

The solutionized samples exhibited higher elongation, yield and tensile strength than the as-cast condition, regardless of the processing route. For the non-treated sample, the increase in the yield and tensile strength was higher than that of the elongation, while the contrary was observed in the US-treated samples. The partial dissolution of the coarse precipitates may underlie this behavior, given that they constitute preferential sites for fracture initiation and act as a continuous easy crack path [39]. In addition, the dissolution of these precipitates occurred to a greater extent in the US-treated sample, leading to the matrix being dopped with aluminum atoms, easing the basal-plane deformation twinning in the magnesium alloys and thereby enhancing the elongation.

In the aged condition, non- and US-treated materials showed a decrease in yield and tensile strength, as well as elongation. Such a loss of the materials' mechanical properties may be promoted by forming precipitates which, contrary to what is observed for aluminum alloys, may harm the mechanical resistance. In fact, under the considered aging conditions, discontinuous precipitates formed at the grain boundaries, showing small interface bonding strength and weakening the bonding force between grains [40]. Under these conditions, the microcracks are suggested to form preferentially in the β-$Mg_{17}Al_{12}/\alpha$-Mg, interfaces, promoting early intergranular fracture due to their growth and propagation along the grain boundaries [41]. In agreement with these findings, the higher yield and tensile strength and elongation values observed in the US-treated material after solution treatment compared to the peak-age properties appear to confirm that the precipitation of the β-$Mg_{17}Al_{12}$ has an overall deleterious effect on the mechanical behavior of the material. Moreover, the elongation decline showed by the non-treated material after aging may be explained by the presence of coarse intermetallic particles that did not fully dissolve and are likely to crack at low strains. This detrimental effect did not significantly compromise the ductility in the solution-treated condition due to the softer matrix, which worked to delay the cracking [42] but played a critical role in decreasing the material's mechanical strength after aging treatment.

According to the obtained results, it can be concluded that AZ91D-1.5%Ca (wt.%) has a limited aging potential under the considered conditions. In fact, no significant improvement in the material's mechanical properties could be achieved through the adopted heat treatment scheme. Such a poor aging response may be explained mainly by two factors: (i) the addition of calcium promotes the formation of the Al_2Ca intermetallic phase during the material solidification and suppresses the formation of the β-$Mg_{17}Al_{12}$ phase. Given its high thermal stability, the Al_2Ca phase did not dissolve during the solution treatment, causing the aluminum content available to precipitate during subsequent aging treatment to be extremely low; on the other hand, (ii) the discontinuous precipitates formed along the grain boundaries consumed the available solute content, promoting the depletion of continuous precipitation, which could further improve the mechanical performance of the material [43].

Optimizing the solution treatment to dissolve the Al_2Ca phase and applying ultrasound treatment during the casting process may constitute a route to enhance the aging response of the AZ91D-1.5%Ca (wt.%) alloy. In fact, applying ultrasonic vibration during the material solidification can promote an excess of vacancies, which act as nucleation sites for continuous precipitates [18]. Such conditions, together with a higher available aluminum content granted by the dissolution of the Al_2Ca intermetallic, can enhance the material's response to the aging treatment by tailoring the morphology of the resultant precipitate.

4. Conclusions

In this research, ultrasound treatment was used during the solidification of an AZ91D-1.5%Ca to modify its microstructure, which was then investigated for its effect on aging kinetics. The following conclusions could be drawn from the performed study:

- Applying ultrasound treatment during AZ91D-1.5%Ca (wt.%) alloy cooling has significantly changed its microstructure, promoting the refinement of β-$Mg_{17}Al_{12}$ and Al_2Ca intermetallic phases.
- The refined microstructure of the US-treated sample yielded a higher hardness than that of the non-treated one in the as-cast condition.
- US-treated samples showed accelerated aging kinetics since precipitation hardening occurred for a shorter heat treatment duration compared to that of non-treated ones.
- The hardness curve of the non-treated material suggests that peak aging was not achieved under the tested conditions, which indicates that aging for periods longer than 4920 min may be required. Conversely, US-treated samples appeared to reach the peak-aging state after 960 min.
- Ultrasound treatment enhanced the ultimate tensile strength and elongation in all the considered conditions—as-cast, solutioned and aged—compared to the absence of treatment. However, the tensile properties showed a decrease in the peak age, possibly due to the formation of precipitates at the grain boundaries that promote the formation of microcracks and intergranular early fracture.

Author Contributions: I.V.G.: Conceptualization, Investigation, Formal analysis, Writing—original draft, Writing—review and editing. F.D.: Resources, Writing—review and editing. J.L.A.: Resources, Funding acquisition, Writing—review and editing. H.P.: Conceptualization, Methodology, Resources, Formal analysis, Writing—review and editing. All authors have read and agreed to the published version of the manuscript.

Funding: This work was supported by Portuguese FCT under the project UIDB/04436/2020 and the doctoral grant PD/BD/140094/2018.

Institutional Review Board Statement: Not applicable.

Informed Consent Statement: Not applicable.

Data Availability Statement: Not applicable.

Conflicts of Interest: The authors declare no conflict of interest.

References

1. Jo, S.; Whitmore, L.; Woo, S.; Aramburu, A.U.; Letzig, D.; Yi, S. Excellent age hardenability with the controllable microstructure of AXW100 magnesium sheet alloy. *Sci. Rep.* **2020**, *10*, 22413. [CrossRef] [PubMed]
2. Han, T.; Huang, G.; Li, H.; Wang, L.; Zhang, H.; Pan, F. Strength-ductility balance of AZ31 magnesium alloy via accumulated extrusion bonding combined with two-stage artificial cooling. *J. Magnes. Alloys* **2021**, *26*, 2043. [CrossRef]
3. Shi, R.; Miao, J.; Avey, T.; Luo, A.A. A new magnesium sheet alloy with high tensile properties and room-temperature formability. *Sci. Rep.* **2020**, *10*, 10044. [CrossRef] [PubMed]
4. Panigrahi, S.K.; Mishra, R.S.; Brennan, R.C.; Cho, K. Achieving extraordinary structural efficiency in a wrought magnesium rare earth alloy. *Mater. Res. Lett.* **2020**, *8*, 151–157. [CrossRef]
5. Bu, D.; Li, T.; Han, X.; Du, Z.; Yuan, J.; Zhang, K.; Li, Y.; Peng, Y.; Pang, Z.; Zhao, C. Enhancing strength and ductility in back extruded WE71 magnesium alloy cylindrical parts by introduction of multi-direction forging process. *J. Rare Earths* **2022**, *5*, 239. [CrossRef]
6. Korgiopoulos, K.; Langelier, B.; Pekguleryuz, M. Mg17Al12 phase refinement and the improved mechanical performance of Mg–6Al alloy with trace erbium addition. *Mater. Sci. Eng. A* **2021**, *812*, 141075. [CrossRef]
7. Zhang, K.; Li, H.; Liang, X.; Chen, Z.; Wang, L. Discontinuous and continuous precipitation characteristics and mechanical properties of a AZ80A magnesium alloy at different aging temperatures. *Mater. Charact.* **2020**, *161*, 110146. [CrossRef]
8. Celotto, S. TEM study of continuous precipitation in Mg ±9 wt%Al ±1 wt%Zn alloy. *Acta Mater.* **2000**, *48*, 1775–1787. [CrossRef]
9. Abd El-Rehim, A.F.; Zahran, H.Y.; Al-Masoud, H.M.; Habashy, D.M. Microhardness and microstructure characteristics of AZ91 magnesium alloy under different cooling rate conditions. *Mater. Res. Express* **2019**, *6*, 86572. [CrossRef]
10. Cepeda-Jiménez, C.M.; Pérez-Prado, M.T. Microplasticity-based rationalization of the room temperature yield asymmetry in conventional polycrystalline Mg alloys. *Acta Mater.* **2016**, *108*, 304–316. [CrossRef]
11. Lee, J.U.; Kim, S.-H.; Kim, Y.J.; Park, S.H. Effects of homogenization time on aging behavior and mechanical properties of AZ91 alloy. *Mater. Sci. Eng. A* **2018**, *714*, 49–58. [CrossRef]
12. Guo, Y.; Quan, G.; Celikin, M.; Ren, L.; Zhan, Y.; Fan, L.; Pan, H. Effect of heat treatment on the microstructure and mechanical properties of AZ80M magnesium alloy fabricated by wire arc additive manufacturing. *J. Magnes. Alloys* **2021**, *32*, 1211. [CrossRef]
13. Fatmi, M.; Djemli, A.; Ouali, A.; Chihi, T.; Ghebouli, M.A.; Belhouchet, H. Heat treatment and kinetics of precipitation of β-Mg17Al12 phase in AZ91 alloy. *Results Phys.* **2018**, *10*, 693–698. [CrossRef]
14. Kim, J.-Y.; Byeon, J.-W. Quantitative relation of discontinuous and continuous Mg17Al12 precipitates with corrosion rate of AZ91D magnesium alloy. *Mater. Charact.* **2021**, *174*, 111015. [CrossRef]
15. Contreras-Piedras, E.; Esquivel-Gonzalez, R.; López-Hirata, V.M.; Saucedo-Muñoz, M.L.; Paniagua-Mercado, A.M.; Dorantes-Rosales, H.J. Growth kinetics of cellular precipitation in a Mg–8.5Al–0.5Zn–0.2Mn (wt.%) alloy. *Mater. Sci. Eng. A* **2010**, *527*, 7775–7778. [CrossRef]
16. Puga, H.; Carneiro, V.H. Light-Alloy Melt Ultrasonication: Shorter T6 with Higher Precipitation Strengthening. *Met. Mater. Int.* **2021**, *27*, 3195–3204. [CrossRef]
17. Puga, H.; Carneiro, V.; Barbosa, J.; Vieira, V. Effect of Ultrasonic Treatment in the Static and Dynamic Mechanical Behavior of AZ91D Mg Alloy. *Metals* **2015**, *5*, 2210–2221. [CrossRef]
18. Hu, Y.; Liu, H.; Fujii, H.; Araki, H.; Sugita, K.; Liu, K. Ultrasonic-induced excess vacancies in friction stir processing and exploration of acoustoplastic effect. *Scr. Mater.* **2020**, *185*, 117–121. [CrossRef]
19. Cong, W.; Ning, F. A fundamental investigation on ultrasonic vibration-assisted laser engineered net shaping of stainless steel. *Int. J. Mach. Tools Manuf.* **2017**, *121*, 61–69. [CrossRef]
20. Khorasani, M.; Gibson, I.; Ghasemi, A.H.; Hadavi, E.; Rolfe, B. Laser subtractive and laser powder bed fusion of metals: Review of process and production features. *Rapid Prototyp. J.* **2023**. *ahead-of-print*. [CrossRef]
21. Ni, J.; Jin, L.; Zeng, J.; Li, J.; Wang, F.; Wang, F.; Dong, S.; Dong, J. Development of high-strength magnesium alloys with excellent ignition-proof performance based on the oxidation and ignition mechanisms: A review. *J. Magnes. Alloys* **2023**, *11*, 1–14. [CrossRef]
22. Xiao, R.; Liu, W.-C.; Wu, G.-H.; Zhang, L.; Liu, B.-L.; Ding, W.-J. Effect of Ca content and rheo-squeeze casting parameters on microstructure and mechanical properties of AZ91−1Ce−xCa alloys. *Trans. Nonferrous Met. Soc. China* **2021**, *31*, 1572–1586. [CrossRef]
23. Gomes, I.V.; D'Errico, F.; Alves, J.L.; Puga, H. Ultrasound-assisted casting of AZ91D-1.5%Ca—Shifting T4 paradigm for downstream processing. *Mater. Lett.* **2023**, *330*, 133305. [CrossRef]

24. Khosro Aghayani, M.; Niroumand, B. Effects of ultrasonic treatment on microstructure and tensile strength of AZ91 magnesium alloy. *J. Alloys Compd.* **2011**, *509*, 114–122. [CrossRef]
25. Zhang, X.; Kotadia, H.R.; Depner, J.; Qian, M.; Das, A. Effect of Ultrasonication on the Solidification Microstructure in Al and Mg-Alloys. In *Light Metals*; Chesonis, C., Ed.; Springer: Berlin/Heidelberg, Germany; New York, NY, USA, 2019; pp. 1589–1595, ISBN 978-3-030-05863-0.
26. Emadi, P.; Ravindran, C. The Influence of High Temperature Ultrasonic Processing Time on the Microstructure and Mechanical Properties AZ91E Magnesium Alloy. *J. Mater. Eng Perform* **2021**, *30*, 1188–1199. [CrossRef]
27. Amir Esgandari, B.; Mehrjoo, H.; Nami, B.; Miresmaeili, S.M. The effect of Ca and RE elements on the precipitation kinetics of Mg17Al12 phase during artificial aging of magnesium alloy AZ91. *Mater. Sci. Eng. A* **2011**, *528*, 5018–5024. [CrossRef]
28. Jun, J.-H. Damping behaviors of as-cast and solution-treated AZ91–Ca magnesium alloys. *J. Alloys Compd.* **2014**, *610*, 169–172. [CrossRef]
29. Huang, X.; Chino, Y.; Ueda, H.; Inoue, M.; Kido, F.; Matsumoto, T. Improvement of mechanical properties of extruded AZX912 magnesium alloy using high-temperature solution treatment. *J. Mater. Res.* **2019**, *34*, 3725–3734. [CrossRef]
30. Suzuki, A.; Saddock, N.D.; TerBush, J.R.; Powell, B.R.; Jones, J.W.; Pollock, T.M. Precipitation Strengthening of a Mg-Al-Ca–Based AXJ530 Die-cast Alloy. *Met. Mat. Trans. A* **2008**, *39*, 696–702. [CrossRef]
31. Bamberger, M.; Levi, G.; Vander Sande, J.B. Precipitation hardening in Mg-Ca-Zn alloys. *Metall. Mater. Trans. A* **2006**, *37A*, 481–487. [CrossRef]
32. Wang, Y.; Liu, G.; Fan, Z. Microstructural evolution of rheo-diecast AZ91D magnesium alloy during heat treatment. *Acta Mater.* **2006**, *54*, 689–699. [CrossRef]
33. Xu, W.; Yu, J.; Jia, L.; Gao, C.; Miao, Z.; Wu, G.; Li, G.; Zhang, Z. Grain refinement impact on the mechanical properties and wear behavior of Mg-9Gd-3Y-2Zn-0.5Zr alloy after decreasing temperature reciprocating upsetting-extrusion. *J. Magnes. Alloys* **2021**, *7*, 672. [CrossRef]
34. Lee, G.M.; Lee, J.U.; Park, S.H. Effects of post-heat treatment on microstructure, tensile properties, and bending properties of extruded AZ80 alloy. *J. Mater. Res. Technol.* **2021**, *12*, 1039–1050. [CrossRef]
35. Zhang, Z.; Wang, L.; Zhang, R.; Yin, D.; Zhao, Z.; Bai, P.; Liu, B.; Wang, F. Effect of solution annealing on microstructures and corrosion behavior of wire and arc additive manufactured AZ91 magnesium alloy in sodium chloride solution. *J. Mater. Res. Technol.* **2022**, *18*, 416–427. [CrossRef]
36. Swetha Chowdary, V.; Dumpala, R.; Anand Kumar, S.; Kondaiah, V.V.; Ratna Sunil, B. Influence of heat treatment on the machinability and corrosion behavior of AZ91 Mg alloy. *J. Magnes. Alloys* **2018**, *6*, 52–58. [CrossRef]
37. Kim, S.-H.; Lee, J.U.; Kim, Y.J.; Bae, J.H.; You, B.S.; Park, S.H. Accelerated precipitation behavior of cast Mg-Al-Zn alloy by grain refinement. *J. Mater. Sci. Technol.* **2018**, *34*, 265–276. [CrossRef]
38. Du, X.; Zhang, E. Microstructure and mechanical behaviour of semi-solid die-casting AZ91D magnesium alloy. *Mater. Lett.* **2007**, *61*, 2333–2337. [CrossRef]
39. Li, F.; Peh, W.Y.; Nagarajan, V.; Ho, M.K.; Danno, A.; Chua, B.W.; Tan, M.J. Development of non-flammable high strength AZ91 + Ca alloys via liquid forging and extrusion. *Mater. Des.* **2016**, *99*, 37–43. [CrossRef]
40. Di, T.; Jiang, Y.; Guan, R.; Chen, M.; Jiang, J.; Gao, F.; Lu, X.; Zhao, Z. The Evolution of Microstructure, Mechanical Properties and Fracture Behavior with Increasing Lanthanum Content in AZ91 Alloy. *Metals* **2020**, *10*, 1256. [CrossRef]
41. Teschke, M.; Koch, A.; Walther, F. Comparison of High-Temperature Compression and Compression-Comp ressionFatigue Behavior of Magnesium Alloys DieMag422 and AE42. *Materials* **2020**, *13*, 497. [CrossRef]
42. Cáceres, C.H.; Davidson, C.J.; Griffiths, J.R.; Newton, C.L. Effects of solidification rate and ageing on the microstructure and mechanical properties of AZ91 alloy. *Mater. Sci. Eng. A* **2002**, *325*, 344–355. [CrossRef]
43. Lai, W.-J.; Li, Y.-Y.; Hsu, Y.-F.; Trong, S.; Wang, W.-H. Aging behaviour and precipitate morphologies in Mg–7.7Al–0.5Zn–0.3Mn (wt.%) alloy. *J. Alloys Compd.* **2009**, *476*, 118–124. [CrossRef]

materials

Article

Numerical Simulation of Slag Entrainment by Vortex Flux during Tapping at Converter

Chengyong Huang [1,2], Ye Sun [1,2], Wei Liu [1,2,*], Jingshe Li [1,2], Shufeng Yang [1,2] and Jianfeng Dong [1,2]

1 School of Metallurgical and Ecological Engineering, University of Science and Technology Beijing, Beijing 100083, China; b20200097@xs.ustb.edu.cn (C.H.)
2 Beijing Key Laboratory of Special Melting and Preparation of High-End Metal Materials, Beijing 100083, China
* Correspondence: liuwei@ustb.edu.cn; Tel.: +86-152-1060-2862

Abstract: In order to improve the yield of steel produced in the converter and the quality of the molten steel, and to understand the distribution of the flow field in the converter and ladle during the steelmaking process, the CFD fluid simulation software Fluent 2020 R2 was used to analyze the flow field of the converter static steelmaking process. The aperture of the steel outlet and the timing of the vortex formation under different angles were studied, as well as the disturbance level of the injection flow in the ladle molten pool. The study revealed that in the steelmaking process, the emergence of tangential vectors caused the entrainment of slag by the vortex, whereas in the later stages of steelmaking, the turbulent flow of slag disrupted the vortex, resulting in its dissipation. When the converter angle increases to 90°, 95°, 100°, and 105°, the eddy current occurrence time is 43.55 s, 66.44 s, 68.80 s, and 72.30 s, and the eddy current stabilization time is 54.10 s, 70.36 s, 70.95 s, and 74.26 s, respectively. When the converter angle is 100–105°, it is suitable to add alloy particles into the ladle molten pool. When the tapping port diameter is 220 mm, the eddy current inside the converter changes and the mass flow rate of the tapping port is "oscillating". When the aperture of the steel outlet was 210 mm, the steelmaking time could be shortened by about 6 s without affecting the internal flow field structure of the converter.

Keywords: converter; vortex; flow field analysis; numerical simulation

Citation: Huang, C.; Sun, Y.; Liu, W.; Li, J.; Yang, S.; Dong, J. Numerical Simulation of Slag Entrainment by Vortex Flux during Tapping at Converter. *Materials* **2023**, *16*, 3209. https://doi.org/10.3390/ma16083209

Academic Editor: Miguel Ángel Sanjuán

Received: 20 March 2023
Revised: 14 April 2023
Accepted: 17 April 2023
Published: 19 April 2023

1. Introduction

With the development of science and technology, the demand for high-quality steel in the market is increasing, which poses higher requirements for metallurgical technology. Improving the cleanliness of steel by reducing the entry of inclusions and then improving the quality of steel is an important direction for the development of the steel industry [1,2]. During the steelmaking process in the converter, as the liquid level gradually decreases, a rapidly rotating free surface vortex is easily formed. Once the free surface vortex is formed, it is prone to roll over the slag. When the height of the steel liquid level is lower than the critical height of the vortex, the phenomenon of vortex-induced slagging will be very serious. This phenomenon occurs in the steelmaking process of the converter, ladle, and tundish. This will lead to a reduction in the yield of steel liquid, a shortening of the service life of the converter, an increase in the content of inclusions, blockage of the nozzle, and other problems [3,4], causing considerable losses to production efficiency [5–7]. Therefore, effectively delaying the occurrence of vortex-induced slagging is a problem that must be solved in the production of clean steel [2,8].

The main reason for the formation of eddy currents is due to the rotational motion of the fluid, where at least one of the three rotational angular velocity components of the fluid is not zero, meaning that the fluid particles exhibit microscopic horizontal rotation. In the actual fluid flow process, due to the viscosity of the fluid, rotation generally occurs. Since the formation of eddy currents during the steelmaking process in the converter is inevitable, it is necessary to have a detailed understanding of the mechanism of eddy current formation in order to mitigate the adverse effects of eddy currents on slag entrapment. So far, the study of eddy currents in steelmaking vessels has mainly focused on the ladle [9,10] and tundish [11–13]. Due to the non-visible process, high-temperature environment, and rotation of the converter during the steelmaking process, there is not yet a complete understanding of the flow phenomena in the converter during the steelmaking process. During the steelmaking process in a converter, once the molten steel enters the tapping hole, it will have a certain initial velocity, and the tapping speed at different tapping angles is completely different. At the end of the converter steelmaking process, alloying treatment needs to be carried out on the molten steel, and the influence of the distribution of the alloy by the molten steel flowing out at different converter angles is also different. Referring to the automatic tapping process of an enterprise, the angle changes in a "step-like" manner over time, so it is particularly important to determine the optimal tapping angle for adding alloying materials. Therefore, the timing of alloy addition needs to ensure that the tapping time is short, the amount of eddy current is small, and the turbulence intensity in the ladle is high enough.

Some researchers have studied factors such as different tapping hole sizes and different steel heights but have not considered the effect of different converter tilts on the steelmaking process. Wang et al. [14] discussed the influence of initial liquid level height on the timing of slag coiling and built a quarter-converter model through numerical simulation. However, because the structure near the tapping port of the converter is more complex, different from the tapping port of the ladle, tundish, and other symmetric structures. Therefore, Pang et al. [15] built a complete model of the converter steel drawing process, but the model ignored the slag layer. Therefore, a full three-dimensional transient multiphase flow model of the converter steelmaking process was established on the basis of the previous studies. The eddy current phenomenon in the converter steelmaking process was studied, and the optimal time for adding alloy materials was determined. The mass flow rate of the wire outlet under these conditions will provide a basis for the study of the melting trajectory and dynamics of alloy particles in the converter steelmaking process in the future. By parametric analysis of the influence of different parameters on the vortex entrain slag, the correlation between the tapping angle of the converter, tapping hole aperture, and alloy adding time was established. Based on existing research, a fully three-dimensional, transient, multiphase flow model is established for mathematical simulation of the converter steelmaking process, studying the eddy current phenomena during converter steelmaking, and determining the optimal timing for adding alloying materials. The quality flow rate of the tapping hole under this condition will be used for the study of the melting trajectory and dynamics of alloy particles in future converter steelmaking processes. By parameterizing the analysis of the influence of different parameters on the eddy current entraining furnace slag, the correlation between the converter inclination angle, tapping hole aperture, and the timing of alloy addition is established.

2. Model Description and Geometric Model

2.1. Computational Model and Meshing

A simplified three-dimensional mathematical model based on the actual size of a 300 t steelmaking converter in a steel plant was established (Figure 1a). During the converter smelting process, a layer of slag floats on the liquid surface, and air fills the remaining space. Based on the characteristics of converter steelmaking, the following assumptions were made when establishing the mathematical model: (1) the experimental environment is relatively stable, and the disturbance caused by air flow on the liquid surface can be ignored; (2) the inner wall and connections of the converter container are smooth, and the frictional effect can be neglected; (3) the temperature change during the experiment is slight, and the influence of temperature on liquid density and viscosity can be ignored; (4) in the turbulent mixing zone, the flow satisfies the basic assumptions of the thin turbulent shear layer theory.

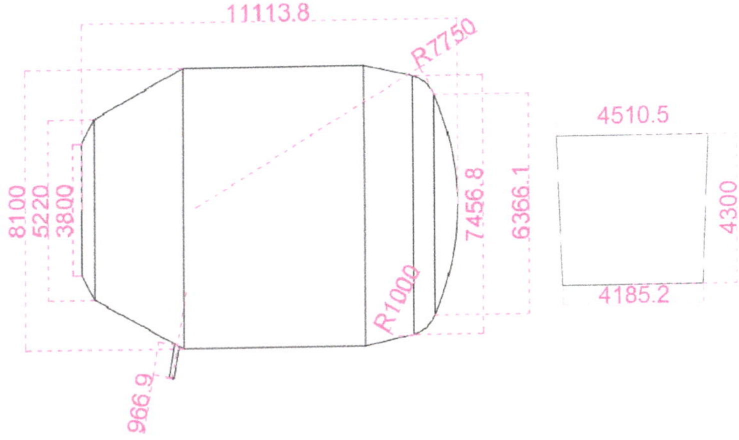

Figure 1. Geometric of the (**a**) converter; (**b**) ladle.

To simulate the converter steelmaking process accurately, a model was built using SolidWorks software 2021 at a 1:1 scale with respect to the actual model. The model includes four computational domains: the converter molten pool domain, the converter air domain, the ladle, and the air domain between the end of the tapping hole and the top of the ladle. The numerical scheme is based on pressure, transient, and three-dimensional calculations. The scheme uses the converter angle (90°, 95°, 100°, 105°) and the tapping hole diameter (200 mm, 210 mm, 220 mm) as variables. The non-structured tetrahedral mesh is created using Workbench, and the Multzone method is used to divide the mesh for the ladle domain. The complete geometric modeling and mesh division of the calculation domain are shown in Figure 2. The calculation mesh consists of 1,600,000 elements, with a maximum positive skewness of 0.823 and a minimum orthogonal quality of 0.177. To more accurately simulate the behavior of the slag–air interface and the steel liquid flow area, local mesh refinement technology was used, as shown in Figure 2c.

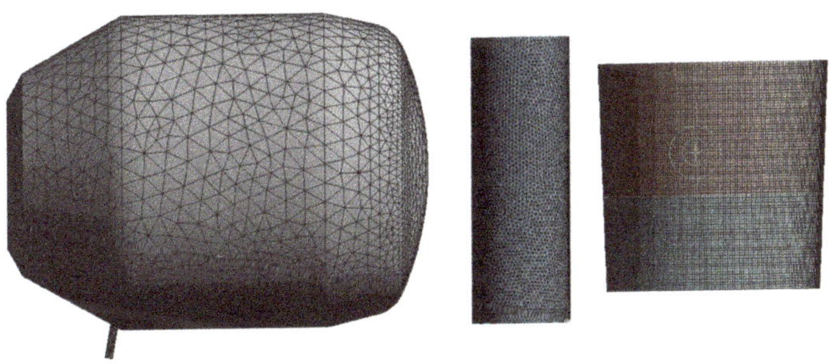

Figure 2. Diagram meshing of (**a**) converter; (**b**) surround; (**c**) ladle.

2.2. Governing Equations

The process of vortex formation in the steel tapping process of the converter can be ignored in the study without considering the effect of temperature, and the control equations used in the simulation of this process are mainly the continuity equation, momentum equation, turbulent kinetic energy equation, and volume equation.

(1) The multiphase VOF (volume of fluid) model can be expressed as follows:

Continuity equation:

$$\frac{\partial(uA_x)}{\partial x}+\frac{\partial(vA_y)}{\partial y}+\frac{\partial(wA_z)}{\partial z}=0 \tag{1}$$

Momentum equation:

$$\frac{\partial u}{\partial \tau}+\frac{1}{V_F}(uA_x\frac{\partial u}{\partial x}+vA_y\frac{\partial u}{\partial y}+wA_z\frac{\partial u}{\partial z})=\frac{1}{\rho}\frac{\partial p}{\partial x}+G_x+f_x \tag{2}$$

$$\frac{\partial v}{\partial \tau}+\frac{1}{V_F}(uA_x\frac{\partial v}{\partial x}+vA_y\frac{\partial v}{\partial y}+wA_z\frac{\partial v}{\partial z})=\frac{1}{\rho}\frac{\partial p}{\partial y}+G_y+f_y \tag{3}$$

$$\frac{\partial w}{\partial \tau}+\frac{1}{V_F}(uA_x\frac{\partial w}{\partial x}+vA_y\frac{\partial w}{\partial y}+wA_z\frac{\partial w}{\partial z})=\frac{1}{\rho}\frac{\partial p}{\partial z}+G_z+f_z \tag{4}$$

where G_x, G_y, G_z are the gravitational acceleration in directions x, y, z, respectively, m/s^2; f_x, f_y, f_z is the viscous force in direction x, y, z; V_F is the volume fraction that can flow; ρ is fluid density, kg/m^3; p is the pressure acting on the fluid element; u, v, w is the component of velocity in direction x, y, z, m/s.

(2) The turbulence modeling is expressed through the standard k-ε model, In the turbulent mixing zone, the flow satisfies the basic assumption of thin, turbulent shear layer theory.

Turbulence kinetic energy equation:

$$\frac{\partial(\rho k)}{\partial t}+\frac{\partial(\rho k u_i)}{\partial x_i}=\frac{\partial}{\partial x_j}[(\mu+\frac{u_i}{\sigma k})\frac{\partial k}{\partial x_i}]+G_k+G_b-\rho\varepsilon-Y_M+S_k \tag{5}$$

Rate of dissipation equation:

$$\frac{\partial(\rho\varepsilon)}{\partial t}+\frac{\partial(\rho\varepsilon u_i)}{\partial x_i}=\frac{\partial}{\partial x_j}[(\mu+\frac{u_i}{\sigma_\varepsilon})\frac{\partial \varepsilon}{\partial x_i}]+C_{1\varepsilon}\frac{\varepsilon}{k}(G_k+C_{3\varepsilon}G_b)+C_{2\varepsilon}\rho\frac{\varepsilon^2}{k}+S_\varepsilon \tag{6}$$

where ρ is the fluid density; t is time; u_i is the fluid flow velocity in direction i; μ_t is the turbulent viscosity coefficient; μ is the turbulent dynamic viscosity; x_i and x_j denote the Cartesian coordinates in directions i and j, respectively; denotes the turbulent energy due to the mean velocity gradient; G_b denotes the turbulent energy due to buoyancy; Y_M denotes the contribution of pulsating expansion to the total dissipation rate in compressible turbulent flow $C_{1\varepsilon}$, $C_{2\varepsilon}$, $C_{3\varepsilon}$, σ_k, and σ_ε are the constants of the k-ε model, which are 1.43, 1.92, 0.09, 1.0 and 1.3, respectively; S_k and S_ε are the user-defined turbulent energy phase and dissipation rate source phase, respectively.

(3) For each phase in the model, introduce a variable called unit phase volume fraction, such that the sum of the volume fractions of all phases in each control volume is equal to 1.

For phase i, the volume fraction equation is:

$$\frac{\partial a_i}{\partial \tau}+\vec{v}\nabla a_i=S_{a_i/\rho_i} \tag{7}$$

The volume fraction equation for the primary phase is:

$$\sum_{i=1}^{n}a_i=1 \tag{8}$$

In the equation, a_i represents the volume occupied by the fluid in the i fluid volume segment, with a value between 0 and 1.

2.3. Boundary Conditions and Numerical Details

The Fluent 20 R2 fluid simulation software under the ANSYS software package was used to perform numerical simulations of the steelmaking process. The physical and chemical properties of the steel and air are shown in Table 1. Since the simulation focuses on the late stage of steelmaking in the converter, 70 t of molten steel were preset in the ladle, and 178 t of liquid steel and 22.24 t of slag were in the converter. Based on the actual flow of fluids in the converter, the boundary conditions of the model were determined as follows: the boundary condition at the converter mouth adopts a pressure inlet, the edge of the air domain above the ladle is a pressure outlet, the inlet and outlet gauge pressure are set to 0 Pa, and the steel liquid flows out freely due to gravity. The gravity acceleration rate $g = 9.81\ m/s^2$, and no-slip wall surface is avoided. The PISO (pressure implicit with splitting of operators) algorithm was used to couple the pressure and velocity terms. In order to obtain more accurate solutions, second-order upwind schemes were used for the discretization of convective terms. The calculation residual was set to 10^{-3}, and iterative calculations were performed after initialization. Calculating the fluxes was carried out using the gradient of each variable evaluated at the cell centroid. The numerical convergence was attained when the sum of the residuals for the flow variables was less than 10^{-3}.

Table 1. Physical parameters of fluid used in simulation.

Property	Air	Steel	Slag
Density, kg/m^3	1.225	7100	2700
Viscosity, pa·s	1.79×10^{-5}	0.0065	0.1998

2.4. Grid-Independency Study

To eliminate the influence of grid resolution on numerical results, a grid-dependency study was conducted to compare the effects of three different grid sizes on the time of slag entrainment by vortex. The results show that the error in the onset and stabilization time of vortex flux on the 160 W and 250 W grids is 0.14% and 0.89%, respectively. The error in the onset and stabilization time of vortex on the 80 W and 160 W grids is 11.2% and 7.8%, respectively. Table 2 summarizes the key parameters of the grid-dependency study. It can be clearly seen that the 160 W grid satisfies the computational requirements of the turbulence model used in this paper.

Table 2. Grid dependency study based on the case.

Number of Grid	80 w	160 w	250 w
Vortex onset time	38.68	43.55	43.49
Vortex stabilization time	49.84	54.10	54.58

3. Results and Discussion

3.1. The Tapping Flow Field of the Converter at 90 Tilt Angles

An analysis was conducted on the steelmaking process of a 200 mm diameter tapping hole with a 90° tapping angle in the converter. Figure 3 shows the distribution diagrams of steel and slag along the vertical section of the converter at 20 s, 50 s, 70 s, and 125 s after tapping. From Figure 3a, it can be seen that at the beginning of steel tapping, the steel and slag are mutually insoluble and clearly stratified, and the flow of the steel inside the converter is relatively stable without the formation of vortex.

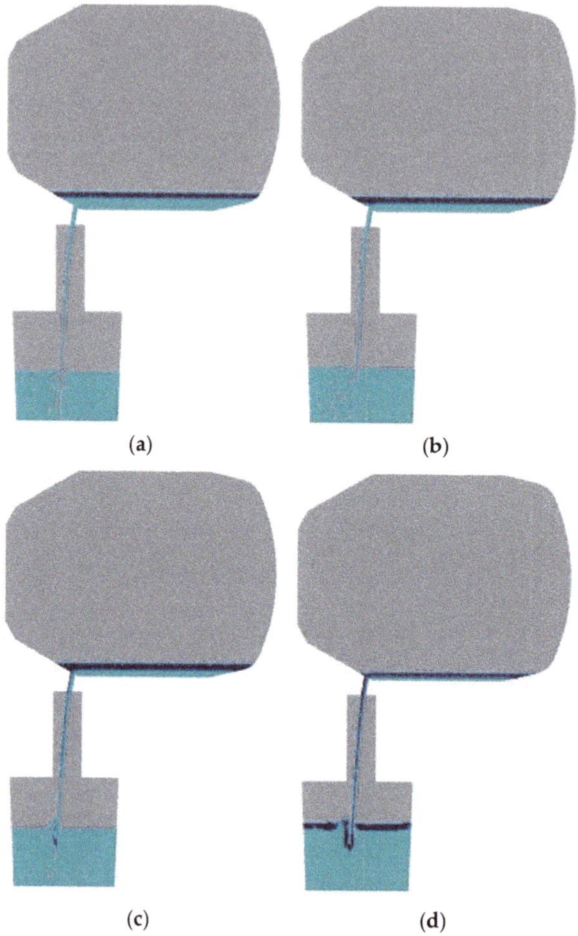

Figure 3. Steel distribution diagram during the 90° tapping process at different times. (**a**) 20 s; (**b**) 50 s; (**c**) 70 s; (**d**) 125 s.

Fluent monitoring shows that the time when slag flows out of the taphole is 43 s, while the time when stable vortex flow appears is 50 s. In the process of steel tapping (Figure 3b), many small vortices disappear as the flow develops [16]. When the fluid in the converter descends to a certain height, a sustained vortex flow will form. The front end of the vortex develops in the direction of the taphole, and the slag is entrained into the taphole, resulting in the phenomenon of vortexing and slag coiling (Figure 3c). At the end of the steelmaking process (Figure 3d), a surface depression can be observed on the slag layer.

During the development of the vortex flow (Figure 3a,b), although the mass flow rate of the steel through the tapping hole decreased, the impact of the jet flow on the ladle increased gradually due to the increase in the depth of the molten pool. This caused greater disturbance to the steel inside the ladle. As the depth of the steel increases, the eddies and vortices generated by the impact of the jet flow will more strongly affect the gas inside the steel. However, if the depth of the steel bath in the ladle is large, the distance between the bottom of the ladle and the impact point of the nozzle will become very large, and the kinetic energy of the impact will be absorbed and dispersed by the steel liquid. At this time, the disturbance effect of the impact will decrease.

During the growth period of a vortex (Figure 3c,d), the amount of air drawn into the steel bath by the descending jet decreases gradually. The stirring effect produced in the steel bath also weakens, because the viscosity of slag is higher than that of steel, and the propagation speed of the jet impact in slag is also slower. According to Figure 4, due to the density difference between the slag and steel, the presence of slag forms a "buffer zone", making it difficult for the impact of the jet flow to directly affect the steel, thereby reducing the disturbance to the steel [17]. As the thickness of the slag layer increases, the disturbance caused by the injection flow to the molten steel in the ladle gradually decreases.

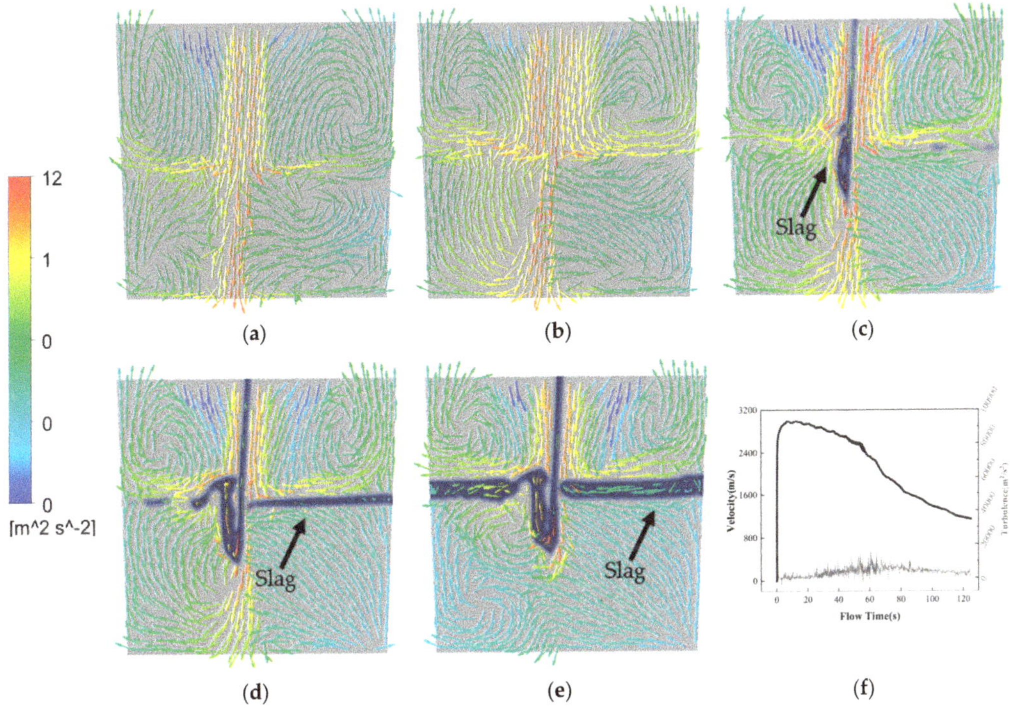

Figure 4. Turbulent flow distribution diagram in the ladle at (**a**) 20 s; (**b**) 50 s; (**c**) 70 s; (**d**) 90 s; (**e**) 125 s during steel tapping when the converter is at 90° and (**f**) velocity vector and turbulence intensity distribution.

3.2. Analysis of Vortex Formation during Converter Tapping

During the process of producing steel in a converter, the molten steel is continuously impacted and stirred against the furnace wall due to unstable and varying flow velocities, resulting in the formation of a rotating vortex. This vortex has both horizontal and vertical rotations. As the rotation process is highly complex, the formation of the converter steel vortex is initially discussed by analyzing the fixed tilt angle. The formation of the vortex is related to the speed and direction of the steel liquid flow. Above the taphole, the flow rate of the steel liquid is faster, and the steel liquid inside the converter needs to pass through a conical position to give the fluid an initial axial velocity. During the tapping process, the turbulence disturbance causes the tangential flow to become unstable and easily form eddies.

Figure 5 shows the velocity streamlines at different moments during the tapping process of the converter at an angle of 100°. The vortex initiation time is 68.8 s, and the stable development time is 70.95 s. As shown in Figure 5a, when the converter begins tapping, the static pressure of the steel liquid near the taphole is no longer balanced due to the effect of gravity. The pressure difference causes the steel liquid to flow out quickly, and the velocity streamlines are relatively uniform and neat. As shown in Figure 5b, although most of the flow directions of the steel liquid streamlines still point to the centerline of the taphole, some flow directions near the mouth begin to align with the tangential direction of the taphole, and there is no vortex above the taphole at this time. As the tapping process continues, two vortices appear above the taphole in the early stage of vortex initiation (Figure 5c), and the centers of the two vortices are connected by rotation in the appropriate tangential direction, causing the flow direction of the fluid streamlines to change, and countless micro-vortices develop into visible vortices. With the further extension of tapping time, stable vortices are formed, and the phenomenon of vortex slag coiling appears (Figure 5d). As shown in Figure 5e,f, after the vortex is stably formed, the vortex gradually becomes smaller and even disappears as the tapping process progresses. On the one hand, this is because the overall flow rate of the taphole decreases in the late stage of tapping. On the other hand, when the slag amount is too large, the flow velocity of the slag liquid will increase, which will intensify the turbulent motion of the slag liquid, destroying the formation and maintenance of the vortex and causing the number and size of the vortices to decrease.

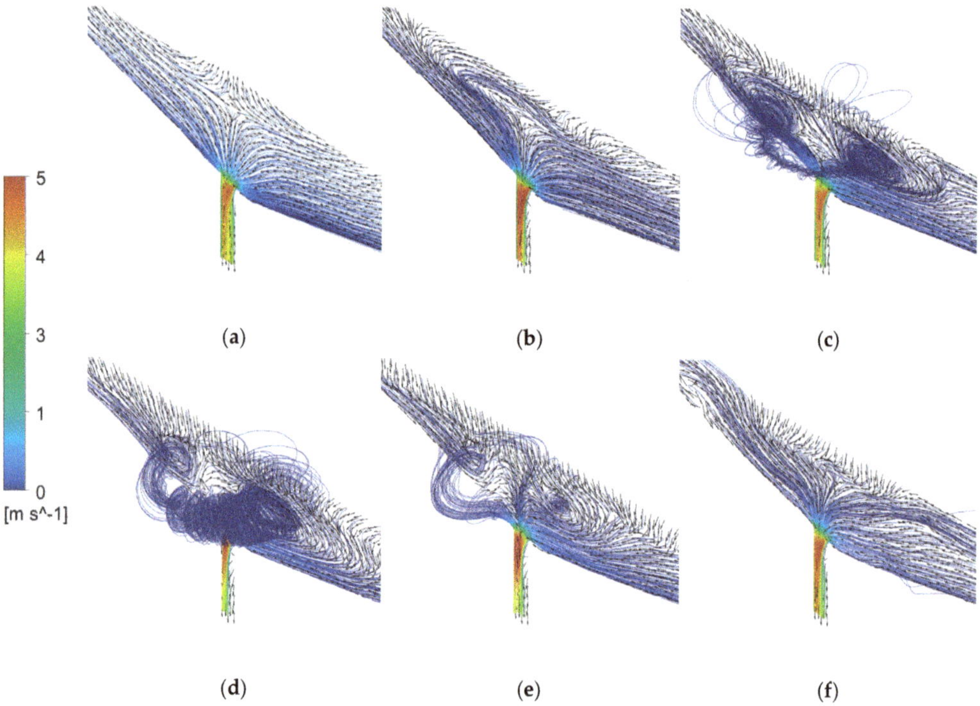

Figure 5. Velocity streamline diagram at a converter angle of 100°: (**a**) 1 s; (**b**) 30 s; (**c**) 67 s; (**d**) 72 s; (**e**) 75 s and (**f**) 80 s.

It can be seen that when a tangential vector appears above the tundish nozzle, the steel liquid vortex begins to form. As the vortex develops, the turbulence of the steel liquid streamlines above the tundish nozzle increases, and the steel liquid vortex becomes stronger and gradually forms a whirlpool, which begins to roll into the slag, forming a phenomenon of swirling slag. With the increase in the slag amount during the steel pouring process, the turbulent motion of the slag liquid destroys the connection of the vortex, resulting in the gradual disappearance of the vortex.

3.3. Comparison of Flow Field with Different Converter Tilt Angles

When the converter tilt angle is fixed, the self-gravity of the fluid in the steel ladle forms a static pressure difference, which is the energy source for the development of vortex motion. Figure 6e–h shows the position enlarged velocity flow diagram of the red box in Figure 6a–d. The red line in Figure 6e–h is the slag layer. Due to the angle of 129° between the tapping hole and the left wall, and the angle of 100° between the tapping hole and the furnace wall, as the tilt angle of the converter increases, the vortex currents in the slag layer gradually shift from above the tapping hole to the right side, reducing the tangential velocity and disturbance of the slag layer. When the converter is tilted at an angle of 105° and the tapping time is 40 s, the angles on both sides of the tapping hole tend to be symmetric, and the tangential vector becomes very small (Figure 6). This suppresses the development of a slag layer vortex, even approaching a converging vortex [18]. In other words, when the tilt angle of the converter is 105°, the vortex in the slag layer has basically stopped rotating.

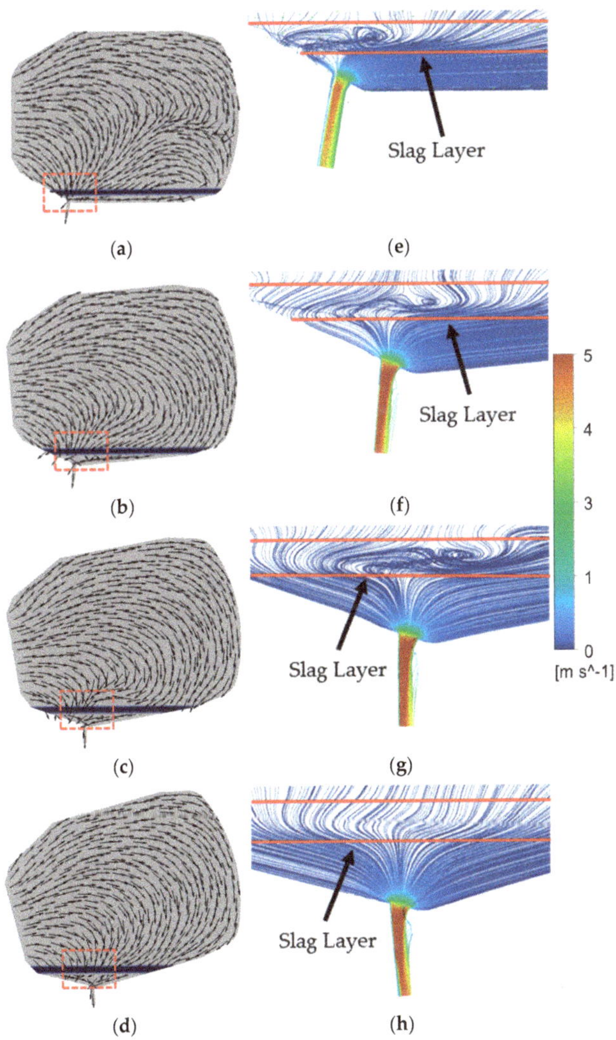

Figure 6. Velocity vector diagram at converter angles (**a**) 90°, (**b**) 95°, (**c**) 100°, (**d**) 105°, and streamlines diagram at angles (**e**) 90°, (**f**) 95°, (**g**) 100°, (**h**) 105°.

The monitoring standard for slag entrapment is to measure the flow rate of slag by monitoring the cross-section of the tapping hole during steelmaking. The onset times of vortex entrainment for converter angles of 90°, 95°, 100°, and 105° are 43.55 s, 66.44 s, 68.80 s, and 72.30 s, respectively. The stable development times of vortex entrainment are 54.10 s, 70.36 s, 70.95 s, and 74.26 s, respectively. Vortexes exist from the beginning of the steel flow, even if the steel–slag interface is not concave, but vortex motion has formed at the bottom. With an increase in converter angle, the tangential vector above the tapping hole tends to be symmetrical, and the duration of stable vortex entrainment decreases. As the initial tangential disturbance is the main factor for the vortex formation [19], the asymmetry of the tangential vector above the tapping hole leads to different onset and development times for the vortex.

According to Figure 7, when slag occurs at different angles of the converter, the remaining liquid steel in the converter is 5.12m^3, 1.45m^3, 0.80m^3 and 0.54m^3, respectively. When the tapping hole is close to the vertical ground, the angle of the converter has little effect on the timing of slagging, but it has a significant impact on the speed of the tapping hole. As the tapping hole is vertical to the ground, the static pressure above the tapping hole caused by gravity is the largest, and the steel liquid flow rate at an angle of 100° is the fastest. Due to the special nature of the converter's tapping hole, the angles of the left and right walls of the tapping hole are inconsistent, so when the angle of the converter is 105°, the tapping speed is slightly reduced, and the timing of slagging is delayed.

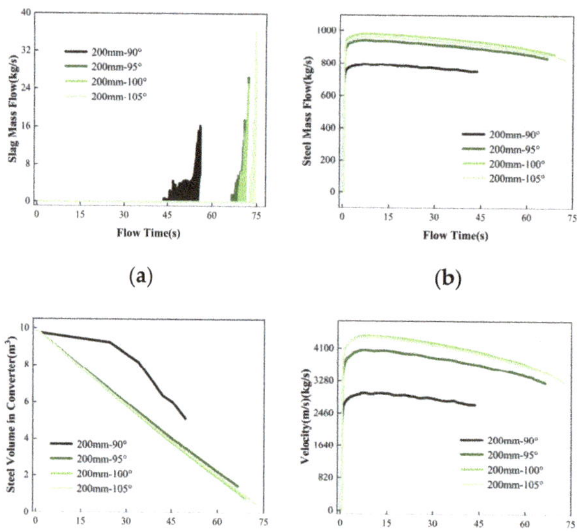

Figure 7. The variation trend of (**a**) slag mass flow; (**b**) steel mass flow; (**c**) steel volume in converter and (**d**) tapping velocity with flowing time when the tilting angle of the converter is 90°, 95°, 100°, or 105°.

When the converter angle is 100° (Figure 8c), the maximum steel pouring velocity is obtained. At this time, the pouring angle is perpendicular to the center of the bottom of the ladle, and the distribution of the velocity vector inside the ladle is nearly symmetrical. This is because the wall above the taphole of the converter is an asymmetric interface, which causes the velocity vector distribution in the pouring of molten steel to be incompletely symmetrical. When the converter angle is 105° (Figure 8d), the stagnant area of the molten pool vector inside the ladle is larger. The pouring velocities of molten steel are relatively close at converter angles of 100° and 105°. However, due to differences in the pouring location and angle deviation, the inner wall of the ladle on one side is eroded when the converter angle is 105°. This is because as the pouring angle of the molten steel increases, the number and size of the vortex will change. When the molten steel flows into the ladle at a larger angle from the nozzle, the eddies generated in the molten pool can more strongly disturb the molten pool, thereby increasing the turbulence level in the molten pool [20]. This helps to mix the chemical components in the molten pool, thus improving the uniformity and quality of the molten steel and reducing the temperature gradient in the molten pool.

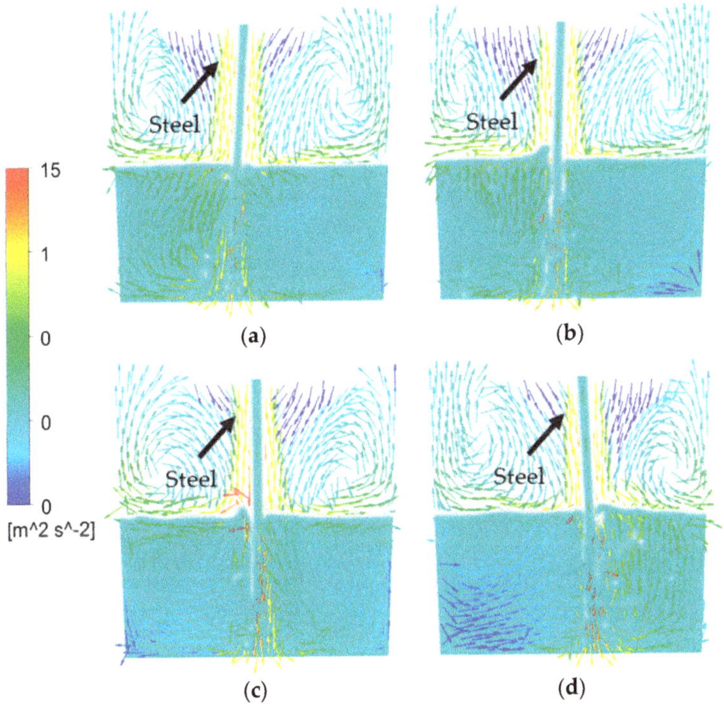

Figure 8. Velocity vector diagram of the ladle at different tapping angles of the converter: (**a**) 90°; (**b**) 95°; (**c**) 100°; (**d**) 105°.

In the late stage of steelmaking, alloy materials are added to the steelmaking furnace. The influencing factors of the movement trajectory and residence time of alloy particles in the molten steel include the physical and chemical parameters of the alloy particles, the steel flow field [21,22], and the location where the alloy particles are added to the steelmaking furnace [20,23,24]. In fact, the alloying process of alloy particles in the steelmaking furnace is mainly carried out by the mechanisms of diffusion and mixing. Diffusion refers to the free diffusion of alloy elements in the steel, and mixing refers to the process in which the turbulent action formed in the steel promotes the mixing of the steel and slag in the molten steel. Therefore, when the angle of the converter is between 100–105°, it is suitable to add alloy particles to the molten steel in the steelmaking furnace. At this time, the flow rate of the steel liquid is relatively high, and the critical height of the molten slag in the converter is relatively low.

3.4. Influence of Different Tapping Hole Sizes on the Flow Field

Figure 9 shows the mass flow rate of steel, slag, and molten steel velocity, as well as the residual volume of molten steel in the converter during the second tapping process with different tapping hole diameters (a) 200 mm, (b) 210 mm, and (c) 220 mm, for 50 s. The time when vortex currents start to roll slag is 68.80 s, 62.64 s, and 57.11 s, respectively. The time when vortex currents start to develop stably is 70.95 s, 64.49 s, and 58.00 s, respectively. As the diameter of the tapping hole increases, the time for the vortex to reach the tapping hole shows a decreasing trend. The residual volume of molten steel in the converter is 0.80 m^3, 0.77 m^3, and 0.73 m^3, respectively, when the different tapping hole diameters start to roll slag, indicating that the critical height of the vortex decreases with an increase in the tapping hole diameter. This is because, during the tapping process, the effect of the tapping hole diameter on the initiation of the vortex is much smaller than the effect of the mass flow rate of the molten steel from the tapping hole. With a larger tapping hole diameter, the mass flow rate of molten steel increases, and the liquid level drops faster. Although the critical height of the liquid level for vortex formation is lower, the development time of the vortex accelerates, and the time for the vortex to reach the tapping hole becomes shorter.

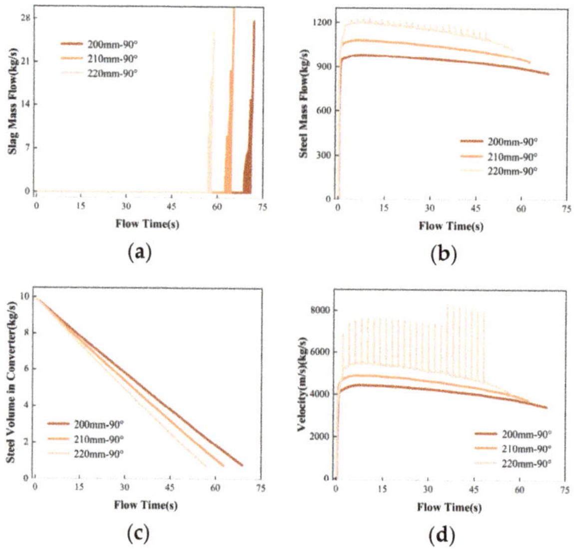

Figure 9. The variation trend of (**a**) slag mass flow; (**b**) steel mass flow; (**c**) steel volume in converter, and (**d**) tapping velocity with flowing time when the tilting hole sizes of the converter are 200 mm, 210 mm, or 220 mm.

However, overall, the effect of increasing the tapping hole diameter on the formation of the vortex is not significant. However, there is a phenomenon where periodic oscillations occur in the flow of molten iron from the taphole when the taphole diameter is increased to 220 mm, which disappears only at the end of tapping. This "oscillation" phenomenon is due to the dynamic instability of the steel in the converter. The occurrence of oscillation tapping can cause fluctuations in the quality of the steel in the furnace, thereby affecting the stability and efficiency of the steel production process. Increasing the taphole diameter of the converter will increase the tapping speed, reduce the surface tension of the molten steel, and reduce the viscosity of the molten steel.

When the diameter of the tapping hole is small, the flow of molten steel is mainly restricted by the hole size, and the flow rate is relatively stable, so it is not easy to cause oscillation. However, when the diameter of the tapping hole increases, the degree of restriction on the flow of molten steel decreases, and the instability of the flow increases, making it easy to cause oscillation. The fluid flow inside the furnace and the velocity and mass flow rate at the tapping hole are closely related. As shown in Figure 10, when the diameter of the tapping hole increases from 200 mm to 220 mm, the symmetric vector position in the converter is closer to the top of the tapping hole. When the diameter of the tapping hole is 220 mm, the internal flow structure of the converter changes after 50 s of flow (Figure 10c), which produces a large vortex, further causing the flow of molten steel to oscillate. When the flow velocity at the tapping hole increases, the kinetic energy of the molten steel also increases, and therefore the mass flow rate also increases accordingly. Figure 10g shows clearly the counteractive effect of two vortices [25]. When the flows meet, the counteractive effect reduces the tangential velocity between the top of the tapping hole, suppresses the formation of the free surface vortex, and reduces the height of the adjacent vortex slag. Therefore, a diameter of 210 mm for the steel tapping hole is the optimal choice, as it can reduce the tapping time of the converter without changing the internal flow field structure of the converter.

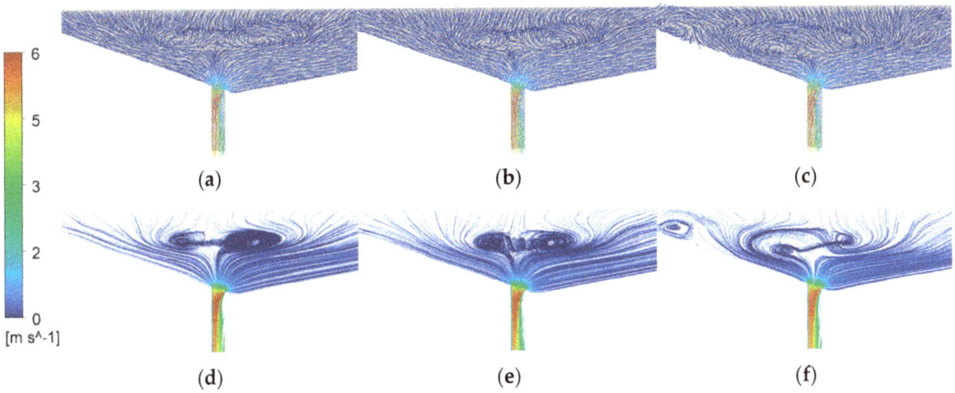

Figure 10. Velocity vector distribution of converter tapping hole with diameters of (**a**) 200 mm, (**b**) 210 mm, and (**c**) 220 mm, as well as (**d**) 200 mm, (**e**) 210 mm, and (**f**) 220 mm velocity streamlines distribution.

4. Conclusions

This paper uses the VOF multiphase flow model and combines it with the 300 t converter steelmaking process to establish a slag entrainment by vortex Flux model for the converter steelmaking process. By discussing the converter steelmaking process at different converter angles and different steel tapping orifice diameters, the following conclusions can be drawn:

1. As the steel pouring process continues, the vortex develops into a whirlpool due to the appearance of tangential vectors above the converter taphole, and the phenomenon of swirling slag occurs. At the end of the steel pouring, as the amount of slag increases, the turbulent motion of the slag liquid disrupts the connection of the vortex, resulting in the gradual disappearance of the vortex.

2. When the converter angle is 90°, 95°, 100°, and 105°, the vortex slagging times are 43.55 s, 66.44 s, 68.80 s, and 72.30 s, respectively. As the converter angle increases, the time for forming a stable vortex becomes shorter. When the converter angle is between 100–105°, the internal flow field of the steelmaking furnace is more suitable for adding alloy particles.

3. When the steel tapping orifice diameter is 200 mm, 210 mm, and 220 mm, the steel liquid residue starts to swirl at 0.80 m^3, 0.77 m^3, and 0.73 m^3, respectively. The initiation times of vortex slag were 68.80 s, 62.64 s, and 57.11 s, respectively. The stable development time of vortex slag was 70.95 s, 64.49 s, and 58.00 s, respectively. When the steel tapping orifice diameter is 210 mm, the steel tapping speed can be increased without affecting the internal flow field structure of the converter. When the steel tapping orifice diameter is 220 mm, the mass flow rate of the steel tapping orifice shows an "oscillating" phenomenon.

Author Contributions: Investigation, Y.S., S.Y. and J.D.; project administration, W.L., J.L. and S.Y.; writing—original draft, C.H. and J.L.; writing—review and editing, C.H., Y.S. and W.L. All authors have read and agreed to the published version of the manuscript.

Funding: This research was funded by the National Natural Science Foundation of China (grant numbers 52104318 and 52074030) and the National Key R&D Program of China (grant number 2021YFB3700402).

Institutional Review Board Statement: Not applicable.

Informed Consent Statement: Not applicable.

Data Availability Statement: Not applicable.

Conflicts of Interest: The authors declare no conflict of interest.

References

1. Wang, Y.; Xu, X.Y.; Li, Z.W.; Fu, J.X. Sulfur segregation and inclusion modification in steel using magnesium addition. *Metall. Res. Technol.* **2022**, *119*, 316. [CrossRef]
2. Riaz, S.; Mills, K.C.; Bain, K.G. Experimental examination of slag/refractory interface. *Ironmak. Steelmak.* **2002**, *29*, 107–113. [CrossRef]
3. Sankaranarayanan, R.; Guthrie, R.I.L. Slag entraining vortexing funnel formation during ladle teeming: Similarity criteria and scale-up relationships. *Ironmak. Steelmak.* **2002**, *29*, 147–153. [CrossRef]
4. Bruna, G.B.; Pereira, J.; Bielefeldt, W.V.; Vilela, A. Assessment of inclusion analysis via manual and automated SEM and total oxygen content of steel. *J. Mater. Res. Technol.* **2015**, *4*, 235–240.
5. Zhang, L.; Thomas, B.G. State of the Art in the Control of Inclusions during Steel Ingot Casting. *Metall. Mater. Trans. B* **2006**, *37*, 733–761. [CrossRef]
6. Tripathi, N.N.; Nzotta, M.; Sandberg, A.; Sichen, D. Effect of ladle age on formation of non-metallic inclusions in ladle treatment. *Ironmak. Steelmak.* **2004**, *31*, 235–240. [CrossRef]
7. Beskow, K.; Tripathi, N.N.; Nzotta, M.; Sandberg, A.; Sichen, D. Impact of slag–refractory lining reactions on the formation of inclusions in steel. *Ironmak. Steelmak.* **2004**, *31*, 514–518. [CrossRef]
8. Ånmark, N.; Karasev, A.V.; Jönsson, P.G. The Effect of Different Non-Metallic Inclusions on the Machinability of Steels. *Materials* **2015**, *8*, 751–783. [CrossRef]
9. Yin, Y.; Yang, J.; Zhang, J.; Tang, L. Physical modeling of slag carryover in the last stage of ladle teeming during continuous casting with dynamic change of slide gate opening. *J. Mater. Res. Technol.* **2023**, *23*, 1781–1791. [CrossRef]
10. Morales, R.D.; Palafox-Ramos, J.; Barreto, J.; Lopez-Ramirez, S.; Zacharias, D. Melt flow control in a multistrand tundish using a turbulence inhibitor. *Metall. Mater. Trans. B* **2000**, *31*, 1505–1515. [CrossRef]
11. Li, Z.; Zhang, M.; Zhou, F.; Lu, Y.; Zhang, X.; Gu, H. Numerical simulation of slag entrapment process during the end of casting in tundish. *Ironmak. Steelmak.* **2022**, *49*, 1039–1047. [CrossRef]
12. Mabentsela, A.; Akdogan, G.; Bradshaw, S.M. Numerical and physical modelling of tundish slag entrainment in the steelmaking process. *J. S. Afr. Inst. Min. Met.* **2017**, *117*, 469–483. [CrossRef]
13. Wang, J.; Zhu, M.; Zhou, H.; Wang, Y. Fluid Flow and Interfacial Phenomenon of Slag and Metal in Continuous Casting Tundish With Argon Blowing. *J. Iron Steel Res. Int.* **2008**, *15*, 26–31. [CrossRef]
14. Wang, J.; Li, J.; Han, Z. Numerical Simulation of Slag Entrainment by Vortex Flux during Tapping at Converter. *Iron Steel Vanadium Titan.* **2012**, *33*, 34–39.
15. Pang, C.; Yuan, F.; Xu, A.; He, D.; Liu, X. Numerical simulation of tapping process in converter steelmaking. *China Metall.* **2022**, *32*, 8–15.
16. Gupta, A.; Kumar, R.; Singh, R.K. Assessment of Critical Vortexing Height to Prevent Slag Entrapment during Tundish Teeming. *Met. Mater. Int.* **2022**, *5*, 1246–1256. [CrossRef]
17. Magnini, M.; Pulvirenti, B.; Thome, J.R. Numerical investigation of the influence of leading and sequential bubbles on slug flow boiling within a microchannel. *Int. J. Therm. Sci.* **2013**, *71*, 36–52. [CrossRef]
18. Li, H.; Wang, Q.; Lei, H.; Jiang, J.; Guo, Z.; He, J. Mechanism Analysis of Free-Surface Vortex Formation during Steel Teeming. *ISIJ Int.* **2014**, *54*, 1592–1600. [CrossRef]
19. Li, H.; Wang, Q.; Jiang, J.L.; Guo, Z.; He, J. Analysis of Factors Affecting Free Surface Vortex Formation during Steel Teeming. *ISIJ Int.* **2016**, *56*, 94–102. [CrossRef]
20. Alfonso, N.B.; Jafeth, R.Á.; Javier, G.G.; Rodolfo, D.M.; Kinnor, C. Changes of Multiphase Flow Patterns during Steel Tapping with Simultaneous Argon Bottom Stirring in the Ladle. *Metals* **2020**, *10*, 1036.
21. Cheng, R.; Zhang, L.; Yin, Y.; Zhang, J. Effect of Side Blowing on Fluid Flow and Mixing Phenomenon in Gas-Stirred Ladle. *Metals* **2021**, *11*, 369. [CrossRef]
22. Cheng, R.; Zhang, L.J.; Yin, Y.B.; Ma, H.T.; Zhang, J.M. Influence of Argon Gas Flow Parameters in the Slot Plug on the Flow Behavior of Molten Steel in a Gas-Stirred Ladle. *Trans. Indian Inst. Met.* **2021**, *74*, 1827–1837. [CrossRef]

23. Jafeth, R.Á.; Morales, R.D.; Nájera-Bastida, A. Numerical Study of Multiphase Flow Dynamics of Plunging Jets of Liquid Steel and Trajectories of Ferroalloys Additions in a Ladle during Tapping Operations. *ISIJ Int.* **2012**, *52*, 814–822.
24. Ren, Y.; Zhang, L.F.; Li, Y.L.; Yu, L.; Li, G.; Li, J.T. Numerical Simulation of Fluid Flow and Alloy Dispersion in an Argon Gas Stirred Ladle. *J. Iron Steel Res. Int.* **2014**, *26*, 28–34.
25. Lee, S.J.; Kim, S.J.; Lee, H.G. *Formation of Free Surface Vortex in the Steelmaking Process (Dual Tapping Hole System in a Converter)*; John Wiley & Sons: Hoboken, NJ, USA, 2013.

materials

MDPI

Article

Numerical Prediction of Microstructure Evolution of Small-Diameter Stainless Steel Balls during Cold Skew Rolling

Jing Zhou [1,2,*], Shengqiang Liu [1], Baoyu Wang [1,2] and Hao Xu [1]

[1] School of Mechanical Engineering, University of Science and Technology Beijing, Beijing 100083, China
[2] Engineering Research Center of Part Rolling, Ministry of Education, Beijing 100083, China
* Correspondence: jingzhou@ustb.edu.cn

Abstract: The wear resistance and hardness of stainless steel (SS) balls formed by cold skew rolling are effectively improved due to the change in internal microstructure. In this study, based on the deformation mechanism of 316L stainless steel, a physical mechanism-based constitutive model was established and implemented in a subroutine of Simufact to investigate the microstructure evolution of 316L SS balls during the cold skew rolling process. The evolution of equivalent strain, stress, dislocation density, grain size, and martensite content was studied via simulation during the steel balls' cold skew rolling process. The corresponding skew rolling experiments of steel balls were carried out to verify the accuracy of the finite element (FE) model results. The results showed that the macro dimensional deviation of steel balls fluctuates less, and the microstructure evolution agrees well with the simulation results, which proves that the established FE model has high credibility. It shows that the FE model, coupled with multiple deformation mechanisms, provides a good prediction of the macro dimensions and internal microstructure evolution of small-diameter steel balls during cold skew rolling.

Keywords: steel ball; skew rolling; microstructure; cold forming; stainless steel

Citation: Zhou, J.; Liu, S.; Wang, B.; Xu, H. Numerical Prediction of Microstructure Evolution of Small-Diameter Stainless Steel Balls during Cold Skew Rolling. *Materials* **2023**, *16*, 3246. https://doi.org/10.3390/ma16083246

Academic Editor: Andrea Di Schino

Received: 17 March 2023
Revised: 13 April 2023
Accepted: 18 April 2023
Published: 20 April 2023

1. Introduction

Steel balls, as one of the critical components of ball bearings, directly influence the dynamic performance, reliability, and life of the bearings. Numerous researchers have conducted in-depth research on the steel ball skew rolling process, which has dramatically deepened the understanding between the microstructure and performance of steel balls. However, most of them are limited to the hot skew rolling process in the formation of steel balls with larger diameters, and there needs to be more research on steel balls with small diameters. The forming of small-diameter steel balls is more complicated than that of large-diameter steel balls, the equipment and process parameters required are stricter and more precise, and the forming defects and dimensions of the steel balls are more challenging to control rigorously. In addition, most of the research on steel ball materials is concentrated on the steel GCr15, while more research needs to be conducted on stainless steel (SS) balls. Compared with GCr15, 316L SS exhibits better rust and corrosion resistance [1]. Bearings manufactured with SS can be used in liquids and have stronger heat resistance and are widely used in medical devices, cryogenic engineering, optical engineering, high-speed motors, and other fields.

Several studies have examined the influence of process parameters on the steel balls' skew rolling forming process. Pater et al. [2] discussed the production process of the multi-wedge spiral rolling process in the formation of steel balls. They analyzed the temperature, stress, strain, force, and movement during the forming process. Gontarz et al. [3] studied the effect of skew groove patterns on obliquely rolled steel balls and analyzed the strain, damage, and temperature distribution during the forming process. Tomczak et al. [4] optimized the spiral indentation of the die during the oblique rolling of steel balls, proposed a design method for the spiral indentation, proposed a die correction method based on it, and verified the effectiveness of the optimized die method by simulation and forming experiments. Liu et al. [5] analyzed the production process of small-diameter steel balls, discussed the influence of process parameters on the forming quality of steel balls, and analyzed the stress, strain, rolling force, and work-hardening distribution of steel balls, and successfully trial-produced small-diameter steel balls with the required dimensional accuracy. The above studies mainly focus on the hot skew rolling process of steel balls, while steel balls with small diameters are more suitable for cold forming due to the small size and fast heat loss during the hot rolling process. Moreover, the macroscopic dimensional fluctuations and internal microstructure evolution during the cold skew rolling of steel balls have an important influence on their performance and service life, so research on the cold skew rolling process of steel balls is urgently needed.

During deformation, the deformation mechanism of austenitic SS is highly dependent on the stacking faults energy (SFE), which is the crucial parameter determining whether mechanical twinning, martensitic transformation, and dislocation slipping dominate the deformation process of the material. The deformation mechanism of metals with low SFE will change from dislocation slip to deformation twinning or martensitic transformation, which is significant for the hardening of the material. For the deformation mechanism of austenitic SS under rolling conditions, Zhang et al. [6] investigated the strain-induced martensite behavior of 316L SS during rolling at different temperatures. After warm rolling deformation, they found that both slate-like martensite and dislocation-type martensite were present in the microstructure, containing both dislocations and deformation twins. Xiong et al. [7] similarly studied the evolution of microstructure and mechanical properties of 316LN SS after cold rolling at different strains. They found that the deformation mechanisms during low-temperature rolling were mainly composed of high-density dislocations, deformation twinning, and martensitic transformation. Eskandari et al. [8] investigated the changes in the deformation-induced martensite volume fraction and mechanical properties of 316L SS under cold rolling process conditions, which showed that decreasing the rolling temperature and increasing the pre-strain could increase the martensite volume fraction and, at the same time, refine the grain size, with a minimum of 30–40 nm grain size. The obtained nanocrystalline SS exhibited excellent tensile strength.

The constitutive model based on the physical mechanism enables reflection of the connection between the macroscopic deformation behavior and the material's microstructure. Huo et al. [9] studied the forming process of bearing steel balls with a diameter of 30 mm during warm skew rolling. They developed an internal state variable constitutive model to predict the microstructure evolution. Then, the experimental and simulation results verified the availability of the developed model. Ran et al. [10] established a multiscale model with coupled damage, grain size, and surface layer models to describe the ductile fracture and deformation behavior during the micro-forming of multiphase alloys and elucidated the relationships between size effects, fracture energy, and expected fracture strain during the micro-forming. Chen et al. [11] developed a constitutive model considering the kinetics of twinning and martensitic transformation using the mixing law to investigate the impact processes during the mechanical abrasion treatment of surfaces. Liu et al. [12] established a constitutive model coupling the electricity and size effect, studied the influences of current density and grain size on the forming process in the electro-assisted forming, and captured the abnormal evolution of surface effect by using the proposed constitutive model.

This work aims to implement a physical mechanism-based constitutive model coupling dislocation slip, martensitic transformation, and grain refinement in the finite element software Simufact 16.0. Establishing the FE model of the small-diameter steel ball skew rolling process is conducive to investigating the macro size and microstructure evolution of cold-rolled steel balls and provides a theoretical basis for the microstructure control in the process of cold skew rolling of steel balls.

2. Development of a Mechanism-Based Constitutive Model of 316L under Cold Deformation

2.1. Establishment of Multiscale Constitutive Equations

As the martensite transformation of 316L stainless steel will occur during cold plastic deformation, the microstructure after deformation consists of austenite and martensite. Therefore, the mixture law describes the macroscopic flow stress of the dual-phase material. The macroscopic flow stress can be expressed in the following formula:

$$\sigma = \sigma_A(1 - f_M) + \sigma_M f_M \tag{1}$$

where σ_A, σ_M are the stress of austenite and martensite, and the stress of martensite can be calculated by:

$$\sigma_M = \sigma_0 + \alpha M G_M b_M \sqrt{\rho_M} \tag{2}$$

where σ_0 is the yield strength of martensite. α, M, G_M, b_M, and ρ_M are the material constants reflecting the dislocation interaction, the Taylor factor, the shear modulus, the magnitude of the burger vector, and the dislocation density of martensite, respectively. The value can be determined by the empirical formula [13]:

$$\sigma_0(MPa) = 461 + 1310 \times \sqrt{(wt.\%C)} \tag{3}$$

Austenitic stress is composed of grain boundary resistance σ_G and dislocation density resistance σ_p [14]:

$$\sigma_A = \sigma_G + \sigma_\rho \tag{4}$$

σ_G can be expressed by the Hall–Petch equation, and σ_ρ is related to mean dislocation density [15]:

$$\sigma_G = \sigma_{y_0} + \frac{K_s}{\sqrt{d}} \tag{5}$$

$$\sigma_\rho = \alpha M G_A b_A \sqrt{\rho_A} \tag{6}$$

where σ_{y_0} is the initial stress, K_s is the parameter indicating the Hall–Petch slope, and d is the grain size. The evolution of dislocation density in austenite can be denoted as:

$$\dot{\rho}_A = \frac{|\dot{\varepsilon}|}{b_A \Lambda_s} - \Omega \rho_A |\dot{\varepsilon}| \tag{7}$$

where $\dot{\varepsilon}$ is the plastic strain rate, and Ω is the content describing dynamic recovery. Λ_s is the mean free path of dislocation density in austenite, which is related to grain size, dislocation density, and martensite volume fraction:

$$\frac{1}{\Lambda_s} = \frac{1}{k_d d} + \frac{\sqrt{\rho_A}}{k_s} + \frac{f_M}{k_M t_M (1 - f_M)} \tag{8}$$

where k_d, k_s, k_M are material constants, and t_M is the thickness of the martensite lath; f_M is the volume fraction of martensite.

In this paper, the martensitic transformation model proposed by Wong et al. [16] is adopted. It is believed that the volume fraction of martensite is related to the martensite nucleation rate $\dot{N}_M$ and the volume of newly formed martensite V_M:

$$\dot{f}_M = (1 - f_M) V_M \dot{N}_M \tag{9}$$

$\dot{N}_M$ can be expressed by the following equation:

$$\dot{N}_M = \dot{N}_0 P_M P_{nsc} \tag{10}$$

where $\dot{N}_0$ is the potential martensitic nucleation density in unit time, P_{nsc} is the probability that cross-slip does not occur. P_M is the probability of the formation of a martensite lath:

$$P_M = \exp[-(\frac{\sigma_{tr}^c}{\sigma_M})^m] \tag{11}$$

where m is the model parameter, σ_{tr}^c is the critical stress of martensitic transformation, and σ_M is the stress applied to the martensite.

$$\sigma_{tr}^c = \sigma_{tr}^0 + \frac{K_{tr}}{\sqrt{d}} \tag{12}$$

where σ_{tr}^0, K_{tr} are the model parameters. The volume of the newly formed martensitic slats can be characterized as:

$$V_M = \frac{\pi}{4}\Lambda_M^2 t_M \tag{13}$$

where Λ_M is the mean free path of the dislocation in the martensite phase.

$$\frac{1}{\Lambda_M} = \frac{c_\gamma}{d} + c_m\sqrt{\rho_M} \tag{14}$$

where c_γ and c_m are the model parameters. The change rate of dislocation density in martensite is given [17]:

$$\dot{\rho}_M = |\dot{\varepsilon}|(\frac{1}{b_M\Lambda_M} - k_a\rho_M) \tag{15}$$

where k_a is the parameter representing the dislocation annihilation. During plastic deformation, the grain of material will be refined, and the grain refinement model can be expressed as:

$$d_\varepsilon = d_0\frac{[(\frac{K}{\sqrt{\rho_A}})^{d_a} + (1-f_M)^{d_b}]}{2} \tag{16}$$

where d_a and d_b are the model parameters, d_0 is the initial grain size, and K is the model parameter.

2.2. *Solution of Model Parameters*

The model is composed of highly coupled differential equations, and there are a total of 24 parameters, so analytical solutions can hardly be obtained. Therefore, the genetic algorithm toolbox in the software package Matlab 2021b is used to solve the model parameters with the goal of residual error between experimental and predicted values. The details of the solution procedure can be found in Ref. [18], and the determined model parameters are given in Table 1.

Table 1. Identified parameters of 316L SS constitutive model.

Parameters	Values	Parameters	Values
σ_{y_0} (MPa)	1.0000×10^1	σ_{tr}^0 (MPa)	7.0746×10^2
K_s (MPa·μm$^{1/2}$)	6.5222×10^2	K_{tr} (MPa·μm$^{1/2}$)	1.1347×10^2
Ω	1.0515×10^{-1}	k_a	4.1500×10^1
k_d	2.1464×10^{-6}	$\dot{N}_0$	3.1500×10^{17}
k_s	3.9676×10^1	d_a	1.5088×10^{-1}
k_M	1.5520×10^3	d_b	4.1151×10^{-1}
m	2.4255×10^2	K	1.0000×10^5
c_m	1.1090×10^{-1}	α	4.0200×10^{-1}

Table 1. *Cont.*

Parameters	Values	Parameters	Values
c_γ	7.7166×10^6	b_M (nm)	1.4700×10^{-1}
σ_0 (MPa)	6.8790×10^2	M	3.06×10^0
b_A (nm)	2.5600×10^{-1}	G_A (MPa)	7.5000×10^4
t_M (μm)	1.1000×10^0	G_M (MPa)	8.0000×10^4

3. Validation of Numerical Simulation of Cold Skew Rolling

3.1. FE Modeling

A skew rolling process FE model considering the physical mechanism-based constitutive model of steel balls is developed in the FE software Simufact 16.0, as shown in Figure 1. The FEM consists of two rolls, two guide plates, a guide pipe, and a workpiece. Due to the little plastic and elastic deformation, the guide plates and rolls are set as rigid bodies. The workpiece used in the simulation is the 316L SS wire with a diameter of 3 mm. The density, Young's modulus, and Poisson's ratio of 316L SS are 7966 kg/m^3, 192 GPa, and 0.3, respectively, and its chemical composition is listed in Table 2. In this work, the workpiece is assumed to be a homogeneous isotropic plastic body, the mesh type is hexahedral, and the number of elements is approximately 19,300. The Coulomb friction model is used, the friction coefficient between the rolls and guide plates is 0.2, while it is 0.1 for the friction coefficient between the rolls and the workpiece [19]. In addition, the initial grain size of 316L SS is 16μm, and the initial dislocation density of martensite and austenite are 1×10^{10} m^{-2} and 1×10^{12} m^{-2}, respectively.

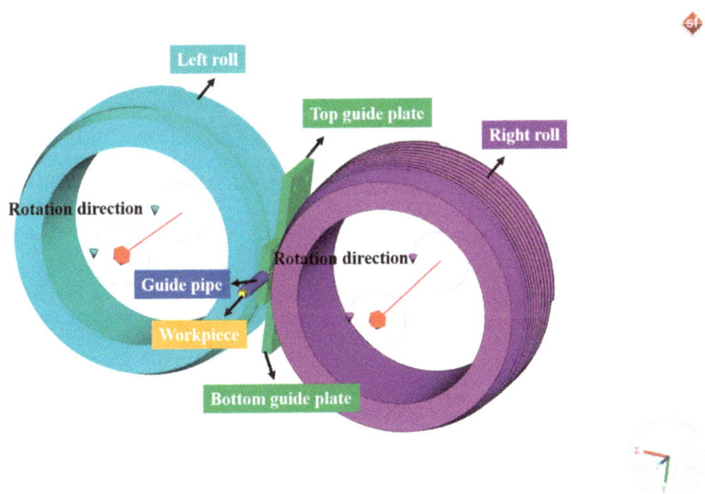

Figure 1. FE model of the cold skew rolling process of steel balls.

Table 2. Chemical composition of 316L SS.

C	Mn	Si	Cr	Ni	Mo	Fe
0.03	1.74	0.27	16.82	10.26	2.08	Bal.

3.2. Experiment Detail

An experiment of steel ball skew rolling was carried out on the newly designed skew rolling mill, as shown in Figure 2. Two servo motor systems drive the rolls of the skew rolling mill. Before the rolling experiment, 2 rolls deflect the same angle around the axis but along the reverse directions, and the deflection angles are nearly 2 and −2 degrees, respectively. During the rolling process, two rolls rotate in the same direction and drive the workpiece forward, and the workpiece gradually forms steel balls under the extrusion of roll grooves.

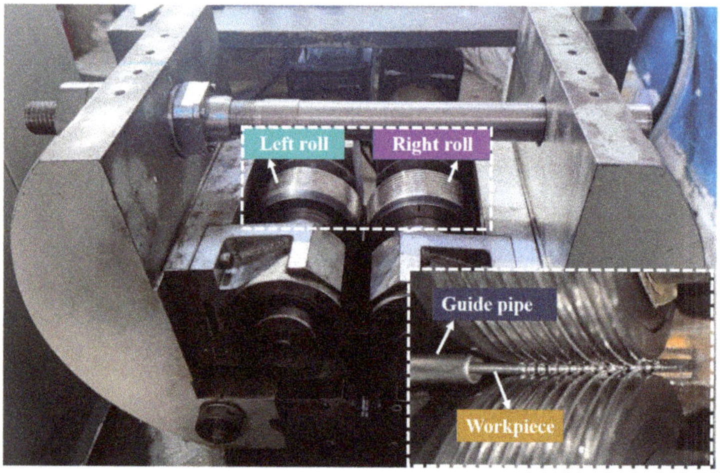

Figure 2. Experimental skew rolling mill and workpiece.

3.3. Comparison of Steel Ball Diameter

A total of 20 steel balls were randomly selected from the finished steel balls and their arbitrary section diameters were measured by the SZM-45T1-560H stereomicroscope; the results are shown in Figure 3. It can be seen from the figure that the diameter of steel balls ranges from 3.114 to 3.126 mm. The mean square error of the selected balls' diameter is calculated to further evaluate the dimensional accuracy of the steel ball.

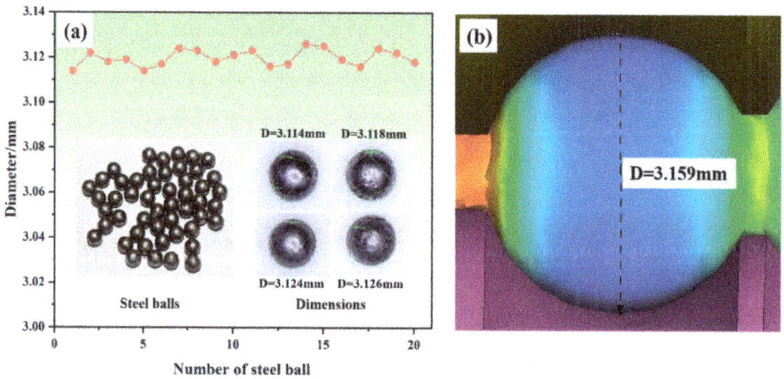

Figure 3. Comparison of the diameter values of steel balls: (**a**) experimental results and (**b**) simulation results.

The standard size of the steel ball diameter is 3.12 mm, and then the mean square deviation can be obtained. The calculated mean square deviation from the selected steel balls is 0.004, which means the diameter has less data dispersion, indicating that the macroscopic size of the steel balls has high accuracy. As shown in Figure 3b, the diameter of the steel balls obtained from the simulation is 3.159 mm, close to the standard size. This indicates that the established FE model has good prediction accuracy for the macroscopic size of cold-skew-rolled steel balls.

3.4. Comparison of Steel Ball Microstructure

In order to verify the predictive accuracy of the developed FE model based on the dislocation density of the microstructure of the steel ball, four points on the transverse section of the steel balls are selected for microstructure characterization, and the exact positions are also taken in the simulation results. Initially, the steel balls were cut along the transverse section, and one half was selected for mechanical polishing, followed by electrochemical polishing in a 25% perchloric acid alcohol solution for 30 s. Finally, EBSD observation was performed on different positions of the steel ball, and the results were analyzed by HKL Channel 5 software. The comparison between the experimental and simulation results is shown in Figure 4. As can be seen from the figure, the grain size gradually decreases from the center to the surface, and the simulation results are in good agreement with the experiment. Due to the large deformation on the surface of the steel ball, the grain refinement phenomenon is more prominent, while the grain in the center is relatively coarse due to the small deformation.

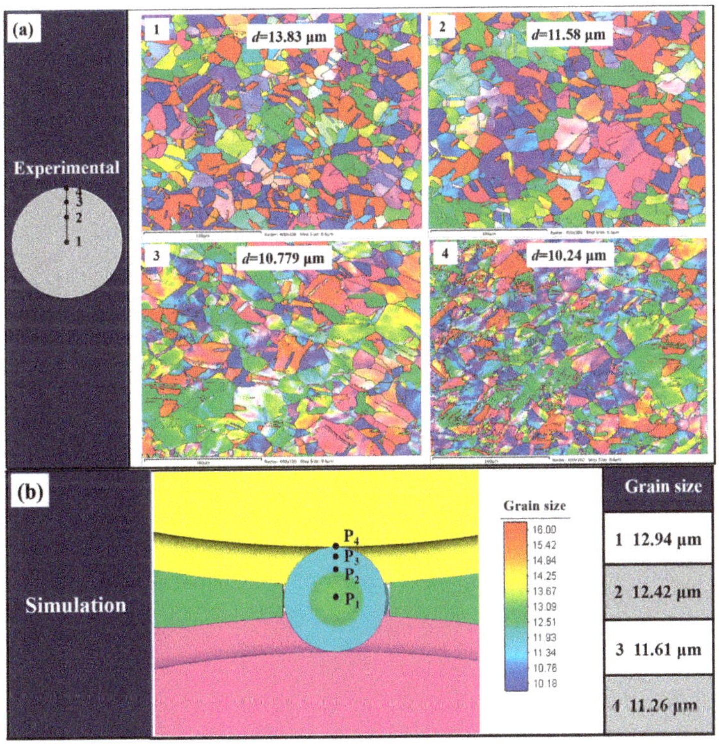

Figure 4. Comparison of grain size at different positions in transverse section of steel balls: (**a**) experimental results and (**b**) simulation results.

Figure 5 shows the comparison of the volume fraction of martensite between the simulation and experiment. Because of the randomness of EBSD measurement results, XRD was used to analyze the martensite content of steel balls quantitatively. The martensite content of the transverse sections of three steel balls was measured to ensure the accuracy of the results. The martensite content was determined by X-ray (Rigaku Ultima IV) equipped with a $Cu-K_{\alpha}$ lamp with a scanning angle range of 30~110° and a step setting of 0.02°. In addition, $(111)_{\gamma}$, $(200)_{\gamma}$, and $(220)_{\gamma}$ diffraction peaks of austenite and $(110)_{\alpha'}$, $(200)_{\alpha'}$, and $(211)_{\alpha'}$ diffraction peaks of martensite were selected to calculate the content of martensite by the following equation [20]:

$$V_{\alpha} = \frac{1/n\sum_{j=1}^{n} I_{\alpha'}^{j}/R_{\alpha'}^{j}}{1/n\sum_{j=1}^{n} I_{\alpha'}^{j}/R_{\alpha'}^{j} + 1/n\sum_{j=1}^{n} I_{\gamma}^{j}/R_{\gamma}^{j}} \tag{17}$$

where n, I, and R are the number of diffraction peaks, intensity factor, and material scattering factor of the corresponding phase, respectively. The results show that the simulation results agree with the experimental results, indicating that the developed FE model has a good prediction accuracy on the martensitic transformation during the cold skew rolling process of steel balls.

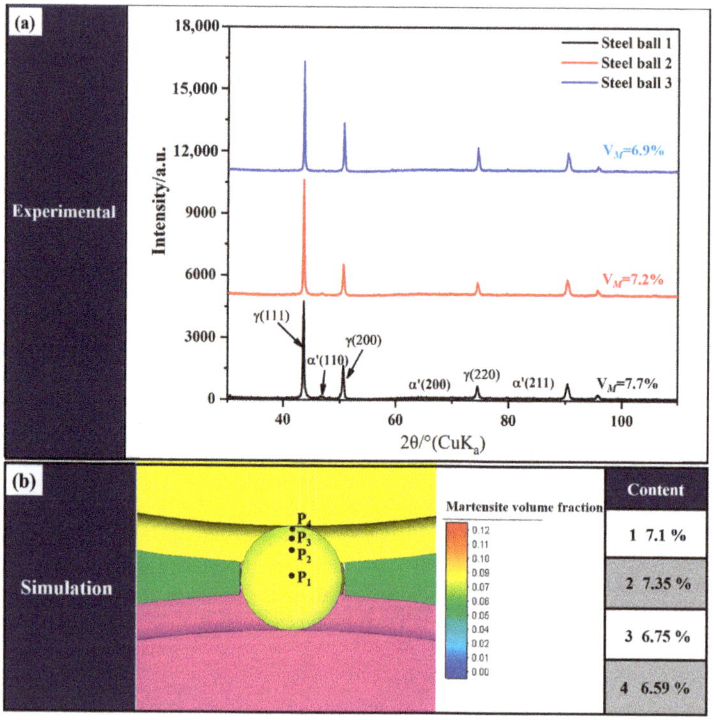

Figure 5. Comparison of volume fraction of martensite: (**a**) experimental results and (**b**) simulation results.

Figure 6 shows the comparison of dislocation density at different positions in the steel ball transverse section between experimental and simulation results. The simulation and experimental measured dislocation densities are of the same magnitude, and the values are close, so the FE model can be considered to have good prediction accuracy. It should be noted that the dislocation density obtained from the simulation is the average dislocation density, which is calculated by mixing the austenite and martensite dislocation densities and the volume fraction of each phase according to the mixing rule. Since the developed model is based on the dislocation density, the comparison results of the dislocation density further demonstrate the model's reliability in predicting the microstructure evolution of steel balls during the skew rolling process.

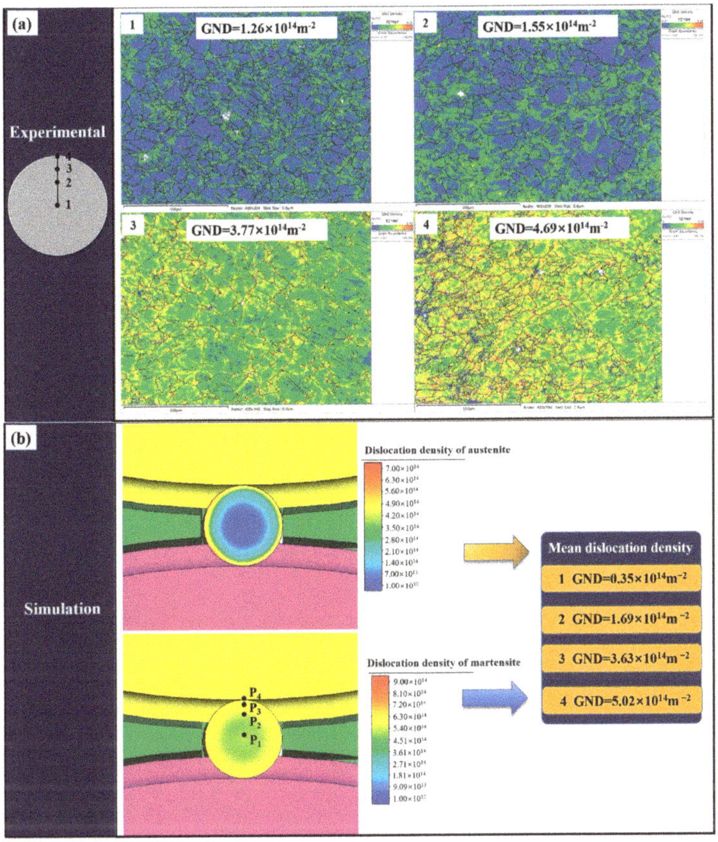

Figure 6. Comparison of dislocation density at different positions in transverse section of steel balls: (**a**) experimental results and (**b**) simulation results.

4. Numerical Simulation Analysis of the Skew Rolling Process

4.1. Equivalent Strain and Stress of Small-Diameter Steel Balls

The equivalent strain and stress in the longitudinal section of the steel ball during skew rolling are illustrated in Figure 7. During the ball-forming process, the radial compression occurs at the linking neck under the action of the roll ridges. The diameter of the linking neck gradually decreases, and the diameter of the steel ball in the groove gradually increases. The equivalent strain at the linking neck is the largest, while the equator part of the steel ball has a minor strain due to the smaller deformation than other parts. When the workpiece is knifed into the roll grooves, the equivalent stress on the part of the steel ball in contact with the convex edge is the most extensive. Due to the different amounts of deformation, the equivalent stress at the pole part is greater than the equator part in the sphere. The equivalent stress decreases gradually from the outer layer to the inner layer, and the overall stress value of the steel ball is moderate.

Figure 8 illustrates the equivalent strain and stress in the transverse section of the steel ball. The outer surface of the steel ball has the most significant strain. In contrast, the ball's core has a minor strain because the ball's surface layer receives the most considerable deformation, and the deformation amount decays from the ball's surface to the core. The deformation penetration effect on the ball core is insignificant, finally showing the gradual decrease in equivalent strain from the ball surface to the center. It can be found from

Figure 8b that the stress distribution in the center is also less than that in the surface, and the area with maximum stress is located at the contact area with the rolls.

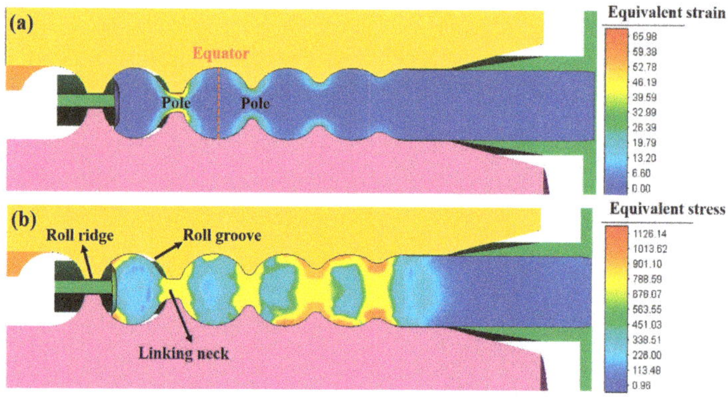

Figure 7. The strain and stress contour along the longitudinal section of the steel ball: (**a**) equivalent strain and (**b**) equivalent stress.

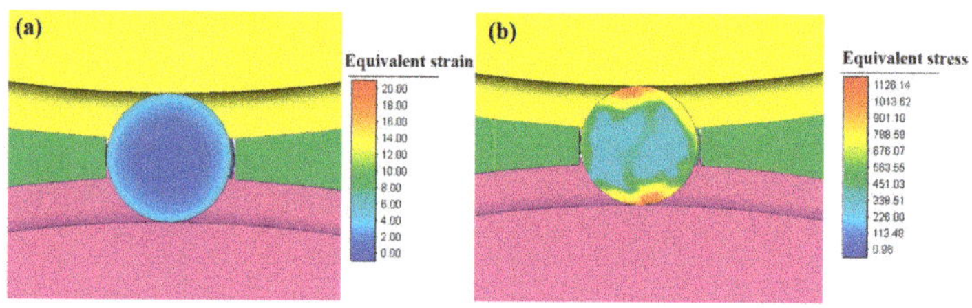

Figure 8. The strain and stress contour along the transverse section of the steel ball: (**a**) equivalent strain and (**b**) equivalent stress.

4.2. *Microstructure of Small-Diameter Stainless Steel Balls*

Figure 9 shows the dislocation density of martensite and austenite of the steel balls during the skew rolling process. Since the dislocation density of austenite is related to the deformation, the distribution pattern of the dislocation density of austenite is identical to that of strain. The linking neck between the balls has the most significant dislocation density due to the most severe deformation. The part close to the linking neck also has an amount of dislocation density. On the transverse section, the austenitic dislocation density gradually increases from the ball center to the surface. The martensitic dislocation density is not solely related to deformation, which is also related to grain size, and the refined grains will have an inhibitory influence on the martensitic transformation [21]. Therefore, the difference in martensite dislocation density from the center and surface of the ball is relatively slight.

A total of five testing points were selected on the transverse and longitudinal sections of the steel ball to measure the microhardness at different positions of the ball. The testing results are shown in Figure 10. It can be seen that the microhardness continuously increases from the ball center to the surface in both the transverse and longitudinal sections. The only difference is that the variation in the longitudinal section exceeds that in the transverse section. The hardness is highest at point C near the linking neck, and the results are consistent with the dislocation density distribution pattern. The high hardness of the steel ball surface can effectively improve the wear resistance of the steel ball, while the softer part of the ball's center can improve the impact resistance.

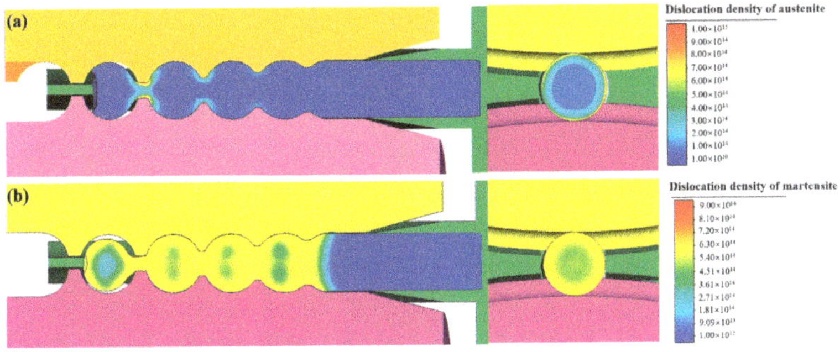

Figure 9. The dislocation density contour of steel balls (**a**) in austenite and (**b**) in martensite.

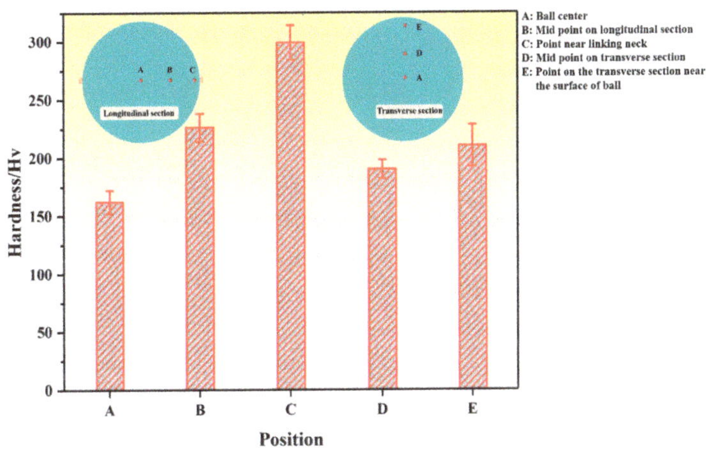

Figure 10. Microhardness of different positions in steel balls.

The grain size of the steel balls during the skew rolling process is shown in Figure 11. Because grain refinement is directly correlated to the amount of deformation, the grain size is smaller in the area with more significant deformation. The refined grains on the surface of the steel balls effectively improve the steel balls' comprehensive mechanical properties and corrosion resistance.

Figure 12 depicts the martensite volume fraction of the steel balls during the skew rolling process. In the transverse section of the steel ball, because the surface grain refinement is more noticeable than the core grain refinement, it will inhibit the formation of martensite. Hence, the martensite content of the surface layer is low, and the center part is slightly high, but the distinction is very slight. In the longitudinal section of the steel ball, due to the deformation of the linking neck part of the steel ball being the largest, the grain refinement is also the most significant, resulting in a reduction in grain size, which will lead to the inhibition of martensitic phase transformation [22]. Meanwhile, because the deformation of the center part is small, the grain size is relatively large. Then, martensite is more easily generated, finally showing that the martensite content of the linking neck part and the ball center part is almost equivalent. The martensite distribution inside the steel ball is relatively uniform.

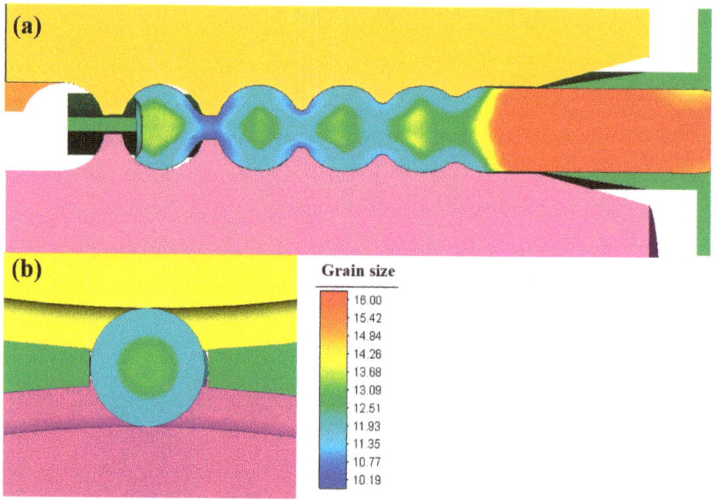

Figure 11. Grain size distribution of steel balls during the skew rolling process in (**a**) the longitudinal section and (**b**) the transverse section.

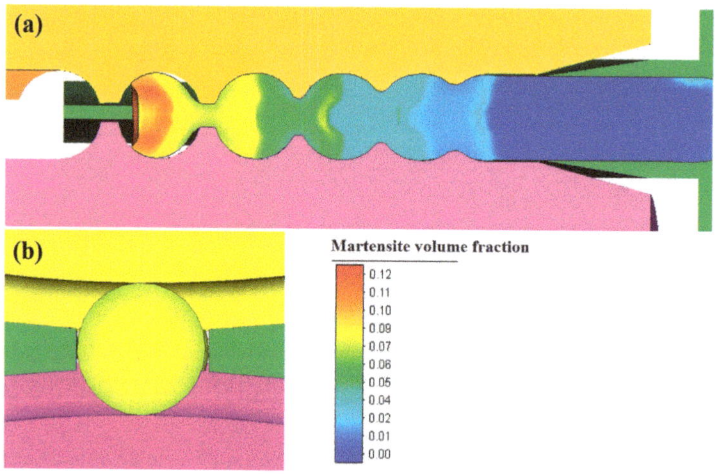

Figure 12. Martensite volume fraction of steel balls during skew rolling process in (**a**) the longitudinal section and (**b**) the transverse section.

5. Conclusions

In this study, we investigated the microstructure evolution of steel balls during the cold skew rolling process via a multiscale constitutive model coupling martensite and grain refinement. The main conclusions drawn are as follows:

1. Based on the mixing rule, a multiscale constitutive model of 316L SS was established by coupling martensitic transformation and grain refinement. By embedding the developed model into the FE software Simufact 16.0, a numerical simulation model of the cold skew rolling process of small-diameter 316L SS steel balls was developed.
2. The simulation results were compared with the experimental results to verify the reliability of the established simulation model. It is found that the diameter size,

dislocation density, grain size, and martensite content of the steel balls are in good agreement, which proves the prediction capability of the established model.

3. The dislocation density in the large deformation of the steel balls is high. Consequently, the grain size in there is relatively small, which was determined through the analysis of the numerical simulation of microstructure evolution. Observing from the transverse section of the steel ball, the martensite content exhibited a slight difference between the areas in the surface and core. The martensite content in the longitudinal section of the steel ball was uniformly distributed in each region.

Author Contributions: Conceptualization, B.W.; methodology, J.Z. and S.L.; validation, J.Z., H.X. and S.L.; writing—original draft preparation, J.Z.; writing—review and editing, J.Z. and S.L.; visualization, H.X. and J.Z.; software, S.L., H.X. and J.Z.; supervision, B.W. All authors have read and agreed to the published version of the manuscript.

Funding: This work is supported by the Fundamental Research Funds for the Central Universities (Grant No. FRF-NP-20-01 and FRF-BD-20-08A).

Institutional Review Board Statement: Not applicable.

Informed Consent Statement: Not applicable.

Data Availability Statement: The data used to support the findings of this study are available from the corresponding author upon request.

Acknowledgments: This work is supported by the Beijing Key Laboratory of Metal Forming Lightweight.

Conflicts of Interest: The authors declare no conflict of interest. We declare that we have no financial and personal relationship with other people or organizations that can inappropriately influence our work.

References

1. Kheiri, S.; Mirzadeh, H.; Naghizadeh, M. Tailoring the Microstructure and Mechanical Properties of AISI 316L Austenitic Stainless Steel Via Cold Rolling and Reversion Annealing. *Mater. Sci. Eng. A* **2019**, *759*, 90–96. [CrossRef]
2. Pater, Z.; Tomczak, J.; Bartnicki, J.; Lovell, M.R.; Menezes, P.L. Experimental and Numerical Analysis of Helical-Wedge Rolling Process for Producing Steel Balls. *Int. J. Mach. Tools Manuf.* **2013**, *67*, 1–7. [CrossRef]
3. Gontarz, A.; Tomczak, J.; Pater, Z.; Bulzak, T. Effect of the Forming Zone Length on Helical Rolling Processes for Manufacturing Steel Balls. *Materials* **2019**, *12*, 2917. [CrossRef]
4. Tomczak, J.; Pater, Z.; Bulzak, T. Designing of Screw Impressions in the Helical Rolling of Balls. *Arch. Civ. Mech. Eng.* **2014**, *14*, 104–113. [CrossRef]
5. Liu, S.; Liu, J.; Xu, H.; Wang, Z.; Shen, J.; Wang, B. Experimental and Numerical Study of Cold Helical Rolling of Small-Diameter Steel Balls. *Int. J. Adv. Manuf. Technol.* **2022**, *119*, 599–613. [CrossRef]
6. Zhang, M.; Chen, H.; Wang, Y.; Wang, S.; Li, R.; Li, S.; Wang, Y. Deformation-Induced Martensitic Transformation Kinetics and Correlative Micromechanical Behavior of medium-Mn Transformation-Induced Plasticity Steel. *J. Mater. Sci. Technol.* **2019**, *35*, 1779–1786. [CrossRef]
7. Xiong, Y.; Yue, Y.; Lu, Y.; He, T.; Fan, M.; Ren, F.; Cao, W. Cryorolling Impacts on Microstructure and Mechanical Properties of AISI 316 LN Austenitic Stainless Steel. *Mater. Sci. Eng. A* **2018**, *709*, 270–276. [CrossRef]
8. Eskandari, M.; Najafizadeh, A.; Kermanpur, A. Effect of Strain-Induced Martensite on the Formation of Nanocrystalline 316L Stainless Steel After Cold Rolling and Annealing. *Mater. Sci. Eng. A* **2009**, *519*, 46–50. [CrossRef]
9. Huo, Y.; He, T.; Wang, B.; Zheng, Z.; Xue, Y. Numerical Prediction and Experimental Validation of the Microstructure of Bearing Steel Ball Formation in Warm Skew Rolling. *Metall. Mater. Trans. A Phys. Metall. Mater. Sci.* **2020**, *51*, 1254–1263. [CrossRef]
10. Peng, L.; Xu, Z.; Gao, Z.; Fu, M.W. A Constitutive Model for Metal Plastic Deformation at Micro/Meso Scale with Consideration of Grain Orientation and its Evolution. *Int. J. Mech. Sci.* **2018**, *138–139*, 74–85. [CrossRef]
11. Chen, Z.; Sun, Z.; Panicaud, B. Constitutive Modeling of TWIP/TRIP Steels and Numerical Simulation of Single Impact During Surface Mechanical Attrition Treatment. *Mech. Mater.* **2018**, *122*, 69–75. [CrossRef]
12. Liu, Y.Z.; Wan, M.; Meng, B. Multiscale Modeling of Coupling Mechanisms in Electrically Assisted Deformation of Ultrathin Sheets: An Example on a Nickel-Based Superalloy. *Int. J. Mach. Tools Manuf.* **2021**, *162*, 103689. [CrossRef]
13. Meng, B.; Liu, Y.Z.; Wan, M.; Fu, M.W. A Multiscale Constitutive Model Coupled with Martensitic Transformation Kinetics for Micro-Scaled Plastic Deformation of Metastable Metal Foils. *Int. J. Mech. Sci.* **2021**, *202–203*, 106503. [CrossRef]
14. Tang, X.; Wang, B.; Huo, Y.; Ma, W.; Zhou, J.; Ji, H.; Fu, X. Unified Modeling of Flow Behavior and Microstructure Evolution in Hot Forming of a Ni-based Superalloy. *Mater. Sci. Eng. A* **2016**, *662*, 54–64. [CrossRef]

15. Benzing, J.T.; Liu, Y.; Zhang, X.; Luecke, W.E.; Ponge, D.; Dutta, A.; Oskay, C.; Raabe, D.; Wittig, J.E. Experimental and Numerical Study of Mechanical Properties of Multi-Phase medium-Mn TWIP-TRIP Steel: Influences of Strain Rate and Phase Constituents. *Acta Mater.* **2019**, *177*, 250–265. [CrossRef] [PubMed]
16. Wong, S.L.; Madivala, M.; Prahl, U.; Roters, F.; Raabe, D. A Crystal Plasticity Model for Twinning- and Transformation-Induced Plasticity. *Acta Mater.* **2016**, *118*, 140–151. [CrossRef]
17. Srivastava, A.; Ghassemi-Armaki, H.; Sung, H.; Chen, P.; Kumar, S.; Bower, A.F. Micromechanics of Plastic Deformation and Phase Transformation in a Three-Phase TRIP-assisted Advanced High Strength Steel: Experiments and Modeling. *J. Mech. Phys. Solids* **2015**, *78*, 46–69. [CrossRef]
18. Liu, S.; Li, W.; Shen, J.; Yang, X.; Wang, B.; Liu, J. Size-Dependent Constitutive Model Incorporating Grain Refinement and Martensitic Transformation. *Arch. Civ. Mech. Eng.* **2023**, *23*, 38. [CrossRef]
19. Cao, Q.; Hua, L.; Qian, D. Finite Element Analysis of Deformation Characteristics in Cold Helical Rolling of Bearing Steel-Balls. *J. Cent. South Univ.* **2015**, *22*, 1175–1183. [CrossRef]
20. Shen, Y.F.; Li, X.X.; Sun, X.; Wang, Y.D.; Zuo, L. Twinning and Martensite in a 304 Austenitic Stainless Steel. *Mater. Sci. Eng. A* **2012**, *552*, 514–522. [CrossRef]
21. Naghizadeh, M.; Mirzadeh, H. Effects of Grain Size on Mechanical Properties and Work—Hardening Behavior of AISI 304 Austenitic Stainless Steel. *Steel Res. Int.* **2019**, *90*, 1900153. [CrossRef]
22. Yeddu, H.K. Phase-Field Modeling of Austenite Grain Size Effect on Martensitic Transformation in Stainless Steels. *Comp. Mater. Sci.* **2018**, *154*, 75–83. [CrossRef]

materials

Article

DSC Analysis of the Effect of Cold Deformation on the Precipitation Kinetics of a Binary Cu-Sc Alloy

Ramona Henle [1,*], Julia Dölling [2], Ulrich Prahl [3], Gerrit Nandi [1] and Andreas Zilly [2]

1 Faculty of Technology, Cooperative State University Heidenheim, Marienstraße 20, 89518 Heidenheim an der Brenz, Germany; gerrit.nandi@dhbw-heidenheim.de
2 Faculty of Technology, Cooperative State University Stuttgart, Lerchenstraße 1, 70197 Stuttgart, Germany; julia.doelling@dhbw-stuttgart.de (J.D.); andreas.zilly@dhbw-stuttgart.de (A.Z.)
3 Institute of Metal Forming, Technische Universität Bergakademie Freiberg, Bernhard-von-Cotta Straße 4, 09599 Freiberg, Germany; ulrich.prahl@imf.tu-freiberg.de
* Correspondence: ramona.henle@dhbw-heidenheim.de

Abstract: The present study aimed to investigate the effect of cold deformation on the precipitation kinetics of a binary CuSc alloy containing 0.4 wt.% scandium using the experimental analysis method of differential scanning calorimetry (DSC). Non-deformed and 75% cross-section-reduced cold-rolled supersaturated specimens were tested in non-isothermal DSC runs at up to five different heating rates. The DSC results showed that cold rolling significantly accelerated the precipitation process in the binary alloy, leading to a decrease in the initial and peak temperatures of the exothermic reactions. The activation energies calculated with the Kissinger method indicated that the precipitation activation energy decreased with increasing cold deformation. The findings of this study provide worthy implications to further optimize the processing of Cu-Sc alloys with improved mechanical properties.

Keywords: copper–scandium CuSc; copper alloy; differential scanning calorimetry DSC; precipitation kinetics; cold-working; cold-rolling; activation energy; Kissinger method

Citation: Henle, R.; Dölling, J.; Prahl, U.; Nandi, G.; Zilly, A. DSC Analysis of the Effect of Cold Deformation on the Precipitation Kinetics of a Binary Cu-Sc Alloy. *Materials* **2023**, *16*, 3462. https://doi.org/10.3390/ma16093462

Academic Editors: Pan Gong, Xin Wang, Maojun Li and Guangchao Han

Received: 12 April 2023
Revised: 25 April 2023
Accepted: 27 April 2023
Published: 29 April 2023

1. Introduction

Microalloyed copper alloys are used in many areas of industry, exhibiting high electrical and thermal conductivity as well as increased mechanical strength [1]. Therefore, they are frequently used for signal cables, connectors, or welding electrodes, particularly in the electrical industry. There are several hardening mechanisms for copper, which is very soft in its pure state [2], that can bring about the desired properties. One of the most important in this context is precipitation hardening, which allows both, the otherwise conflicting mechanical strength of an alloy and the electrical conductivity to be increased by optimizing the microstructural properties during specific heat treatments [2–5].

However, a conflict arises from the fact that precipitates can act as scattering centers for current flow. The smaller and more numerous the precipitates are, the more they impede the flow of electric current and reduce the conductivity of the material [6]. To balance the trade-off between mechanical strength and electrical conductivity, it is essential to carefully control the size, distribution, and morphology of precipitates. This can be achieved through a combination of processing conditions such as heating rate, aging temperature, and prior cold working, as well as the alloy composition [2,3,6,7].

Recently published studies by Franczak et al. [8] and Dölling et al. [9] have shown that a combination of copper and scandium has potential in terms of electrical conductivity and mechanical strength, in addition to having advantages in terms of recrystallization behavior and grain refinement [9–11]. In particular, cold working prior to hardening of the precipitates increases the density of dislocations and strain-induced defects in the material, creating more nucleation sites for fine precipitates to form.

Figure 1 shows the phase diagram of binary Cu-Sc alloy. The maximum solubility of scandium in copper is reported to be 0.35 wt.% at 865 °C [12–14] and decreases with decreasing temperature, leading to the possibility of precipitation formation in binary Cu-Sc alloy [15].

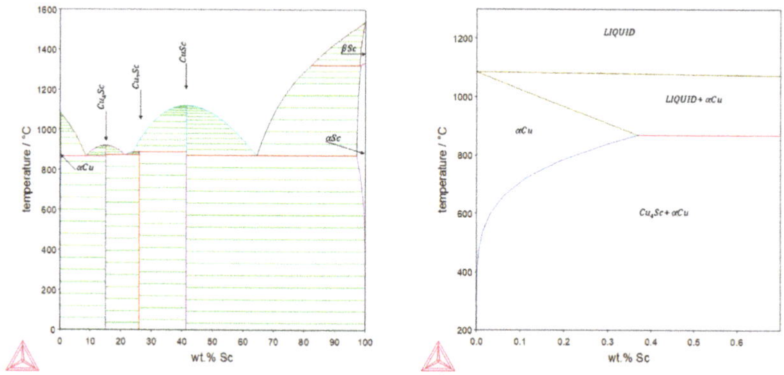

(**a**) (**b**)

Figure 1. Phase diagram of the binary Cu-Sc system calculated with the Thermo-Calc SGTE database (2022a): (**a**) overall; (**b**) magnification of the copper-rich area [9].

Hao et al. investigated the precipitation behavior and hardening effects of a highly deformed and cryorolled CuSc0.4 wt.%. The precipitation reaction started initially from the supersaturated solid solution, forming Sc-rich atomic clusters. Subsequently, these formed coherent lamellar precipitates and became increasingly detached from the matrix structure to become tetragonally-oriented lamellar Cu_4Sc precipitates [11,16]. The precipitation strengthening that occurs at this point is described by Hao et al. as a combination of coherent strengthening with significant matrix distortion and presence of the incoherent Orowan bypass mechanism at larger precipitate sizes [11,17].

To optimize precipitation treatments for materials with desired properties, differential scanning calorimetry (DSC) can be used to develop a deeper understanding of the underlying precipitation mechanisms. DSC analysis is a common thermal analysis technique that measures the heat flow into or out of a material as a function of temperature and time. Therefore, it can detect reactions within the microstructure as exothermic (precipitation and recrystallization) or endothermic (dissolution) processes [18,19]. This analysis provides valuable information about the thermal behavior of a material, such as the melting point, crystallization kinetics and phase transitions [18,20–22]. In the context of precipitation kinetics, DSC analysis can be used to study the evolution of precipitates in a material during aging or heating. By monitoring the heat flux, DSC provides information on precipitation kinetics such as the nucleation and growth rate of precipitates, as well as the temperature dependence of precipitation reactions [21,23,24].

An important parameter characterizing the kinetics of a reaction is the activation energy, which can be obtained from DSC data using one of several available methods, including the Kissinger [25,26], Ozawa [27,28], and Boswell [29] methods. This can be used to predict the reaction behavior at different temperatures, such as the time required to complete the reaction to a certain degree, or to predict the temperature at which the reaction will occur. The Kissinger method is often the preferred approach for determining the activation energy of precipitation reactions using DSC, as it assumes a single reaction mechanism and only requires a single DSC measurement at each heating rate, making it a simple and convenient approach that can provide valuable information for materials science and engineering applications [23,30].

This study aims to determine the activation energy required for the precipitation reaction in an undeformed alloy and a 75% cold-worked CuSc0.4 alloy.

2. Materials and Methods

The composition of the alloy was analyzed using a specially calibrated optical emission spark spectrometer (Spectrotest, SPECTRO Analytical Instruments GmbH, Kleve, Germany) and determined to be Cu with a content of 0.40 wt.% Sc. From this, an ingot was cast on a VC400 casting machine (Indutherm Blue Power Casting Systems, Walzbachtal, Germany) utilizing the raw materials Cu-OFE and CuSc23 as a master alloy. Melting was performed in a boron nitride-covered graphite crucible up to 1300 °C, which along with the casting process was performed under vacuum conditions. After pouring in a graphite mold (heated at 250 °C), the 5 mm thick bar was solution annealed at 870 °C for 120 min in a preheated furnace (ME65/13, Helmut ROHDE GmbH, Prutting, Germany) and then quenched to room temperature in circulated water. The as-quenched plate was divided and one part was longitudinally cold rolled on a duo-roll stand (Bühler, Pforzheim, Germany) with 110 mm diameter rolls driven with a speed of 27 min^{-1}.

Differential scanning calorimetry (DSC) analysis of Cu-Sc alloys was performed using a Netzsch STA 449 C. The calorimeter calibration was performed thermally with Al_2O_3 crucibles by melting In, CsCl, Ag, and Au to obtain a baseline. The mass of the samples ranged from 40 to 202 mg. Several measurements of the same heating rate with higher sample mass resulted in identical curves, though with an increased signal-to-noise ratio (SNR). During the heating process, a protective argon atmosphere (20 mL/min) and purge gas (30 mL/min) were utilized, and an empty crucible was used as a reference. The precipitation experiments were carried out using continuous heating rates (5, 10, 15, 20, and 40 K/min) with a temperature ranging from room temperature (RT) to 750 °C. Data output was performed with an accuracy of 0.5 °C; the error for the DSC measurements as compared to the calibration measurements is shown in Table A1 of Appendix A.

The raw data from the DSC measurements were smoothed using a locally-weighted linear regression with the second-degree polynomial (LOESS) function in MATLAB with a span of 5%. Then, the second derivative of the function was calculated in order to determine the initial and final temperatures of the exothermic peak (T_i and T_f), which can be seen as inflection points in the curve. Estimation of the baseline of each curve was performed by spline interpolation of the smoothed DSC signal, with the maximum difference value between the smoothed DSC signal and the baseline indicating the peak temperature of the exothermic reaction (T_p).

The activation energy of the precipitation of the Cu_4Sc phase was calculated based on the dependence of previously determined T_p temperatures on the heating rates using the Kissinger equation [25,26] provided by

$$\ln\left(\frac{V}{{T_p}^2}\right) = C - \frac{E}{RT_p} \quad (1)$$

where V is the heating rate, T_p is the peak temperature, C is a constant, E is the apparent activation energy, and R is the molar gas constant. By plotting $\ln(V/{T_p}^2)$ as a function of $1/T_p$ and fitting a linear regression line y to the data, the activation energy E can be calculated from the slope of the line using the relationship

$$E = -slope \times R. \quad (2)$$

For microscopic analysis, one of the non-deformed CuSc0.4 specimens were heated using an identical heating process with a heating rate of 10 K/min up to 610 °C and cooled with the same temperature gradient. Thus, the microscopic evolution is directly comparable to measured exothermic peaks during the precipitation reaction in the DSC. Microstructure characterizations were observed with a scanning electron microscope (SEM) (Gemini Sigma VP with the used NTS BSD (Carl Zeiss Microscopy Deutschland GmbH,

Oberkochen, Germany)) operating at 12 kV and Bruker XFlash 6 | 30 detector (Bruker Nano GmbH, Berlin, Germany).

3. Results and Discussion

The following chapter presents and discusses the results obtained from investigating the effect of cold deformation on the precipitation kinetics of a binary Cu-Sc alloy. This chapter is divided into three sections: DSC analysis, calculation of the activation energy, and microscopic analysis. Each section provides a comprehensive examination of the experimental findings and their implications for understanding the alloy's precipitation behavior with and without prior cold deformation.

3.1. DSC Analysis

During the DSC analysis, precipitation reactions with a clearly visible exothermic peak appeared. Depending on the chosen heating rate, the location of this peak was slightly different.

The DSC scans at different heating rates (10, 15, 20, and 40 K/min) for the non-deformed CuSc0.4 specimen (Figure 2a) all show an exothermic peak (Exo) between 760 K and 950 K. A closely related study by Dölling et al. [31] showed highly comparable curves for a non-deformed CuSc0.3 alloy at a 10 K/min heating rate with a peak temperature of 842.1 K. Furthermore, there was no recrystallization detected for non-formed Cu-Sc specimens, resulting in only a single peak indicating the precipitation reaction.

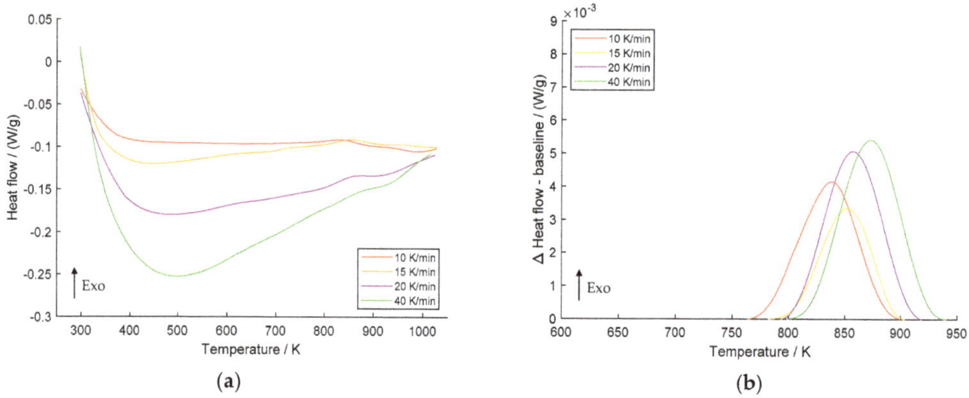

Figure 2. DSC scans of non-deformed CuSc0.4 alloy at heating rates of 10 K/min, 15 K/min, 20 K/min, and 40 K/min (**a**); difference between heat flow and the baseline of the precipitation reaction (**b**).

The difference between the smoothed DSC signal and the baseline and spline interpolations is displayed in Figure 2b. Obviously, the maximum temperature of the Cu_4Sc precipitation shifts to higher values as the heating rate increases, implying that the precipitation reaction is thermally activated. Similar findings have been reported for recrystallization in cold-rolled pure copper [32], Cr clustering in equal-channel angular pressing (ECAP) processed CuCrZr [1], and recrystallization phenomena in Cu–Ni–Si alloy processed by high-pressure torsion (HPT) [33]. Due to the baseline shape extracted from the experiment, the determination of the 20 K/min curve using spline interpolation resulted in an exaggeratedly large area for the peak. At this point, manual adjustment of the baseline calculation parameters was necessary. Despite this adjustment, the peak temperature T_p remained unchanged, thereby keeping any potential impacts on the calculation of the activation energy unaffected. Comprehensive insights pertaining to the interpolated baselines are provided in Appendix A, Figure A1.

With an increase in the heating rate, the temperature range at which the precipitation reaction takes place is widened. In addition, the initial (T_i), peak (T_p), and final (T_f) temperatures shift to higher values, which can be attributed to the kinetics of the reaction. Reactions and transitions, such as precipitation or recrystallization, need time to transform, resulting in a narrower time range at lower heating rates due to the longer duration. On the other hand, at higher heating rates the time required to complete the reaction may not be sufficient because of the limitations imposed by the kinetics. For this reason, the peak expands at higher temperatures and widens with increasing heating rates.

However, scans of the non-deformed specimen at a heating rate of 5 K/min did not show a significant peak, which is why it is not mentioned in this study.

Figure 3 shows the DSC scans of the 75% cold-rolled Cu-Sc specimen at all heating rates (Figure 3a) and the differences between the precipitation peak curve and baseline (Figure 3b). Again, the maximum temperature shifts to higher temperatures at higher heating rates. However, in direct comparison to the non-deformed specimen, the precipitation reaction of the cold-rolled specimen starts about 100 K earlier (660–840 K).

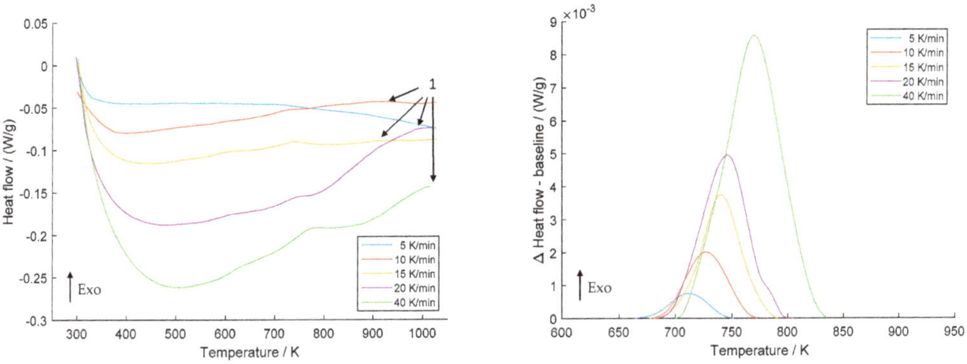

(a) (b)

Figure 3. DSC scans of 75% cold-rolled CuSc0.4 alloy at heating rates of 5 K/min, 10 K/min, 15 K/min, 20 K/min, and 40 K/min (**a**); difference between heat flow and baseline of the precipitation reaction (**b**).

In addition, the curves show the peak of the second exothermic reaction (indicated by the number 1 with arrows), which can be attributed to recrystallization of the microstructure. This investigation agrees with the findings of Dölling et al. [31] obtained with a directly comparable experimental setup and raw materials.

The obtained peak temperature values related to Cu_4Sc precipitation as a function of heating rates are shown in Table 1. The temperature peaks ranged from 837.7 K to 872.3 K without prior cold deformation, while the peak temperatures of the 75% cold-rolled specimens ranged from 711.7 K to 769.0 K. It is obvious that the temperature peaks of the precipitation reaction decrease when increasing deformation is applied. This can be attributed to the fact that cold rolling introduces various lattice defects such as dislocations and vacancies into the material, which provide additional nucleation sites for precipitation, leading to an accelerated reaction. This effect was shown during isothermal aging experiments for comparable alloys (CuSc0.15 and CuSc0.3) and identical experimental processing conditions by Dölling et al. [9]. The mechanical and physical properties were analyzed under varying degrees of cold rolling during isothermal heat treatments, and the precipitation reactions occurred reproducibly earlier with higher degrees of cold rolling prior to aging at different temperatures. This phenomenon was evident due to the evolution of the material properties, namely, the concurrent increase in electrical conductivity and hardness.

Furthermore, several previous investigations have shown that a high dislocation density significantly improves the kinetics of precipitation [5,34–36] and recrystallization [33,37,38].

Table 1. Values of maximum temperatures T_p [K] of the Cu_4Sc precipitation reaction in non-deformed (0%) and 75% cold-rolled CuSc0.4 wt.% alloy.

	V [K/min]	0%	75%
Tp [K]	5	-	711.7
	10	837.7	726.3
	15	851.4	739.7
	20	856.8	745.2
	40	872.3	769.0

3.2. *Determination of Precipitation Activation Energies*

The activation energy for scandium precipitation was calculated using the Kissinger method [25,26] using Formula 1. For this purpose, the values of the peak temperature T_p and the corresponding heating rate V are used for the equation and then $\ln(V/T_p{}^2)$ is plotted versus 1000/K. Using Formula 2, the activation energy is derived from the slope of the linear fitting curve of the calculated points multiplied by the molar gas constant R.

Figure 4 shows the Kissinger plots of the non-deformed (Figure 4a) and cold-rolled specimens (Figure 4b) versus $1000/T$ for the precipitation reaction in the utilized CuSc0.4 alloys. All plots show straight lines, with high Pearson correlation coefficients of $r^2 = 0.991$ for the undeformed specimen and $r^2 = 0.982$ for the cold-rolled specimen. The activation energies were calculated from their slopes, and are listed in Table 2. The mean values of the activation energy for precipitation of the Cu_4Sc phase are 232.02 kJ/mol and 151.51 kJ/mol, respectively.

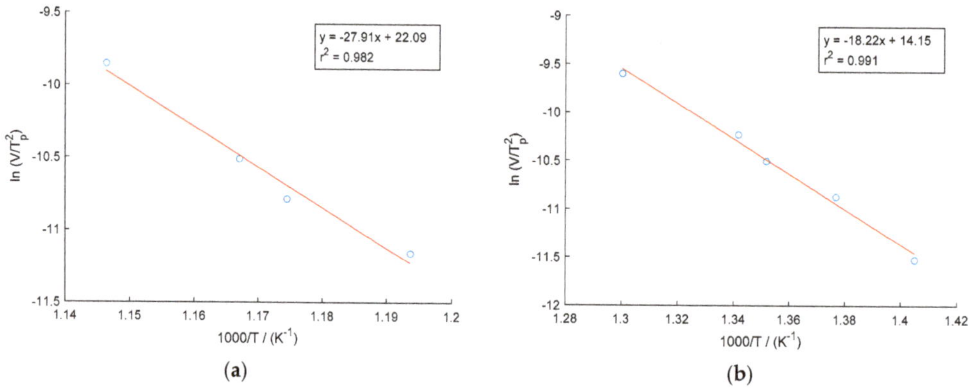

Figure 4. Kissinger plots of $\ln(V/T_p{}^2)$ against $1000/T$ of Cu_4Sc precipitation in Cu-Sc alloy for non-deformed (**a**) and 75% cold-rolled (**b**) specimens.

Table 2. Mean values of estimated activation energies of the Cu_4Sc precipitation reaction in non-deformed (0%) and cold-rolled (75%) CuSc0.4 wt.% alloy using the Kissinger method.

	0%	75%
E [kJ/mol]	232.02	151.51
r^2	0.9913	0.9816

However, temperature errors can cause inaccuracies in determination of the peak temperature during DSC measurements, potentially having a significant influence on the calculated activation energy. An error of just a few degrees between high and low heating rates can lead to an activation energy error of 10–20% [30]. Thus, careful temperature calibration and control to minimize such errors are crucial in DSC measurements.

Considering the temperature deviations during the temperature calibration shown in Appendix A, Table A1, a range of activation energy of 226.91–244.25 kJ/mol can be determined for the non-deformed specimen and 147.82–156.24 kJ/mol for the cold-rolled specimen.

The estimated activation energy for the cold-rolled specimen in the present study is lower than the value of self-diffusion through the lattice in copper (~197 kJ/mol) [39], whereas the activation energy for the non-deformed specimen is slightly higher. This may be due to residual scandium atoms in the copper matrix. During aging, not all scandium atoms are transferred to the Cu_4Sc phase; residues remain in the copper matrix depending on the temperature and duration of the aging treatment. Because the alloying element has a larger atomic radius than copper, distortions in the lattice of the copper matrix result, which act as impediments to dislocation movement and thereby hinder diffusion [3]. Therefore, a higher amount of energy needs to be applied to overcome this barrier, resulting in higher activation energy.

3.3. Microstructure Analysis

Changes in microstructure that occur during thermal treatment can be visually investigated by metallographic analysis. An SEM backscatter detector can be used to identify differences in chemical composition and distinguish between the phases of different elemental compositions [40]. The material contrast is determined by the atomic number of the elements. As the atomic number increases, the degree of backscattering increases as well, resulting in higher brightness in the SEM image of regions containing elements with higher atomic numbers. [41].

Figure 5 shows the microstructure of the non-deformed CuSc0.4 specimen after heating with 10 K/min up to 610 °C followed by cooling with 10 K/min. The selected temperature can be deduced from the DSC curve shown in Figure 2b, which shows that the selected temperature of 883.15 K is just below the precipitation's final peak temperature.

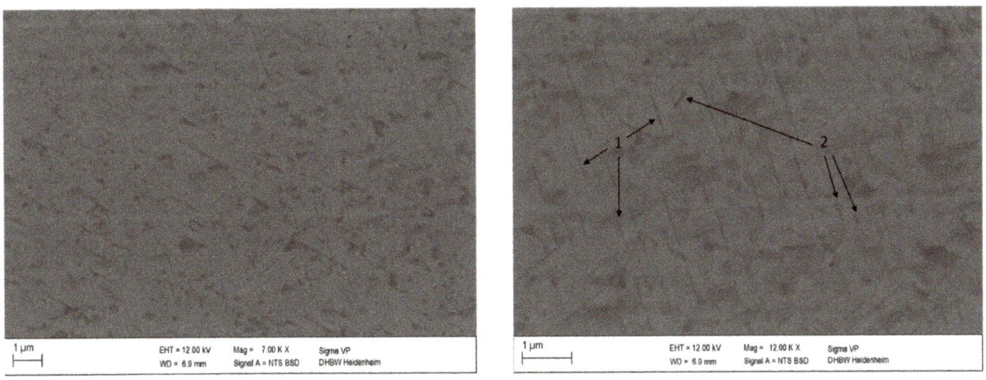

(**a**) (**b**)

Figure 5. Microstructure of non-deformed CuSc0.4 specimen after DSC analysis heated up to 610 °C at a heating rate of 10 K/min with magnitude of 7000 (**a**) and 12,000 (**b**) showing homogeneously distributed lamellar Cu_4Sc phases (1) and coarsened phases detached from the structure (2).

Because the light metal Sc has a lower atomic number than copper [6], the Cu_4Sc phase containing one-fifth scandium atoms appears darker compared to the copper matrix. The fine lamellar structures (dark grey) are homogeneously distributed within the copper matrix (light grey).

The characteristics of the Cu_4Sc precipitates are comparable to those reported by Dölling et al. [31], which were analyzed at the same heating rate and a slightly lower final temperature (600 °C). Energy Dispersive X-ray spectroscopy (EDS) proved that the darker periodically-appearing structures are directly associated with enhanced scandium content (Appendix A, Figure A2). This observation is further supported by the EDS images obtained by Dölling et al. [9,31] with a slightly lower scandium content of 0.3 wt.%.

In direct comparison to the microstructural images of the non-deformed specimen, the 75% cold-rolled specimen with identical heat treatment already shows the first signs of partially recrystallized areas within the microstructure (Figure 6a). This observation fits the analysis of the 10 K/min DSC curve depicted in Figure 3a. The DSC measurement shows that the precipitation reaction has already been completed at the same maximum temperature of 883.15 K and that the recrystallization of the microstructure has started. However, it is notable that the Cu_4Sc precipitates exhibit a significant decrease in size (Figure 6b), making them discernible only at higher magnifications.

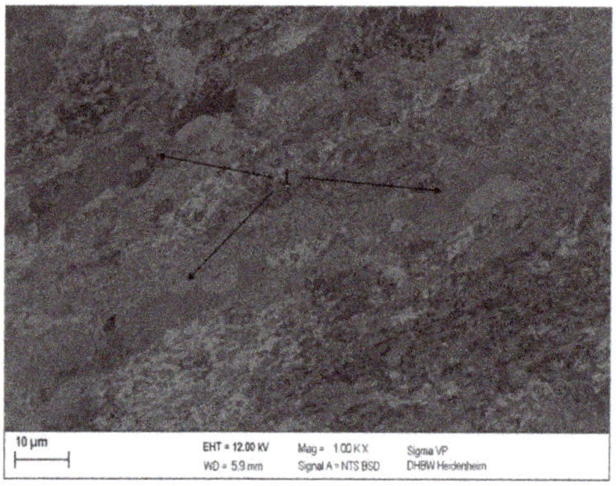

(a)

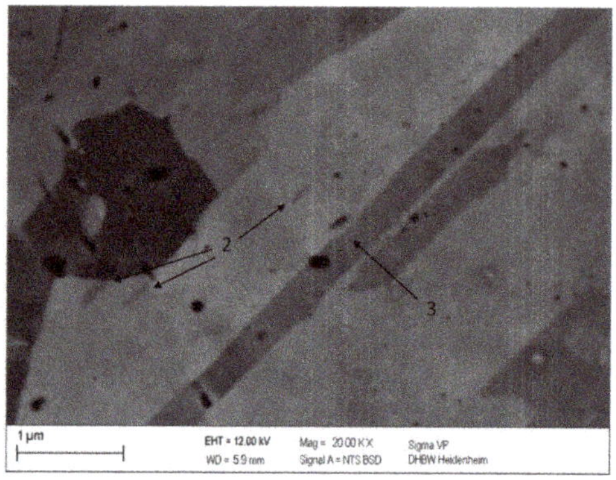

(b)

Figure 6. Microstructure of 75% cold-rolled CuSc0.4 specimen after DSC analysis heated up to 610 °C at a heating rate of 10 K/min with a magnitude of 1000 (**a**) and 20,000 (**b**), showing partly recrystallized areas within the strongly deformed microstructure (1), Cu_4Sc precipitates (2), and twins (3).

4. Conclusions

The present paper describes the kinetics of Cu_4Sc precipitation of a CuSc0.4 wt.% alloy with and without prior cold working using differential scanning calorimetry (DSC). The results showed that a prior 75% cross-section reduction by cold working can accelerate the precipitation reaction and lower the reaction's starting temperature by about 100 K. The activation energies needed for the reaction were determined by the Kissinger method and resulted in values of 226.91–244.25 kJ/mol for the non-deformed specimen and 147.82–156.24 kJ/mol for the 75% cold-rolled specimen. The differences observed in the activation energies provide new insights into the effects of cold deformation on precipitation kinetics in binary alloys, and can help in the design and optimization of such materials. The results of this study highlight the importance of carefully considering the effects of processing on precipitation behavior in order to achieve the desired material properties.

Author Contributions: R.H. is the principal author of this article, and carried out most of this study for her doctoral research. J.D. was responsible for alloy production, specimen preparation, and SEM analysis. U.P. and G.N. supervised the research project. J.D., U.P., G.N. and A.Z. helped in scripting and finalizing the article. All authors have read and agreed to the published version of the manuscript.

Funding: The APC was funded by the Baden-Wuerttemberg Ministry of Science, Research, and Culture and the Baden-Wuerttemberg Cooperative State University Heidenheim through the Open Access Publishing funding program.

Institutional Review Board Statement: Not applicable.

Informed Consent Statement: Not applicable.

Data Availability Statement: Not applicable.

Conflicts of Interest: The authors declare no conflict of interest.

Appendix A

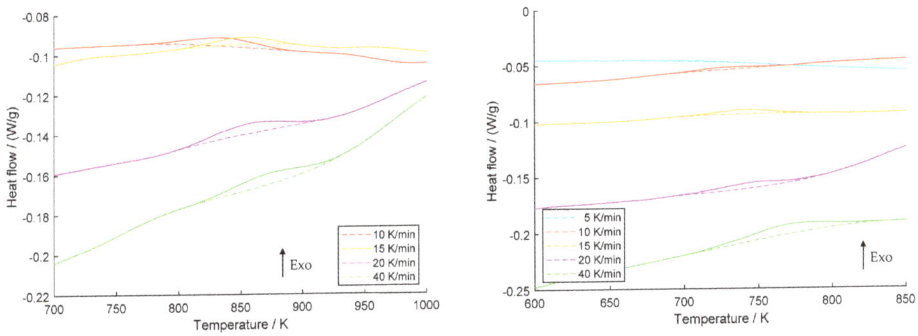

(**a**) (**b**)

Figure A1. Overview of peak areas of non-deformed (**a**) and 75% cross-section-reduced (**b**) CuSc0.4, with DSC signal (straight lines) and interpolated baselines (dashed lines).

Table A1. Maximum deviation of the experimentally determined melting temperature [*dT*] of the calibration sample materials (In, CsCl, Ag, and Au) for heating rates of 5, 10, 15, 20, and 40 K/min.

V [K/min]	+*dT* [K]	−*dT* [K]
5	0.3	0.8
10	0.3	0.1
15	0.4	0.8
20	0.3	0.1
40	0.7	1.5

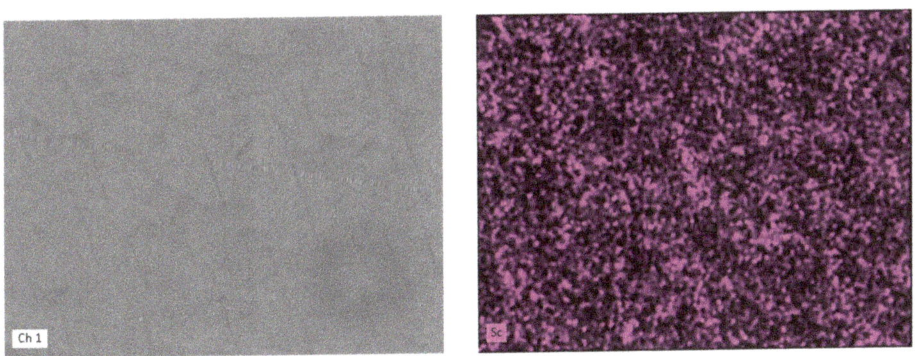

(**a**) (**b**)

Figure A2. Lamellar distribution of Cu_4Sc precipitates without prior cold deformation after DSC analysis heated up to 610 °C at a heating rate of 10 K/min, as analyzed with EDS: (**a**) backscatter detection and (**b**) EDS distribution of Sc.

References

1. Bourezg, Y.I.; Abib, K.; Azzeddine, H.; Bradai, D. Kinetics of Cr clustering in a Cu-Cr-Zr alloy processed by equal-channel angular pressing: A DSC study. *Thermochim. Acta* **2020**, *686*, 178550. [CrossRef]
2. Copper and Copper Alloys. Davis, J.R., Ed.; ASM International: Materials Park, OH, USA, 2008; ISBN 9780871707260.
3. Gottstein, G. *Materialwissenschaft und Werkstofftechnik: Physikalische Grundlagen*; Springer: Berlin/Heidelberg, Germany, 2014; ISBN 978-3-642-36602-4.
4. Roos, E.; Maile, K.; Seidenfuß, M. *Werkstoffkunde für Ingenieure: Grundlagen, Anwendung, Prüfung*; Springer: Berlin/Heidelberg, Germany, 2017; ISBN 9783662495322.
5. Bonvalet Rolland, M.; Borgenstam, A. Modeling precipitation kinetics in multicomponent alloys during deformation. *Front. Mater.* **2022**, *9*, 1–12. [CrossRef]
6. Dies, K. *Kupfer und Kupferlegierungen in der Technik*; Springer: Berlin/Heidelberg, Germany, 2014; ISBN 978-3-642-48932-7.
7. Freudenberger, J.; Heilmaier, M. *Materialkunde der Nichteisenmetalle und -Legierungen*; Wiley-VCH Verlag GmbH & Co. KGaA: Weinheim, Germany, 2020; ISBN 9783527822546.
8. Franczak, K.; Kwaśniewski, P.; Kiesiewicz, G.; Zasadzińska, M.; Jurkiewicz, B.; Strzępek, P.; Rdzawski, Z. Research of mechanical and electrical properties of Cu–Sc and Cu–Zr alloys. *Archiv. Civ. Mech. Eng.* **2020**, *20*, 1–15. [CrossRef]
9. Dölling, J.; Henle, R.; Prahl, U.; Zilly, A.; Nandi, G. Copper-Based Alloys with Optimized Hardness and High Conductivity: Research on Precipitation Hardening of Low-Alloyed Binary CuSc Alloys. *Metals* **2022**, *12*, 902. [CrossRef]
10. Arzhavitin, V.M.; Korotkova, I.M.; Sytin, V.I. Grain-boundary internal friction of yttrium- or scandium-microalloyed copper. *Russ. Metall.* **2016**, *2016*, 229–234. [CrossRef]
11. Hao, Z.; Xie, G.; Liu, X.; Tan, Q.; Wang, R. The precipitation behaviours and strengthening mechanism of a Cu-0.4 wt% Sc alloy. *J. Mater. Sci. Technol.* **2022**, *98*, 1–13. [CrossRef]
12. Shubin, A.B.; Shunyaev, K.Y. Copper-scandium system: Thermodynamic properties of intermetallics and liquid alloys. *Russ. Metall.* **2010**, *2010*, 672–677. [CrossRef]
13. Bo, H.; Liu, L.B.; Jin, Z.P. Thermodynamic analysis of Al–Sc, Cu–Sc and Al–Cu–Sc system. *J. Alloy. Compd.* **2010**, *490*, 318–325. [CrossRef]
14. Goncharuk, L.V.; Sidorko, V.R. Thermodynamic properties of scandium-copper compounds. *Powder Metall. Met. Ceram.* **2006**, *45*, 72–75. [CrossRef]
15. Porter, D.A.; Easterling, K.E.; Sherif, M.Y. *Phase Transformations in Metals and Alloys*, 4th ed.; CRC Press, Taylor & Francis Group: Boca Raton, FL, USA, 2022; ISBN 9781003011804.
16. Zhao, Z.; Li, Z.; Lv, L. Quantum chemical calculations of thermodynamic and mechanical properties of the intermetallic phases in copper–scandium alloy. *J. Theor. Comput. Chem.* **2017**, *16*, 1750056. [CrossRef]
17. Gladman, T. Precipitation hardening in metals. *Mater. Sci. Technol.* **1999**, *15*, 30–36. [CrossRef]
18. Milkereit, B.; Starink, M.J.; Rometsch, P.A.; Schick, C.; Kessler, O. Review of the Quench Sensitivity of Aluminium Alloys: Analysis of the Kinetics and Nature of Quench-Induced Precipitation. *Materials* **2019**, *12*, 4083. [CrossRef] [PubMed]
19. Deschamps, A.; Hutchinson, C.R. Precipitation kinetics in metallic alloys: Experiments and modeling. *Acta Mater.* **2021**, *220*, 117338. [CrossRef]
20. Adeli, M.; Seyedein, S.H.; Aboutalebi, M.R.; Kobashi, M.; Kanetake, N. Implementation of DSC analysis in reaction kinetics during heating of Ti–50 at.%Al powder mixture. *J. Therm. Anal. Calorim.* **2017**, *128*, 867–874. [CrossRef]
21. Lang, P.; Povoden-Karadeniz, E.; Falahati, A.; Kozeschnik, E. Simulation of the effect of composition on the precipitation in 6xxx Al alloys during continuous-heating DSC. *J. Alloy. Compd.* **2014**, *612*, 443–449. [CrossRef]
22. Luo, A.; Lloyd, D.J.; Gupta, A.; Youdelis, W.V. Precipitation and dissolution kinetics in Al-Li-Cu-Mg alloy 8090. *Acta Metall. Mater.* **1993**, *41*, 769–776. [CrossRef]
23. Starink, M.J. Analysis of aluminium based alloys by calorimetry: Quantitative analysis of reactions and reaction kinetics. *Int. Mater. Rev.* **2004**, *49*, 191–226. [CrossRef]
24. Heugue, P.; Larouche, D.; Breton, F.; Martinez, R.; Chen, X.G. Evaluation of the Growth Kinetics of θ′ and θ-Al2Cu Precipitates in a Binary Al-3.5 Wt Pct Cu Alloy. *Metall. Mat. Trans. A* **2019**, *50*, 3048–3060. [CrossRef]
25. Mittemeijer, E.J. Analysis of the kinetics of phase transformations. *J. Mater. Sci.* **1992**, *27*, 3977–3987. [CrossRef]
26. Kissinger, H.E. Reaction Kinetics in Differential Thermal Analysis. *Anal. Chem.* **1957**, *29*, 1702–1706. [CrossRef]
27. Ozawa, T. A New Method of Analyzing Thermogravimetric Data. *BCSJ* **1965**, *38*, 1881–1886. [CrossRef]
28. Koga, N. Ozawa's kinetic method for analyzing thermoanalytical curves. *J. Therm. Anal. Calorim.* **2013**, *113*, 1527–1541. [CrossRef]
29. Boswell, P.G. On the calculation of activation energies using a modified Kissinger method. *J. Therm. Anal.* **1980**, *18*, 353–358. [CrossRef]
30. Vyazovkin, S.; Burnham, A.K.; Criado, J.M.; Pérez-Maqueda, L.A.; Popescu, C.; Sbirrazzuoli, N. ICTAC Kinetics Committee recommendations for performing kinetic computations on thermal analysis data. *Thermochim. Acta* **2011**, *520*, 1–19. [CrossRef]
31. Dölling, J.; Kracun, S.F.; Prahl, U.; Fehlbier, M.; Zilly, A. A Comparative Differential Scanning Calorimetry Study of Precipitation Hardenable Copper-Based Alloys with Optimized Strength and High Conductivity. *Metals* **2023**, *13*, 150. [CrossRef]
32. Benchabane, G.; Boumerzoug, Z.; Thibon, I.; Gloriant, T. Recrystallization of pure copper investigated by calorimetry and microhardness. *Mater. Charact.* **2008**, *59*, 1425–1428. [CrossRef]

33. Azzeddine, H.; Bourezg, Y.I.; Khereddine, A.Y.; Baudin, T.; Helbert, A.-L.; Brisset, F.; Kawasaki, M.; Bradai, D.; Langdon, T.G. An investigation of the stored energy and thermal stability in a Cu–Ni–Si alloy processed by high-pressure torsion. *Philos. Mag.* **2020**, *100*, 688–712. [CrossRef]
34. Sha, G.; Wang, Y.B.; Liao, X.Z.; Duan, Z.C.; Ringer, S.P.; Langdon, T.G. Influence of equal-channel angular pressing on precipitation in an Al–Zn–Mg–Cu alloy. *Acta Mater.* **2009**, *57*, 3123–3132. [CrossRef]
35. Härtel, M.; Frint, P.; Abstoss, K.G.; Wagner, M.F.-X. Effect of Creep and Aging on the Precipitation Kinetics of an Al-Cu Alloy after One Pass of ECAP. *Adv. Eng. Mater.* **2018**, *20*, 1700307. [CrossRef]
36. Härtel, M.; Wagner, S.; Frint, P.; Wagner, M.F.-X. Effects of particle reinforcement and ECAP on the precipitation kinetics of an Al-Cu alloy. *IOP Conf. Ser.: Mater. Sci. Eng.* **2014**, *63*, 12080. [CrossRef]
37. ABIB, K.; LARBI, F.H.; RABAHI, L.; ALILI, B.; BRADAI, D. DSC analysis of commercial Cu–Cr–Zr alloy processed by equal channel angular pressing. *Trans. Nonferrous Met. Soc. China* **2015**, *25*, 838–843. [CrossRef]
38. Rodríguez-Calvillo, P.; Ferrer, N.; Cabrera, J.-M. Thermal analysis of CuMg alloys deformed by equal channel angular pressing. *J. Therm. Anal. Calorim.* **2021**, *146*, 1393–1403. [CrossRef]
39. Neumann, G.; Tölle, V. Monovacancy and divacancy contributions to self-diffusion in face-centred cubic metals reanalysis for copper, silver, gold, nickel and platinum. *Philos. Mag. A* **1986**, *54*, 619–629. [CrossRef]
40. Kejzlar, P.; Švec, M.; Macajová, E. The Usage of Backscattered Electrons in Scanning Electron Microscopy. *Manuf. Technol.* **2014**, *14*, 333–336. [CrossRef]
41. Goldstein, J.I. *Scanning Electron Microscopy and X-ray Microanalysis: A text for Biologists, Materials Scientists, and Geologists*; Plenum Press: New York, NY, USA, 1984; ISBN 030640768X.

MDPI
St. Alban-Anlage 66
4052 Basel
Switzerland
Tel. +41 61 683 77 34
Fax +41 61 302 89 18
www.mdpi.com

Materials Editorial Office
E-mail: materials@mdpi.com
www.mdpi.com/journal/materials

www.ingramcontent.com/pod-product-compliance
Lightning Source LLC
Chambersburg PA
CBHW061619220926
51850CB00005B/546
9783036578040